Research Methodology in Plant Science

2nd Edition

About the Authors

P. Suresh Narayana obtained his M. Sc. degree (1971) in Botany from Sri Venkateswara University. Later, he joined the Department of Botany, S. S. B. N. Junior College, Anantapur (1971) as a Junior Lecturer and served the institution for more than 37 years. He rose to become the Principal of the same institution (2006). He took active interest in popularizing Botany, among College and High School students. Many a time he participated in District Science Fairs and Children`s Science Congresses as an adjudicator and encouraged the students to develop the scientific temper. He participated in several National Symposia in Botany, Environmental Sciences and Biotechnology. He gave many radio talks on plants, environment and biotechnology over All India Radio. He actively participates in the social activities among the elite of Anantapur. He is a member of a few scientific societies of India. He also organized a couple of symposia in Plant Sciences and Biotechnology, Quiz programmes and elocution and essay writing competitions. During his tenure as Principal, he established the NSS unit in the college and conducted many programmes such as `Clean and Green`, `Planting` and `Environmental Awareness`. He conducted many botanical tours and accompanied students for collections. His hobbies include reading, correspondence, photography and collection of books and stamps. He is also an Executive Member of a few local educational institutions and organizations. He retired from active service in June 2008. His recent publications include: *Eminent Indian Botanists – Past and Present – (Biographies and Contributions), Text Book of Pharmacognosy – Volume I, II and III and Research Methodology in Zoology* etc.

D. Varalakshmi obtained her M. Sc., degree (1972) in Botany from Sri Venkateswara University, Tirupathi, and M. Phil., degree from Sri Krishnadevaraya University, Anantapur (1986). She joined the Department of Botany, Government Junior College, Pendurthi, Visakhapatnam (1975) as Junior Lecturer and later served in different Government Junior Colleges of Visakhapatnam, Kurnool and Anantapur districts for 18 years. She was promoted as a Lecturer and joined the Department of Botany, Government Degree College, Anantapur (1993). She was a Lecturer in the Department of Biotechnology since its inception (1998) in the college and was kept in-charge of the Department in the same institution (2001) where she established a full working laboratory and library in the department which was a model for the establishment of biotech laboratories in other colleges. She took active interest in the popularization of Botany among students and many a time participated in the District Science Fairs and Children`s Science Congresses as an adjudicator. She participated in several National Symposia in Botany, Biotechnology and Environmental Sciences. She also gave radio talks on different topics of botany and biotechnology over All India Radio. She actively participated in the social activities among the elite of Anantapur. She organized a couple of Regional and National Symposia in Biotechnology and Quiz programmes. She conducted many botanical tours and educational tours in Biotechnology and accompanied the students for collections. Her hobbies include reading and collection of books. She retired from active service in August 2008. She is the Co-author of 5 publications: *A Text Book of Pharmacognosy – Volume I, II and III and Research Methodology in Zoology* etc.

Prof. T. Pullaiah obtained his M. Sc. (1973) and Ph. D. degrees in Botany from Andhra University. He was a Post-doctoral Fellow at Moscow State University, Russia, during 1976-1978. He travelled widely in Europe and visited Universities and Botanic Gardens in about 17 countries. He joined Sri Krishnadevaraya University as Lecturer (1979) and became Professor (1993). He held several positions in the University as Dean, Faculty of Life Sciences; Head, Department of Botany; Chairman, Board of Studies in Botany; Head, Department of Sericulture; Co-ordinator and Chairman, Board of Studies in Biotechnology; Principal, Sri Krishnadevaraya University College. He retired from active Service in 2011 and was selected by UGC as UGC-BSR Faculty Fellow in the Department of Botany, Sri Krishnadevaraya University. He has published 75 books, 330 research papers and 35 popular articles. His books include: *Global Biodiversity* (4 volumes, Apple Academic press), Ethnobotany of India (5 volumes Apple Academic press), Red Sanders: Silviculture and Conservation (Springer), *Monograph on Brachystelma and Ceropegia in India* (CRC Press), *Flora of Andhra Pradesh* (5 volumes), *Flora of Eastern Ghats* (7 volumes), *Encyclopaedia of World Medicinal Plants* (7 volumes), *Biodiversity in India* (8 volumes), *Taxonomy of Angiosperms* (presently in 4th edition), *Plant Development, Eminent Indian Botanists – Past and Present* – (Biographies and Contributions), *Text Book of Pharmacognosy* – Volume I, II and III etc. He was the Principal-Investigator of 20 major Research Projects totalling more than three Crores of Rupees funded by DBT, DST, CSIR, UGC, BSI, WWF, GCC etc. Under his guidance, 54 students obtained their Ph.D. degrees and 34 students their M.Phil. Degrees. He is the recipient of Best Teacher Award from the Government of Andhra Pradesh, Prof. P. Maheshwari Gold Medal of the Indian Botanical Society and Prof. Y.D. Tiagi Gold Medal of the Indian Association for Angiosperm Taxonomy. He was President of Indian Association for Angiosperm Taxonomy (2013) and President of the Indian Botanical Society (2014).

Research Methodology in Plant Science

2nd Edition

P. Suresh Narayana
Principal (Rtd.).
S.S.B.N. Junior College
Anantapur 515001, A. P.

D. Varalakshmi
Lecturer in Botany (Rtd.).
Government Degree College
Anantapur 515001, A. P.

Prof. T. Pullaiah
Professor (Rtd.).
Department of Botany
Sri Krishnadevaraya University
Anantapur 515003, A. P.

Published by:

SCIENTIFIC PUBLISHERS (INDIA)
5A, New Pali Road, P.O. Box 91
Jodhpur 342 001, India
E-mail: info@scientificpub.com
Website: www.scientificpub.com

ISBN: 978-93-89832-26-6
eISBN: 978-93-89832-28-0

Printed in India

PREFACE

For many years of observation during the last decade, it is our common observation that our students who approached us for the necessary suggestions about their future after post-graduation, and those who wish to carry out research in plant science, expressed that they find it very difficult to procure relevant material to take up their M. Phil, or pre-Ph. D, examination as the different chapters of research methodology cannot be located all at one place, and it was very difficult to procure the same. And some of them have also requested us to prepare a source book on research methodology in plant science which can be of use to the researchers and teachers where all the topics are at one place. Thus, the impetus for the publication of this book on `*Research Methodology in Plant Science*' began with an effort to provide a source book on research methodology wherein all the topics lie at one place for the benefit of the students.

This publication is a practical, down-to-earth text book for those people who wish to carry out research and also who want to appear for their examinations for M. Phil, and Pre Ph.D., as no book exclusively dealing with *Research Methodology in Plant Science* is available in the market. This book is designed to provide M.Phil. and Ph.D, students, with a concise introduction to conducting research not only in organizations but also in Universities. The main aim of the book is to impart knowledge for making the appropriate methodological choice and to provide readers a practical guide to design and execute a research problem. The book is aimed at those who are new to research and have no prior knowledge of the issues covered herein. It will also provide more about the research methods that are available to those people who have already conducted some research and interested in finding out more. This book is more than a text on research methods in plant science, and it is an introduction to the theoretical concepts as well as the descriptive and analytical research methods that are used by different disciplines associated with research in plant science.

These days' people, students and teachers frequently talk of research, both in the academic institutions as well as outside. Many research studies are undertaken and accomplished year after year, but in most cases, very little attention has been paid to an important dimension related to research, that of research methodology. Thus, a great deal tends to be futile as the significance lies in its quality and not in its quantity. Therefore, an urgent need is to pay due attention to designing and adhering to appropriate methodology for improving the quality of research. The methodology may always differ from problem to problem, yet the basic approach towards research remains the same.

Keeping all these points in view, the present book has been written with three clear objectives, namely, to enable the students as well as the researchers in plant science, to prepare well and to be successful in the M.Phil. and Pre Ph.D. examination, irrespective of their branch/field; when once they get into research, in the development of an appropriate methodology for their research studies; and also to make them familiar with an art of using different methods and techniques. It is our ambition that this humble effort in taking out this book will definitely assist the students and the researchers in the accomplishment of result-oriented studies.

Regarding the organization of the content, the structuring of chapter scheme of the book is designed to be more practical in nature. The content of this revised edition is organized into nineteen chapters, well arranged in coherent manner. More details have been added and some new chapters are also incorporated keeping in mind the need of advanced techniques in plant science research. The book provides the basic as well as advanced level discussion on research methodology so that the students and/or the researchers may become familiar with the art of different advanced methods and techniques. This is also a useful source for those writing a thesis or dissertation or a monograph and/or a research paper. The key concepts have been explained so that they can be easily understood by those with little or no research experience. Though the book focuses specifically on issues of research methodology, we have included one chapter that focuses on statistics (biostatistics) wherein we have provided only a simple overview of some of the basic concepts in the field, and for those who intend to conduct detailed analysis, expert consultation is recommended. For some of the chapters where necessary, early history has been added to enable the readers to follow the time line of development.

Writing this book has been an enjoyable journey that we have been encouraged by a number of our students and colleagues teaching botany in and around Anantapur (State of Andhra Pradesh) along the way which served as an impetus for this work. This has been a much bigger project than we originally anticipated, and we hope that we have managed to provide what is required for research in plant science. We have done our best to incorporate all the details as far as possible and the book has run longer than we have expected. We wish to state that every care has been taken to bring out an error-free volume and some errors and omissions, if any, might have been crept in without our knowledge, and if any present, they will be rectified in future editions. Any suggestions for further improvement of the quality of this volume are warmly invited.

Our sincere thanks are due to the authors of Wikipedia, and various other sources of web, books, periodicals, monographs, bulletins, research reports/papers etc. Finally, we express our deep sense of gratitude to the members of our families for their continued cooperation, encouragement and best support in times of stress during the preparation of this volume. Last but not the least, we express our sincere thanks to our Publisher, Scientific Publishers, New Delhi, for their continued cooperation in bringing out this publication in a highly presentable form within a short time.

P. Suresh Narayana
D. Varalakshmi
T. Pullaiah

Anantapur
Dated: 26 January, 2020

CONTENTS

Preface v

1. **RESEACH METHODOLOGY** **1-23**

Introduction 1
Research - What it is? 1
Aim and Purpose of Research 2
Early History of Research 3
Types of Research 5
Possible Motives for Doing Research 9
Methodology - What it is? 10
Types of Research Methodology 11
Research Design 13
Research Plan 14
Concepts and Context of Research 14
Research Logistics 14
Research Methods versus Research Methodology 14
Research Proposal 15
Different Stages of Research 15
Criteria for Good Research 21
Objectives of Research 22
Why do people carry out Research? 22
Importance of Research 22
Significance of Research 22

2. **CENTRIFUGATION** **24-57**

Introduction 24
Centrifugation - Definition 25
Types of Centrifugation 25
Preparative Ultracentrifugation - Separation Methods 26
Centrifuge - The Instrument 33
Centrifuge - Components and Parameters 34
Centrifuge - Operation 34
Common Centrifuges Used in Biological Sciences 36

	Centrifuges - Types	37
	Centrifuges - Classification	39
	Ultracentrifuge - Basic Principle	42
	Preparative Centrifugation - Types of Rotors	42
	Rotors - Selection	45
	Rotors and Accessories - Care	46
	Centrifuge - Containers	47
	Density Gradient Separation - Formation and Choice of Density Gradients	48
	Gradient Material - Nature and Use	49
	Suitable Density Gradient Medium - Selection	49
	Fractions - Collection	51
	Sample Application to the Gradient	51
	Recovery and Monitoring of Gradients	52
	Analysis of Sub-cellular Fractions	52
	Presentation of the Results	53
	Analytical Ultracentrifuge - Applications	53
	Centrifuges - Safety Aspects in Use	55
	Centrifugation - Applications in Biological Research	55
	Many Uses of Centrifuge	56
3.	**PLANT TISSUE CULTURE**	**58-86**
	Introduction	58
	Plant Tissue Culture - Materials and Methods	59
	Explant	59
	Aseptic Environment	59
	Tissue Culture Media or Nutrient Media	60
	Nutritional Composition of Most Common Plant Cell/Tissue Culture Media (Angiosperms)	61
	Composition or Components of Tissue Culture Media	62
	General Technique of Plant Tissue Culture	68
	Plant Tissue Culture Protocol (General)	73
	Plant Tissue Culture - Laboratory and Equipment	74
	Methodology for the Establishment of Cultures	77
	Types of Plant Tissue Culture	82
	Applications of Plant Tissue Culture	83
4.	**ELECTROPHORESIS**	**87-118**
	Introduction	87
	Definition	88
	How Separation Occurs in Electrophoresis?	88

	Requirements for Electrophoresis	88
	Basic Principles of Electrophoresis	88
	Electrophoresis - Method	89
	Factors that affect Electrophoresis	91
	Physical Background of Electrophoresis	92
	Instrumentation for Electrophoresis	92
	Electrophoresis – Apparatus	93
	Theory of Electrophoresis	94
	Types of Electrophoresis	96
	Specific Techniques of Electrophoresis Commonly Used in Biochemistry	100
	Detection Methods – Blotting	116
	Electrophoresis – Applications	117
	RECENT ADVANCES IN ELECTROPHORESIS	118
	Introduction	118
	Non-linear Frictiophoresis or Non-linear Electrofrictiophoresis	119
	ZIFE (Zero Integrated Field Electrophoresis)	119
	Pulsed - field Gel Electrophoresis (PFGE)	120
	Isotachophoresis (ITP)	121
	Transient Isotachophoresis (tITP)	121
	Multi-junction IEF System	122
	Dielectrophoresis (DEP)	122
	Optical Dielectrophoresis	123
	Affinity Electrophoresis	123
	Difference Gel Electrophoresis (DIGE)	124
	Discontinuous Electrophoresis (colloquially Disc Electrophoresis)	124
	Microfluidic Chip-based Electrophoresis	125
	Iontophoresis	126
	Temperature Gradient Gel Electrophoresis (TGGE) and Denaturing Gradient Gel Electrophoresis (DGGE)	127
	Free Flow Electrophoresis (FFE) or Carrier-Free Electrophoresis	127
	New Technique to Separate Microsatellite Alleles	128
	Conclusion	129
5.	**HERBARIUM TECHNIQUE**	**130-152**
	Herbarium - What it is?	130
	Plant Collection	131
	Why Preserve Plant Specimens?	132
	Preservation of the Plant Specimens	132
	Equipment Needed for Collection of the Plant Specimens	132

Method of Plant Pressing 138
Method to keep Wet Plant Material in Adverse Conditions 140
Methods of Drying the Plant Specimens 141
What Makes a Good Specimen? 141
The Data to be recorded in the Field Note Books 142
Mounting of the Dry Plant Specimens 147
Labeling the Plant Specimens 148
Filling or Arrangement of the Plant Specimens 149
Special Care of the Plant Specimens 149
Safe Storage of the Sheets / Specimens 150
Safe Preservation of the Specimens (Plants) 151
Use and Significance of the Herbarium 152
Role of Herbarium in Modern Plant Taxonomy 152

6. CHROMATOGRAPHY 153-189

Introduction 153
Definition 154
Principle 154
Chromatography - Early History 155
Classification of Chromatographic Methods 156
Special Techniques of Chromatography 160
Hyphenated (Hybrid) Techniques and Alternative Methods of Detection in Chromatography 161
Chromatography - Basic Principle 161
Types of Chromatography 161
- Adsorption Chromatography 162
- Partition Chromatography 163
- Planar Chromatography 163
- General Procedure - Paper Chromatography (PC) 165
- Thin Layer Chromatography (TLC) 168
- High Pressure Liquid Chromatography (HPLC) or High Performance Liquid Chromatography 171

General Procedure - High-Performance Liquid Chromatography (HPLC) 172
Chromatography - Applications 175
Chromatography - Applications in Biology and Chemistry 177
Chromatography - Uses in Pharmacognosy 177
Chromatography - Applications for the Analysis of Herbal Drugs 178
Chromatography - Latest and Advanced Techniques 179
- Evolution of Chromatography 179

Chromatography - Improvements - Multi-dimensional Analysis 180
Dye-Ligand Chromatography 181
Hydrophobic Interaction Chromatography (HIC) 182
Pseudo-affinity Chromatography 183
Immobilized Metal Affinity Chromatography (IMAC) 184
Perfusion Chromatography (PC) 184
Supercritical Fluid Chromatography (SFC) 185
Expanded Bed Adsorption (EBA) Chromatography 185
Nanoflow Liquid Chromatography (NLC) 185
Micellar Liquid Chromatography (MLC) 185
Electrochromatography 186
Capillary Electrochromatography (CEC) 186
Weak Affinity Chromatography (WAC) 187
High Performance Liquid Chromatography (HPLC) 187
Conclusion 188
Hy-phe-nated Techniques or Hybrid Techniques 188
Hy-phe-nated HPLC (High-performance Liquid Chromatography) Methods 188

7. BOTANICAL MICRO-TECHNIQUE 190-208

Introduction 190
Killing and Fixation 191
Embedding the Material in Paraffin 197
Stains and Staining Methods 199
Procedure of Staining 202
Dehydration of the Material 203
Clearing the Sections 204
Dewaxing the Sections 204
Mounting the Material on to the Slide 204
Adhering Sections to the Slides 204
Ringing the Slide / Sealing the Cover slip 205
Slide Label - Preparation 205
Free Hand Sections or Sectioning by Hand 206
Free Hand Sectioning - Procedure 206
Epidermal Peeling - Preparation 207

8. MICROSCOPY 209-233

Introduction 209
Microscopy - Definition 209

Types of Microscopes 209
Optical (Light) Microscope 209
Electron Microscope 211
Scanning Electron Microscope (SEM) 211
Transmission Electron Microscope (TEM) 211
Scanning Probe Microscope (SPM) 211
Optical or Light Microscopy 212
The Compound Light Microscope 214
Electron Microscope 215
Cytophotometry or Microfluorometry 217
Phase Contrast Microscopy (PCM) 218
Fluorescence Microscopy (FM) 220
Scanning Electron Microscopy (SEM) 226
Transmission Electron Microscopy (TEM) 230
Electron Microscopy (EM) 232

9. SMEAR AND SQUASH TECHNIQUE 234-245

Introduction 234
Smear and Squash 235
Squash Technique for Root Tips - Protocol 236
Preparation of the Onion Root Tip Squash 238
Chromosome Squash Technique 243
Preparation of the Anther Smear 244

10. SPECTROPHOTOMETRY 246-260

Introduction 246
History and Spectrophotometry 246
Spectrophotometer - Definition 248
Spectrophotometer - Components 248
Spectrophotometer - Design of the Device 249
Spectrophotometer - Basic Principle 251
Using Spectronic-20 Spectrophotometer 253
Spectrophotometers - Different Types 254
Spectrophotometer - Source of Light 257
Spectrophotometer - Methods of Measurement 258
Spectrophotometer - Uses 259
Spectrophotometer - Applications 259
Colorimeter and Spectrophotometer - Differences 260

11.	**COLORIMETRY**	**261-273**
	Introduction	261
	History and Colour/Colorimetry	261
	Colorimetry - Definition	262
	Colorimeter - The Instrument	263
	Principle of Colorimetry	265
	Chemistry of Colorimetry	266
	Colorimeter - Operational Procedure (Experimental Method)	268
	Colorimetry - Chemical Method	269
	Absorption Colorimeter	269
	Photo-Electric Colorimeter - Operation	269
	Colorimetric Assays	270
	Precautions to be taken in Colorimetry	270
	Factors that affect/influence Colorimetric Measurements	270
	Advantages and Disadvantages of a Colorimeter	271
	Uses of Colorimeter	271
	Colorimetry - Applications in Plant Sciences	272
	Colorimetry - Applications in Other Fields	272
12.	**HYDROGEN-ION CONCENTRATION (pH)**	**274-302**
	Introduction	274
	Earliest pH Meter - Litmus	274
	Hydrogen Ion Concentration (pH)	276
	Measurement of pH	279
	pH Meter	280
	pH Meters - Types	284
	Accurate Measurement of pH	284
	pH Indicators	285
	Buffers and Buffer Solutions	286
	Determination of pH	288
	Care and Use of pH Meter	290
	pII Elcctrode / pH Meter	290
	pH Scale	293
	Uses of pH Meters	294
	pH Meters - Advantages and Disadvantages	294
	pH Controller	294
	Calibration of pH Meter - Why?	296
	Care of pH Electrodes	296
	pH Measurement - Applications	298
	pH in Nature	300
	pH and Life	301

13. MICROTOMY AND MICROTOME 303-317

Introduction 303
History and Microtomy 303
Parts of a Microtome 305
Proposed Classification (Types) of Microtomes 305
Microtome - Knife Design and Cut Types 312
Sharpening of the Knives 314
Method of Sectioning by a Microtome 315
Maintenance of a Microtome 315
Applications of a Microtome 316
Conclusion 317

14. BIOSTATISTICS 318-356

Introduction 318
History and Biostatistics 318
Scope of Biostatistics 320
Principles of Experimental Design and Statistical Inference 321
Tabular Design 324
Tabular Presentation of the Raw Data 327
Tabular Presentation of Data 327
Types of Tables 328
Best Practice of Presenting the Table 328
Graphical Design 329
Diagrammatic Representation of Data 333
Statistical Method - Types 336
Data (Information) - Types 336
Central Tendencies - Measures 337
Averages 338
Standard Deviation (SD) 341
Correlation Coefficient 341
Standard Error of Mean (SEM) and its Applications 342
Standard Deviation (SD) and Standard Error (SE) 342
Confidence Interval (CI) or Fiducial Limits 342
Distribution - Types 343
Null Hypothesisμ 344
Errors - Types 344
Level of Significance 345
Outliers 345
One - tailed and Two - tailed Test 345
Determination of Sample Size - Importance 346
Power of Study 347
Choosing an Appropriate Statistical Test 348

	Common Problems Faced by the Researcher	351
	Software for Biostatistics	353
	Statistical Applications in Biological Research	353
	Biostatistics - Applications	355
	Recent Development in Modern Biostatistics	355
15.	**COMPUTERS IN LIFE SCIENCES**	**357-372**
	Introduction	357
	Computer - Basic Principles	357
	Computer - Parts	359
	Computers - Types according to Size and Power	362
	Brief History of Computing	362
	Computers - Applications in Life Sciences	365
	Computers in Agricultural Research	367
	Computers and Plant Sciences	368
	Importance of Computer Literacy and Useful Software for Students and Researchers in Life Sciences	370
16.	**APPLICATIONS OF RADIOISOTOPES IN BIOLOGY/PLANT SCIENCES**	**373-387**
	Introduction	373
	Radioactive Isotopes	373
	Half-Life ($t_{1/2}$)	374
	Half-life in Biology and Pharmacology	374
	Radioactive Tracers	374
	Tracer Studies / Tracer Techniques	375
	Instruments to Measure Ionizing Radiation	376
	Geiger-Muller Counter (GM Counter)	376
	Scintillation Counter	379
	Radioactivity - Applications in Life Sciences	380
	Radioisotopes - Applications in Plant Sciences	381
	Radioisotopes and Agriculture	381
	Radioisotopes - Applications in Agriculture	383
	Stable Isotopes - Use in Agriculture	386
	Radioisotopes - Uses in Molecular and Cell Biology	387
	Conclusion	387
17.	**BIOINFORMATICS**	**388-403**
	Introduction	388
	Definition	389
	History and Bioinformatics	389
	Bioinformatics - Concept and Goal	391

	Bioinformatics - Relevance to Biotechnology	392
	Bioinformatics - Important Thrust Areas	392
	Bioinformatics - Uses	393
	Goals of Bioinformatics	394
	Approaches of Bioinformatics	394
	Bioinformatics in relation to other fields	395
	Conclusion	403
18.	**POLYMERASE CHAIN REACTION (PCR)**	**404-430**
	Introduction	404
	PCR - What it is?	405
	PCR - Principle	405
	PCR - General Protocol	406
	PCR - Different Schemes and Types	413
	Applications of PCR in Different Disciplines	424
	PCR - Benefits, Drawbacks and Limitations	429
	Conclusion	429
19.	**PHOTON COUNTING AND PHOTO-MULTIPLIER TUBES (PMTs)**	**431-452**
	Introduction	431
	Photon Counting	432
	Photomultiplier Tube (PMT)	432
	Photomultipliers - Properties	436
	Photomultipliers - Modification	439
	Photon Counting: SPCMs, CPMs and CCDs: Which Technology Works Best?	440
	The 3 - Technologies	440
	SPCMs (Single Photon Counting Modules)	440
	CPMs (Channel Photo Multipliers)	441
	CCDs (Charged Coupled Devices)	442
	Photon Counting Mode (Analog and Digital)	446
	Photon Counting Methods	446
	Detectors for High Speed Photon Counting	446
	Photon Counting Techniques	446
	Detector Principles	450
	Photon Detection - Uses	451
	Photon Counting - Applications	451
	Photon Counting - Benefits	451
	Conclusion	452
	REFERENCES	**453-458**

	Common Problems Faced by the Researcher	351
	Software for Biostatistics	353
	Statistical Applications in Biological Research	353
	Biostatistics – Applications	355
	Recent Development in Modern Biostatistics	355
15.	**COMPUTERS IN LIFE SCIENCES**	**357-372**
	Introduction	357
	Computer – Basic Principle	357
	Computer – Parts	359
	Computers – Types according to Size and Power	362
	Brief History of Computing	362
	Computers – Applications in Life Sciences	365
	Computers in Agricultural Research	367
	Computers and Plant Science	368
	Importance of Computer Literacy and Useful Software	370
16.	**APPLICATIONS OF RADIOISOTOPES IN BIOLOGY/PLANT SCIENCES**	**373-387**
	Introduction	373
	Radioactive Isotopes	373
	Half-Life ($t_{1/2}$)	374
	Half-life in Biology and Pharmacology	374
	Radioactive Tracers	374
	Tracer Studies / Tracer Techniques	375
	Instruments to Measure Ionizing Radiation	376
	Geiger-Muller Counter (GM Counter)	376
	Scintillation Counter	379
	Radioactivity – Applications in Life Sciences	380
	Radioisotopes – Applications in Plant Science	381
	Radioisotopes and Agriculture	381
	Radioisotopes – Applications in Agriculture	381
	Stable Isotopes – Use in Agriculture	386
	Radioisotopes – Uses in Molecular and Cell Biology	387
	Conclusion	387
17.	**BIOINFORMATICS**	**388-403**
	Introduction	388
	Definition	389
	History and Bioinformatics	389
	Bioinformatics – Concept and Goal	391

	Bioinformatics – Relevance to Biotechnology	392
	Bioinformatics – Important Thrust Areas	392
	Bioinformatics – Uses	393
	Goals of Bioinformatics	394
	Approaches of Bioinformatics	394
	Bioinformatics in relation to other fields	395
18.	**POLYMERASE CHAIN REACTION (PCR)**	**400-430**
	Introduction	404
	PCR – What it is?	405
	PCR – Principle	405
	PCR – General Protocol	406
	PCR – Different Schemes and Types	413
	Applications of PCR in Different Disciplines	424
	PCR –Benefits, Drawbacks and Limitations	428
	Conclusion	429
19.	**PHOTON COUNTING AND PHOTO-MULTIPLIER TUBES (PMTs)**	**431-452**
	Introduction	431
	Photon Counting	432
	Photomultiplier Tube (PMT)	432
	Photomultipliers – Properties	436
	Photomultipliers – Modification	439
	Photon Counting: SPCMs, CPMs and CCDs: Which Technology Works Best?	440
	The 3 - Technologies	440
	SPCMs (Single Photon Counting Modules)	440
	CPMs (Channel Photo Multipliers)	441
	CCDs (Charged Coupled Devices)	442
	Photon Counting Mode (Analog and Digital)	446
	Photon Counting Methods	446
	Detectors for High Speed Photon Counting	446
	Photon Counting Techniques	446
	Detector Principles	450
	Photon Detection – Uses	451
	Photon Counting – Applications	451
	Photon Counting – Benefits	451
	Conclusion	452
	REFERENCES	**453-458**

CHAPTER - 1

RESEARCH METHODOLGY

Introduction

In any profession, education is fundamental to the establishment and ongoing vitality. In the definition of any profession, the mastering of a body of knowledge and its practical application is an essential element. The importance of learning in defining a profession and its practice has long been recognized in the debates over definition of education. The main purpose of this chapter is to review the role of both the research process in general and research techniques and skills in particular as integral components of professional education. According to Richard Fox, '*specialized knowledge or systematic theory is the basic attribute of a profession*' and it was suggested that '*the profession's specialized knowledge is developed through systematic research*'. According to Robert Warner, '*a firm commitment to research*' is a major component of the profession.

Research – What it is?

Etymologically, the word, `research' is derived from the Middle French word, *recherchér* meaning `to go about seeking`, and the term itself being derived from the Old French term, `*recerchier*`, which is a compound word from `*re*` + `*cerchier*` (searcher) meaning `to search`. The earliest recorded use of the term was in 1577, and the first known use of the word, *research* is 1593. Research is *careful or diligent search* or *studious inquiry, especially investigation or experimentation aimed at the discovery and interpretation of facts, revision of accepted theories or laws in the light of new facts, or practical application of such new or revised theories or laws or in simple terms refers to collection of information about a particular subject.* Research is *systematic work to create new knowledge or devise new applications of knowledge.* Scientific research is the *investigation of phenomena or acquisition of knowledge using scientific methods.* Research comprises `*creative work undertaken on a systematic basis in order to increase the knowledge, including knowledge of man, culture and society, and the use of this knowledge to devise new applications*`. This consists of three steps, namely, pose a question, collect the data to answer the question, and present an answer to the question. Thus, research is an original contribution to the existing stock of knowledge making for its advancement, and it is the pursuit of truth with the help of study, observation, comparison and experiment. In short, research is the search for knowledge through objective and systematic method of finding the solution to a problem.

Aim and Purpose of Research

The ultimate aim of research is to generate measurable and testable data, gradually adding to the accumulation of human knowledge. Philosophers of ancient times were of the belief that all the answers could be achieved through deduction and reasoning rather than the measurement. At present, science uses established research methods and standard protocols to test the theories thoroughly. It is always important to remember that science and philosophy are intertwined together and are the essential elements of human advancement, both contributing to the way we view. The gradual accumulation of knowledge in this way dictates the overall direction of both the science and philosophy. The aim of research is to establish `what are the underlying causes and relationships between the different processes fueling the trend?` and in majority of the cases, it is always necessary to review the earlier research and try to separate the better quality sources from the inaccurate or poorly designed studies. It is also equally important to take into account any opposing points of view and accept that they may be equally valid. The explanation with viable reasons that comes up as a result must be objective and unbiased as far as possible.

The first stage of any research is to observe the world around us and to ask questions about why things are happening, because every phenomenon in the universe has a reason behind it and the aim of research is to understand and evaluate what is happening? The scientific research demands rigorous testing of a truth to be accepted however simple the phenomenon or however easy it appears to be to generate logical and intuitive answers. Whether it is a case study or a full-blown true experimental design, describing the overall behavior of the subject is the first stage of any research. This is where one can make a statement of intent and develop a strong hypothesis, and this must be testable, with the aim of research being to prove or disprove the statement. At this stage, one can express his/her personal opinion that favours one side or the other, and he/she can make a statement predicting what final answer is to be expected. The researcher must keep an open mind and understand that there is a chance that one may be wrong, but the research is never about right or wrong but about arriving at an answer, which improves the knowledge of natural processes.

The purpose of the research can be a complicated issue and varies across different scientific fields and disciplines. At the most basic level, the science can be split, loosely, into `*pure research*` and `*applied research*'. Both of these follow the same structure and protocols for propagating and testing hypotheses and predictions, but vary slightly in their ultimate purpose. The *pure scientific research* often referred to as `*pure science*`, is about the explaining of the world around us and trying to understand how the universe operates, and it is about finding out what is already there without any greater purpose of research than the explanation itself. It is definitely a direct descendant of philosophy, where both the scientists and philosophers try to understand the underlying principles of existence. Though pure research offers no direct benefits, it often has many indirect benefits that can contribute greatly to the advancement of humanity (for example, in the structure of the atom, pure research has led to x-rays, nuclear power and silicon chips etc.). The *applied scientific research* looks for answers to specific questions that help humanity (*medical research* or *environmental studies* for example), and such research generally takes a specific question and tries always to find a definitive and comprehensive answer. The purpose is about testing the theories, often generated by pure science, and applying them to the real situations and addressing more than just abstract principles. This research can be about finding out the answers to specific problems (like `*is global warming avoidable*?` or `*does a new type of medicine really help the patients*?` etc.).

Early History of Research

Our present mode of thinking is a product of our past, as the discoveries and theories of the previous generations have influenced us much. For example, the people of the ancient ages gazed at the stars with wonder, and also considered that the earth was flat. But today, we look at the stars with a yearning for the adventure, and believe the images shown to us from the space of a circular planet are very much true. Our fundamental core beliefs and values go on changing with time, learning and the exposure and assimilation of new ideas. We thought our universe was stable and constant fifty years ago. Albert Einstein has been proved wrong and it is not stable. New discoveries are being made constantly, and we are constantly exploring our world and our surroundings more than ever before than in our past history, and to a very large extent, our research ancestors are responsible for this. It is not easy to create a framework of our past, and there have been many great thinkers who have changed the course of thinking markedly and influenced the thought for generations to come. We can easily identify some of them and to understand their influence, one must understand the time in which they lived, for their ideas are a rebellion of the main stream thoughts that persisted at that time.

Socrates, a Greek philosopher and a stone-cutter by trade (470 – 399 BC), while roaming the markets got engaged himself in discourse by asking the people many questions of ethics and virtue that provoked the discussion, as his motto was `*know thyself*. This was the beginning of research. Plato and Aristotle were his students. He turned our attention inward, away from the heavens towards ourselves, and to our relationships and responsibilities to other people around us. He was convicted of atheism, treason and corruption of the young and was forced to drink the poison to die at the age of 70.

Socrates was followed by Plato (428 – 347 BC) born to an Athenian household, got engaged in poetry as a young man and he believed that the formation of a noble character was to be before all else, and how we think and what we take to be real have an important role in how we act. He came to believe that a philosophical compartment toward life would lead one to being just and ultimately happy (*the theory of forms*).

Later, Aristotle, a Greek philosopher, a student of Socrates and a tutor to Alexander the Great (384 – 322 BC), proposed the concept that the earth was the centre of the universe consisting of four elements, earth, water, air and fire. He also advocated that the heavenly bodies were perfect and divine and made of a fifth element, the ether.

The view that the earth was the centre of the universe and that the heavenly bodies were the dwelling places of Gods remained relatively unchallenged for 1000s of years. With the beginnings of industrialization (including the rapid means of transport), when the cities started to develop, this view began to be challenged. This was referred to as the beginning of scientific revolution. Nicolas Copernicus, a Polish astronomer (1473 – 1543) held the view that the earth was not at the centre of the universe and proposed the *heliocentric theory* that placed the sun at the centre, and also introduced the idea that all the planets including the earth orbited the sun. This theory could accurately predict the planetary position.

Later, Galileo Galilei (1564 – 1642) discovered the laws associated with the falling bodies (any two objects will fall at the same speed irrespective of their size) and thus helped to establish the hypothesis of Copernicus as a valid scientific theory. He also devised a number of ingenious experiments to measure things, constructed the first telescope (1609) and directly challenged Aristotle`s idea of the centrality of the universe and confirmed the view of Copernicus, When he saw the threat by the Roman Catholic Church at 70 years of age, he recanted the theories and lived the remainder of his life under house arrest. He showed that observation and careful measurement are the important tools in reaching an understanding of the world around us.

This was followed by Sir Francis Bacon, Lord Chancellor of English (1561 – 1626), considered as one of the *founders of modern scientific thinking*, set forth the *empirical method* (the word, *empirical* is derived from the Greek word for `experience`, and it is concerned with that which can be measured or observed) using the experiments and drawing conclusions from them. According to him, *inductive reasoning* played an important part in any investigation. He was also responsible for the King James Version of the Bible. Inductive reasoning means looking at what some items have in common, and then generalizing that commonality to other items (extrapolating a general principle from a limited set of data – for example, if it is true of A and B and C, then it could be true of D – inductive reasoning using the colours).

This was followed by the work of Renes Descartes, a French mathematician, philosopher, and anatomist, considered as *the Founder of Modern Philosophy*, who is famous for the *rational and deductive approach*. The key to understanding was mathematics and he founded the field of analytic geometry, and also wrote on the difference between man and animals, which even today influences the field of artificial intelligence. It is believed that at the age of 23, he experienced an illuminating vision and believed in the certainty of scientific knowledge. The main theme is that of radical doubt and for Descartes, one must doubt everything except one`s own existence. According to him, the essence of human nature lies in thought, and knowledge is achieved through intuition and deduction. Descartes used the analytical approach and suggested a breaking up of a problem into smaller pieces and then arranging those pieces in a logical order. With such an approach, one could begin to understand the world around us, and this is a systems approach to research, based on the presumption that ` *the sum of the parts is the whole*`. Such a mode of thinking of Descartes is called *Cartesian*, and this proposes a basic division between the mind and the matter, and `there has to be a rational explanation for this'. The deductive reasoning assumes that a specific case is true (a particular property of an item) because it is a part of a generally accepted principle that is widely accepted as true (extrapolating a specific instance from a general principle). Deductive reasoning is logically valid (for example, if it was widely accepted principle that people drive on the left-hand side of the road, you could deduce that a particular driver drove on the same side of the road as the general population).

Later, it was Sir Isaac Newton, an English scientist and a mathematician (1642 - 1727), who is responsible for developing the idea that the universe is one huge mechanical system, and to explain the motion of bodies, he developed *differential calculus*. He used this to derive the orbit of the moon as around 27 days, but was so concerned about others stealing his new brand of mathematics, that he rewrote the proof in the conventional mathematics of that time. He was struck on the head by an apple around 1665/1666, an event that has been attributed to help him to develop the *theory of gravity*. He firmly dispelled the earlier views of the world, and his mechanistic model of the universe shattered views espoused by the church. He also investigated the properties of colour using prisms.

Later, Jean Baptiste de Lamarck, A French naturalist (1744 – 1829), proposed that all living beings evolved from earlier simple forms. He was the *founder of modern invertebrate zoology* and classified the invertebrates. He also claimed that man was a class of animal, and that animals passed on acquired characteristics to their offspring, and this theory is the beginning of the theory of evolution. Later, it was suggested by Charles Darwin that the passing of the characteristics from one generation to another occur by the process of adaptation and natural selection, and as of now, we know that the genes (DNA) control this.

Types of Research

These refer to various procedures, schemes and algorithms used by a researcher during a research study, and include theoretical procedures, experimental studies, numerical schemes, statistical approaches etc. These help the researcher to collect samples, data and also find a solution to a problem. In science, these methods call for explanations based on the collected facts, measurements and observations (not reasoning alone), and accept only those explanations which can be verified by the experiments.

1. *Descriptive Research*

This varies widely in scope and complexity and is often used for a literature review (survey). The purpose of literature reviews according to Borg and Gall (1997) is – further refinement of the research question, discovery of explicit recommendations for further research, prevention of the repetition of work that has already been done, to provide insight into the potential research strategies and methodologies, and to highlight research possibilities that have been overlooked in the past. The sources for the descriptive research based on a qualitative approach include: text books, journal, magazine and newspaper articles, thesis and dissertations, information brochures, conference proceedings, personal interviews and articles on the internet.

Descriptive research includes surveys and fact-finding enquiries of many kinds. The characteristic of this method is that the researcher has no control over the variables, and he/she can only report what is happening or what has happened. The methods of research are represented by the survey methods of all kinds, which include comparative and co-relational methods. In *analytical research*, the researcher has to use the facts or information already available, and analyze to make a critical evaluation of the material.

2. *Pure and Applied Research*

Research can be either applied research or fundamental (basic or pure) research. The *applied research* generally aims at finding a solution for an immediate problem facing the society, while the *fundamental research* is concerned with generalizations and with the formulation of a theory. The pure or basic research is gathering of knowledge for knowledge's sake (example: concerned with natural phenomena or relating to pure mathematics and human behaviour). The main aim of applied research is to discover a solution for a pressing practical problem, while basic research is directed towards finding information that has broad base of applications, and adding to the existing scientific knowledge.

a. Basic Research/Pure Research/Fundamental Research

Basic research (*fundamental research* or *theoretical research*) refers to an investigation on the basic principles and reasons for the occurrence of a particular process or phenomenon (natural phenomenon related to pure science) which may sometimes lead to an immediate application (or use) and is original or basic in character. It always provides a systematic and deep insight into a problem and facilitates extraction of scientific and logical explanation and conclusion on it.

b. Applied Research

Applied research helps to solve certain problems by employing well known and accepted theories and principles. This is a research the outcome of which has an immediate potential application and of practical use to current activity (for example, increasing efficiency of a machine, increasing factor of material production, control of pollution, preparation of vaccines etc).

3. *Quantitative Research*

Quantitative research is based on the measurement of quantity or amount, and is applicable to phenomena that can be expressed in terms of quantity. This usually involves collecting and converting the data into numerical form so that the statistical calculations can be made and conclusions drawn. In order to find the answers to the hypotheses, the researchers will also have various instruments and materials (like paper or computer tests, observation check lists etc.) and a clearly defined plan of action. The data is collected by different means following a strict procedure and prepared for statistical analysis which is carried out with the help of sophisticated statistical computer packages. The data collection methods include online surveys, paper surveys, mobile surveys, face-to-face interviews, telephonic interviews, longitudinal studies, website interceptors, online polls and systematic observation. The final results of the analyses are presented in journals in a standard way (the end result being a p value). For those researchers who are not familiar with scientific research jargon, the discussion at the end of the articles in peer reviewed journals describe the results of the study and explain the implications of the findings in straight forward terms. The main emphasis of this type of research is on deductive reasoning which tends to move from the general to the specific (this is sometimes referred to as a top down approach). The validity of the conclusions is shown to be dependent on one or more premises (prior statements, findings or conditions etc.) being valid (best examples come from the deductive reason of Aristotle and Sherlock Holmes). The quantitative designs may be *descriptive, correlational, quasi-experimental* and *experimental.*

4. Qualitative Research

Qualitative research is the approach usually associated with the emphasis of the socially constructed nature of reality. It is concerned with phenomena relating to or involving the quality of kind, and this is especially important in the behavioural sciences where the main aim is to discover the underlying motives of human behavior. This is about the recording, analyzing and attempting to uncover the deeper meaning and significance of human behaviour and experience, including contradictory beliefs, behaviours and emotions. The researcher is interested in gaining a rich and complex understanding of people's experience and not in obtaining the information which can be generalized to other larger groups. The approach of the researcher tends to be inductive meaning that he/she develops a theory or look for a pattern of meaning on the basis of the data that he/she has collected. This always involves a move from the specific to the general (called a bottom-up approach). The researchers do not base their research on the pre-determined hypotheses. The approach to the collection of data and analysis is methodical but allows for a greater flexibility than in quantitative research. The data is collected in textual form on the basis of observation and interaction with the participants (through participant observation, in-depth interviews and focus groups) and is not converted into numerical form and is not statistically analyzed. The sample size typically small and the respondents are selected to fulfill a given data.

5. Conceptual Research

The *conceptual research* is related to some abstract ideas or theories, and this is generally used by philosophers and thinkers to develop new concepts or to reinterpret the existing ones. The *empirical research* (data-based research) relies on the experience or observation alone, without regard for system and theory. In this research it is always necessary to get the facts first hand (at their source) and to do certain things to stimulate the production of desired information. He/she must provide himself/herself with a working hypothesis and guess the probable results, set up experimental designs which will manipulate the material concerned to bring forth the desired information.

6. Other Types of Research

Other types of research are variations of one or more types of the above approaches. From the point of view of time, research may be ***one-time research*** or ***longitudinal***

research, and in the former type, the research is confined to a single time-period, while in the latter, the research is carried on over several time-periods. The research can be field-setting research or laboratory research or simulation research depending upon the environment in which it is to be carried out. Research can be clinical research or diagnostic research in which case-study methods (in-depth approaches) are followed to reach the basic causal relations, which usually go deep into the causes of things (events) that interest us using very small samples and very deep probing data gathering devices. The research may be ***exploratory research*** in which the objective is the development of hypotheses rather than testing whereas the studies of ***formalized research*** are those with substantial structure and with specific hypotheses to be tested. ***Historical research*** utilizes sources like documents, remains etc, to study events or ideas of the past, including the philosophy of persons and groups at any remote point of time. In ***conclusion-oriented research***, a researcher is free to pick up a problem, and redesign the enquiry as he/she proceeds and is prepared to conceptualize as he/she wishes, while the ***decision-oriented research*** is for the need of a decision maker and the researcher is not free to embark upon research according to his/her own inclination. The ***operations research*** is a decision oriented research since it is a scientific method of providing executive departments with a quantitative basis for decisions regarding operations under their control.

There are also other types of research, namely: ***survey research*** – developed since 1930s, this refers to the amount of error (expressed as plus and minus a certain percentage) is determined by the size of the sample (the number of the people surveyed); ***basic biomedical research*** – which is conducted to increase the understanding of fundamental life processes (discovery of the molecular structure of DNA) – one half of the genetic code of life – or investigating the genetics of the lipid disease etc; ***clinical research*** – which addresses the important questions of normal function and disease using human subjects; ***directed research*** or ***targeted research*** – which is conducted by an investigator in response to an outside request to explore a specific scientific area or a question; ***investigator-initiated research*** – which investigates a question or a hypothesis that the researcher has defined; ***outcomes research*** – which focuses upon the end results of health care, the tangible and quantifiable manifestations of the disease upon patients and society and the determinants of such outcomes; ***population health research*** – the science and art of studying the distribution and determinants of health status as influenced by social, economic and physical environments, human biology, health policy and services and of preventing the disease, prolonging life and promoting health at the population levels; ***strategically focused research*** – which focuses on science areas that the association has determined are important to achieving its mission and strategic objectives; ***translational research*** – which takes a result from basic or fundamental science and studies its applicability in the clinical or human situation or addresses the adoption of prevention and treatment strategies that have been demonstrated to be effective through clinical research in the care of patients and in population-based prevention of conditions like heart disease and stroke; ***observational research*** - this includes case studies, ethnographic studies, ethological studies etc. and the primary characteristic of each of these types of studies is that phenomena are being observed and recorded, and often the studies are qualitative in their nature; ***correlational research*** - which can be accomplished by a variety of techniques which include the collection of empirical data, and examines the co-variation of two or more variables; ***primary research*** – defined as factual, firsthand accounts of the study written by a person who was a part of the study which typically follow a scientific method and that it is typically original research and consists of a collection of original primary data collected by the researcher and analyzing the previously collected primary data; ***secondary research*** - is an analysis and interpretation of primary research and is to collect primary research that is relevant to a writing the topic and interpret what the primary research found was the summary, collation and/or synthesis of existing research where data is collected from research subjects or experiments; ***self-generated research*** - refers to the research generated by one`s own merits and has originality to it and could simply be seen as one`s own forms of media and that is correct, but it becomes self-generated research when he/she looks back at the produced content for the purpose of either own or someone else; ***audience research*** – defined as any type of communication research that is conducted on specific audience segments to gather information about their attitude, knowledge, interest, preferences or behavior with respect to the prevention issues; ***market*** or ***marketing research*** - refers to the process of assessing the viability of a new product or service through techniques like surveys, product testing and focus groups, which allows a company to discover who their target market is and what the consumers think about a product or service before it becomes available to the public; and ***production research*** – this is the research to help give the information on the characteristics of the product and is always needed when developing a new product, and it focuses on the production of a product, in other words, how it is made?

Possible Motives for Doing Research

The possible motives for doing research are as follows: desire to get a research degree along with consequential benefits; desire to face challenge in solving the unsolved (practical) problems; desire to get joy of doing some creative work; desire to be of some service to the society; and desire to get some respect. Other factors include: directives by the government, the conditions of employment, curiosity about identifying new things, and desire to understand causal relationships etc.

Research is a logical and systematic search for new and useful information on a particular topic. It is an investigation of finding solutions to scientific problems through objective and systematic analysis. It is a search for knowledge or a discovery of truths. A research can lead to new contributions to the existing knowledge and only through research it is possible to make progress in any field of science and determines the development of a nation. Research is not confined to science and technology only, and there are vast areas of research in other disciplines as well like languages, literature, history and sociology. Irrespective of the subject, research has to be an active, diligent and systematic process of inquiry in order to discover, interpret/revise the facts, events, behavior and theories. Research is normally done with the help of study, experiment, observation, analysis, comparison and reasoning.

Research is used to establish or confirm facts and reaffirm the results of the previous work, and solve new or existing problems, support the theorems, or develop new theories. A research problem may also be an expansion on the past work in the required field, and is also used to test the validity of the instruments, procedures, or experiments, may also replicate the elements of the earlier projects or the project as a whole. The main purpose of the basic research (as opposed to applied research) is documentation, discovery, interpretation, or the research and development of methods and systems for the advancement of human knowledge. The approach to research depends on epistemology which varies considerably both within and between humanities and sciences. The research is of several forms and may be scientific, humanities, artistic, economic, social business, marketing etc.

Scientific research is a systematic way of gathering data and a harnessing of curiosity. This provides scientific information and theories for explanation of the nature and properties of the world and also makes practical applications possible. This can be subdivided into different categories according to the academic and application disciplines and is widely used as a criterion for judging the status of an academic institution (business schools). But, it is also argued that it is an inaccurate assessment of the institution, because the quality of research does not tell about the quality of teaching.

The main objective of research is to discover answers to questions through the application of scientific procedures. The important aim is to find out the hidden truth which has not yet been discovered as yet. The objectives of research fall into following broad groups, namely, *exploratory* or *formulative* research studies, *descriptive* research studies, *diagnostic* research studies and *hypothesis testing* research studies.

Methodology – What it is?

Methodology is a set of practices and this term may be used to refer to practices which are widely used across a scientific discipline, the techniques used in a particular research study, or the techniques used to accomplish a particular project, and some people may also use the term, `*methodology*` to refer to the study of such methods, rather than the methods themselves. Research methodology is a science of studying how research is done and is a systematic way to solve a problem. Research methodology essentially refers to *the procedures by which researchers go about their work of describing, explaining and predicting phenomena.* It is also defined as *the study of methods by which knowledge is gained* and aims at giving the work plan of research. A clear methodology is important especially in the sciences. The clearly outlined directions and procedures tend to increase the consistency, and to create work which can be repeated elsewhere is an important characteristic of rigorous scientific research. Researchers who conduct the studies attempt to use rigorous methods so that their results will be respected by other people, and the write-ups often include an extensive discussion along with the errors and issues which they have identified.

Methodology normally refers to *ways of obtaining, organizing and analyzing the data* (Polit and Hungler, 2004). The decisions of methodology generally depend on the nature of the research question, and methodology in research can be considered to be *the theory of correct scientific decisions* (Karfman in Mouton & Marais, 1996). It also refers to *how the research was done and its logical sequence.* Mouton (1996) describes methodology as the *means or methods of doing something*; and according to Burns and Grove (2003), methodology *includes the design, setting, sample, methodological limitations, and the collection of data and analysis techniques in a study*. Methodology is described as *coherent group of methods that complement one another and that have the ability to fit to deliver the data and findings that will reflect the research question and suit the purpose of the researcher* (Henning, 2004). According to Holloway (2005), methodology means *a framework of theories and principles on which methods and procedures are based.*

TYPES OF RESEARCH METHODOLOGY

This is a way to find out the result of a given problem on a specific matter or problem that is also referred to as research problem. In methodology, the researcher uses different criteria for solving and/or searching the given problem and different sources use different types of methods for solving the said problem. In research methodology, the researcher always tries to search the given question systematically in his/her own ways to find out all the answers till conclusion.

The research methods refer to particular way in which one can verify or discover a set of data. The methodologies define different types of research and are concerned with the overall strategy the researcher chooses (Leedy and Ormrod, 2005). The research methods can be broadly classified into two types, namely, *Quantitative* and *Qualitative* methods. The *qualitative research* will, yield more specific and in-depth results while the quantitative research will give a more general picture, perhaps on a larger scale. The former refers to the quality of data while the latter deals with numbers. The quantitative research involves the generation of data in quantitative form which can be subjected to rigorous quantitative analysis in a formal and rigid fashion, which can be further sub-classified into *inferential, experimental* and *simulation approaches* to research. The purpose of *inferential approach* to research is to form a database from which to infer characteristics or relationships of population, which usually means survey research where a sample of population is studied or observed or questioned to determine its characteristics, and is then inferred that the population has the same characteristics. The *experimental approach* is generally characterized by much greater control over the research environment, and some variables are manipulated to observe their effect on other variables. The *simulation approach* involves the construction of an artificial environment within which relevant information and data can be generated, which permits an observation of the dynamic behavior of a system (or its sub-system) under controlled conditions. The term `simulation` refers to the operation of a numerical model that represents the structure of a dynamic process, and given the values of initial conditions, parameters and exogenous variables, a simulation is run to represent the behavior of the process over time. This can also be useful in building models for understanding future conditions. The qualitative research is normally concerned with the subject assessment of attitudes, opinions and behaviour. The research is a function of his/her insights and impressions, and such an approach generates results either in non-quantitative form or in the form which is not subjected to rigorous quantitative analysis. The techniques used are group interviews, projective techniques and depth interviews which are explained at length.

During discussion of research methodology, it is always important to distinguish between applied and basic research. The former examines a specific set of circumstances and its ultimate goal in relating the results to a particular situation, and uses the data directly for real world application. The goal of applied research is to predict a specific behavior in a very specific setting. The latter focuses on the fundamental principles and testing theories, and it is sometimes implied that basic research doesn't have practical applications. Both the applied and basic research is important to the scientific process.

Basic research generally focuses on the fundamental principles and testing theories, and it does not have practical implications. The classical principles were developed mostly from experimenting on non-human subjects. Both the applied and basic research is important to the scientific process. Applied principles and findings are of use immediately, and there is nothing as practical as a general and accurate theory.

a. Quantitative Methodology

As the name suggests, this is concerned with trying to quantify things, and it asks questions like `*how long*`, `*how many*` or the `*degree to which*`. These methods look to quantify data and generalize the results from a sample of the population of interest. These measure the incidence of various views and opinions in a chosen sample, for example, or aggregate results. Quantitative research is always based on the measurement of quantity (amount) wherein a process is generally expressed (described) in terms of one or more quantities, and the result is essentially a number or a set of numbers. The important qualities of quantitative research are: it is numerical, non-descriptive and applies statistics/mathematics and uses numbers; it is iterative whereby evidence is evaluated; the results are presented in tables and graphs; it is conclusive; and it investigates the *what*, *where* and *when* of decision making. The quantitative research begins with the collection of data based on a theory/hypothesis/experiment followed by the application of descriptive statistical methods.

This is dialectic and interpretive, and according to De Vos (2002), during the interaction between the researcher and the research participants, the world of the participants is discovered and interpreted by means of qualitative method. According to Burns and Grove (2003), phenomenologists are of the opinion that the clear definition of methodology tends to limit the creativity of the researcher.

b. Qualitative Methodology

This is concerned with a quality of information, and the methods attempt to gain an understanding of the underlying reasons and motivations for actions, and establish how people interpret their experiences and the world around them. These methods provide insights into the setting of a problem, generating ideas and/or hypotheses. The qualitative research is concerned with phenomena involving quality and the important characteristics of qualitative research are: it is non-numerical, descriptive and applies reasoning by using words, it gets to the meaning and describes the situation, the data cannot be graphed, it is exploratory, and it investigates the *why* and *how* of decision making. The detection of a particular compound is called *qualitative analysis* (the determination of a particular compound present in a volume) and can be done by carrying out physical or chemical tests (*volumetric, gravimetric* or *colorimetric* or *instrumental* methods).

This refers to inductive, holistic, emic, subjective and process–oriented methods used to understand, interpret, describe and develop a theory on a phenomena or setting. According to Burns and Grove (2003) and Morse and Field (1996, 1999), this is a systematic, subjective approach used to describe life experiences and give them meaning. This is also mostly associated with words, language and experiences rather than measurements, statistics and numerical figures. According to Holloway (2005), this is also characterized as developmental and dynamic, and does not use formal structured instruments, regarding the generation of knowledge. This also involves the systematic collection and analysis of subjective narrative data in an organized and intuitive fashion to identify the characteristics and the significance of human experience (Holloway, 2005). The researcher normally tries to examine from the participant's point of view in order to interpret his/her words, as such he/she becomes involved and immersed in the phenomenon to become familiar with it. The immersion helps to provide dense descriptions from the narrative data gathered from the participants, in order to interpret and portray their experiences, and also to generate empathetic and experiential understanding, and the immersion cannot be obtained without a researcher-participant trusting relationship, which is built through basic interviewing and inter-personal skills.

Quantitative and Qualitative Methodology - Differences

The *aim* in quantitative methodology is to count things in an attempt to explain what is observed, while in qualitative methodology, the aim is complete, detailed description of what is observed. The *purpose* in the former is generalisability, prediction and causal application, while in the latter; it is contextualization, interpretation and understanding perspectives. The *tools* in quantitative methodology used by the researcher include surveys to collect numerical data, while in qualitative methodology, researcher gathers the data. The *collection of the data* in the former is structured while in the latter it is unstructured. The *output* of the data in the former is in the form of numbers and statistics while in the latter, it is in the form of words, pictures or objects. In the former, the *sample* is usually a large number of cases representing the population of interest (randomly selected respondents) while in the latter, it is usually a small number of non-representative cases (respondents selected on their experience). The *objective* seeks precise measurement and analysis in the former, while in the latter, the *subjective* is the individual's interpretation of events which is important. The *role of the researcher* in the former tends to remain objectively separated from the matter of the subject, while in the latter, he/she tends to become subjectively immersed in the subject matter. The *analysis* in quantitative methodology is more statistical while in qualitative methodology, the analysis is interpretive.

RESEARCH DESIGN

No research has meaning without proper validation. Research design is the *overall approach to the research process, from the theoretical underpinning to the collection and analysis of data* (Hussey and Hussey – 1997). According to Saunders, Lewis and

Thornhill (1997), this helps the researcher – to *make an informed decision about the research methodology* (to decide how data is to be collected and analyzed); *to adapt the design to cater for limitations and constraints* (include limited access to data or insufficient knowledge of the subject or an inadequate understanding of the subject or time constraints); and *to determine which methods would be appropriate for the particular study* (proper methods should help to explain the why's, how's and what's of the subject). According to Saunders *et al* (1997), the systematic research encompasses specific methods to collect data, deliberation on the significance of the results obtained, and an explanation of any limitations experienced. The research primarily focuses to increase knowledge of a particular topic in order to help solve relevant problems.

RESEARCH PLAN

This consists of two general areas, namely, *research concepts and context* and *research logistics*. How well the project is planned and how well the steps in the plan are integrated can make the difference between the success and failure. The writing of a research plan should start as soon as possible.

CONCEPTS AND CONTEXT OF RESEARCH

Primarily, become familiar with literature to identify a problem and to explore the areas of biology, and explore the ways that enhance the integration of the study into broader biological disciplines and ways to collect, analyze and present the data useful to others (for comparison or to use in models). Secondly, state the problem carefully and clearly in the form of questions or as working or null hypotheses and make a list of possible scenarios related to most probable research questions. Next, discuss all the aspects of proposal with colleagues and investigators that work in the same or closely related fields for sources of valuable information. Finally, write a detailed research proposal.

RESEARCH LOGISTICS

Basically, identify the information which needs to be collected and how to be collected by – developing a logistical and quality control plan for the collection of data, handling and storage procedures; design the data sheets for each measurement, with data recording, as computer files and editing data analysis; try to walk through as many techniques and procedures as possible to detect the problems with protocols and equipment prior to collection of data; and evaluate the data for recurring errors; order the supplies in advance with the available funds; and also plan for the unexpected failure of the equipment, accidents, illness and unusual weather.

Research Methods versus Research Methodology

Research methods or techniques refer to the methods the researchers use in performing research. All the methods which are used by the researcher during the study of his problem are referred to as the research methods. These can be divided into three groups, namely, those methods which are concerned with the collection of the data (where the already available data is not sufficient to arrive at the required solution); the statistical techniques which are used to establish the relationships between the data and the unknown; and the methods which are used to evaluate the accuracy of the results obtained.

Research methodology is a way to systematically solve the problem of research. Researchers need not only the research methods but also the methodology. They need not only to know how to develop some indices (tests), how to calculate the mean, the mode, the median or the standard deviation or chi-square, how to apply the particular technique, and also need to know which of these methods are relevant and which are not, and what would they indicate and why. They also need to understand the assumptions underlying various techniques and the criteria by which they can decide that certain procedures will be applicable to certain problems while others will not. It is also necessary that the researcher should design his own methodology for his problem as the same may differ from problem to problem. The research methodology has many dimensions and the techniques do constitute a part of methodology as the scope of methodology is much wider than that of the methods. *When one talks of research methodology, he/she not only talks of the research methods but also considers the logic behind the methods which he/she uses in the context of his/her research study and explains why he/she is using a particular method and why he/she is not using others so that the research results are capable of being evaluated either by the researcher himself/herself or by the others.* When one talks of research methodology related to a research problem or research study, he/she tries to answer a host of questions like - why the study has been undertaken, how the problem has been defined, in what way and why the hypothesis has been formulated, what data have been collected, and what particular method has been adopted, and why particular technique of analyzing data has been used etc.

The scientific method is normally based on some basic postulates which are as follows: The scientific method mainly relies on empirical evidence and utilizes relevant concepts. This method is committed to only objective considerations and presupposes ethical neutrality (aims at only making adequate and correct statements). The results are probable predictions and the methodology is made known to all concerned for a critical scrutiny and can be used in testing the conclusions through replication. The method mainly aims at formulating the most general axioms (which can be termed as scientific theories).

RESEARCH PROPOSAL

This is a document that is typically written by a scientist or an academician which describes the ideas for an investigation on a certain topic. The proposal outlines the process from beginning to end and may be used to request financing for the project, certification for performing certain parts of research of the experiment, or as a required task before beginning a dissertation.

DIFFERENT STAGES OF RESEARCH

There are several important steps to follow whenever a scientific problem is to be solved. During this process, the problem must be very clear and must include any simplifying assumptions and this should be developed into a mathematical statement involving the use of one or more mathematical procedures. Different techniques and procedures are to be learnt through advanced text books or review articles, and to arrive at a decision, the results have to be interpreted. This requires experience and understanding of the sequence of components of research, which are as follows: *Selection (choosing) of a research topic, Definition of a research problem, Literature survey and reference collection, Assessment of the current status of the chosen (selected) topic, Formulation of a hypothesis, Research design, Actual investigation, Data analysis, Interpretation of the result and the Result.*

1. Selection (choosing) of a Research Topic

The selection of a topic and the problem is the starting point of research and in choosing a problem of research, the significance of certain problems for the advancement of science must be taken into consideration. The identification of a suitable research topic for research work is one of the most difficult parts of a research and before choosing a topic and a problem, the researcher should keep the following points in his/her mind – the topic must be suitable for carrying out research, the investigator should have interest in it, and the topic should not be chosen by compulsion by someone else. The research topic and the problem must be fixed in consultation with the guide (research supervisor). The problem suggested in a broad view must be narrowed down and defined in an operational form. It is not necessary that the topic of a Ph.D., should be different from a project of his/her PG or M.Phil., dissertation (thesis), as it saves sometime in the duration of his Ph.D. work. After choosing a topic of his/her own interest, he/she must find a research supervisor working in that topic or interested in guiding him/her in a proper manner. As there are many limitations in this method, it is always better to choose a topic on which he/she has to work, the goal of which is already defined by a research supervisor who is working in that topic or interested in guiding him/her in a proper manner. A research topic may also be chosen by a researcher him/herself which has many advantages such as: he/she can pursue his/her own interest to the farthest limits, there is an opportunity to spend a long time on something, and the results would prove better in terms of growth and the quality of the work but this has its own limitations. The researcher should not choose his/her topic because it is fascinating, and he/she should always take care of the possibility of collection of data, the size of the topic (it should not be too narrow) etc. He/she has to study text books of advanced level as well as latest articles of research to identify the most reliable problems.

2. Definition and Formulation of a Research Problem

The research problem after identification and selection has to be properly defined and formulated in order to solve it. This can be executed as follows: The problem must be converted in the form of a questionnaire or its equivalent, it must be specified in detail in precise terms, the assumptions made should be listed out, the ambiguities in the statement if any should be removed and the feasibility of a particular solution should be properly examined. Defining the problem is important than its solution and this should not be done in a hurry. Hence, proper care must be taken in defining the topic. The very first definition of the problem may not be appealing, and as such redefinition is required in order to make it a good problem. Hence, it can be made to meet the criteria of a good problem by suitably rewording or reformulating the chosen problem. Before initiating research work on a problem, a rough estimate on costs and the time required to complete the work in every respect must be made by the researcher. The researcher must make sure that the necessary experimental setup and materials to perform the actual work are available in the department where the research work is to be carried out.

3. Literature Survey and Reference Collection (Bibliography)

The investigator has to do literature survey connected with the problem after defining the research problem. Literature survey is *a collection of research publications, books and other documents related to the defined problem*. It is highly essential to know whether the problem has already been solved, its status, the techniques that are useful to investigate and other details. The investigator can carry out his/her survey of the journals which regularly publish the abstracts of papers (published) in different journals, the review articles related to the chosen topic, journals that publish articles of research, books (advanced level) on the selected topic, conference proceedings and workshops etc, reprint and preprint collections available with the research supervisor and experts working on the same topic, internet and free e-print service providers.

Before attempting to plan the study, review of literature in the area of research is a preliminary step which helps the investigator in sharpening the problem, for reformulation and even defining other related problems, to have proper understanding of the problem, to acquire proper theoretical and practical knowledge to investigate, to show how the problem is related to previous studies and to know whether the problem has already been solved. The investigator can collect the relevant information about the problem and also can acquire clarity of ideas through literature survey. The literature connected with a similar problem helps to formulate the problem in a clear-cut way and it also helps the researcher to design methodology for the present work and helps familiarize with the precepts, contexts as well as interpretations and finally to formulate a satisfactory structure of the research. Under no circumstances, a work already carried out by another investigator is to be duplicated. A good library of an institution or a university will be of a great help during the survey.

At present, the internet has become an important source of knowledge and an effective medium for research, as it provides many new opportunities for the collection of information, networking, for conducting the research, for the collection of data and the dissemination of the results. The latest tools provided by the internet include e-mail, e-journals; submission of articles on-line to journals, video conferencing etc, and the internet also provides the details about various scientific and non-scientific topics, full published articles of the past, preprint papers of the investigators, information about the scientific meetings, contact details of other investigators, databases for reference material, and the places from which the help can be sought. The internet is also useful as the collection of data can be made within a short span of time, identification of data unknown to the researcher etc.

The preparation and collection of references (bibliography) with annotations should be undertaken as soon as the literature survey begins. During the investigation, keeping systematic records (in files) helps a researcher to preserve the data for a future use. *Current Contents* (journal) is the most important source of reference collection as it provides titles (of articles), authors (names), dates of publication, numbers of the volumes, page numbers of the articles and the addresses (of the authors) from whom a reprint of the article can be collected. The useful articles can be collected from a local library or from a centre of document delivery service (INFLIBNET – India) or purchase the articles on payment from the publishers or a free reprint from the author by contact through e-mail or letter. With reference to a research paper, its title, name of the journal, volume number, pages (starting and ending) of it, and the year of publication should be noted; and for a book, name of the publisher, place of publication and year of publication are necessary. These details of references can be stored in a computer as well as in two to three compact discs. The collected references can also be classified, for example, the sources that deal with the theory, with experimental techniques, with numerical methods etc. Similarly, cross references should also be collected and classified to provide useful information. The collection should be organized with utmost care to retrieve it when it is needed.

4. Assessment of the Current Status of the Chosen (selected) Topic

The current status of the chosen topic can be easily identified by going through relevant journals and recent published papers, conference discussions, seminars and workshops etc. The letter sections of many international journals also publish important articles and these can also be referred to. The scientific papers accepted by journals for publication can also be downloaded from the internet.

5. Formulation of a Hypothesis

Scientific research is normally carried out with some aim or expectation which forms the hypothesis. *A hypothesis is a scientifically reasonable prediction.* The hypothesis refers to certain outcome from the study before the commencement of actual research work (experiment, theoretical calculation or numerical analysis) which can be stated in the form of sentences or certain logical terms. The proposed solutions on the basis of the available data and their tentative explanation constitute the hypothesis. The formulation of hypothesis is generally done after exclusive survey of relevant literature and learning the status of the research topic. A good hypothesis should have clarity and a theoretical orientation and should be stated in a suitable way so that it can be tested by further investigation. If this withstands the investigation by experiment and provides the necessary facts to make it acceptable (to researchers as well as others) and sufficiently reinforced by continuous verification, then alone it becomes a theory.

6. Research Design

The manner in which the investigation is carried out refers to the mode of approach, and this should always *keep the investigator on the right track and make him/her successfully complete the planned work.* Every job is to be done with maximum care. A research design is very helpful in order to make a steady progress in research and to assess the progress of it. According to Napolean Hill, *`plan your work and work your plan`* is the important suggestion to a researcher. The research design should indicate the different approaches to be used in solving the problem, sources and information related to the problem, and time frame as well as the budget. The research design helps to create the foundation of the entire work and will help to perform the chosen task easily and also in a systematic way. The very first step in the actual research work is to learn the facts associated with the problem, such as the theoretical methods, numerical techniques, experimental methods and other relevant data as well as the tools that are necessary for the study. An investigator has to identify and select the materials which are useful for his/her selected problem, and the validity and the utility of information gathered should necessarily be tested before using it. The sources of mathematical, numerical and experimental methods on which the scientific research is based on, have to be properly studied and judged before their application to the problem of interest.

7. Mode of Approach

The most important requirements for scientific research are absolute honesty, patience, stamina, precision and devotion to the subject matter with imagination and analytical ability. The investigator should not do the entire work in collaboration with others and should have the *ability to do all by himself/herself independently.* The researcher is advised to perform all the work starting from identification of the problem to report preparation by himself/herself under the guidance of the supervisor, and this helps to acquire enough knowledge, confidence and training to carry out the research independently after getting the Ph.D. degree. The thesis/ dissertation should reflect the efforts of the researcher and hence, one should avoid collaboration with others as far as possible in the first stage of his/her investigation. The investigator should avoid embarrassing the supervisor and senior researchers by asking doubts often and a good supervisor/senior researcher does not provide any answers to the questions but gives appropriate directions to clarify the doubts. He/she should not allow the personal life to interfere with research and should make research a part of his/her everyday life. A researcher should have persistence, tolerance and self-control over the unpleasant outcomes like not getting an expected result, not recognized by the supervisor, and the rejection of his article from a journal. One should maintain an excellent good relationship with the supervisor and also with his/her seniors and colleagues throughout the period of association during research. Always remember to follow the professional etiquette, and take the supervisor's criticism in the right spirit and respond appropriately, and there should be no emotional outbursts with anybody during the work. The most crucial period for a research scholar doing his full-time Ph.D. is the last year of the programme and during this period, one should always concentrate on the completion of the final work for his thesis (dissertation) and writing of various chapters, by avoiding diversions to other activities. The researcher should always be motivated by a noble goal and work gets the first, second and third priority. The attributes of a good research scholar can be summarized as: *self-confidence, dedication, concentration, determination, analytical mind, scientific discipline, global outlook, innovative approach, originality, intellectual curiosity, freedom from the obsessions of clock* (time) *and calendar* (days), *flexibility, keen observation, intelligence, passion for knowledge, questioning attitude, spirit of enquiry, insight, precision and accuracy, resilience to withstand temporary setbacks, persistence, patience, social skills, presentation skills,* and *writing skills*.

8. Actual Investigation and Analysis of Data

A researcher should always aim at doing good research. To identify a best research, the quality of research publications, the number of citations of his/her publication, the projects completed, the books published, and the contribution made to science and society etc, can be considered. The work that has been published in reputed international journals, cited by other researchers working in the same (similar) topic, and the work which added new information to the existing knowledge on a particular topic is generally considered good research.

An investigation should always lead to an original contribution (it should not involve objectionable duplication of some other's work) and this is basic credit point in any research. He/she should develop always new ideas and a deep insight into the problem in order to get novel and new ideas so that it will be a good research. One should always use recently introduced theories, experimental techniques and numerical algorithms instead of outdated methods. The work should be such that it should invite others to read it, use it and cite it in their work, and finally should lead to recognition and respect. The different steps of the work should be scientifically taken and the work should be free from mistakes. According to Prof. Devanathan, *a researcher should check, recheck, cross-check... all the results before submitting a research paper to a journal*. Before beginning to write a part of the work done and the results obtained, check and recheck the data and the results by repeating the experiment once again, rerunning the programs and going through the theoretical derivations and arguments if any. During the analysis of the data, appropriate tools have to be employed, as many statistical tools as possible have to be employed, and the number of data used, different units of the data, the error bars and other important details must be noted in the graphs if any. The validity, efficiency and applicability of special cases of the theory (method) identified if any, along with merits and demerits must be demonstrated with examples. As mere measurement (certain quantities) alone is not enough, the interpretation of the observed data and explanation for the particular pattern must be made on the basis of which the general principles underlying the process to be formulated. The following are some of the common errors generally in research: Selective and inaccurate observation, over-generalization and made-up information, and illogical reasoning and premature closure of inquiry.

9. Interpretation of the Result, Discussion and Conclusion

After the completion of the actual research work on the chosen topic, next step is to prepare the results and conclude the work, because the ultimate goals of research are predictions, results and conclusion. The investigator should analyze cause and effect, and pay greater attention to minute details of the work as the keenness in observation is the hall mark of any scientific research. Interpretation refers to the task of drawing inferences from the actual research work, and also means drawing of conclusion which is based on the study and helps to bring out relations and processes that underlie the findings. The interpretation of the result is important because: it links the present work (completed) to the previous (already completed), leads to identification of future problems, opens up new avenues and stimulates the quest for more knowledge, helps others to understand the importance of the research findings, and also suggests a possible experimental verification. During interpretation, under no circumstances, baseless statements should be made by the researcher and he/she should not yield to fabricate the results and interpretation, and should not try to report the already reported findings of others as their own. One should always acknowledge and cite the original sources which enhance the credibility of his/her work.

10. Write-up or the Preparation (Presentation) of the Report/Thesis/ Dissertation and Research Paper

The researcher finally has to prepare the report (thesis) of what has been done by him/her. The *presentation* or writing of a report must be done with great care keeping in view the following steps. The *general layout* of the report should contain the preliminary pages, the main text and the end matter. The *preliminary pages* should necessarily carry the *title* and the *date* followed by *acknowledgements* and *foreword.* There should be a *table of contents* followed by a *list of tables and graphs and charts*, if any available. The *main text* of the report should have *introduction, summary of findings, main report* and *conclusion*. The *introduction* should contain a clear statement of the *objective* of the investigation and an *explanation* of the methodology adopted along with many *limitations* stated therein. The summary of *findings* and *recommendations* in non-technical language should follow introduction (if extensive, they should be summarized). The *main body* of the report should be generally presented in *logical sequence* as readily *identifiable sections*. The main text should be followed by conclusion wherein the researcher should put down clearly and precisely the results (final summing up) of his/her research. The *appendices* in respect of all *technical data* should be enlisted at the end of the report. This should be followed by *bibliography* (the list of books, journals, reports etc, consulted). Finally an *index* should also be given in a published research report. The report should be written in a *concise and objective style in a simple language* and vague expression like `*it seems*`, `*there may be*` etc, should be avoided. The *charts and illustrations* contained in the main report should be used only if they present the information more clearly. It is also mandatory to mention the *calculated `confidence limits` and many constraints* experienced by the researcher during the conduct of research.

CRITERIA FOR GOOD RESEARCH

Any scientific research should satisfy the following criteria: The purpose of the investigation should be clearly defined and common concepts should be used. The procedure used should be described in sufficient detail to permit others to repeat the research for further advancement. The design of the research should be carefully planned to yield good results. The investigator should report flaws in the procedure and their effects upon the final findings with complete frankness. The validity and the reliability of the data should be checked carefully during analysis. The final conclusions should be confined to those well justified by the data, and a greater confidence is warranted if the investigator is experienced and has a good reputation in research and is a person of integrity. Thus, the qualities of a good research are: a good research is *systematic*, a good research is *logical*, a good research is *empirical* and a good research is *replicable*.

OBJECTIVES OF RESEARCH

The main objectives of research are: to discover the new facts, to verify and test the important facts, to analyze a process or phenomenon, to identify the cause and effect of relationship, to develop new scientific tools, concepts and theories to solve and understand and find the solutions to the scientific and non-scientific problems, and to overcome or solve the problems that occur in our everyday life.

Why do people carry out Research?

No investigator would like to do research unless there are some motivating factors like: to get a research degree (like Ph.D.) along with it benefits such as better employment, promotion, increment etc; to acquire a research degree and to get a teaching position in a college/university or become a scientist in a research institution; to get a research position abroad and settle there; to solve the unsolved and challenging problems; to get recognition and acquire respectability; and curiosity to find new things and to find out the unknown facts of an event and thereby serve the society by solving the social problems.

Importance of Research

The importance of research is: it calls for a thorough understanding and possible solution for a theoretical or a practical situation; it helps to identify the range and applications of existing theories and concepts; it provides the necessary guidelines for solving the problems; it is important for higher gain and productivity and to improve the quality of the products in industry; to optimize the problems in industry (mathematical and logical research); to lead to the identification and characterization of new materials; to invent new and novel phenomena and processes; and to lead a new style of life.

Significance of Research

The significance of research can well be understood in the famous maxim - `all progress is born of inquiry; doubt is often better than overconfidence, for it leads to inquiry, and inquiry leads to invention`. Research inculcates scientific and inductive thinking and it promotes the development of logical habit of thinking and organization. Research provides the basis for nearly all government policies in our system. Research has its special significance in solving various operational and planning problems of business and industry. Research is equally important for social scientists in studying the social relationships and also in seeking answers to various social problems.

Research is significant to students who write a masters or Ph.D. thesis and it may mean a way to attain a higher position in the social structure; to the professionals it may mean a source of livelihood; to the philosophers it may mean the outlet for new ideas and insights; to literary persons it may mean the development of new things and creative work; and to analysts and intellectuals, it may mean the generalizations of new theories.

■ ■ ■

CHAPTER - 2

CENTRIFUGATION

Introduction

Cell organelles refer to the structures such as plastids, mitochondria, nuclei, dictyosomes, ribosomes, peroxisomes, glyoxysomes and even vacuoles. The membranes include plasmalemma, tonoplast, endoplasmic reticulum and the membranes associated with cell organelles. In order to understand a particular set of metabolic reactions and its control, we must devise means of separating the essential cellular entities from other parts of the cell. The most useful tool for this is a combination of tissue homogenization and differential centrifugation. The common methods of fractionation include: perfusion, tissue slicing, homogenization and centrifugation.

The techniques of centrifugation are based upon the particle behaviour in an applied centrifugal field, assumed that the parameters under investigation like the relative molecular mass, shape of the particles, density of the particles etc, may be related to the behaviour of those molecules in the gravitational field. When a solution of large particles is left to stand for a longer time, they tend to sediment under the influence of gravity. For any given particle, the rate of velocity at which it sediments is proportional to the force applied, so that the particles sediment more rapidly when the applied force is greater than the earth`s gravitational force. So, the basis of these techniques is to exert a larger force than does the earth`s gravitational field, thus increasing the rate at which the particles sediment. The particles normally are suspended in a specific liquid medium held in tubes or bottles located in the rotor. The *rotor* is centrally positioned on the *drive shaft* of the centrifuge. The particles differing in density, shape or size get separated because they sediment at different rates in the centrifugal field, each sedimenting at a rate that is directly proportional to the applied centrifugal field.

Centrifugation is one of the most important and most widely applied research technique in biochemistry, cellular and molecular biology and in medicine. The current research and clinical applications of centrifugation rely on the isolation of cells, sub-cellular organelles and macro-molecules often in higher yields. Many particles or cells in a liquid suspension, given time, will eventually settle at the bottom of a container due to gravity, but, the time required for such settling is impractical. Other particles which are extremely small in their size will not separate at all in solution unless they are subjected to a high centrifugal force. When such a suspension is rotated at a certain speed or RPM (revolutions per minute), the centrifugal force causes the particles to move radially away from the axis of rotation. This force (compared to gravity) on the particles is called RCF (relative centrifugal force). For example, an RCF of 500 x g indicates that the centrifugal force applied is 500 times greater than Earth`s gravitational force.

Centrifugation is a process which involves the use of centrifugal force for the sedimentation of heterogeneous mixtures with a centrifuge. The process is generally used to separate two immiscible liquids both in the industry and laboratory. The more dense components of the mixture migrate away from the axis, while less dense components migrate towards the axis. The effective gravitational force on a test tube may be increased so as to more rapidly and completely cause the *pellet* (precipitate) to gather on the bottom of the tube. The *supernate* or the *supernatant* is the remaining solution which is quickly decanted from the tube without disturbing the pellet, or withdrawn with a Pasteur pipette.

The rate of centrifugation is specified by the angular velocity measured in RPM (*revolutions per minute*) or acceleration expressed as *g*. The conversion factor between RPM and *g* depends on the radius of the sample in the rotor. The settling velocity of the particles is a function of their size and shape, the acceleration, the volume fraction of the solids present, the difference in the density between the particle and the liquid and the viscosity. The ultra-centrifuge can spin in excess of 70,000 rpm and thus it is well suited for the isolation of small particles (DNA or viruses).

Centrifugation is a process by which a centrifuge is used to separate the components of a complex mixture by spinning the laboratory samples at very high speeds, the components of a given mixture are subjected to centrifugal force, which causes more dense particles to migrate away from the axis of rotation while the lighter particles move toward it. These can sediment at the bottom of the tube into what is known as the pellet, and this isolated specimen or the remaining solution as the supernatant can be further processed or analyzed.

Centrifugation - Definition

Centrifugation (syn. Centrifugalization) is *"the process of separating fractions of systems in a centrifuge, the most basic separation is to sediment a pellet at the bottom of the tube, leaving a supernatant at a given centrifugal force"*.

TYPES OF CENTRIFUGATION

These techniques are mainly of two types, namely, Preparative and Analytical centrifugation techniques.

1. Preparative Centrifugation Techniques

These techniques are concerned with the actual preparation, isolation and purification of the whole cells, sub-cellular organelles, plasma membranes, polysomes, ribosomes, chromatin, nucleic acids, lipoproteins and viruses for further investigation. Very large amount of the material gets involved in these techniques (for example, cells of microbes from culture media, plant cells/animal cells from tissue culture and plasma from blood). Larger amounts of cellular particles are also to be isolated to study their morphology, composition and biological activity. Similarly it is possible to isolate macromolecules (biological – like nucleic acids and proteins) from the preparations with some measure of purification (by fractional precipitation)

2. Analytical Centrifugation Techniques

These techniques are mainly associated with the study of pure particles or macromolecules. These are primarily concerned with the study of the characteristics of sedimentation of biological macromolecules rather than with the collection of specific fractions. These methods require only smaller amounts of the material. These also utilise specially designed rotors and detector systems for continuous monitoring of the process of sedimentation. These studies often yield much information from which the purity, relative molecular mass and shape of the particles can be easily deduced.

PREPARATIVE ULTRACENTRIFUGATION - SEPARATION METHODS

a. Differential Centrifugation / Differential Pelleting

Sometimes called as *differential pelleting*, this is the simplest form of separation. The particles of different densities or sizes in a suspension will sediment at different rates, with larger and denser particles sedimenting fastest followed by less dense and smaller particles. The rates can be increased by using centrifugal force. A suspension of cells subjected to a series of increasing centrifugal force will yield a series of pellets containing cells of decreasing sedimentation rate. In this method, separation is done based on the size and density, with larger and denser particles pelleting at lower centrifugal forces (for example, unbroken whole cells will pellet at low speeds and at short intervals like 1000g for 5 min). The smaller cell fragments and organelles remain in the supernatant as they require more force and greater times to pellet out. During the process, a tissue sample is first homogenized to break the cell membranes and mix up the cell contents. The homogenate is then subjected to repeated centrifugations, each time removing the pellet and increasing the centrifugation force. Finally, purification may be done through equilibrium sedimentation, and the desired layer is extracted for further analysis.

Differential centrifugation is commonly used in simple pelleting and in obtaining partially pure preparations of sub-cellular organelles and macro-molecules. This method is commonly used for harvesting cells or producing crude sub-cellular fractions from the tissue homogenate. This is a common procedure in microbiology and cytology and is used to separate organelles from whole cells for further analysis of specific parts of cells. It is unusual to use more than four differential centrifugation cycles for a normal tissue homogenate. The process suffers from contamination and poor recoveries due to heterogeneity in biological particles. The contamination can be addressed by re-suspension and repeating the steps (washing the pellet). This serves as the preliminary step for further purification using other types of centrifugal separation like density gradient separation.

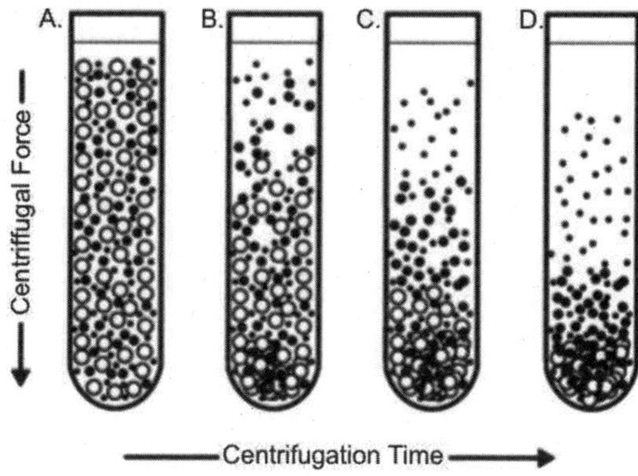

Differential Centrifugation

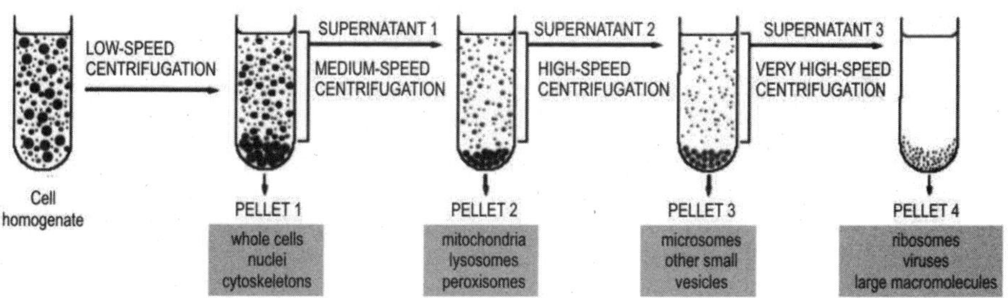

Differential Centrifugation – Flow Sheet Diagram

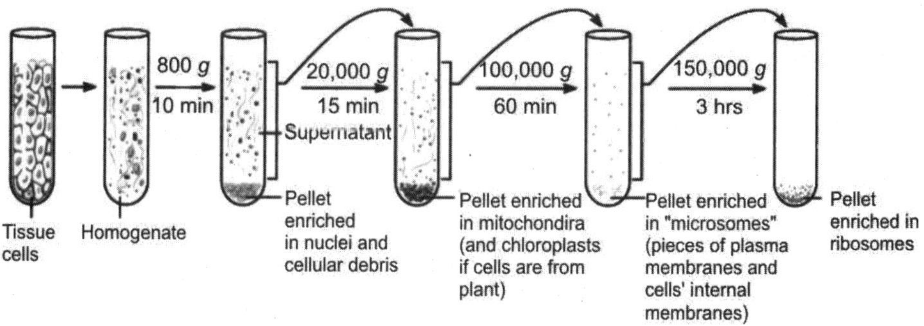

Differential Centrifugation – Different steps

b. Density Gradient Centrifugation / Isopycnic Centrifugation / Equilibrium Sedimentation

This is also called as isopycnic (= equal density) centrifugation or equilibrium sedimentation, and this procedure is widely used in biochemistry to separate molecules based on their isopycnic point (buoyant density). A self-generating density gradient is established via equilibrium sedimentation, and then the analyte molecules concentrate as bands where the molecule density matches that of the surrounding medium. This is achieved by spinning biological preparations at high g-force over long periods of time in buffers or solutions containing a variable amount of a viscous molecule (a 20 to 50% linear sucrose gradient is used in the purification of clathrin coated vesicles [CCVs]).

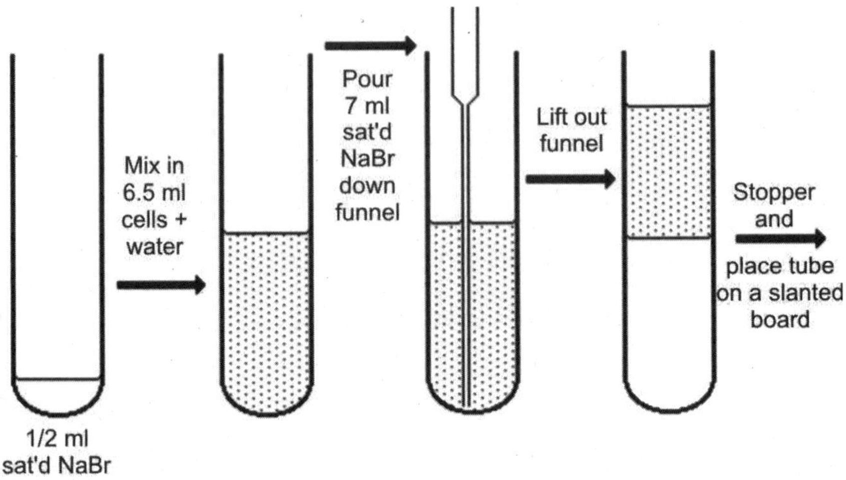

Density Gradient Centrifugation – Flow sheet diagram

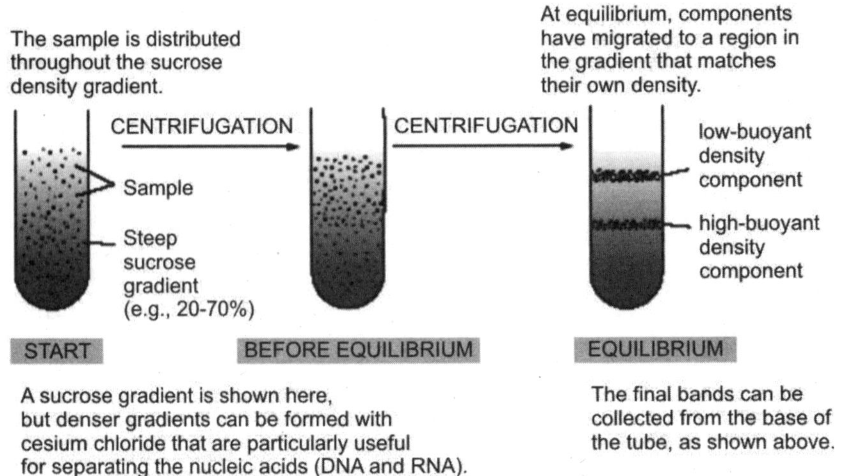

Equilibrium Sedimentation – Method

Density gradient centrifugation is the preferred method to purify the sub-cellular organelles and macromolecules. The density gradients can be generated by placing layer after layer of gradient media like sucrose in a tube with the heaviest layer at the bottom and the lightest layer at the top in either a discontinuous or continuous mode. The cell fraction to be separated is placed on top of the layer and centrifuged. This can be classified into 2 categories, namely, rate zonal (size) separation and isopycnic (density) separation

1. Rate Zonal (Size) Separation

Rate zonal separation takes advantage of particle size and mass instead of particle density for sedimentation. The common applications include the separation of cellular organelles like endosomes or separation of proteins like antibodies. The antibody classes all have very similar densities but different masses. Hence, separation based on mass will help separate the different classes whereas the separation on the basis of density will not be able to resolve the antibody classes. The important criteria for a successful rate zonal centrifugation include: the density of the sample solution must be less than that of the lowest density portion of the gradient, the density of the sample particle must be greater than that of the highest density portion of the gradient, the path length of the gradient must be sufficient for the separation to occur. The time is the most important criterion. The rate zonal separations are dynamic rather than static. If the centrifugation is continued long (long runs) enough, all the zones end up as one pellet at the bottom of the tube. In this method, the problem of cross-contamination of particles of different sedimentation rates may be avoided by layering the sample as a narrow zone on the top of a density gradient. In this way, the faster sedimenting particles are not contaminated by the slower particles as in differential centrifugation. However, the narrow load zone limits the volume of sample (typically 10%) that can be accommodated on the density gradient. The gradient stabilizes the bands and provides a medium of increasing density and viscosity. The speed at which the particles sediment depends primarily on their size and mass instead of density. As the particles in the band move down through the density medium, the zones containing the particles of similar size form as the faster sedimenting particles move ahead of the slower ones. Because the density of the particles is greater than the density of the gradient, all the particles will eventually form a pellet if centrifuged long enough.

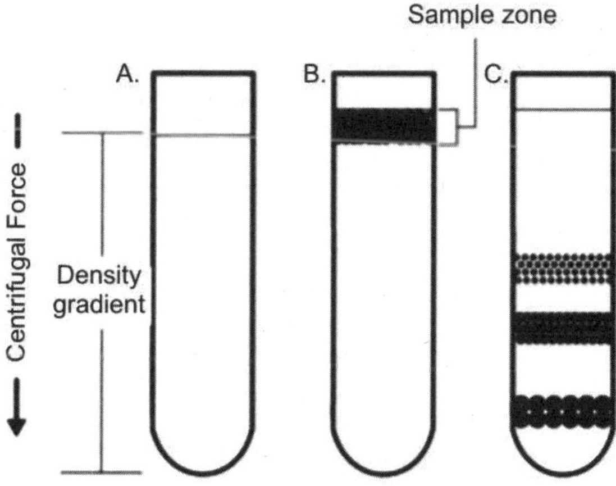

Rate-zonal Centrifugation

2. Isopycnic (Density) Separation

Isopycnic separation is also called as *buoyant* or *equilibrium separation* and in this method, the particles are separated solely on the basis of their density. The particle size only affects the rate at which particles move until their density is the same as the surrounding gradient medium. The density of the gradient medium must be greater than the density of the particles to be separated. Through this method, the particles will never sediment to the bottom of the tube, no matter how long the centrifugation time. The particles with specific density sediment until they reach the point where their density is the same as the gradient media (the equilibrium position). The gradient then is said to be isopycnic and the particles are separated according their buoyancy. Since the density of biological particles is sensitive to the osmotic pressure of the gradient, isopycnic separation may vary significantly depending on the gradient medium used. The preparative techniques commonly use a discontinuous gradient in which the particles band at the interface between the density gradient layers although a continuous gradient may be more suited to analytical purposes. This tendency makes harvesting certain biological particles like lymphocytes easier.

In this method, a particle of a particular density will sink during centrifugation until a portion is reached where the density of the surrounding solution is exactly the same as the density of the particle. When the quasi-equilibrium is reached, the length of centrifugation does not have any influence on the particle migration. The best example is the separation of nucleic acids in caesium chloride gradient. A variety of gradient media can be used for this separation. For example, sugars like sucrose are good for viruses and organelles; polysaccharides like ficoll are good for some cells, some viruses and some organelles; colloidal silica like percoll is good for cells and organelles; iodinated media like nycodenz is excellent for cells, organelles and nucleoproteins; and alkali metal salts like CsCl is good for some viruses and nucleoproteins and excellent for macromolecules. The important criteria for this separation include: the density of the sample particle must fall within the limits of the gradient densities, any gradient length is possible, and the run time must be sufficient for the particles to band at their isopycnic point and excessive run times have no adverse effect. The analysis of the fractions is done by standard biochemical procedures such as: Spectro-photometric or Radio-isotopic methods to characterize their biochemical activities.

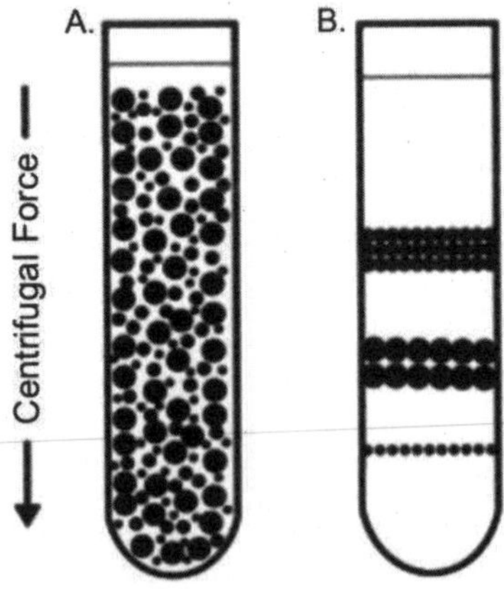

Isopycnic Centrifugation

c. Counter-flow Centrifugation Elutriation (CCE)

Counter-flow centrifugation elutriation is a cell separation technique that enables the scientists to separate different cells with different sizes. As the cell size is correlated with different stages of the cell cycle, the method allows the separation of cells at different stages of the cell cycle. The main concept is that the larger cells tend to stay within the flowing buffer solution while the smaller cells will be washed away following buffer solution, due to different sedimentation property.

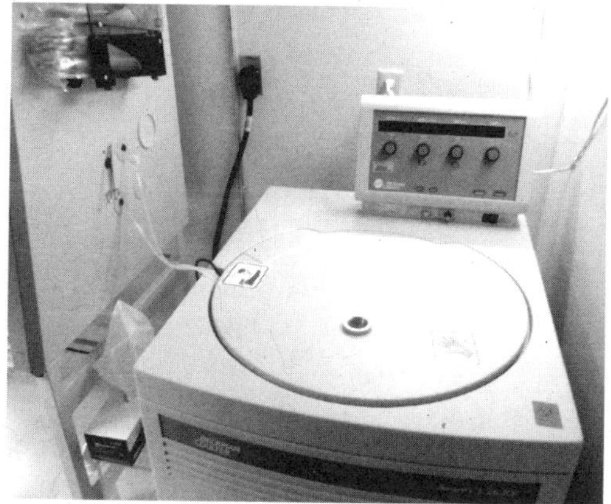

CCE or Elutriation System

d. Sucrose Gradient Centrifugation

Sucrose gradient centrifugation is a type of centrifugation which is often used to purify the enveloped viruses and ribosomes, and also to separate the cell organelles from the crude cell extracts. This method is also used to purify exosomes. In this method, a sucrose density gradient is created by gently overlaying lower concentrations of sucrose on higher concentrations in a centrifuge tube. This is of two types, equilibrium centrifugation and non-equilibrium centrifugation.

In the former method, a sucrose density gradient is created gently by overlaying lower concentrations of sucrose on higher concentrations (layers extending from 70% sucrose to 20% sucrose in 10% increments) in a centrifuge tube. Sometimes, gradient mixers of gradient makers can be used to form linear, concave, convex and exponential gradients. The sample containing the particles is placed on the top of the gradient and centrifuged at forces in excess of 150,000 x g. The particles travel through the gradient until they reach the point in the gradient at which their density matches that of the surrounding sucrose. This fraction is removed and analyzed.

The latter method is similar to the former except that the experiment is only run until a particular point (gradients are called velocity gradients). The run ends before the equilibrium is reached, and the desired particle will be at a known distance from the surface. When the run ends, this band (fraction) is collected by several methods. The easiest way is to elute (drain) the sucrose solution by puncturing at the bottom of the tube while the desired protein or material is kept behind. It is also possible to suction the sucrose gently (not mixing the layers formed) and dividing the material into successive fractions, which can be assayed by a number of methods to determine the distribution of the desired material.

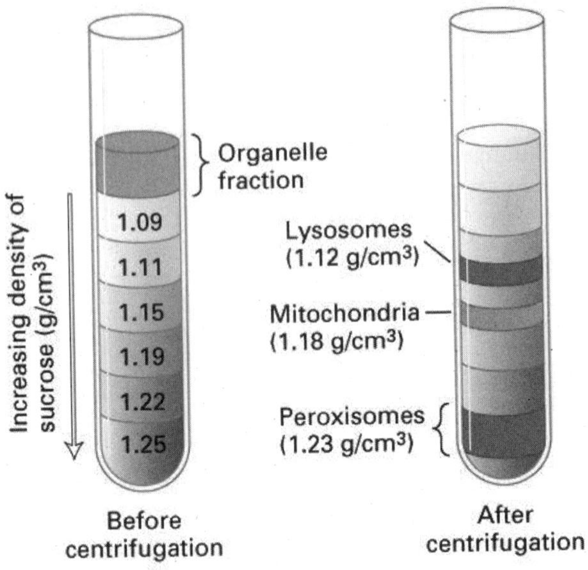

e. Buoyant Density Centrifugation

Buoyant density centrifugation uses the concept of a buoyant force to separate the molecules. Usually, caesium chloride solution approximately of the same density as the molecules is used. The sample is put on the top of the solution and the tube is spun at a very high speed for a longer period of time (sometimes even days). The molecules of caesium chloride get densely packed towards the bottom and these result in even layers of different densities. These layers go to a level where their density and density of caesium chloride are the same, and form a sharp distinctive band. This method separates the molecules sharply and it can even separate different molecular isotopes from one another.

f. Ultracentrifugation

In ultracentrifugation, the cell or a tissue homogenate is prepared by a glass homogenizer or a Waring blender. The homogenization is generally carried out at 0°C – 4°C so that the cell components may not break enzymatically. The tissues, such as animal liver is homogenized in 0.25 molar sucrose solutions before subjecting to differential centrifugation. The pH of the isolation medium is also kept at the required level. The ultracentrifuge can spin in excess of 70,000 rpm and thus it is well suited for the isolation of small particles (DNA or viruses).

Ultra-centrifugation may be: Preparative or Analytical.

1. Preparative Ultracentrifugation

This method involves the preparation of the samples. The Preparative Ultra-centrifugation is again of 2 types, namely:

a. Zone Ultra-centrifugation

Sedimentation Velocity or Zone Ultra-centrifugation involves the separation of particles of same density but of different size.

b. Isopycnic Ultra-centrifugation.

Equilibrium Density or Isopycnic Ultra-centrifugation involves the separation of the particles of same size but of different densities.

2. Analytical Ultracentrifugation

Analytical ultracentrifugation involves homogeneity or purity test as well as the determination of the S - value. During analytical ultra-centrifugation, the homogeneity test is done by using quartz cubebs, using white light as well as ultra-violet light, as quartz is transparent to both the lights. With the help of a Photographic Recording Unit (PRU) and Spectroscopic Recording Unit (SRU), the purity of the fraction is confirmed. If there is one band in both the cases, the same is considered to be pure. The sedimentation coefficient (S) is calculated in terms of Svedberg`s unit. It is the sedimentation velocity per unit of force (**$1S=1x10^{-13}$ cm/sec/dyne**).

CENTRIFUGE – THE INSTRUMENT

Centrifuge is an instrument used in every bio-medical research lab across the globe. This is a device used to increase the rate of sedimentation of suspensions and is used to separate the solid, liquid or two immiscible liquids. The word centrifuge has its origin in 1795 – 1895 and is derived from French word, *centrifuge* meaning `centrifugal `and Neo-Latin word, *centrifugus* meaning `centre-fleeing`. The first known use of centrifuge is in 1887 or in 1895 and is not definitely known. Various definitions for a centrifuge are in use.

Portable Centrifuge

Centrifuge is an apparatus for separating the components of a suspension or to separate two or more substances of different density, that is, two liquids or a liquid and solid. It consists of a fixed base or a frame and a rotating part in which the mixture is placed and then spun at high speed. The liquid with the suspended particles is kept in test tubes (centrifuge tubes) which are set in the rotating part in holders which is whirled or spinned horizontally in a circle. The holders are arranged so that when the rotary motion begins, the test tubes swing into a slanted to horizontal position with their open ends toward the axis of rotation. The heavier or the solid part of the specimen is thrown outward into the bottom of the tube (pellet) while the lighter liquid part comes to the top, into the supernatant.

Centrifuge – Components and Parameters

The most important component of a centrifuge is the *rotor*. This is the moving part that spins at high speeds. The rotors can be fixed in position, or a centrifuge can use multiple rotors fixed atop a part called the *spindle*. Usually the rotor will have a lid that is screwed down tightly to prevent the samples from flying out during spinning. Many centrifuges will have a refrigeration unit that allows the internal temperature to be controlled during the spin. These also have some knobs or buttons for inputting the parameters for each run, such as the duration of the spin, the temperature, and the magnitude of the spin in terms of speed or relative force. The spin speed is measured as *revolutions per minute* (RPM). This is a tricky value, because it is not the speed of the centrifuge that causes the particles to separate from a mixture, but the force acting on these particles. This *force is related to the radius of the centrifugal rotor*, and since different centrifuges have different rotor sizes, different forces can be applied at the same speed or RPM. The strength of centrifugation can also be quantified as *relative centrifugal force* (RCF), and this is generally presented as a multiple of earth`s gravitational acceleration. RCF is expressed as *the product of the radius of the rotor and the square of the angular velocity divided by earth`s gravitational acceleration*. RCF can be related to RPM by the equation, mv^2/r, where r stands for the radius of the centrifuge measured in centimetres. Centrifuges often come with nomograms that can help convert RCF to RPM quite easily. A ruler is used to connect the radius of the centrifuge and a given RPM value, in order to obtain RCF.

Centrifuge - Operation

To begin the spinning the samples with a centrifuge, one must consider the temperature. If a refrigerated centrifuge is being used, the internal temperature of the machine should reach the desired value before starting the spin, or one must find other ways to cool the rotor. Before every spin, one has to ensure that all of the caps on the tubes are tight and secure. It is to be made sure that each sample is counterbalanced with another sample directly across from it when loading the tubes. If there is only one tube, one must make sure that another tube will act as a counterweight. In the case of 3 tubes, they can be arranged in a triangle so that they are equidistant from each other. In a centrifuge, the balancing weights are critical as the rotors reach high speeds and possess a lot of kinetic energy. If they are improperly balanced, the entire unit can be propelled from its place resulting in serious damage. Once the rotor and the lid are secure, the centrifuge is started and one has to wait until it reaches the desired speed. In case of a problem, an experienced technician is called for, for rectification of the defect. The supernatant can be removed by decanting (poured off) or aspirated (removed by suction). The pure specimen can be returned to a solution by re-suspending it. All the cells (spinning cells, aspirating cells and washing cells [re-suspension in buffer]) are finally washed.

The biochemical experiments are usually conducted with dissolved or suspended particles in solution. As such, the rate of sedimentation of a particle depends not only upon the centrifugal field applied but also upon the mass of the particle. This can be expressed as the product of the volume and density of the particle, and the density and viscosity of the medium in which it is sedimenting, and also the extent to which its shape deviates from spherical. It is also clear that a mixture of heterogeneous particles can be separated on the basis of their densities and/or their size, either by the time required for their complete sedimentation or by the extent of their sedimentation after a given time. These form the basis for the separation of biological macromolecules and also of cell organelles from tissue homogenates. The order of separation is generally the whole cells and cell debris first, followed by the nuclei, chloroplasts, mitochondria, lysosomes (or other microbodies), microsomes (the fragments of the smooth and rough ER) and finally the ribosomes.

The sedimentation rate or the velocity (**υ**) of a particle can also be expressed in terms of sedimentation rate per unit of centrifugal field, called as its sedimentation coefficient, **s**. As the sedimentation rate studies can be performed by the use of a wide variety of solvent-solute systems, or at different temperatures, the experimental value of the sedimentation coefficient (affected by temperature, viscosity of the solution and the density of the solution) by convention is corrected to the sedimentation constant theoretically obtainable in water at 20°C.

The sedimentation coefficient for many macromolecules (nucleic acids and proteins) usually decreases in its value with an increase in the concentration of the solute. This becomes more severe with an increase both in the relative molecular mass and the degree of extension of the molecule. The sedimentation coefficients of most biological particles are very small. As such, for the sake of convenience, its basic unit is taken as $10^{-13}s$, which is termed as *Svedberg Unit* (S), in recognition of T. Svedberg`s pioneering work in these types of analyses. In general, the larger the molecule (or particle), the larger is its *s* and hence, the faster is its sedimentation rate. For example, the value of *s* for the enzymes, peptide hormones and soluble proteins are 2 to 25s; for nucleic acids is 3 to 100s; for ribosmes and polysomes is 20 to 200s; for viruses is 40 to 1000s; for lysosomes is 4000s; for biological membranes is 100 to 100 x 10^3s; for mitochondria is 20 x 10^3s to 70 x 10^3s; and for nuclei between 4000 x 10^3s and 40,000 x 10^3s.

COMMON CENTRIFUGES USED IN BIOLOGICAL RESEARCH

In biological research, the use of the following centrifuges is in practice.

1. Microcentrifuges

These are normally used to process small volumes of biological molecules, cells or nuclei. The tubes generally hold 0.5 to 2.0 ml of liquid and are spun at maximum angular speeds of 12,000 to 13,000 rpm. These instruments are small enough to fit on a table top and have the rotors that can quickly change their speeds. They may or may not have a refrigeration system.

2. High-speed Centrifuges

These super-speed centrifuges can handle larger sample volumes, from a few tens of ml to several litres. Larger centrifuges can also reach higher angular velocities (around 30,000 rpm). The rotors come with different adapters to hold various sizes of test tubes, bottles or micro-titer plates.

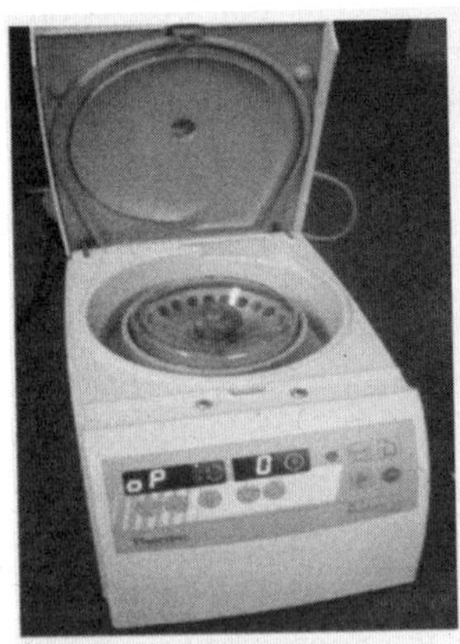

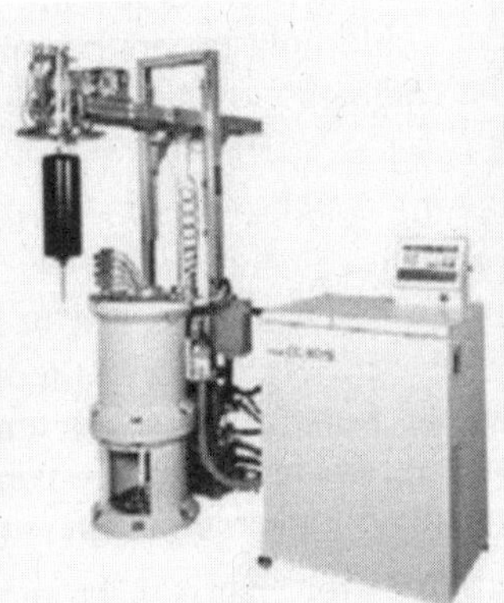

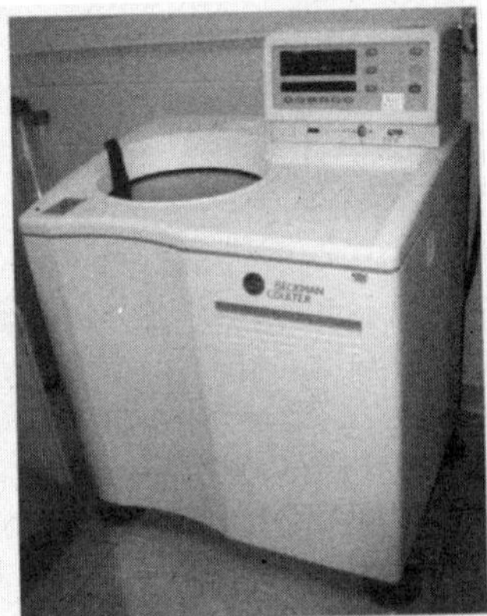

Microcentrifuge High Speed Centrifuge Ultracentrifuge

3. Ultracentrifuges

These can help isolate much smaller particles including ribosomes, proteins and viruses and can also be used in the studies of membrane fractionation. These instruments can reach maximum angular velocities in excess of 70,000 rpm. The molecules can be separated in batch or continuous flow systems. Analytical ultracentrifugation (AUC) can also be used for the determination of the properties of macromolecules like shape, mass, composition and conformation in addition to purification. The samples are centrifuged with a high density solution (sucrose, caesium chloride or iodixanol). The solution may be at a uniform concentration throughout the test tube (cushion) or of a variable concentration (gradient). During the run, the particles or molecules will migrate through the test tube at different speeds depending on their physical properties and the properties of the solution, and finally form a pellet at the bottom of the tube or bands at various heights. The molecular properties can also be modelled through sedimentation velocity analysis or sedimentation equilibrium analysis.

CENTRIFUGES - TYPES

The centrifuge is a motorized piece of laboratory equipment which spins the liquid samples to separate their components. Generally, they come in two sizes, namely a table-top version (often called as micro-centrifuge) and a larger floor model. The centrifuges can be classified on the basis of their intended use or by their rotor design. The types of centrifuges by their rotor design include: *Fixed-angle Centrifuges* (these are designed to hold the sample containers at a constant angle relative to the central axis); *Swinging-head or Swinging-bucket Centrifuges* (these have a hinge where the sample containers are attached to the central rotor, which allows the samples to swing outwards as the centrifuge spins); and *Continuous Tubular Centrifuges* (these do not have individual sample vessels and are generally used for high volume applications).

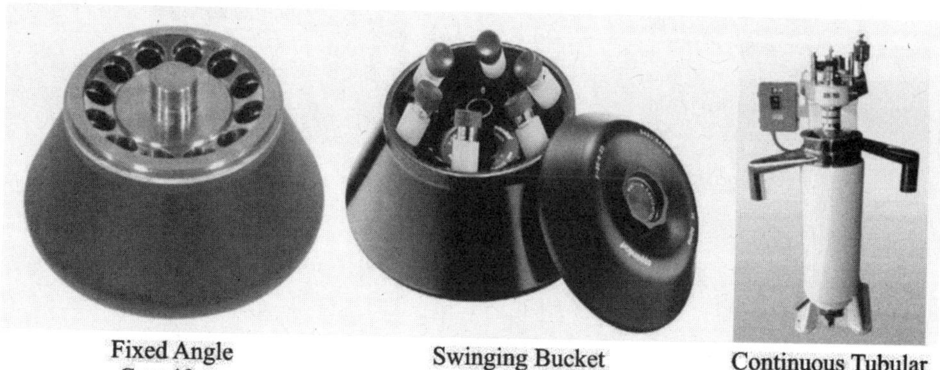

Fixed Angle Centrifuge | Swinging Bucket Centrifuge | Continuous Tubular Centrifuge

The types of centrifuges by their intended use include: *Ultracentrifuges* (these are optimized for spinning a rotor at very high speeds, and are popular in molecular biology, biochemistry and polymer science, and may be preparative or analytical and fixed-angle or swing-head types); *Haematocrit Centrifuges* (these are used to measure the percentage of RBCs in the whole blood in pathological investigations); *Gas Centrifuges* (this is a device that performs isotope separations of gases, which relies on the principle of centripetal force accelerating molecules so that the particles of different masses are physically separated in a gradient along the radius of a rotating container. A prominent use of gas centrifuge is for the separation of uranium-235 from uranium-238 and this replaces use of gaseous diffusion method of uranium-235 extraction. These also include Zippe-type centrifuges. The industrial centrifuges can be classified according to the type of separation of the high density fraction from the low density fraction); *Screen Centrifuges* (in these instruments, the centrifugal acceleration allows the liquid to pass through a screen, through which the solids cannot pass through due to granulometry larger than the screen gap or due to agglomeration. The common types of screen centrifuges are: screen/scroll centrifuges, pusher centrifuges and peeler centrifuges); *Decanter Centrifuges* (in these instruments, there is no physical separation between the solid and liquid phase, but, an accelerated settling due to centrifugal acceleration occurs); *Continuous Liquid Centrifuges* (the common types of these instruments are: solid bowl centrifuges and conical plate centrifuges); and *Airfuge* or *Air Centrifuge* [an airfuge (manufactured by Beckman & Coulter Inc., Spinco Division, Palo Alto, CA, USA), is a table-top preparative ultracentrifuge which is an air-driven ultracentrifuge and is a miniature air turbine, which is provided with a small rotor that operates at 90,000 to 100, 000 rpm and generates a maximum RCF of 178,000 x g within a short time. This is capable of accelerating a magnetically suspended 3.7 cm diameter rotor. It accommodates 6 x 175 mm^3 tubes on a virtually friction-free cushion of air in a non-evacuated chamber. This instrument has applications in biochemical research and clinical research where only small volumes of samples require high centrifugal forces, such as macromolecule/ligand binding-kinetic studies, steroid hormone receptor assays, separation of major lipoprotein fractions from plasma and deproteinisation of physiological fluids for amino acid analysis. This is used to separate chylomicrons from serum allowing accurate analysis on the clear *infranatant*].

Ultracenrtifuge

Haematocrit Centrifuge

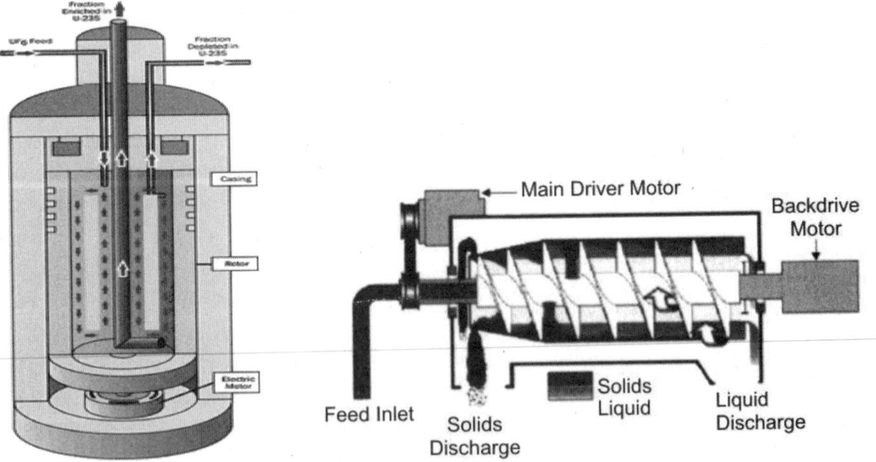

Gas Centrifuge (Uranium enrichment)

Decanter Centrifuge

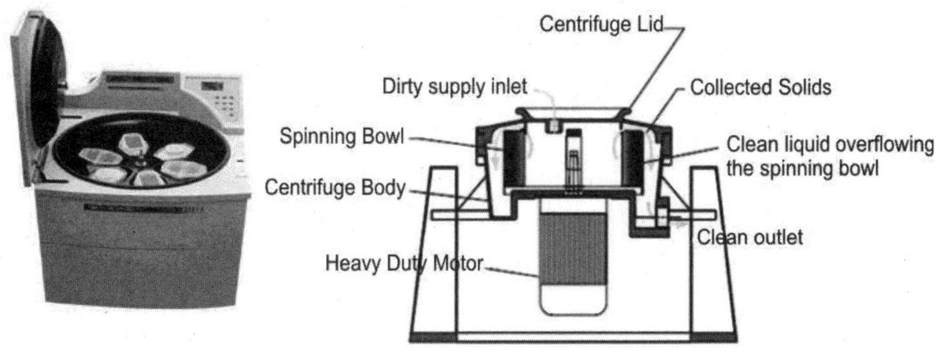

Refrigerated centrifuge

Continuous flow Centrifuge

CENTRIFUGES - CLASSIFICATION

The centrifuges are classified into 4 major groups, namely, small bench centrifuges, large capacity refrigerated centrifuges, high speed refrigerated centrifuges, continuous flow centrifuges and ultracentrifuges which are of 2 types, namely, Preparative ultracentrifuges and Analytical ultracentrifuges.

1. Small Bench Centrifuges or Table-top Centrifuges or Portable Centrifuges

These centrifuges are the simplest and least expensive. The design exists in many types. These are generally used to collect small amounts of material that sediments rapidly (for example, cells of yeast, RBCs and coarse precipitates). These possess a maximum speed of 4000 to 6000 RPM with maximum relative centrifugal fields of 3000 to 7000g. Majority of these centrifuges operate at ambient temperature. However, the latest designs of these centrifuges incorporate a refrigeration system to keep the rotors cool to prevent protein denaturation. Small microfuges provide virtually instant acceleration to maximum speeds of 8000 to 130,000 RPM and develop relative centrifugal fields up to 10,000g. These are extremely useful for sedimenting small volumes of material (250 mm^3 to 1.5 cm^3) very quickly (with in 1 or 2 min). These are best applied for the rapid sedimentation of samples of blood and of synaptosomes (used to study the effect of drugs on the uptake of biogenic amines).

2. Large-capacity Refrigerated Centrifuges

These centrifuges have a maximum speed of 6000 RPM and generate a maximum relative centrifugal field of about 6500g. These possess a refrigerated rotor chamber and vary in their maximum carrying capacity. All are capable of utilising a variety of interchangeable swinging-bucket and fixed angle rotors and enabling separation in 10, 50 and 100 cm^3 tubes. Large capacity centrifuges (4 to 6 dm^3) capable of holding bottles (each of 1.25 dm^3 capacity) in addition to small tubes are also available. In this type, the rotors are mounted on a rigid suspension so that the tubes and their contents are balanced accurately (within 0.25 g of each other). The rotors should never be loaded with an odd number of tubes. When the rotor is partially loaded, the tubes must be located diametrically opposite to each other in order that the load is distributed evenly around the axis of the rotor. These centrifuges are used most often to collect (or compact) the rapidly sedimenting substances (like RBCs, coarse or bulky precipitates, cells of yeasts, nuclei and chloroplasts).

3. High-speed Refrigerated Centrifuges

These centrifuges are available with maximum rotor speeds about 25,000 RPM and generate a relative centrifugal field of about 60,000g. These have a total capacity up to 1.5 dm^3 and a range of interchangeable fixed-angle and swinging-bucket rotors. These are often used to collect microbes, cellular debris, larger cell organelles as well as proteins precipitated by ammonium sulphate. These cannot effectively sediment viruses or smaller organelles like ribosomes.

4. Continuous Flow Centrifuges

These are relatively simple high speed centrifuges. The rotor is long, tubular and non-interchangeable. This rotor helps the particles suspended in medium flow continuously (1 to 1.5 dm^3 min^{-1}). As the medium enters the rotating rotor, the particles are sedimented against its wall and the excess clarified medium overflows through an outlet port. This is mainly used in harvesting cells of bacteria or yeast from larger volumes of culture medium (10 to 500 dm^3). Some centrifuges can also be adapted to function in continuous flow mode with a specially designed rotor.

5. Ultracentrifuges - Preparative Ultracentrifuges

These instruments are capable of spinning their rotors to a maximum speed of 80,000 RPM and produce a relative centrifugal field of up to 600,000g. The chamber containing the rotor is refrigerated, sealed and evacuated to minimise excessive rotor temperatures generated by the frictional resistance between the air and the spinning rotor. The temperature monitoring system of the centrifuge is more sophisticated and employs an infra-red temperature sensor that can continuously monitor the temperature of the rotor and also control the refrigeration system. An over speed control system in the instrument prevents the operation of the rotor above its maximum rated speed and the rotor imbalances are detected by the electronic circuits within. The instrument is also fitted with a flexible drive shaft system in order to minimise vibration caused by slight rotor imbalance due to unequal loading of the tubes. The tubes and their contents must be accurately balanced to within 0.1 g of each other. The rotor chambers (high-speed and ultracentrifuges) are enclosed in heavy armour plating for safety reasons.

6. Analytical Ultracentrifuges

These centrifuges operate at high speed of about 70,000 RPM and produce a maximum relative centrifugal field of about 500,000g. These consist of a motor and a rotor contained in a protective armoured chamber that is refrigerated and evacuated. An optical system contained within enables the observation of the sedimenting material throughout the process of centrifugation to determine the concentration distribution in the sample at any given time during the process. The ultracentrifuge has 3 types of optical system within, a light absorption system and alternative Schlieren system and Rayleigh interferometric system, both of which detect changes in the refractive index of the solution. There is a solid rotor with holes to hold the cells that contain the samples. The rotor is suspended on a wire that comes from the drive shaft of a high-speed motor which allows the rotor to find its own axis of rotation. The rotor tip contains a thermistor to measure the temperature.

High Speed Refrigerated Centrifuge

Analytical Ultracentrifuge

Many types of rotors are available for use. The simplest type of rotor incorporates 2 cells – an analytical cell and the counterpoise cell. The latter counterbalances the former. The latter possesses 2 holes drilled through it (at distances calibrated from the centre of rotation). These holes facilitate the calibration of distances in the analytical cell, and the cells have a capacity between 0.4 and 1.0 cm^3. The analytical cells are used with the UV light absorption optical system and Schlieren optical system (a single 2° or 4° sector shape and with a 12 mm optical path length centre piece) to prevent convection. The interference optical system uses double-sector cells (with two 2.5° sectors and an optical length that can vary from 12 – 30 mm). The absorption optical system is used with photo-electric scanner while Schlieren optical system is used when a baseline is required. The double-sector cells allow to take account of absorbing components in the solvent, and also to correct for the redistribution of the components of the solvent, preferably at high g values. The double-sector cells also help to facilitate the measurements of differences in sedimentation coefficients and of diffusion coefficients. The centre piece when aligned correctly in the rotor, gives ideal conditions for sedimentation (behaving as the swinging-bucket rotor). During centrifugation, this also ensures that there will be no accumulation of material against the wall of the analytical cell. These are provided with upper and lower plane windows of optical grade quartz or synthetic sapphire. The latter is used with interference optics as they have a tendency to distort under a high gravitational field.

The rotor chamber has an upper condensing lens and a lower collimating lens. The former is provided with a camera lens (to focus light on to a photographic plate) while the latter collimates light as such the sample is illuminated by parallel light. In more advanced centrifuges, an electronic scanning system replaces the photographic detector system.

ULTRACENTRIFUGE - BASIC PRINCIPLE

The principle involved in the method, is the sedimentation of the particles in a suspension by the centrifugal force. In a centrifuge, the particles when they are subjected to an accelerating force sediment at different rates. The rate of sedimentation depends on the size of the particles, their shape and the density. For the sedimentation of smaller particles, the speed must increase. The nuclei require a sedimentation force of 500 *g*, the mitochondria of 10,000 *g*, and the ribosomes of 100,000 *g*, and broadly speaking, various cell organelles and macromolecules sediment in the following order:

Nucleus --- Chloroplast --- Mitochondrion --- Ribosome --- DNA --- m-RNA --- t-RNA etc.

The process of centrifugation is based on Stoke`s law and the particle sedimentation velocity increases with – an increasing diameter, an increasing difference in density between two phases and a decreasing viscosity of the continuous phase.

PREPARATIVE CENTRIFUGATION - TYPES OF ROTORS

A certain amount of load or stress is created by the centrifugal force on the material with which the rotor is made of. To overcome this stress, the rotors are generally made of brass, steel or Perspex. The higher amount of stress forces generated during high-speed centrifugation necessitates the use of aluminium alloy or titanium alloy in making the rotors. The rotors made of the latter alloy possess a greater strength-to-weight ratio and as such they are capable of withstanding nearly twice the centrifugal force made from aluminium alloy. These are also more resistant to chemical corrosion and are less subject to metallic fatigue. The rotors made of aluminium alloy though less expensive, are more susceptible to metallic corrosion as they are readily attacked by the acids, alkalis and highly concentrated salt solutions like Sodium chloride and Potassium bromide (used in lipoprotein fractionations, ammonium sulphate used in protein precipitation and salts of caesium or rubidium used in the preparation of density gradients). Hence, all the rotors are given a protective durable coating on their metallic surface either by anodising (aluminium rotors) or by applying black epoxy paint. Recently, carbon fibre which produces not only a lighter rotor but also a rotor with a better strength-to-weight ratio is being used in the production of rotors.

The rotors are of many types, namely, Swinging-bucket rotors, Fixed-angle rotors, Vertical tube rotors, Zonal rotors and Elutriator rotors.

a. Swinging-bucket Rotors or Swing-out Rotor or Horizontal-head Rotor

These are used for pelleting, isopycnic studies and rate zonal studies. These are best applied for rate zonal studies in which maximum resolution of sample zones are needed or pelleting runs where it is desirable for the pellet to be in the exact centre of the tube bottom. Gradients of all shapes and sizes can be used.

b. Fixed-angle Rotors or Angle Head Rotors

These are general purpose rotors especially useful for pelleting sub-cellular particles and for short-column banding of viruses and sub-cellular organelles. The tubes are generally held at an angle (usually 20 to 45 degrees) to the axis of rotation in numbered tube cavities. The tube angle shortens the particle path length compared to swinging bucket rotors and as such pelleting results in reduced runtimes.

c. Vertical Tube Rotors

These rotors hold samples in a vertical position within the instrument. These are not suitable for pelleting centrifugation but good when used for rate-zonal sedimentation and excellent with isopycnic sedimentation. These rotors hold the tubes parallel to the axis of rotation, as such the bands separate across the diameter of the tube rather than down the length of the tube. These are useful for isopycnic and in some cases, rate zonal separations where run time reduction is important.

d. Near Vertical Tube Rotors

These rotors are designed for gradient centrifugation when there are components in a sample mixture that do not precipitate in the gradient. The reduced tube angle of these rotors significantly reduces run times from the more conventional fixed angle rotors, while allowing components that do not band under separation conditions to either to pellet to the bottom or float to the top of the tube.

e. Zonal Rotors

In density gradient centrifugation, the capacity may be insufficient for certain applications and in such cases; zonal rotors provide a larger internal volume for a given radius. These are designed to contain a larger volume of sample in a single central cavity rather than in tubes/bottles. Somc are capable of dynamic loading and unloading of samples while the rotor is spinning at high speed.

f. Elutriator Rotors

This is a type of continuous flow rotor which contains recesses to hold a single, conical shaped separation chamber. The apex of this chamber points away from the axis of rotation. A by-pass chamber on the opposite side of the rotor serves as a counterbalance and also to provide the fluid outlet. The separation chamber is of two types which differ in their capacity and with different shapes. The centrifuge door has a window along with a stroboscopic lamp. The flash is synchronised with the speed of the rotor. The rotor apertures allow the contents of the rotor chamber to be observed during the process. The particles which are suspended in a uniform low density medium are pumped into the chamber at the peripheral edge through a rotating seal assembly. The rotor spins at a pre-selected speed (between 1000 and 3000 rpm).

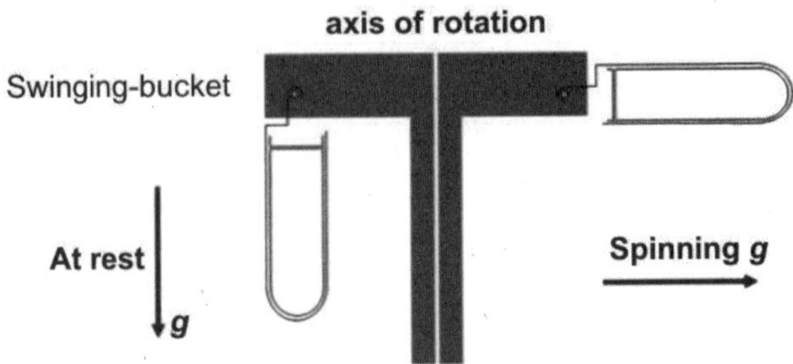

Swinging Bucket Rotor – Orientation (Diagrammatic)

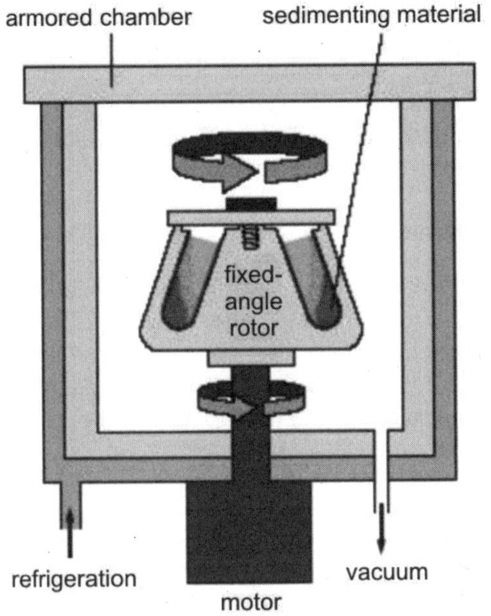

Fixed Angle Rotor – Orientation (Diagrammatic)

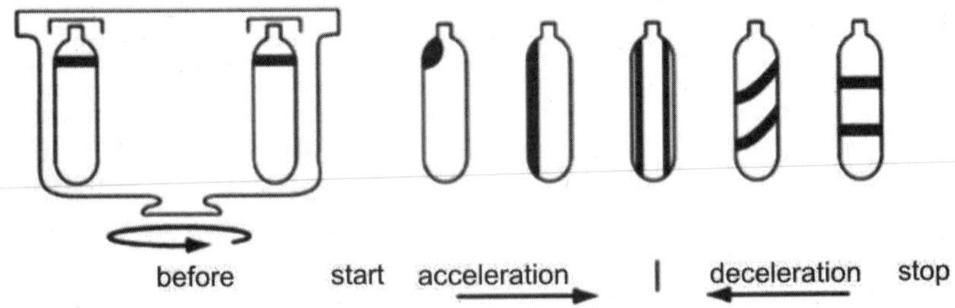

Vertical Rotor – Movement and Orientation (Diagrammatic)

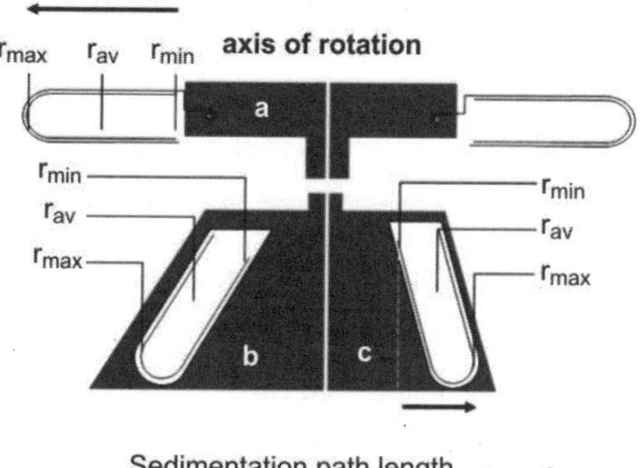

NVT rotor – Orientation (Diagrammatic)

Zonal Rotor

Elutriator Rotor

ROTORS - SELECTION

Centrifuge rotors mainly fall into three categories, namely, swinging bucket rotors, fixed angle rotors and vertical rotors. Each category is designed to account for three key factors, namely, type of centrifugation (either differential, rate zonal or isopycnic), speed and volume range. Of these, the first two are the most common types for bench-top and high-speed floor-model centrifuge applications while the vertical rotors are mainly used in ultracentrifugation. The rotors can be fabricated from a range of materials such as carbon fibre, aluminium and titanium and each material has different characteristics that lend themselves to particular applications.

The swinging bucket type rotors are ideal for separation of large volume samples at low speeds. The common bench-top and super speed centrifuge applications include: batch harvesting of whole cells from growth media, high capacity processing of blood collection tubes and large volume tissue culture processing. An ultracentrifuge with swinging-bucket rotor supports rate zonal separation (based on mass or size) or isopycnic separation (based on density). This rotor is advantageous because the path length of the gradient is long enough for separation to occur, and the position of buckets (during static and dynamic states) helps the sample to retain its orientation which minimizes any pellet or band disturbance.

The fixed angle rotors are the most common rotors used in the process. The majority of these are used for basic pelleting applications (differential sedimentation) either to pellet particles from a suspension or discard the excess debris, or to collect the pellet. The type of the rotor required is determined by two factors, namely, desired g-force (RCF) and the desired volume. In general, the size of the rotor is inversely proportional to its maximum speed capability (the larger the rotor, the lower the maximum speed). K-factor is another specification during the selection of the fixed angle rotor. A low K-factor indicates a higher pelleting efficiency. Hence, this can be a useful metric for comparing the speed at which particles will pellet across a range of rotors.

The vertical rotors which are fairly specialized are most commonly used during ultracentrifugation for isopycnic separation, specifically for DNA banding in caesium chloride. In this type, the density range of the solution contains the same density as the particle of interest. Hence, the particle will orient within this portion of the gradient. The rotors will have a very low K-factor (in the range of 5 to 25) which indicates that the particle must only travel a short distance to pellet.

ROTORS AND ACCESSORIES - CARE

The care of rotor involves careful attention to: regular cleaning, decontamination, and/or sterilization as required; frequent inspection of the rotors; prevention of corrosion, and regular and proper lubrication.

Sharp tools should not be used on a rotor, as the surface can get scratched. Corrosion begins in scratches and may open fissures in the rotor with continued use. The processes of corrosion accelerate with speed-induced stress. The potential for damage from corrosion is greatest in aluminium rotors and components. Rotors and rotor components are washed immediately if salts or other corrosive materials are used or if spillage has occurred. See that corrosive materials are not allowed to dry on the rotors. With normal usage, the rotors are to be washed frequently to prevent corrosion that can begin in scratches. The plastic or wooden tools are to be used to remove the rings or gaskets for cleaning. The metal tools should not be used as they may result in scratches on the anodized surfaces. A mild detergent diluted 10 to 1 with water and a soft brush are to be used for cleaning the rotors and components. After cleaning, it is to be thoroughly rinsed with water. The body or the buckets are air-dried upside down. Acetone should not be used to dry rotors. The rings or the gaskets are to be wiped clean regularly and lubricated after cleaning. They are to be replaced about twice a year or as required. All the surfaces are frequently cleaned. The rotor threads are cleaned regularly with a non-metal brush and a small quantity of concentrated detergent and dried thoroughly and lubricated. If a rotor or its accessories get contaminated with radio-active material, it should be decontaminated using a solution that will not damage the anodized surfaces. If the rotor and its accessories get contaminated with toxic or pathogenic materials, the appropriate decontamination procedures are to be followed by selecting an agent that will not damage the rotor. Periodical inspection of the rotor at least a month depending on the use, is essential. If rough spots, cracks, pitting, white powder deposit on aluminum rotors or heavy discolouration are noticed, do not run the rotor until after proper inspection and rectification of the defects. Proper lubrication is essential to obtain specified torque values and to minimise the thread wear. Proper care of the tubes and bottles is essential. These are not to be washed in dishwashers, and too harsh detergents and temperatures. A detergent with a pH less than 9.0 is to be used for washing polycarbonate bottles and tubes. A brush with exposed metal should not be used. The tubes, bottles or accessories should not be dried in an oven. The laboratory ware should be air-dried.

CENTRIFUGE - CONTAINERS

Selection of Centrifuge Tubes

Depending on the sample, size of the rotor and the speed of centrifugation, different types of tubes can be used during the process. A proper selection of the tubes helps to ensure that leakage does not occur, and none of the sample is lost and the chemical properties of the sample and tube do not conflict (chemical compatibility) and that the sample is recovered with a little effort easily. The major factor in the selection of a tube material (plastic) is clarity, chemical resistance and sealing mechanism if needed. For biological work, the tubes are divided into functional groups, made of regular glass, Corex glass, nitrocellulose or polyallomer. Regular glass tubes can be used at speeds below 3,000 rpm in a standard centrifuge. Above this speed, the centrifugal forces will shatter the glass. For centrifugation in the higher speed ranges, the tubes made of plastic or nitrocellulose which can withstand speeds up to 20,000 rpm are used. The preparative centrifuge tubes made of polypropylene or polyethylene can withstand speeds up to 20,000 rpm. The tubes made of nitrocellulose are inexpensive and are used for ultracentrifugation. These are to be used once and then discarded as repeated use increases the chance of tube collapse due to internal molecular stress. With age, the nitrocellulose also becomes less flexible and hence, the tubes older than one year should be discarded. Polyallomer tubes are more chemically inert, re-usable, more expensive and slippery. The molecules slide down the walls of these tubes more easily and thus the tubes are used only in precipitation centrifugation.

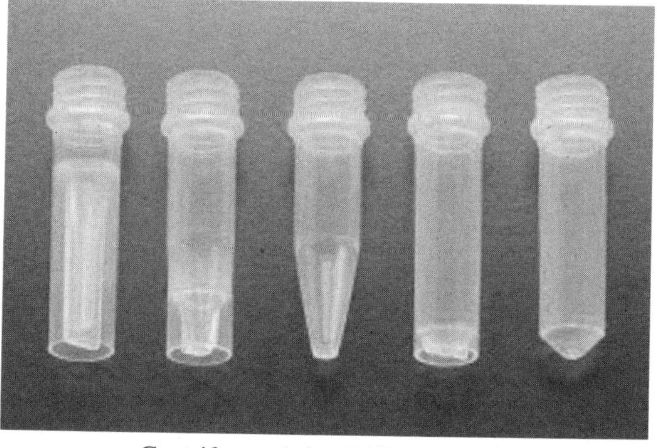

Centrifuge tubes – Different types

Centrifuge bottles – Different types

Conical Bottom Tubes

DENSITY GRADIENT SEPARATION – FORMATION AND CHOICE OF DENSITY GRADIENTS

The gradients are used not only for particle separation in the tubes of conventional rotors, but also in instruments for special purpose like zonal rotors, continuous flow rotors and stay-put devices. The basics of density gradient preparation are the same in all the applications. According to preparation, the density gradient can be subdivided into two main categories, namely, step gradients and continuous gradients.

1. Step Gradients

These are prepared by successive layering solutions of different density in the tube followed by layering the sample to be fractionated on the top of the last step. These are useful in density gradient centrifugation because the abrupt density steps that persist can be used as surfaces on to which the particles can sediment during the process which results in discrete particle layers at each step.

The production of step gradients does not require special density gradient generating equipment. The gradient may be built up in the tubes by careful layering of one step on another. This begins with the densest step by using a pipette. Alternatively, the gradient can be generated beginning with the least dense step by depositing each layer at the bottom of the tube through a narrow cannula.

2. Continuous Gradients

These are the gradients in which the density changes smoothly and continuously from one limit or extreme to another. These can be produced from step gradients by allowing sufficient time for diffusion to smooth out the steps. The gradients are normally prepared by the use of special devices called gradient makers or gradient engines.

GRADIENT MATERIAL - NATURE AND USE

The ideal solute used to produce a density gradient for most of the biological work should satisfy the following criteria: The solute should be sufficiently soluble in aqueous media to produce the range of different densities needed. The solute dissolution should not result in high viscosity solutions in the desired range. This is because the viscous solutions pose special mixing problems for gradient makers and handling difficulties. Moreover, the sedimentation rate of a particle decreases directly in relation to the viscosity of the surrounding medium. The solution should not also exert too high osmotic pressure in the desired range of the density gradient. If the osmotic pressure of the surrounding medium changes markedly, the size, shape and the density of the whole cells, membrane-enclosed sub-cellular organelles, and even macromolecules can be altered as these particles sediment through a density gradient. The solutions of the gradient solute should be adjustable to the pH and ionic strengths that are compatible with the particles being sedimented. The solute should not react with the particles being sedimented and should not interfere with the methods of analysis of the collected fractions. The solutions of the gradient solute should exhibit a property that can be conveniently used as a measure of concentration or gradient density. The solute should be readily removablc from the collected gradient fractions (should not be necessary or desirable). Finally, the solute should be affordable in the needed quantities.

SUITABLE DENSITY GRADIENT MEDIUM - SELECTION

The main function of density gradient centrifugation is to separate particles, either on the basis of their buoyancy density or their rate of sedimentation. The function of the gradient for rate-zonal separations is to provide a gradient of viscosity which improves particle resolution while stabilizing the column from convection currents. The important feature for isopycnic separations is that the maximum density of the gradient media is higher than that of the particles.

An ideal density gradient medium has the following properties:

- The medium should possess sufficient solubility to produce the range of required densities

- It does not form solutions of high viscosity in the desired density range
- It should not be hyper-osmotic when the particles to be separated are osmotically sensitive
- The solutions of the gradient should be adjustable to the pH and the ionic strengths that are compatible with the particle being separated
- It does not affect the biological activity of the same
- It should be non-toxic and not metabolized by the cells
- It does not interfere with assay procedures or react with the centrifuge tube
- It should exhibit a property that can be used as a measure of concentration
- It should be autoclavable and
- It should be available at a reasonable cost.

As no single compound can satisfy all of the above criteria, a wide range of gradient media are used for different types of samples. Most of the media are capable of producing the range of the required densities and being easily removed from the particles of interest.

The effect of osmolarity on biological particles needs special consideration. Most mammalian fluids have an osmolarity between 290 – 300 mOsm. This is the osmolarity of balanced salt solutions (for example, NaCl – 0.85 – 9.0%) and most common media. High osmolarity solutions not only remove water from the interior of membrane-bound particles, they also remove water bound to macromolecules (like DNA). Such a loss of water from cells will reduce their size and increase their density, thereby affecting the buoyancy and also their rate of sedimentation. The osmotic effect on the cells and macromolecules may be reversible, though it is a possible source of error that should be avoided.

A variety of different compounds have been developed over the years as density gradient media in order to enhance the separation process and to overcome the problems of osmolarity and viscosity. The density gradient media are mainly of five classes, namely:

Polyhydric (sugar) alcohols like Sucrose (used for sucrose organelles, membrane vesicles, viruses, proteins, ribosomes and polysomes), Glycerol (used for mammalian cells and proteins) and Sorbitol (used for non-mammalian sub-cellular particles);

Polysaccharides like Ficoll, Polysucrose and Dextrans (used for mammalian cells [sometimes in combination with iodinated density gradient media] and mammalian sub-cellular particles);

Inorganic salts like Caesium chloride (used for DNA, viruses and proteins), Caesium sulphate (used for DNA and RNA) and Potassium bromide (used for plasma lipoproteins);

Iodinated gradient media like Diatrizoate (used mainly as a component of commercial lymphocyte isolation media), Nycodenz and Histodenz (used for mammalian cells, organelles, membrane vesicles and viruses) and Iodixanol (used for mammalian cells, organelles, membrane vesicles, plasma lipoproteins, proteins and DNA); and

Colloidal silica media like Percoll (used for mammalian cells, organelles and membrane vesicels).

FRACTIONS - COLLECTION

At the conclusion of the process, the density gradient and particle zones are usually removed from the tubes (rotors) and collected as a series of fractions. This can be achieved in different ways.

One of the oldest and the simplest methods of collecting the fractions are to puncture the bottom of each centrifuge tube and allow the contents to slowly drip out. The gradient may also be removed by lowering a narrow cannula carefully to the bottom of the tube and withdrawing the gradient by using a peristaltic pump. In both of these procedures, the gradient exits the tube denser end first, and the separated particles are collected in order of decreasing sedimentation rate or density. Another option is to puncture the tube at the location in the gradient of the material of interest and extract it manually.

Measurement of Gradient Density

The profile of the gradient is revealed by the measurement of the density of a solution and its mass volume ratio in successive aliquots of the collected gradient. Such an approach is time consuming and is rarely used in connection with density gradient centrifugation. Otherwise, the refractive index of small drops of liquid collected from different regions of the gradient is determined, and the corresponding densities obtained from tables or curves relating refractive index to density.

SAMPLE APPLICATION TO THE GRADIENT

Each gradient has a defined sample capacity which is a function of its density slope, and if exceeded this will result in a poor resolution of the sample. Hence, before the sample is applied to the density gradient, its optimum value and concentration should be determined. The volume of the sample to be applied to a centrifuge tube is a function of the cross sectional area of the gradient exposed to the sample. The sample volumes in the range of 0.2 to 0.5 cm^3 may be added to tubes 1.0 to 1.6 cm in diameter, and sample volumes of up to 1.0 cm^3 to tubes having a diameter of approximately 2.5 cm. The effective separation of the particles in a multi component sample would not be achieved with larger sample volumes due to insufficient radial distance in the tube. If the sample concentration on the gradient is too high or too low, then either the gradient may become over-loaded (resulting in a broadening of the separated zones and loss of resolution) or difficulty may be encountered in the identification of the separated bands.

The correct method of application of the sample to the gradient is of paramount importance for rate zonal separations. This is because the resolving power of the gradient is greater, the narrower the initial sample zone. Therefore, the sample is applied as slowly and as gently as possible by using a narrow bore pipette to reduce mixing with the gradient. The tip of the pipette is touched against the meniscus at the wall of the tube.

The pipette is moved slightly upward to leave a thin channel of liquid between the pipette and the meniscus. The sample is allowed to run slowly out of the pipette and gently on to the surface of the gradient to form a sharply defined sample layer to minimise the mixing with the gradient. The separation in isopycnic separation is independent of the initial distribution of the sample. The volume of the sample and its initial distribution through the gradient are unimportant and the sample can be carefully layered on the top of a pre-formed gradient. In the case of self-forming gradients, the sample is mixed with the gradient medium. Sometimes, it gets mixed with a dense solution of the gradient medium and layered under the density gradient.

RECOVERY AND MONITORING OF GRADIENTS

It is usually necessary to remove the gradient solution after particle separation in order to isolate the bands of separated material. If the bands can be detected visually, recovery from the gradient can be easily achieved without having to unload the gradient by Pasteur pipette placing it into the band and withdrawing slowly the required material from the band. The band may also be recovered through the side of the centrifuge tube using a hypodermic syringe. This is possible through a number of techniques. For most applications, the method of choice is that of upward displacement. The denser solution of the gradient medium or a solution for unloading available in the market (like Maxidens – an inert, non-viscous, water-immiscible organic liquid with a density of 1.9 g cm^{-3} that does not react with the medium) is either pumped through a long needle passing through the gradient to the bottom of the tube or introduced via a hollow needle piercing the bottom of the tube. The gradient is displaced upwards and the fractions removed in sequence using a syringe or a pipette, or preferably by being channelled out through an unloading cap to which is attached a collection pipe leading either to a fraction collector or directly into a flow cell of an UV spectrophotometer. A variation to this is to fit a cap to the top of the tube and pump either air or a less dense solution of the gradient medium into the top of the tube, the gradient then being displaced through a tube that leads from the bottom of the tube.

If a pellet is not present, the tube may be punctured at its base by the use of a fine hollow needle. The drops of the gradient that pass through the needle may be collected by using a fraction collector. The analysis of the displaced gradient (to identify and locate the separated components) can be achieved by UV spectrophotometry, measurements of the refractive index, scintillation counting and enzymatic or chemical analysis.

ANALYSIS OF SUB-CELLULAR FRACTIONS

Homogeneity Assessment

The evaluation of the purity of the sub-cellular fractions is very essential. Light microscopic and electron microscopic examination have been used for analysis of a particular homogenisation technique and to determine the degree of contamination of a sub-cellular fraction after centrifugation. Absence of any visual contamination is not a conclusive proof of purity, because it is difficult to detect lower levels of contaminating material in the fraction. Hence, quantitative determination of purity has to be obtained by chemical analysis. The organelles and molecules lacking assayable enzyme activities can be located by their light absorption or by radio-active labelling techniques.

PRESENTATION OF THE RESULTS

In the whole homogenate and in each sub-cellular fraction that has been isolated, the enzyme activity as well as the protein content is to be determined. The sum of enzyme activity and the protein content of the fractions should not differ from that in the initial homogenate, and this represents the total recovery. The enzyme activity and the protein content in each fraction as a percentage of the total recovery are calculated. The results can also be represented by a graph where the enzyme distribution patterns are presented in the form of histograms which provide a visual diagram of the net result. The histograms of the fractions are to be presented separately on the ordinate scale. This is done by the relative specific activity which is a measure of the degree of the purification achieved. Each of the fractions are represented cumulatively, from left to right, on the abscissa, in their order of isolation by its percentage to the total protein. Then, the area of each rectangle is equal to the percentage of the enzyme activity in the isolated fraction.

ANALYTICAL ULTRACENTRIFUGE - APPLICATIONS

In biology, especially in protein and nucleic acid chemistry, the analytical ultracentrifuge has many applications in providing much information from which the sedimentation coefficient, relative molecular mass, purity and shape of the particles can be easily deduced.

a. Relative Molecular Mass - Determination

The relative molecular mass of a macromolecule can be determined by two important methods, namely, the sedimentation velocity method and sedimentation equilibrium method.

In the former method, the sedimentation coefficient of the molecule is first determined either by boundary sedimentation or band (zonal) sedimentation. The particles are uniformly distributed through the solution in analytical cells at the beginning of the experiment in the boundary sedimentation method. The instrument is operated at high speeds which cause the randomly distributed particles to migrate through the solvent radially outwards from the centre of rotation. A sharp boundary, the *plateau region*, is formed between the solvent (that has been cleared of the particles) and the solvent (still containing the sedimenting material). The boundary movement with time is a measure of the particle sedimentation. This is followed by the use of UV absorption optical system or Schlieren optical system.

In the latter method, a small amount of material (about 15 mm^3) is layered on the top of a denser solvent, which generates its own density gradient during the process. Thus, the solute band is stabilised against convection disturbances. The migrating band movement is followed by the use of the absorption optical system using a special band forming centre piece in the analytical cell.

The relative molecular mass of a molecule can be calculated, and such calculations are complicated by difficulties encountered in the determination of the diffusion coefficient of the particle and differences in viscosity and temperature. The relative molecular mass of a macromolecule determined by sedimentation velocity analysis is less accurate and time consuming than determination by sedimentation equilibrium.

Sedimentation equilibrium methods are more versatile and most accurate. These can be used to determine the values of relative molecular mass (ranging from a few hundred to several million). This is due to the larger range of centrifugal fields in the ultracentrifuge. The field varies with the square of the rotor speed (700-fold range at speeds of 800 to 68,000 rpm) of the analytical ultracentrifuges. In this method, the instrument is operated until a balance is established between sedimentation (under the influence of speed) and diffusion of the material in the opposite direction (until there is no net migration of solute throughout the length of cell). For a single and ideal solute component, at equilibrium, the resulting concentration distribution increases exponentially with the square of the radial position, invariant with time, and is a function of the relative molecular mass of the sedimenting solute. The relative molecular mass can be calculated from the concentration gradient of the solute that is set up. A major disadvantage of low speed sedimentation equilibrium is the long periods (from several days to weeks) that are necessary for achieving the equilibrium.

b. Macromolecules - Estimation of Purity

In the investigation of purity of DNA preparations, viruses and proteins the analytical ultracentrifuge has been used extensively. Purity of the sample is extremely important if an accurate estimation of the relative molecular mass of the molecule is required. The widely used methods of determination of the homogeneity of a preparation include the analysis of sedimenting boundary using sedimentation velocity technique. Many criteria have been devised for the assessment of homogeneity of a preparation. It must be remembered that homogeneity can be presumed only through the absence of detectable heterogeneity. Homogeneity is best recognised by a single sharp symmetrical sedimenting boundary throughout the duration of the experiment. The impurities present in the preparation are generally displayed as additional peaks, shoulders on the main peak or asymmetry of the main peak. The absence of the heterogeneity is no guarantee that all the molecules have the same biological activity or electric charge. For example, partial deamidation of a protein sample, while increasing the negative charge on the molecule at a neutral pH would have no significant effect on the size, shape or relative molecular mass of the molecule. This sample fails to show heterogeneity in relative molecular mass or sedimentation coefficient (but it shows multiple zones in capillary electrophoresis).

c. Macromolecules - Detection of Conformational Changes

The changes in the conformation in macromolecules may be ascertained by examining the differences in the sedimentation velocity of the sample. The more compact the molecule is, the lower would be its frictional resistance in the solvent. The more disorganised the molecule is, the greater is the frictional resistance, and the sedimentation occurs more slowly. Therefore, the changes in the conformation can be detected by differences in sedimentation rates of the sample both before and after treatment. In allosteric proteins, these changes may accompany combination of the protein with the substrate and/or small ligands. If the protein is treated with reagents like urea, it may result in disaggregation of the protein into its protomers. Such changes may readily be studied by analytical ultracentrifugation.

CENTRIFUGES - SAFETY ASPECTS IN USE

If not maintained properly or if used incorrectly, the centrifuges can be extremely dangerous. By the fitting of effective lid locks (to prevent access to the rotor chamber while it is spinning), imbalance detectors, rotor over-speed devices etc, the operator is safe-guarded against accidents. During the process of filling and emptying the zonal rotors, and in the operation of continuous flow rotors, every care must be taken to ensure that the moving rotor is not touched, to prevent possible physical injury. Similarly, longer hair and loose clothing (such as neck-ties) of the operator should not be allowed to get caught in any rotating part of the instrument (in the case of older centrifuges where the lid can be opened before the rotor has stopped spinning). The samples of hazardous materials (microbial pathogens, infectious viruses, carcinogenic, and corrosive or toxic chemicals and radio-active material) are kept in air-tight and leak-proof containers, to prevent the formation of aerosols from accidental spillage of the sample which can contaminate the rotor, centrifuge and the whole laboratory.

CENTRIFUGATION - APPLICATIONS IN BIOLOGICAL RESEARCH

The most common type of centrifuges encountered in the laboratories is fixed angle rotor centrifuges like table-top centrifuges. The tubes sit in a fixed and angled position in these instruments, and are used in differential centrifugation protocols. In these protocols, a series of centrifugations at different speeds can be used to purify biological specimens (animal cells), and these involve several cell washing steps.

In contrast to these (fixed angle rotors), swing rotors have flexible tube holders and allow the samples to rotate outward. These are beneficial in density gradient centrifugation, where the biological samples migrate to distinct layers of gradient media. This is particularly useful for quick isolation of one cell type from another or for the isolation of individual organelles.

The micro-centrifuges are used to process small volumes of biological molecules such as cells or nuclei. The tubes generally hold 0.5 to 2.0 mL of liquid, and are spun at maximum angular speeds of 12,000 to 13,000 rpm. These are small enough to fit on a table-top and have the rotors that can quickly change the speeds. They may or may not have a refrigeration function. These can separate particles in batches (limited volumes of samples must be handled manually in test tubes or bottles).

The high-speed or super-speed centrifuges can handle larger volumes of the sample, from a few 10s of ml to several litres, and larger instruments can also reach higher angular velocities around 30,000 rpm. The rotors come with different adapters to hold various sizes of test tubes, bottles or microtiter plates. These can separate particles in batches.

Ultracentrifugation makes use of high centrifugal force for studying the properties of biological particles. These can isolate smaller particles including ribosomes, proteins and viruses. These can also be used in the study of membrane fractionation because these can reach maximum angular velocities in excess of 70,000 rpm. These can also separate molecules in batch or continuous flow systems.

Analytical ultracentrifugation (AUC) can be used in addition to purification, for the determination of the properties of macro-molecules like shapes, mass, composition, and conformation. The samples are normally centrifuged with a high density solution like sucrose, caesium or iodixanol which may be at a uniform concentration throughout the test tube (cushion) for a varying concentration (gradient). The molecular properties can be modelled through sedimentation velocity analysis or sedimentation equilibrium analysis. During the run, the particles of molecules will migrate through the test tube at different speeds depending on their physical properties and the properties of the solution and finally form a pellet at the bottom of the tube or bands at various heights.

Other applications of centrifugation include: the removal of fat from milk to reduce skimmed milk, separating the textiles, removal of water from lettuce after washing it in a salad spinner, separation of particles from an air-flow using cyclonic separation, clarification and stabilization of wine, separation of water from clothes while spin-drying in washing machines and separation of components of urine and blood in forensic and research laboratories. Centrifuges are also used in a variety of scientific and technical applications and in some cases, the solid is a waste. In cases like chemical manufacturing, the solid is the final product that is being harvested. Centrifuges are also used in many industries.

Centrifugation is used quite often in the dairy industry and some uses include: clarification (the removal of solid impurities from milk prior to pasteurization), skimming (the separation of cream from skim milk), standardization, separation of whey (of whey cream or fat from whey), bactofuge treatment (separation of bacteria from milk), quark separation (separation of quark curd from whey) and butter oil purification (separation of serum phase from anhydrous milk fat). The widest use of centrifuges is for the concentration and purification of materials in suspension or dissolved in fluids. The idea of centrifugation is also applied in washing machines for drying clothes. Centrifuges are also used for simulating the gravitational fields in space and for drying laundry, usually with a water outlet. The centrifuges find an extensive use in waste water management (to dry sludges), in the oil industry and in the processing of sugar and milk. Geotechnical modelling is another area where centrifuges are used to simulate gravitational stresses in prototypes.

MANY USES OF CENTRIFUGE

Many uses of centrifuges include:

The simple centrifuges are used in chemistry, biology and biochemistry laboratories for the isolation and separation of suspension. The simple centrifuges vary in their speed as well as their capacity. They usually have a rotor that contains 2, 4, 6 or many more numbered wells within which the samples may be placed.

Other types of centrifuges like Zippe-type centrifuges are useful in the separation of isotopes. These find their use in nuclear power and nuclear weapon programmes.

Gas centrifuges are normally used in uranium (Uranium-238) enrichment. To enrich uranium needed for use in a nuclear reactor (around 3.5% enrichment), it requires many 1000s of centrifugations. Similarly, to enrich it to weapons-grade (above 90% enrichment) for use in nuclear weapons, it requires many 1000s more centrifugations.

The human centrifuges are the exceptionally large centrifuges that are used to test the reactions and tolerance of the pilots and astronauts to acceleration above that is experienced in the earth`s gravity. The United States Air Force at Holloman Air Force Base, New Mexico, operates a human centrifuge, which is operated by the department of aerospace physiology for training and evaluating the prospective fighter pilots for high-g-flight in fighter aircraft. Erasmus Darwin, grandfather of Charles Darwin, used the first centrifuges for human research, and the first large-scale human centrifuge was created in Germany (1933) which was designed for aeronautical training.

Centrifuge acceleration is also applied to scale models in geo-technical centrifuge modelling, to scale the gravitational acceleration so as to enable the prototype scale stresses These include many problems like building foundations, bridge foundations, earth dams, earth tunnels, slope stability which include many effects such as blast loading and earthquake shaking etc.

The commercial applications of centrifuges include: to separate sugar crystals from the mother liquor (in sugar industry); for drying hand-washed clothes usually provided with a water outlet (in a washing machine); a combination of a centrifuge and a motion simulator which propels riders to simulate the feeling of going into space (at EPCOT in Walt Disney World); to match the soil stresses in a scale model to those found in reality (in soil mechanics); in water treatment and waste water treatment where the large industrial centrifuges are used to dry the sludges (the dry product is termed as cake, while the water after the most of the solids have been removed is called centrate); in oil industry, large industrial centrifuges are used to remove the solids from the drilling fluid; disc-stack centrifuges are used to separate small amounts of water and solids from bitumen (in oil sands industry); and to separate the cream (removal of fat) from the milk in dairy industry.

■■■

CHAPTER – 3

PLANT TISSUE CULTURE

Introduction

Plant tissue culture is a technique with which "the plant cells, tissues or organs are grown on an artificial nutrient medium, either static or liquid, under aseptic and controlled conditions". The sterile medium usually contains a nutrient solution (of salts and vitamins), sucrose, hormones and antibiotics (optional) and a solidifying or gelling agent (agar). Small parts of the plant (called explants – such as the single cells of the leaf, shoot tips, pieces of leaf, pieces of root, lateral buds, sections of the stem, pollen grains, anthers, ovaries, embryos etc.) can be induced to produce 100s of plantlets by the use of plant hormones and other growth regulators which can later be grown either in a green house, or in a field or as house plants.

To understand plant cell culture and regeneration, two concepts are central, *plasticity* and *totipotency*. Due to their sessile nature and longer life span, plants have developed a greater ability to endure not only extreme environmental conditions but also predation than the animals. Many physiological processes associated with plant growth and development too adapt to the variation in environmental conditions. This *plasticity* allows the plants to alter their metabolism, growth and development to suit best to the changes in the environment. As far as plant tissue culture and regeneration are concerned, the important aspects of such an adaptation are the ability to initiate the process of cell division from almost any cell/tissue/organ of the plant, and to regenerate any lost organ, or undergo different developmental pathways in response to a particular external stimulus. When the cells, tissues or organs of a plant are cultured *in vitro*, they generally exhibit a very high degree of plasticity, which allows one type of cell or tissue or organ to be initiated from another type, and as such the whole plant can be regenerated subsequently. Such regeneration of the whole organism depends upon the concept that all plant cells can, given the correct stimuli, express the total genetic potential of the parent plant, and such maintenance of the genetic potential is called *totipotency*. The cell culture and regeneration do provide the most compelling evidence for totipotency.

The plant cells (or other isolated parts of the plant) can be grown into a complete plant due to *cellular totipotency*. Tissue culture can be initiated from almost any part of a plant, and the parent plant must be healthy and free from signs of disease or decay. The younger tissues contain a high proportion of actively dividing cells and are more responsive to the initiation of callus. The conditions required for initiation and sustenance of cells in culture or to regenerate intact plants from cultured cells are different for each plant species, and these conditions have to be identified for a successful culture.

PLANT TISSUE CULTURE – MATERIALS AND METHODS

The technique of plant tissue culture mainly relies on the three conditions, namely, explants, aseptic environment and the nutrient medium.

1. Explant

Explant is a small tissue from any part of the plant which is the starting point for the culture. The process can be initiated from any part of the plant – a root, stem, petiole, leaf or flower, and the choice of the explants varies with species. The meristems are more responsive and give better success as they are actively dividing. The physiological state of the plant also has an influence on its response to initiate tissue culture, and hence, the parent plant must be healthy and free from signs of disease or decay. Any part obtained from any plant species (explant) can be employed to induce the callus tissue, and the successful production of callus depends upon the nature of the plant species and its quality. Generally, dicotyledons are rather amenable for callus induction rather than monocotyledons, and the callus of woody plants grows slowly. Stems, leaves, roots, flowers and seeds and any other part of the plant can be used as an explant, but younger and fresh explants are preferable. The explants after collection must be surface-sterilized using ethanol, sodium hypochlorite and/or other chemicals to remove all organisms by typical sterilization procedure.

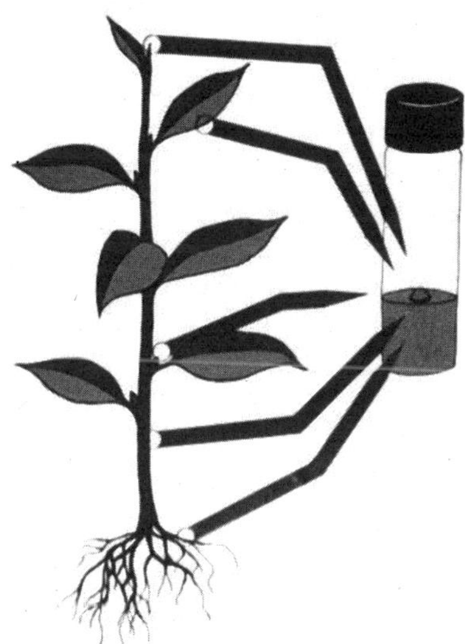

Explants from the body of a plant – stem tip, leaf, axillary bud, stem and root

2. Aseptic environment

During culture, aseptic environment is required to avoid contamination from microbes. The contaminants like bacteria, fungi and minor organisms can easily overgrow the plant tissue as cell division is slow in plant tissue when compared to the growth of the microbes. All the materials like glassware, instruments, medium, explants etc., to be used in the technique must be freed of microbes through several techniques. Laminar air flow is a mandatory prerequisite for any tissue culture technique for contamination free work.

3. Tissue Culture Media or Nutrient Media

The type and composition of culture media very strongly govern the growth and morphogenesis of plant tissues. The choice of culture medium largely depends upon the plant species to be cultured, as some species are sensitive to high salts or have different requirements for plant growth regulators, and some tissues show better response on solid medium while others prefer a liquid medium. Hence, the development of a culture medium formulation is a result of systematic trial and error experimentation depending on specific requirements of a particular culture system.

The cells or organs or tissues when cultured *in vitro*, all their physical and chemical needs have to be met by the culture vessel, the growth medium and the external environment (light, temperature etc.). The growth medium has to supply all the essential mineral nutrients required for growth and development. In addition, it must also supply other additional organic supplements (amino acids, vitamins etc.). A vital component that must be supplied is water, the main biological solvent, and many physical factors like temperature, pH, gases, light and osmotic pressure also have to be maintained within acceptable limits for successful growth and development of the cell/tissue/organ.

In vivo plant cells, tissues and organs get their appropriate nutrients and growth requirements from the intact plant body for their organized growth and development. Isolated cells, tissues and organs also need nutrients for their *in vitro* growth and development. These nutrients are supplied through artificial media (culture media) formulated by several workers. No single medium is capable of maintaining optimum growth of all the plant tissues. Hence, the most suitable medium for a particular cell/tissue must be determined by trial and error method. The composition of a culture medium has often been modified to stimulate the growth of a particular plant tissue or organ.

The inorganic and organic nutrients needed should be either analytical (AR) or guaranteed (GR) reagents. On the basis of the constituents, culture media are either chemically defined (of known composition) or chemically undefined (exact composition is not known). The former is prepared from inorganic salts and organic chemicals of known concentration, while in the latter some natural products (like coconut milk) are added as a result of which the exact composition and the concentration of the constituents is not known. The medium may be either solid (6-8% agar is dissolved in the liquid nutrient medium) or semi-solid (partially solidified medium – with a lesser amount of agar added to it) or liquid (medium without agar). Generally, the solid and semi-solid media are used for callus culture while the liquid media are used for cell suspension culture.

To induce a callus from an explant and also to cultivate the callus, various kinds of media have been designed. One of the most commonly used media for plant tissue culture is that developed by Murashige and Skoog (1962 – popular as MS medium) for tobacco tissue culture. The significant feature of this medium is its very high concentration of nitrate, potassium and ammonia. Many researchers also use B-

5 medium established by Gamborg *et al*., (1968) wherein the levels of the inorganic nutrients are lower than in MS medium. It is not always necessary to test different basal media for callus or multiple shoot induction. It would be always better to use only one or two kinds of basal media in combination and different concentrations of phytohormones. The composition of the most suitable medium should be optimized afterwards in order to obtain higher level products as well as growth rate.

NUTRITIONAL COMPOSITION OF MOST COMMON PLANT CELL/TISSUE CULTURE MEDIA (Angiosperms)

A. Nutrient Composition of Murashige-Skoog`s (MS) Medium (1962) [mg/L]

Macro-nutrients:

$MgSO_4 \times 7H_2O$ (Magnesium sulphate)	: 370
$CaCl_2 \times 2H_2O$ (Calcium chloride)	: 440
KNO_3 (Potassium nitrate)	: 1, 900
NH_4NO_3 (Ammonium nitrate)	: 1, 650
KH_2PO_4 (Potassium dihydrogen phosphate)	: 170

Micro-nutrients:

$FeSO_4 \times 7H_2O$ (Ferrous sulphate)	: 27. 8 (*Iron source*)
Na_2EDTA (Disodium etheylene diamine tetra-acetate)	: 37.25 (*Chelating agent*)
$MnSO_4 \times 4H_2O$ (Manganese sulphate)	: 22. 3
$ZnSO_4 \times 7H_2O$ (Zinc sulphate)	: 8. 6
$CuSO_4 \times 5H_2O$ (Copper sulphate)	: 0. 025
$CoCl_2 \times 6H_2O$ (Cobalt chloride)	: 0. 025
KI (Potassium iodide)	: 0. 83
H_3BO_3 (Boric acid)	: 6. 2
$Na_2MoO_4 \times 2H_2O$ (Sodium molybdate)	: 0. 25

Carbohydrate source:

Sucrose/Glucose	: 30, 000

Vitamins:

Myo-inositol	: 100
Niacin	: 0. 5
Pyridoxine HCl	: 0. 5
Thiamine HCl	: 0. 1

Amino acid source:

Glycine	: 2
pH	: 5. 8

B. Nutrient Composition of Gamborg *et al* (B-5) Medium (1968) [mg/L]

Macro-nutrients:

$(NH_4)_2SO_4$ (Ammonium sulphate)	: 134
$MgSO_4$ x $7H_2O$ (Magnesium sulphate)	: 500
$CaCl_2$ x $2H_2O$ (Calcium chloride)	: 150
KNO_3 (Potassium nitrate)	: 3, 000
NaH_2PO_4 x H_2O (Sodium di-hydrogen phosphate)	: 150

Micro-nutrients:

$MnSO_4$ x $4H_2O$ (Manganese sulphate)	: 10
$ZnSO_4$ x $7H_2O$ (Zinc sulphate)	: 2
$CuSO_4$ x $5H_2O$ (Copper sulphate)	: 0. 025
$CoCl_2$ x $6H_2O$ (Cobalt chloride)	: 0. 025
H_3BO_3 (Boric acid)	: 3
Na_2MoO_4 x $2H_2O$ (Sodium molybdate)	: 0. 25
Fe (as sequence)	: 28 (*Iron source*)

Carbohydrate source:

Sucrose	: 20, 000

Vitamins:

Myo-inositol	: 100
Nicotinic Acid	: 1
Thiamine HCl	: 10
pH	: 5. 5

4. Composition or Components of Tissue Culture Media

A number of media have been devised for specific tissues and organs. White's (1953) medium is one of the earliest plant tissue culture media. White's medium is originally formulated for root culture. To induce organogenesis and regeneration of plants in cultured tissues, Murashige and Skoog (MS) and Linsmaier and Skoog (LS) media are widely used. B_5 medium, originally designed for cell suspension or callus cultures, has with modifications, proved valuable for protoplast culture. Another medium, designated as N_6 has been developed for cereal anther culture and other tissues, though in experiments on anther culture the medium devised by Nitsch and Nitsch (1969) is frequently used. Most of these media contain inorganic salts of major and minor elements, vitamins and sucrose. A medium with these components is known as Basal Medium (BM). Growth regulators like auxins, cytokinins and gibberellins may be added to this medium. Natural plant extracts such as coconut milk, yeast extract, casein hydrolysate and various fruit juices are also added to the medium for promoting growth in some tissues.

A. Inorganic nutrients (Inorganic salts)

In vitro growth of plants also requires a combination of macro- and micro-nutrients like *in vivo* growth. A variety of mineral elements (salts) supply the needed macro- and micro-nutrients required in the life of a plant. Elements required in concentration greater than 0.5 m mol/L are referred to as macronutrients and those less than 0.05m mol/L as micronutrients.

i) Macro-nutrients

The macro-nutrients (macro-elements) include six major elements: Nitrogen (N), Phosphorous (P), Potassium (K), Calcium (Ca), Magnesium (Mg) and Sulphur (S) (and carbon which is added separately) present as salts that constitute various media. All are essential for plant cell and tissue growth.

ii) Micro-nutrients

Inorganic elements required in small quantity (trace amounts) but essential for plant cell and tissue growth and development constitute microelements (micro-nutrients). These are: Iron (Fe), Manganese (Mn), Zinc (Zn), Boron (B), Copper (Cu), Iodine (I), Cobalt (Co) and Molybdenum (Mo) in addition to Nickel (Ni) and Aluminium (Al) found in some media. Of these, iron seems to be more critical since chelated forms of iron and zinc are commonly used in preparing culture media. Iron is usually added as iron sulphate. Iron tartrate and citrate are difficult to dissolve and frequently precipitate in the medium. This problem may be overcome by using Ethylene-diamine-tetra-acetic acid (EDTA) – an iron chelate in place of iron citrate, particularly for embryo induction. EDTA complexes with the iron allow the slow and continuous release of iron into the medium, and un-complexed iron precipitates out of the medium and ferric oxide.

B. Carbon and Energy source

Sugar is the most important part of nutrient medium as an energy source, since most cultures are unable to photosynthesize effectively owing to inadequately developed cellular and tissue development, lack of chlorophyll, limited gas exchange and carbon-dioxide in tissue culture vessels etc. So, tissue cultures lack autotrophic ability and need external carbon for their energy. The most preferred carbon source in plant tissue culture is sucrose at a concentration of 20-60 g/L. Sucrose is cheap, easily available, readily assimilated and relatively stable, and is the most commonly used carbon source. Glucose supports equally good growth, while fructose is less efficient. Sucrose or glucose at 2-4% is the most suitable carbon source which is added to the basal medium. Other sugars like fructose (is toxic if autoclaved), maltose, galactose, raffinose and sorbitol also support the growth of various plant cells, and these may sometimes prove superior to sucrose. These are used depending upon the plant species. During autoclaving the medium, sucrose is hydrolyzed to glucose and fructose which are then used up for growth. The most suitable carbon source and its optimal concentration should be chosen to establish the efficient production of useful metabolites and these factors depend on plant species and products.

C. Vitamins

Vitamins are organic substances required for metabolic processes as co-factors or parts of enzymes and as such for optimum growth; the medium should be supplemented with vitamins. Plants synthesize vitamins endogenously and they are used as catalysts in various metabolic processes. When plant cells and tissues are grown *in vitro*, some essential vitamins are synthesized but only in suboptimal quantities. Hence, it is necessary to supplement the medium with required vitamins and amino acids to achieve best growth of the tissue. The vitamins stimulate the growth of the cells. Thiamine (vitamin-B_1), Nicotinic acid (vitamin-B_3), Pyridoxine (vitamin-B_6), Calcium pantothenate (vitamin-B_5), and myo-inositol or meso-inositol are used more often, and of these thiamine is the basic vitamin required by all cells and tissues. Nicotinic acid and pyridoxine are usually added to culture media but may not be essential for cell growth in many species. Other vitamins are premonitory.

D. Amino acids

Cultured tissues are normally capable of synthesizing the amino acids necessary for various metabolic processes. In spite of this, the addition of amino acids to media is important for stimulating cell growth in protoplast cultures and establishing cell lines and inducing and maintaining somatic embryogenesis. Unlike inorganic nitrogen, amino acids are taken up more readily by plant cells. Amino acids provide a source of reduced nitrogen whose uptake causes acidification of the medium. Casein hydrolysate (relatively a cheaper source of a mix of amino acids), L-glutamine, L-aspargine, L-glycine, L-arginine, L-cysteine, aspartic acid, alanine, glutamic acid and proline are common sources of organic nitrogen used in culture media as mixtures because individually they inhibit cell growth. Tyrosine should be used only when agar is added to the medium. Only L-isomers of amino acids should be used. The most commonly used amino acid is glycine.

E. Other Complex Organic Supplements

In order to stimulate the growth of the cells, the culture media are often supplemented with a variety of organic extracts. These undefined organic supplements are often used when no other combination of known defined components produce the desired growth. Some plant tissues grow well in the presence of complex additives like coconut milk, casein hydrolysate, peptone, yeast extract, ground banana, watermelon juice, malt extract, potato extract (for anther culture), ripe tomato extract, orange juice etc. Yeast extract is a good source of organic nitrogen as well as vitamins. These natural extracts are avoided as their composition is not known and vary from lot to lot and also with age affecting the results.

Casein hydrolysate (edamin) is an undefined protein mixture that is occasionally used in media as a non-specific source of organic nitrogen which contains all amino acids. This has given significant results in tissue cultures. Casein, a phosphor-protein in milk is hydrolyzed (treated with water) to form this weak acid. It is obtained from milk by removing the cream and acidifying the skimmed milk which causes casein to precipitate without any decomposition. The extracts of potato and ripe tomato and juices of orange and water-melon contribute a number of essential nutrients and vitamins.

Coconut milk is the liquid endosperm of coconut. Coconut milk is also known as a supplier of growth regulators. This undefined medium was used with success in early plant tissue culture of *Datura* embryos. Caplin and Steward (1948) grew carrot phloem explants using coconut milk and casein hydrolysate as basal salt supplements. Steward achieved 80-fold multiplication of carrot roots in 3 weeks using coconut milk. Coconut water has been shown to stimulate shoot proliferation in many species of plants. Coconut water should be used at a concentration of 10-15% v/v, and diphenyl-urea, a growth factor in coconut milk exhibits cyto-kinin-like responses.

F. Plant Growth Regulators (PGRs)

Growth regulators are the critical media components in determining the developmental pathway of the plant cells. These stimulate cell division and regulate the growth and differentiation of shoot and the root on explants and embryos in semi-solid and liquid medium cultures. The most commonly used growth regulators are phyto-hormones and their synthetic analogues. Phyto-hormones are required to induce callus tissue and also to promote the growth of many cell lines. Since each plant species requires different kinds and levels of phyto-hormones for the induction of the multiple shoots, callus and its growth and metabolite production, it is important to select the most appropriate growth regulators and to determine their optimal concentrations. The concentration and ratio of hormones may vary from plant to plant and should be standardized for a particular plant tissue. Four broad classes of growth regulators, namely auxins, cytokinins, gibberellins and abscisic acid or their synthetic counterparts (either singly or in combination) are important in tissue culture. The growth regulators used in cultures fall into 5 main classes, namely, auxins, cytokinins, gibberellins, abscissic acid and ethylene.

1. Auxins

Auxins induce and promote cell division, cell elongation, apical dominance, adventitious root formation and somatic embryogenesis and growth. An auxin is required for the induction of cell division and root initiation in cultured tissues. At low auxin concentrations, adventitious root formation predominates, whereas at high auxin concentrations, root formation fails to occur and callus formation takes place. The commonly used auxins include IAA (indole-3-acetic acid), 2, 4-D (2, 4-dichloro-phenoxy-acetic acid), BTOA (2-benz-thizoly-acetic acid), NAA (alpha-naphthalene acetic acid), and IBA (3-indole butyric acid). The most important naturally occurring auxin is indole-3-acetic acid (IAA), but its use in culture media is limited because it is unstable to both heat and light. Hence, amino acid conjugates of IAA (indole-acetyl-L-alanine and indole-acetyl-L-glycine) which are more stable are used. It is more common to use stable chemical analogues of IAA as a source of auxin in culture media. 2, 4-D is the most commonly used auxin which is extremely effective in most circumstances. IAA and IBA are photo-sensitive and as such the stock solutions must be stored in the dark. The auxins are mostly used in combination with cytokinins. 2, 4-D and 2, 4, 5-T are mostly used for callus induction whereas IAA,

IBA and NAA are used for root induction. The concentration of auxins in the medium is generally between 0. 1 – 50 micromoles. 2, 4-D is used to induce and regulate somatic embryogenesis.

2. Cytokinins

Cytokinins promote cell division and stimulate initiation and growth of shoots *in vitro*, and all naturally occurring cytokinins are structurally related purine derivatives. Of these, two have some use in culture media, namely, zeatin and 2iP. The use of these is not widespread as they are expensive and are also relatively unstable. Hence, the synthetic analogues (kinetin and BAP) are used more frequently. Substituted phenyl-ureas are also used as cytokinins (in some systems, these also substitute for auxins). The cytokinins (Kn, BAP, 2iPA and Zeatin) are adenine derivatives and they promote cell division, shoot proliferation, organogenesis and somatic embryogenesis. They have essential role in differentiation and micropropagation of most plant species. The most frequently used cytokinins are kinetin (6-furfuryl-aminopurine), BA (benzyl adenine) or 6-BAP (6-benzyl-amino purine), zeatin, and 2iPA (2-isopentyl-adenine). Zeatin and 2-iPA are naturally occurring cytokinins while BAP and kinetin are synthetically derived cytokinins. These are often used to stimulate growth and development. They modify apical dominance by promoting apical shoot formation, and when used in high concentration, inhibit root formation and induce adventitious shoot formation. The ratio of auxin and cytokinin decides morphogenesis in the culture. If this ratio is high, it leads to embryogenesis, callus initiation and root initiation. They usually promote cell division if added together with an auxin. At higher concentrations (1 to 10 mg/L), adventitious shoot formation is induced but root formation is generally inhibited. Kinetin or benzyladenine as a cytokinin is occasionally required. The ratio of auxins and cytokinins is important with respect to morphogenesis in the culture system. For embryogenesis, callus initiation and root initiation the requisite ratio of auxins to cytokinin is high, while the reverse leads to axillary and shoot proliferation.

3. Gibberellins

Gibberellins (numerous, naturally occurring and structurally related compounds) are lesser used growth hormones involved in the regulation of cell elongation and are agronomically important in determining the height of the plant (mostly used for intermodal elongation and meristem growth) and fruit-set. Gibberellins are infrequently or rarely used in plant tissue cultures as they can inhibit callus growth. Gibberellins are normally used in plant regeneration. GA_3 is essential for meristem culture of some species. In general, gibberellins induce elongation of internodes and the growth of meristems or buds *in vitro*. Gibberellins usually inhibit adventitious root as well as shoot formation. The effect of GA_3 in promoting cell growth in general is conflicting, although in some cases, stimulation of cambial activity in shoot segments and cell division in cell suspension culture is noted.

4. Abscissic acid

ABA inhibits cell division and is most commonly used in tissue culture to promote distinct developmental pathways like somatic embryogenesis (in woody species). Abscissic acid (ABA) is an important growth regulator for induction of embryogenesis. It is also known to promote morphogenesis in *Begonia* cultures. Ethylene is produced by cultured cells, but its role in cell and organ culture is not known. It may be promotory or inhibitory for the same process in different systems. For example, it promoted somatic embryogenesis in maize, but the same process was inhibited in *Hevea brasiliensis*.

5. Ethylene

This is a gaseous, naturally occurring growth regulator, associated with controlling the ripening of fruit in climacteric fruits, and its use in cultures is not widespread. Some cultures generate ethylene, which, if it builds up sufficiently, can inhibit the growth and development of the culture.

G. Activated Charcoal

The activated charcoal acts both in the promotion and inhibition of culture growth depending upon plant species being cultured. Charcoal has been reported to stimulate growth and differentiation in orchids, carrots, ivy and tomato whereas it inhibits tobacco and soya bean etc. It absorbs brown-black pigments and oxidized phenolics produced during culture and thus reduce the toxicity. It also absorbs other organic compounds like plant growth hormones, vitamins etc., which may cause the inhibition of growth. The activated charcoal causes darkening of the medium and thus helps in root formation and growth.

H. Gelling Agents or Solidifying Agents

The culture media for plant tissue culture *in vitro* can be used in either liquid or solid form, depending on the type of culture being grown. Gelling agents are used for preparing the semi-solid media to enable the explants to be placed in right contact with nutrient media (on the surface or slightly embedded but not submerged) to provide aeration to the tissue under growth. For any culture type that requires the cells or tissues to be grown on the surface of the medium, it must be solidified (more appropriately *gelled*). In static cultures, if liquid medium is used, the tissue would get submerged and die due to lack of oxygen. A gelling agent is generally used to circumvent this problem. The most desirable property of a gelling agent is that it should withstand sterilization by autoclaving, and the medium should be liquid when hot but form a semisolid gel when cool. Plant tissues have been shown to do better when cultured on a semi-solid support than otherwise. Agar, a high molecular weight polysaccharide from the seaweeds which can bind water, is the most commonly used gelling agent which is ideal for routine applications. This is preferred over other gelling agents because it is inert; neither does it react with media constituents nor digested by the plant enzymes. The quality of the agar from supplier to supplier and from batch to batch may vary because it is a natural product. It is added to the medium in concentration ranging from 0.5% to 1% (w/v). Agarose, a purified extract of agar is used for protoplast culture. Alternate gelling compounds (agar substitutes) like gelrite, gellan and phytagel or agargel (a blend of agar and phytagel) at 3.5-5 grams/liter, form clear gels (unlike agar which is translucent) as such it is easier to detect contamination during culture growth. Mechanical support for the cell or tissue can also be provided by Filter Paper Bridge, perforated cellophane and polyurethane foam etc., in the absence of a gelling agent.

I. pH (Hydrogen Ion Concentration)

Hydrogen ion concentration (pH) affects the absorption of ions and also solidification of gelling agent. Optimum pH for culture media is 5.8 before sterilization. The values of pH lower than 4.5 or higher than 7.0 greatly inhibit the growth and development *in vitro*. The pH of the culture media generally drops by 0.3 to 0.5 units after autoclaving and keeps on changing though the period of culture due to oxidation and also differential uptake and secretion of substances by the growing tissue.

GENERAL TECHNIQUE OF PLANT TISSUE CULTURE

Many techniques have been adopted for tissue culture among which there are some general techniques that are to be essentially followed. These include: the preparation of the nutrient medium, sterilization, aseptic manipulation and maintenance of culture.

a. Preparation of the Culture Medium

The preparation of the medium is a very crucial step for the culture experiment to be successful. While making the media taking individual constituents, each ingredient is separately weighed and dissolved before putting them together. After making up the volume by water, pH is adjusted and then the medium is autoclaved. Preferably, four stock solutions are prepared, namely: Macro-nutrients, Micro-nutrients (20X concentration), Iron (200X concentration) and Organic nutrients (200X concentration). Separate stock solutions for each growth regulator are prepared. Appropriate quantities are taken from the stocks and mixed together to constitute the basal medium. Required quantity of agar, sucrose and organic supplements are added separately if needed.

Stock solutions of micronutrients, vitamins and growth hormones are prepared separately and stored in a refrigerator for 2 months. Iron stock is prepared separately as chelated form as the sodium salt of the ferric-ethylene-diamine-tetra-acetic acid (Na or Fe–EDTA) and stored in amber coloured bottle in a refrigerator. Cytokinin stocks can be stored for several months in the refrigerator. Cytokinins are thermostable during autoclaving in media.

The macronutrients are prepared in concentrations of 5X, 10X, 20X etc., depending upon the frequency of usage. While preparing macronutrient stocks, in order to avoid the precipitation, calcium (supplied in the form of nitrate or chloride) is prepared separately. All the constituents of the medium after weighing properly are dissolved in double-distilled water. Appropriate quantities of stock solutions are mixed together to prepare required volume of basal medium. Hormonal stocks are added in required concentrations either individually or in various combinations. After the thorough mixing up of the contents in water, sugar and other organic supplements are added. Finally, the volume is made up to 1 liter with double distilled water. The pH of the medium is adjusted around 5.6 to 5.8 (slightly acidic) by adding 0.1 N NaOH (Sodium hydroxide) or 0.1N HCl (Hydrochloric acid). In order to make the medium semi-solid, agar in appropriate quantities is added to the liquid medium and boiled till the agar melts totally. Now, equal quantities of the medium are dispensed into culture vials (test tubes, flasks or petri plates). The culture vials, containing medium are plugged either with non-absorbent cotton wrapped in cheese cloth, aluminium foils or plastic closures and autoclaved. Culture medium is allowed to cool and stored at 4°C until use.

While preparing a solid medium in culture tubes, it is desirable to make slants by keeping the tubes in tilted position during cooling as such slants provide larger surface area for tissue growth.

Now-a-days, the dry powdered media with required concentration of the different constituents are commercially available, and this powder is dissolved in distilled water (10% less than the final volume) directly to get the solution. To this solution, agar, sugar and other desired supplements are added and the final volume is made up by adding the distilled water. The pH of the medium is adjusted to 5.6 to 5.8 and the medium is autoclaved. If the standard medium is not satisfactory, the required quantities of different components are weighed and dissolved in distilled water.

If un-autoclavable plastic culture vials are used, the medium is autoclaved in 250 or 500 ml flasks, and is allowed to cool to 60°C before pouring it into the vials. In both cases, the vessels are stored at 4°C and used according to the need.

b. Sterilization of Glassware and Instruments

The glassware, the instruments, the medium and other articles are to be sterilized to be free from the microbes and contamination. This is done by autoclaving under steam at 121°C and at a pressure of 15 lb/in^2 for 15 minutes. If the medium contains thermo-labile compounds, it is to be sterilized at the room temperature or in cold by passing through a bacterial filter. An alternate method of sterilization of glassware and the instruments is by heating them in a hot-air oven at 150°C for 3-4 hours. Care should be taken during autoclaving the screw-capped glass vessels. The caps are not to be closed too tight to prevent any explosion.

c. Sterilization of Plant Material or Explant

The plant material or the explants to be cultured should be surface sterilized to remove the microbes on the surface. This is to be done in front of the laminar air flow or inside the inoculation chamber prior to inoculation. The plant material or the explant is thoroughly washed in tap water and is immersed in a solution of mild liquid detergent (Teepol – 5% v/v) for 10-15 minutes. Later, it is washed in tap water followed by one-time distilled water. The remaining steps are done in front of a laminar air flow or in pre-sterilized inoculation chamber.

The plant material or the explant is dipped in to 70% ethyl alcohol for 60 seconds. It is immediately transferred into an autoclaved bottle to which is added 0.1% Mercuric chloride or 5-10% Sodium hypochlorite solution for 10-15 minutes. The bottle is to be frequently shaken so that all the surfaces come into contact with the sterilizing solution equally. Later, the solution is decanted and the plant material or the explant is washed thoroughly many a time with autoclaved distilled water to remove all the traces of the sterilizing solution, to make it ready for culture.

d. Preparation of Aseptic Plants

As most of the solutions used for surface sterilization are toxic, they may cause deleterious effects on plant tissues except the seeds as they resist the effect due to the presence of the seed coat. The seeds are directly surface sterilized and cultured on simple basal medium. The seeds germinate in culture to give rise to aseptic seedling. The explants from these seedlings are the most suitable material for culture which need no further surface sterilization.

The seeds to be grown in culture are washed thoroughly with tap water and dipped into 5% v/v Teepol solution for 10-15 minutes. Later, the solution is decanted and seeds are washed again with tap water followed by one-time distilled water. The seeds are then rinsed with 70% Ethyl alcohol for 1 minute. Then they are transferred into an autoclaved bottle in front of a laminar air flow to be immersed in 0.1% w/v Mercuric chloride solution for 10-15 minutes. The bottle is to be stirred frequently. Finally, the seeds are transferred to an autoclaved petri dish with the help of a sterile forceps. The medium is transferred very quickly from the culture vial followed by the seed transfer and the petri plate is incubated in dark either at room temperature or at 25-28°C.

e. Aseptic Technique

Following the surface sterilization and inoculation, all the remaining operations should be necessarily carried out under aseptic conditions. All the sterilized glassware, instruments and the medium are kept ready in the inoculation chamber or laminar air flow. Ultra-violet lamps are put on one hour before, and so also the power switch for air flow in the laminar air flow at least 15 minutes before the work. UV lamp is put off after entry. The top of the work bench is swabbed with alcohol prior to starting the work. Use a clean apron and a mask and clean the hands with alcohol and dry it. Dip all the instruments in alcohol contained in a coupling jar. Light the spirit lamp to flame the neck of culture vial. Transfer the explants or the plant material on to the medium quickly.

The important precautions to be taken while carrying out this work include: keep away the hands moistened with alcohol from the spirit lamp; always make it a point to enter the chamber 15-30 minutes after putting off the UV lamp to prevent the exposure to ozone; do not dip the hot instruments into alcohol and do not use hot instruments either for cutting the material or holding it; and do not apply excessive heat to the neck of the culture vial.

f. Incubation of the Culture

All the cultures are to be incubated in a culture room where light, temperature and humidity are controlled. The control of temperature is necessary as higher temperature may lead to the dissociation of the culture medium as well as tissue damage, while low temperatures decrease the growth of the tissues. Similarly, the intensity and the duration of light are to be regulated as some tissues grow well in the dark while others need both light and dark. Humidity has its own influence, and low humidity results in the quick desiccation of the culture medium while the higher humidity favours contamination.

After inoculation, the cultures are incubated on the culture racks at a constant temperature of 20-25°C. The tubes are always placed at an angle of 30-45° in inclined (slanting) position with the help of a support. The cool, white fluorescent tubes are placed about 18 inches above the culture for a light intensity of $4 - 10 \times 10^3$ Lux for 16 hours. If the culture is to be grown in dark, the lights are put off and the rack is completely covered with a black cloth.

g. Formation of Callus

During incubation, the explant absorbs the exogenously supplied growth hormones from the medium (*initiation medium*) along with other nutrients available which results in a continuous nutrient gradient among the different cells of the explant according to their location. This triggers off a growth sequence during which cell enlargement and cell division predominate, and the cells divide asynchronously to form an *unorganized, amorphous, proliferative aggregate (mass) of loose parenchyma cells called a callus*, around the cut edges or from the ruptured surface of the explant. This gradually increases in mass as the new cells are added by mitosis. The cellular composition is extremely heterogeneous ranging from small cells (with dense cytoplasm which divide actively) to large cells (with vacuolated cytoplasm which do not divide). The shape of the cells also varies, from spherical to elongated cells. On the basis of its texture, the callus may be a *soft callus* (friable) or *hard callus* (compact and brittle). The colour of the callus is either white or creamy-yellow. If white callus is grown in dark, it becomes pigmented. The pigmentation may be either uniform or patchy, and green (due to the development of chloro-plastids) or yellow (due to the synthesis of carotenoids). In some cases, the callus may become brown, due to the excretion of phenolics, where the growth of the tissue is inhibited. The callus has no organized meristems.

Callus culture in a tube

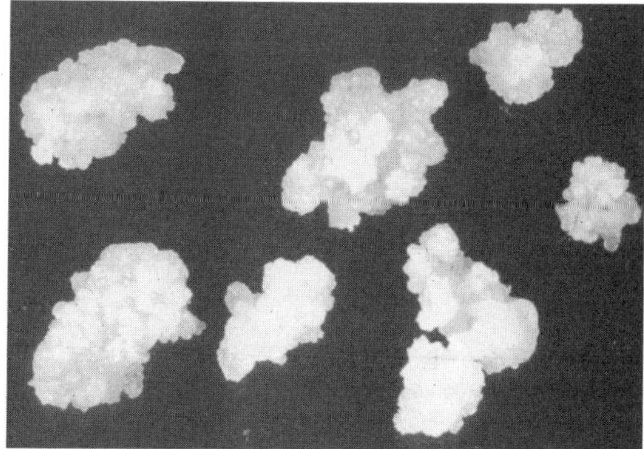

Callus - separated

h. Organogenesis or Morphogenesis or Regeneration of the Plant

The callus has the potentiality to produce normal roots and embryoids which in turn develop into plantlets. *Organogenesis* is the development of adventitious organs or embryoids from the callus in tissue culture. This is generally controlled by the ratio of auxin and cytokinin in the culture medium. A relatively high ratio of auxin:cytokinin induces the root formation (*rhizogenesis*), whereas a low ratio of auxin:cytokinin induces the shoot formation (*caulogenesis*). Sometimes, anomalous structures develop from the callus during organogenesis, called *organoids*. The localized meristematic cells on a callus capable of giving rise to shoots and roots are termed as *meristemoids*. After 6 weeks, the rootless shoots can be excised and placed on to the root inducing medium (MS medium with 0.2 mg/L IAA and 0.02 mg/L cytokinin). Recently, the cut ends of the shoots are treated with a suitable auxin solution (or powder mix) and transplanted in pots containing vermiculite or potting mix and kept under high relative humidity and low light intensity. Rooting takes place in about 10-15 days depending on the species. The plantlets with 0.5 – 1.0 cm are usually transplanted into pots as longer roots tend to get damaged.

The aseptic procedures are not required for the transplantation of the plantlets to soil. The plantlets are carefully removed from the culture vessels and every care should be taken not to damage the root or the shoot system. They are carefully washed with tap water to remove the residual agar medium sticking to their roots. The individual plantlets are separated and transplanted into plastic cups or pots containing a suitable potting mix or seedling compost. The mix or compost is watered and the whole pot is covered with a small inverted polythene bag to reduce the amount of water lost by transpiration. The plants are kept in high humidity (90% or more) and initially low light intensities. High humidity can be attained by fog, mist or a clear plastic sheet to cover the individual or groups of plants. The humidity is gradually decreased to ambient level after 7-15 days and the light intensity is increased. Finally (after 3 weeks), the plastic bags are completely removed and the plants are exposed to green house conditions to be grown to maturity.

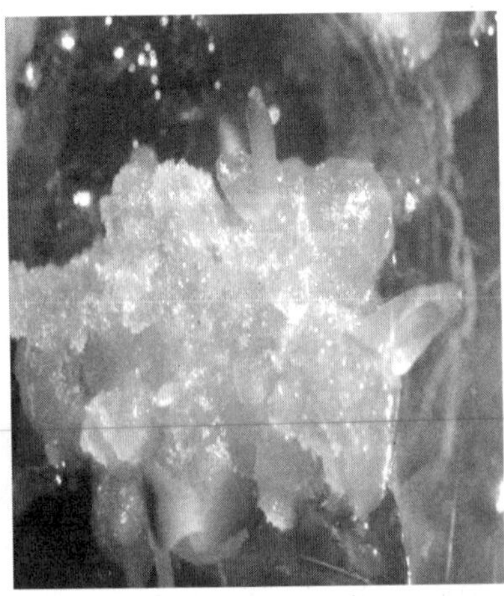
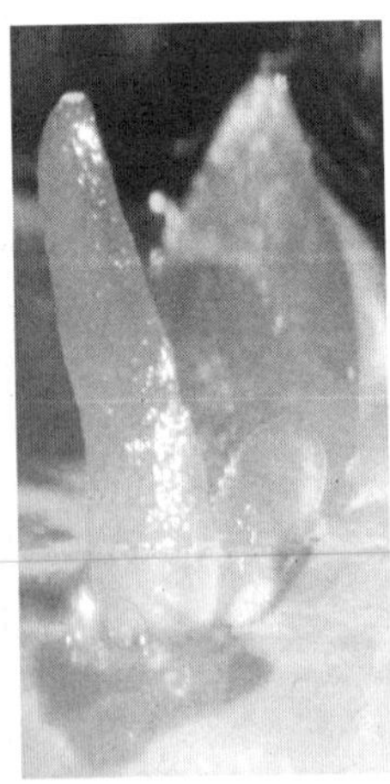
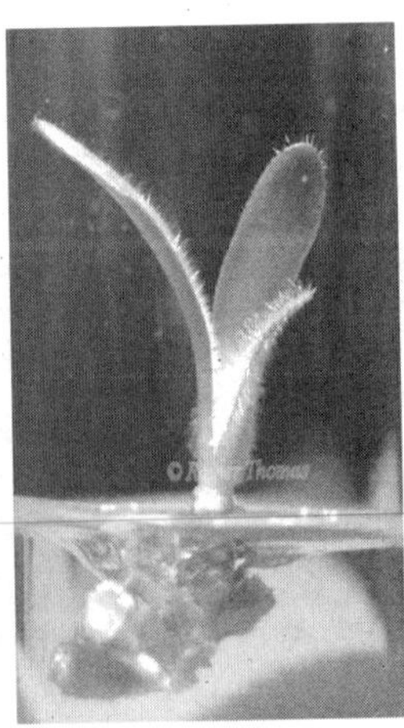

Organogenesis – Different stages

Development of the shoots from the callus – shown separately

PLANT TISSUE CULTURE PROTOCOL (GENERAL)

Most methods of plant regeneration from isolated plant cells or tissue is conducted *in vitro* so that the environment and growth medium can be manipulated to ensure a high frequency of regeneration. The primary aim is to produce, as easily and as quickly as possible, a large number of regenerable cells. Hereunder is considered some basic issues concerned with plant tissue culture *in vitro.*

The culture can be sustained as a mass of undifferentiated cells for an extended period of time before they regenerate into whole plants. The starting point for all tissue cultures is plant tissue, called an *explant.* It can be initiated from any part of a plant, root, stem, petiole, leaf or flower, although the success of any one of these varies between species. It is essential that the surface of the explant is sterilized to remove all microbial contamination. Plant cell division is slow compared to the growth of bacteria and fungi, and even minor contaminants will easily over-grow the plant tissue culture. The explants are then incubated on a sterile nutrient medium to initiate the tissue culture. The composition of the growth medium is designed in such a way so as to both sustain the plant cells, encourage cell division, and control development of either an undifferentiated cell mass (*Callus*), or particular plant organs. The concentration of the growth regulators in the medium (auxins and cytokinins), seems to be the critical factor for determining whether a tissue culture is initiated, and how it subsequently develops?

The explant should initially form a callus, from which it is possible to generate multiple embryos and then shoots, forming the basis for plant regeneration. The first stage of tissue culture initiation is vital for information on what combination of media components will give a friable, fast-growing callus, or a green chlorophyllous callus, or embryo, root or shoot formation. At present there is no way to predict the exact growth medium, and growth protocol, to generate a particular type of callus or organ. These have to be determined through a carefully designed experiment for each new plant species, and also for each new variety of the species. The basis of the experiment will be media and protocols that give the desired effect in other plant species, and experience.

PLANT TISSUE CULTURE - LABORATORY AND EQUIPMENT

It is not always necessary to design a special laboratory for plant cell culture, and general microbiology laboratory can be used and the aseptic conditions are a prerequisite for the incubation of plant cells as well as microbial cultures. An ideal laboratory for tissue culture should have at least three rooms, two big rooms (one for general laboratory work i.e., for the preparation of the media, autoclaving, water distillation etc., other big room for keeping the cultures under controlled light, temperature and humidity etc,) and one small room (for carrying out aseptic work and for keeping the autoclaved articles).

1. Laboratory (for aseptic inoculation): The larger room should be without any windows or ventilators to make it dust-free. The room should be provided with a double door with an automatic door closer. The floor should be covered with a rubber mat to facilitate easy cleaning. The shoes should be left outside before entry into the room. A large wooden chamber measuring 4` x 4` x 7` should be made available to carry out short term aseptic work. The upper half of the side walls of the chamber should be made of large glass sheets. The chamber should be fitted with two ultra-violet sterilizing lamps (one bigger and the other smaller) and a fluorescent lamp, with the operating switches fitted on the outside so that they can be safely switched on and off. The working area and the tables inside the chamber should be made of thick glass sheets.

For routine work (aseptic seed germination, harvesting the cultured tissue, cytological work, microtome work), a small inoculating hood can be used, and this is to be placed on a small table in one corner of the room.

2. Washing area: This is very important and should be provided with a sink, running hot and cold tap water, brushes of different sizes, detergents and a bucket with one-time distilled water to rinse the washed glass articles finally. A number of plastic buckets are necessary for soaking the glass articles to be washed well. A separate bucket with a lid is essential for disposing off the used/infected media before cleaning, and this is to be always kept outside the room/cleaning area and should be thoroughly cleaned twice a week to avoid contamination inside the laboratory.

The equipment required for a general culture laboratory is as follows:

3. Laminar Air Flow Cabinet: These cabinets are available in different sizes commercially in the market. The cabinets are always not necessary when there is a sterile room in the laboratory. The cabinets can be directly purchased. This is the most suitable, most convenient and the most reliable instrument for aseptic work in the tissue culture laboratory. This is because it allows one to work for longer periods which is not so with inoculation chamber. Long hours of work in the inoculation chamber may result in suffocation and as such the work is to be interrupted. But, on the table of laminar air flow cabinet one can work easily and openly for a longer period. This cabinet has a number of small blower motors to blow air which passes through a number of *high efficiency particulate air* (HEPA) filters which help to remove particles larger than 0.3 micro metres. The ultra-clean air free from bacterial and fungal contaminants flows at the velocity of about 27 (+ or -) 3 m/minute through the working area. The instrument is put on for about 1-15 minutes before starting the work. The flow of air will not put out the flame of a spirit lamp; as such it can be used conveniently during the work.

4. Autoclave: These are available commercially in the market in different sizes. Suitable autoclaves can be purchased and the experiments are conducted. Autoclave is essential for the sterilization of the nutrient media, glassware, instruments etc. Sometimes, if autoclave is not available, sterilization can be carried out by using a pressure cooker.

5. Refrigerator: A refrigerator is essential to store various thermo-labile chemicals like vitamins, growth hormones, amino acids, casein hydrolysate, yeast extract, coconut milk etc. The stock solutions of the nutrients (salts) are also kept inside the refrigerator to prevent contamination.

6. Hot air oven (for dry sterilization): An autoclave can be used for dry sterilization but, a hot air oven is very useful for the sterilization of instruments like scalpels and glassware after washing like petri-dishes, pipettes etc. A number of enameled trays of different sizes are essential for keeping the wet glassware inside the oven.

7. Sterilization Equipment (by filtration): By using the autoclave the media containing the carbon source and growth regulators are simultaneously sterilized, Sometimes, aseptic filtration is favourable to avoid the decomposition of some unstable chemicals.

8. Pure Water De-mineralizer or Wet Distillation Apparatus or Distillation Plant: For the preparation of the medium, distilled water or deionized water is generally used. A single distillation or a double distillation water plant (sometimes both) is indispensable for the laboratory for the preparation of the required medium. Two plastic containers of a larger size are also needed to store the distilled water (distilled water and double distilled water separately) for a future use.

9. Weighing Balance: A pan balance, a chemical balance and an electronic balance are essential to weigh different substances needed for culture experiment accurately.

10. pH Meter: A pH meter is an essential requirement for measurement and adjustment of pH of the nutrient medium.

11. Vacuum Pump: This is essential for filtering the liquid media, sugar solutions etc. using air suction.

12. Culture Room: For the successful cultivation of plant cells, culture rooms under different temperatures are very essential. The temperature and the light intensity as well as the duration of lighting in the room are to be controlled under optimal conditions. The culture room refers to the room or the place where the cultures are kept or incubated under controlled conditions of temperature, light and humidity. The room is also generally fitted with a double door in order to make it completely dust-free and also to maintain a constant room temperature. One must enter the room without shoes. Air coolers are generally used to maintain the temperature around 25 (± 2°C) inside the room. The room is also provided with specially designed shelves made of glass or plywood to keep the culture vessels therein for incubation. Other glassware such as flasks, bottles, jars, petri-plates etc., can be placed directly on the shelves. The culture tubes are kept on a support (such as empty paper cover of fluorescent lamps). The cultures can be grown either in light or in dark. Each culture rack is fitted with fluorescent lamps which are photo-periodically controlled by an automatic timer. The racks are generally covered with black curtains for dark incubation of the culture. A thermometer to check the temperature and a hygrometer to check the relative humidity are to be fixed on the wall at the safe corner of the room periodically. The relative humidity of the room is to be always maintained at above 50%. Some more shelves of smaller size are to be placed in the culture room for keeping the autoclaved articles and the culture vials with the medium temporarily.

13. Shelf: Shelves made from a rigid wire mesh to allow the maximum air movement and minimize shading should necessarily be used in the culture room.

14. Rotary Shaker: To raise the suspension cultures either a rotary shaker or a reciprocal shaker is necessary. The speed of revolution of the shaker is to be controlled each time and its platform is provided with light. The clips for holding the conical flasks (150 ml and 200 ml) are to be fitted to the platform of the shaker.

15. Working Tables: These are required for the preparation of the medium and also to carry out other work.

16. Heater: This is required for heating up the medium or for warming up the medium to dissolve agar or sometimes to melt the solid agar medium.

17. Microscope: Simple microscope, compound microscope and an inverted binocular dissecting microscope are needed to carry out the microscopic work. Another essential feature is that some microscopes should be fitted with a camera for photo-micrographic work.

18. Microtome: This is required for sectioning the cultured tissues when needed.

19. Wooden (metal) racks: These are required to keep the various chemicals and solutions properly labeled for use as per the necessity in the laboratory.

20. Glassware to be used: To culture the plant tissues different types of glassware and apparatus are used, which are both conventional and also specific. The glassware should necessarily be of Pyrex or Corning or Boro-silicate glass. The glassware is represented by the measuring cylinders, conical flasks, pipettes, beakers etc. Plant tissues are generally grown in wide-necked Erlenmeyer conical flasks (100 ml or 150 ml or 200 ml), culture tubes (25 mm in diameter and 150 mm in length), petri-plates (50 mm or 90 mm or 140 mm in diameter) and screw-capped universal bottles (of 20 cm^3 capacity). Sometimes, jam bottles or milk bottles may also be used. Proper care must be taken to ensure that the glass ware is properly cleaned (either new or dirty before use. The traditional way of cleaning the glassware is to soak them in soap water for some time, followed by brushing and washing them well with tap water and finally rinsing them with one-time distilled water. Later, they are dried in a hot-air oven and then stored in a dust-proof cup-board or a drawer. Prior to autoclaving, all the culture vessels (flasks and tubes) must be fitted with cotton plugs (tightly rolled plugs of non-absorbent cotton wrapped in gauge) to allow free gaseous exchange but not to allow the entry of microbes. The exposed parts of the plug as well as the rim of the culture vessel should be covered by a brown paper or an aluminum foil which will keep the plug as well as the rim of the vessel free from dust and also protect the plug not to become wet during autoclaving. Sometimes, pre-sterilized, disposable, autoclavable, plastic ware is used to culture plant tissues. In case of sterilization of the medium containing thermo-labile compounds or enzymes (for protoplast isolation), a specially designed glass made of bacterial filter or an autoclavable plastic-made bacterial filter is used. For the flame sterilization of the instruments using methylated spirits, a small spirit lamp made of glass is essential.

21. Instruments to be used: The instruments which are used in routine for culture work include different sized scalpels, forceps, spatula, scissors etc., made of stainless steel. The following is the list of instruments of common use with their measurement and use:

1. Scalpel – for collecting the plant material or the explants (20 cm or 32 cm or 25.4 cm in length).
2. Surgical scalpel – with a holder (size no. 3 or no. 4) and arrow-headed disposable blade (size no. 11) – for collecting the plant material or the explant.
3. Forceps – without a corrugated marking at the pointed end (plain tip) – for holding the material (20 cm or 32 cm or 25.5 cm in length).
4. Jewellery fine forceps – for peeling off the epidermis of the leaf (size no. 5).
5. Arrow-headed sharp needle – for dissecting out the anthers from the flower (25.4 cm in length).
6. Spatula (with one end spoon-headed and the other end flat) - for transferring or sub-culturing the material (25.4 cm in length).
7. Metal cork borers (1 cm in diameter) - for making tissue cylinders (potato or carrot etc.).
8. Scissors – for cutting the roots or shoots from the aseptic culture (25.4 cm length).

METHODOLOGY FOR THE ESTABLISHMENT OF CULTURES

i. Selection of a Suitable New Medium

A suitable tissue culture medium may be devised by trying many combinations of different concentrations of major components like salts, sucrose and plant growth regulators. Various combinations of low, medium and high concentrations of minerals, organic nutrients, auxins and cytokinins can also be evaluated to get a suitable medium. The exact conditions required to initiate and sustain the plant cells in culture or to regenerate intact plants from the cultured cells are different for each plant species, and each variety of species will have a particular set of culture requirements. All these conditions have to be identified for each variety through experimentation.

A simple approach for selecting a suitable medium is to first test several concentrations of an auxin and cytokinin (like zero, 0.5, 2.5, 5 and 10 micromoles/ L^{-1}) to identify the suitable combination of the two. The different auxins and different cytokinins available may be tested to identify the best of each. Later, using these growth hormones, the different standard media may be evaluated. Then check the half, full and even higher salt concentration of the selected medium as well as different sucrose concentrations (2-6%). Alternately, the worker may evaluate the various combinations of low, medium and high concentrations of minerals, organic supplements, and growth hormones, to arrive at a suitable composition.

ii. Methods of Sterilization

The application of plant cell, tissue and organ culture require sterile technique. The maintenance of aseptic or sterile conditions is essential for successful tissue culture procedures. Sterilization is a procedure used for elimination of microorganisms.

The most important factor in the selection of suitable working area is the possible flow of unfiltered air over the disinfected working area; air currents must be avoided, because air-borne spores or contaminating microorganisms are carried along with it.

For aseptic manipulation, the instruments such as forceps, scalpels and scissors are sterilized by dipping in ethanol, followed by flaming and cooling, which is referred to as flame sterilization.

The methods of sterilization (physical and chemical) can be described as follows:

a. Sterilization by Heat

Fire and boiling water have been used for sterilization and disinfection since the time of Greeks, and heating is still one of the most popular ways to destroy microorganisms. Either moist heat or dry heat does this. Moist heat is the most effective and efficient. It kills the organisms by coagulating and denaturing their enzymes and proteins; and this is helped a lot by the presence of water. Spores are killed by exposure to moist heat at 121°C for 10-30 minutes. Dry heat kills organisms by oxidative destruction of the cell constituents. Spores are killed by exposure to dry heat at 160-180°C for 1-3 hours. Following factors influence sterilization by heat: (1) temperature and time of exposure, (2) the number of vegetative organisms and spores present, and (3) the nature of the organism.

b. Sterilization by Dry Heat

This method is used only for glassware, Teflon plastic-ware, metal instruments, and other material that are not charred by oven temperatures. Strips of tape should not be present on any glassware, and prolonged oven heat may dull the cutting edges of surgical blades. Cotton, paper, and plastic should never be placed in the oven for sterilization. It is highly important not to use a laboratory oven that has previously been used for paraffin embedding. Objects to be sterilized are wrapped in heavy-duty aluminum foil before being placed in the oven. Care is taken not to pack the foil-wrapped packages too tightly. After sterilization, the wrapped packages are shifted to the laminar flow cabinet. More recently, glass bead sterilizers (300°C) are being used for the sterilization of the instruments like forceps, scalpels, needles etc.

c. Flame Sterilization

Normally the metallic instruments are sterilized by dipping in 95% ethyl alcohol, followed by flaming and cooling, a technique called *flame sterilization.* The instruments are repeatedly sterilized during the process to avoid contamination. It is customary to flame the mouths of the culture vessels prior to inoculation or subculture.

d. Sterilization by Moist Heat

This procedure employs an autoclave operated with steam under pressure. All culture vessels, both empty and those containing the medium are generally sterilized by heating in an autoclave or a pressure cooker to 121°C at 15 psi (pounds per square inch: 1.06 kg/cm^2) for 15 (20-50 ml medium) to 40 (2L medium) minutes, the time being longer for larger volumes. The plastic-ware and instruments like micro-pipettes are also autoclavable. Proper care should be taken to properly stopper all the vessels and to open the autoclave only when its pressure gauge indicates zero pressure.

e. Filter Sterilization

Filter sterilization is the process of removing contaminants in the range of 0.025 – 10μm from fluids (solutions) by passage through a micro-porous medium, such as a membrane filter (0.45 micro-metres or lower pore size). The filter is held in a suitable assembly, and the assembly together with the filter is sterilized before use by autoclaving. This is necessary for media components that are degraded during autoclaving like gibberellic acid, zeatin, abscissic acid, urea, certain vitamins and enzymes which are thermo-labile) and must be sterilized at room temperature. A relatively small volume can be sterilized by passage through the filtration unit attached to a hypodermic syringe. Most of the membrane filters are screens consisting of a uniform continuous mesh of polymeric material with pore size precisely determined by the manufacturer. Although majority of workers use a 0.45 μm pore size, the pore diameter should measure at least 0.22 μm for the complete removal of all bacteria, yeast, and molds.

f. Infra-red radiation

The infrared rays are directed from an electricity heated element on to objects like glass syringes etc., at 180°C. Surgical instruments are sterilized above 200°C in vacuum chamber in which N_2 is passed during cooling time to prevent oxidation.

g. Wiping with 70% Ethanol

The surfaces that cannot be sterilized by other techniques such as the platform of the laminar air flow cabinet, hands of the operator etc. are sterilized by wiping them thoroughly with 70% ethanol (alcohol) and allowed to dry.

h. Sterilization by Radiation

Ultraviolet (UV) radiation around 260 nm is quite lethal but does not penetrate glass, dirt films, water and other substances very effectively. Because of this disadvantage, UV radiation is used as a sterilizing agent only in a few specific situations. UV lamps are sometimes placed on the ceilings of rooms in biological safety cabinets to sterilize the air and any exposed surfaces.

UV radiations are harmful to mankind; therefore, skillful installation is required during the practical use. UV radiation burns the skin and damages eyes. People working in such areas must switch off the UV lamps when the areas are in use. Commercial UV units are available for water treatment. Pathogens and other microorganisms are destroyed when a thin layer of water is passed under the lamps. Ultraviolet rays kill many of the vegetative forms of microorganisms, present in the air, but they vary in their susceptibility. These radiations are being used in sterilizing the surgical area, instruments etc. The application of ultraviolet radiation is dependent when it makes direct contact with the particles carrying the organisms, because they have very little penetrating power. Mercury vapour lamps produce it. It is used to sterilize plastic material.

iii. Maintenance of Aseptic Conditions

Prior to commencing work, the hands are washed with aseptic soap followed by swabbing with ethanol. The laminar airflow chamber is sprayed with 95% ethyl alcohol and all paraphernalia used in inoculation such as medium, forceps, scalpel, petri plates etc., are transferred to it and UV light is switched on for 30 minutes. Since UV radiation is harmful to the eyes, care should be taken to avoid direct exposure to UV. UV light is switched off and air flow is switched on which flows at the velocity of 27 ± 3 m/min. Work in laminar is started after 15 minutes of air flow, so as to remove the O_3 (ozone) which was built up in the chamber during UV radiation. In order to minimize the contamination during inoculation, hands and inoculating area are swabbed with alcohol frequently, followed by flame sterilization of forceps and scalpel.

Tie back your hair, roll your sleeves up and remove your watch and other jewellery. Wash your hands thoroughly with the disinfectant solution suitable for skin application and start inoculation.

iv. Surface Sterilization of the Explant

The explant (*the excised piece of tissue or organ of the plant used for culture*), which is used for inoculation generally, is loaded with many microorganisms. The growth of explant will be hampered because microorganisms with brief life cycle grow very fast on the medium. Hence, for the development of explant without any problem, the plant material is cleaned and surface sterilized (treated with an appropriate sterilizing agent to inactivate the microbes present on their surface).

The protocol for surface sterilization will depend mainly on the source and the type of tissue of the explant which will determine the contamination load and tolerance to the sterilizing agent. The material is first washed with a detergent followed by a sterilant. The sterilizing agents generally used for surface disinfection include Sodium hypochlorite (0.5-2% from 5-30 min.), Calcium hypochlorite (9-10% from 5-30 min.), Hydrogen peroxide (10-12% from 5-15 min.), Mercuric chloride (0.1-1.0% from 2-10 min.), Ethyl alcohol (75-95%), Benzalkonium chloride (0.01-0.1% from 5-20 min.), Silver nitrate (1% from 5-30 min.) or Bromine water (1-2% from 2-10 min.) and antibiotics (4-50 mg/L). Of these, the first two give good results and mercuric chloride gives a satisfactory result. Hence, their use is most common. The duration of the treatment varies from 15-30 minutes. Since all the agents used are toxic to plant tissues, the duration and concentration used should be such as to cause minimum effect and the rinsing after treatment should remove them as completely as possible. Hence, the treated material is rinsed with sterilized double distilled water several times to completely wash off the sterilant.

Most of the explants are larger enough to be treated directly. But, explants like ovules, embryos, pollen etc., are rather small for efficient handling. So, the organs containing them like ovaries (for ovules), developing seeds (for embryos), mature anthers (for pollen) are surface sterilized following which the desired explants are dissected out aseptically.

The entire process is carried out in an aseptic area. If the explants are obtained from the field, they should first be thoroughly washed under running water to remove the dirt and epiphytic microbes, followed by the treatment with 1% Cetavlon upto 10-15 minutes to reduce their bacterial contamination.

v. Transfer of Explant

Inoculation is carried out near spirit lamp. Sterilized plant material is transferred on to sterilized petri plate containing filter paper. With the help of forceps and scalpel the explants are dissected and damaged ends are removed and blotted on a sterile filter paper. Then explants are inoculated on to culture medium vertically to maintain polarity. The mouth of culture tubes is covered with sterilized aluminum foil and tied with rubber bands or plugged with non-absorbent cotton. Culture tubes are labeled giving the details of the experiment, name of the explant, date of inoculation etc.

vi. Aeration

The tissue or organs can be cultured either on a solid medium or liquid medium. Agar-agar (a polysaccharide obtained from sea weeds) is used at a concentration of 0.8 1% to solidify the medium. When liquid medium is used, aeration is provided by shaking the vials on an automatic shaker. The advantage of growing tissues in agitated liquid media is that it facilitates gaseous exchange and reduces nutrient gradient.

vii. Growth Measurements

Under appropriate light, temperature, aeration and nutrient medium, the growth of suspension culture follows a predictable pattern or growth curve. The growth of suspension culture can be monitored very easily by simply counting the cell number per unit volume of culture in relation to days of culture.

TYPES OF PLANT TISSUE CULTURE

The plant cells cultured on a solid medium will result in friable, pale brown, un-organized mass of cells called callus. The cells and tissues cultured in a liquid medium aerated by agitation will grow as suspension of single cells and clumps of cells. The cells need to divide for growth, while the broken up cells from the explants are non-dividing and mature. So, the differentiated tissue (parenchyma and collenchyma) undergo modifications to become meristematic for cell division. This phenomenon of a mature cell reverting back to a meristematic cell to produce undifferentiated callus is called *de-differentiation*. Initially, the divisions are confined to the cut ends, and subsequently it covers the entire explant and the resulting cell mass is organized ordinarily and often consists of fibres and vascular elements.

A. Callus Cultures

Callus culture is the culture of undifferentiated mass of cells (*callus*) on agar medium produced from explants of a seedling or any other part of a plant. Auxins and cytokinins are required for the formation of a callus. This can be sub-cultured indefinitely by transferring a small piece of the same to fresh agar medium. This process needs to be done every 3-5 weeks in view of cell growth, depletion of the nutrients and drying up of the medium. The calli are easy to maintain and are the most widely used. On solid agar medium, the rate of growth of the callus is relatively slow. The new cells are generated on the periphery of the existing mass, and as such the callus consists of cells which vary considerably in their age. A vertical nutrient gradient is formed as the nutrients are gradually depleted from the agar. The usefulness of the callus in experimental system is limited because of low degree of uniformity among the cells, slower growth rate and development of nutrient gradients. This method is only useful for maintaining the cell lines as well as morphogenesis.

B. Suspension Cultures

Suspension culture is the culture of cells and tissues in a liquid nutrient medium resulting in suspension of single cells and cell clumps of few to many cells. The liquid cultures must be constantly agitated, generally by a gyratory shaker at 100-250 rpm (revolutions per minute) to facilitate aeration and dissolution of cell clumps into smaller pieces. In this method, callus mass is transferred to liquid medium and the culture vessel is incubated. By subculture, gradually over several weeks, the cells dissociate and a liquid suspension culture is obtained. These suspensions are maintained by sub-culturing in the early stationary phase being transferred to a fresh medium. The cell growth is much faster than the callus culture, and as such they are to be sub-cultured more frequently, from 3-14 days. When fully established, the suspension cultures consist of a homogeneous population of cells. The advantage of this system is that the nutrients can be continuously adjusted, and also allow a more accurate determination of the nutritional requirements of cells. These experiments are useful for the large scale production of cells and somatic embryos.

Due to the depletion of the nutrients and also the drying up of the medium after a period of time, it becomes necessary to transfer the cells, tissues or organs on to the fresh media. A portion of the tissue is used to inoculate new culture vessels, and this is known as *sub-culturing*. This process is taken up every 4-6 weeks in general, and every 3-14 days in suspension cultures, to maintain the cultures indefinitely. The volume of the inoculums in suspension cultures should be 20-25% of the volume of the fresh medium. Similarly, the initial density just after inoculation should be 5 x 10^4 cells/ml^{-1} or higher. Otherwise, the cells fail to divide.

C. Single Cell Culture

Single cell culture refers to a method where in the free cells isolated from different plant organs or cell suspensions are grown as single cells under *in vitro* conditions resulting in a clone of identical cells.

1. Bergmann's Plating Technique

In this technique, the free cells are suspended in a liquid medium. The culture medium with added agar (1%) is cooled and maintained at 35^0C in a water bath. Equal volumes of the liquid medium and agar medium are mixed and immediately spread in a petri dish. The single cells remain embedded in the soft agar medium and the cells are observed under inverted microscope. The cell colonies after development are isolated and cultured separately. The single cells should be cultured in dark as light has detrimental effect on the proliferation of the cells.

APPLICATIONS OF PLANT TISSUE CULTURE

The methods of tissue culture offer advantages over traditional methods of propagation such as: production of copies of plants that produce flowers, fruits etc., with desirable traits, quick production of mature plants, production of large numbers of plants in the absence of seeds or required pollinators, regeneration of genetically modified whole plants from cells, production of plants in sterile condition which allows the reduced chances of transmission of diseases, pests and pathogens, production of plants whose seeds have low chances of germination and growth (like *Nepenthes* and Orchids), and a clean stock of plants without infection for use in horticulture and agriculture. Tissue culture is possible due to the fact that many plant cells have the ability to regenerate a whole (complete) plant, called *totipotency*. Thus, individual cells, protoplasts, pieces of leaves, stem tips, root tips, pollen grains, ovaries, embryos etc., can be used to generate a new plant on culture media with the required nutrients and plant hormones. Tissue culture is the only way to propagate the more uniform plants vegetatively, the plants often grow faster and show improved vigour, often mature earlier than those formed from the seed, and this is the only way to produce disease-free plants in bulk.

Tissuc culture techniques are used widely in plant science, and also have a number of commercial applications too. The important applications include: Micro-propagation is widely used in floriculture and forestry and can also be used to conserve rare and endangered plant species. Tissue culture can also be used to screen cells for advantageous characters like resistance and tolerance to herbicides. This is also used in bio-pharmaceuticals where plant cells are grown in large scale in bioreactors for the production of secondary metabolites and recombinant proteins.

These methods are useful to cross distantly related species of plants (protoplast fusion) and regeneration of the novel hybrids. The technique is useful to carry out the cross pollination of distantly related species and to culture the resulting embryo (embryo rescue). This is also useful for the production of dihaploid plants from haploid cultures (colchicine is used to double the chromosome number) to generate homozygous lines more rapidly during the breeding programmes. It is also used in the regeneration of transgenic plants. Meristem tip culture can be used to generate clean plant material from virus-infected stock (potatoes and soft fruits). Culture of meristems and shoots helps to generate large number of identical individuals. The production of identical sterile hybrid species is also possible through tissue culture.

The difficulties that are increasingly encountered in procuring ample supply of raw plant material because of drastic decrease in natural resources due to human disturbance to the environment, ruthless exploitation of the available resources, increasing labour costs, and technical as well as economic difficulties in cultivating wild plants have prompted the adaptation of *in vitro* technology for the commercial production of substances of medicinal importance. Besides providing an alternative technology to bypass the above difficulties, the plant tissue culture (used in a broad sense to include cell, tissue and organ culture) offers some other advantages (Tabata, 1977) such as: (i) Useful compounds could be produced under controlled environmental conditions independent of climatic changes or soil conditions, (ii) Cultured cells would be free of microbes and insects, (iii) Cells of any plant, tropical or alpine, could easily be multiplied to yield their specific metabolites, and (iv) The automated control of cell growth and rational regulation of metabolic processes leading to improved productivity of the desired products.

In addition to these, the *in vitro* technology also facilitates: (i) novel means of conserving the genetic diversity of the germplasm of medicinal plants through cryopreservation, and (ii) production of novel compounds through biotransformation, somatic hybridization and selective gene transfer through recombinant DNA technology. *In vitro* propagation is useful in many ways. Plants derived from nucellus and shoot tips are free from virus and can be used for raising virus free clones. For example, Citrus plant propagation from embryogenic callus of nucellar origin is free from virus.

The Pfizer Company made the first attempt for the industrial production of secondary metabolites *in vitro* during 1950-60 and Routien and Nickell obtained the first patent in 1956.

Mitsui Petrochemical Industries (Tokyo) is commercially producing Shikonin, used both as a dye and pharmaceuticals from *Lithospermum erythrorhizon* cell suspension cultures. Shikonin is also used as herbal lipstick. Similarly, is has been possible to establish cell cultures for industrial production of berberine from *Coptis japonica* and ginseng extracted from roots of *Panax ginseng*.

Cell cultures are advantageous in that the screening of spontaneous mutations at the cellular level effectively inhibit chimeric formation which is a drawback in mutation breeding. Plants cell cultures are known to produce a variety of secondary metabolites hitherto not known to occur in the intact plant, as for instance, *Andrographis paniculata* (Sesquiterpenes), *Morinda jasminoides* (anthraquinones) and *Ruta graveolens* (furocoumarin) which are used in pharmacy, medicine and industry. The cell cultures can be effectively utilized for the production of these chemicals on a commercial scale for enhanced yield and better production control. Some of the important secondary compounds produced through plant tissue culture include: Stevioside from *Stevia rebaudiana* (a widely used sweetener), Jasmine from *Jasminum* species (a perfume), Codeine from *Papaver* species (an analgesic), Atropine from *Atropa belladonna* (a muscle relaxant), Morphine from *Papaver somniferum* (an analgesic), Quinine from *Cinchona officinalis* (an antimalarial), Quinones from *Cinchona succirubra* (antimalarials), Nicotine from *Nicotiana tabacum* and *N. rustica* (an insecticide), Catharanthine from *Catharanthus roseus* (an anti-cancerous substance), Shikonin from *Lithospermum erythrorhizon* (a cosmeceutical), Saiko-saponin from *Bupleurum falcatum* (a photo-protective), and Diosgenin from *Dioscorea deltoidea* (a widely used female contraceptive) etc.

Tissue culture is also useful in *micropropagation* of Horticultural plants, Forest trees, and tree crops. Several companies in India are propagating banana plants through tissue culture. A large number of companies in India and abroad are propagating ornamental and horticultural plants through tissue culture. Orchids, which are of high value, are difficult to propagate in garden. They can be propagated through tissue culture and in a very short period a large number of plants can be produced.

The success of plant tissue culture techniques has made possible large-scale cloning of plant species. With a relatively small investment in space, technical support and materials, it is possible to produce literally millions of high-quality, genetically uniform plants. Micropropagation is also now being used extensively in the production of forest tree species. Here, the propagules are generated primarily from cultures of axillary and adventitious buds; callusing and differentiation of new buds is rarely used. A similar approach has been applied successfully to cultivars of apple (*Malus*), peach (*Pyrus*) and pear (*Pyrus*). Because most of the temperate fruits are heterozygous, they do not breed true from seed but are propagated by vegetative cuttings. Rooting of micro-cuttings in culture is now a routine procedure in many commercial laboratories. By 1983, growers in Netherlands were producing more than 21 million plants by micropropagation.

Tissue culture is one the best methods for producing disease free plants. Disease free plants can be produced through culture of meristems and through nucellus culture. Through somatic hybridization, it is possible to transfer disease resistance from the wild relatives to crop plants. *Nicotiana repanda*, *N.nesophila* and *N. stocksonii* are resistant to a number of diseases but are not sexually crossable with tobacco (*N. tabacum*). However, fertile hybrids have been reported in combination *N. tabacum* + *N. nesophila* and *N. tabacum* + *N. stocksonii* by protoplast fusion. Somatic hybridization of di-haploid and tetraploid potato protoplasts with isolated protoplasts of *Solanum brevidens*, *S. phureja* and *S. pennellii* resulted in the synthesis of fertile, partially amphi-euploid plants possessing important agricultural traits, e.g., resistance to potato leaf virus, potato virus Y and Erwinia soft rot. Interspecific somatic hybridization involving species that are sexually incompatible with brinjal (*Solanum melongena*) has also resulted in the production of amphidiploids with traits resistant to verticillum wilt. Somatic hybridization, between *Brassica napus* and *B. nigra* cultivar, possessing the gene for resistance to *Phoma lingam*, yielded amphidiploid plants, carrying this gene. These hybrids possess all the three *Brassica* genomes (A, B and C) and are now incorporated in breeding programmes.

The culture of excised embryos can be advantageously exploited by the plant breeder. Very often interspecific crosses fail due to arrested development of embryo, either due to abortion of embryo at an early stage, or due to degradation of endosperm. A number of hybrids have been successfully raised through embryo culture. These include *Abelmoschus esculentus* x *A. manihot*, *A. esculentus* x *A. moschatus*, *A. tuberculatus* x *A. moschatus* (Gadwal *et al.*, 1968), *Hordeum vulgare* x *Secale cereale* (Kreuse, 1974), *Hordeum vulgare* x *Agropyron repens*, *Hordeum vulgare* x *Triticum* sp., *Lycopersicon esculentum* x *L. peruvianum*, and *Brassica chinensis* x *B. pekinensis*.

■■■

CHAPTER - 4

ELECTROPHORESIS

Introduction

In molecular biology, the separation of charged compounds of a mixture by means of the difference in their rates of migration in an electrical field is one of the important methods of separation. The compounds possessing different charges can be successfully separated as their migration depends upon the sign and magnitude of the charge as well as their size. This method is valuable for labile macromolecules as the separation can be carried out in aqueous media under mild conditions of temperature, pH and ionic strength. Earlier, such separations were carried out in homogeneous solutions as such they were restricted to frontal analysis only. But, with the introduction of porous media (to prevent the convection of the solute boundary) it became possible to carry out the zonal separation. It was Tiselius who pointed out that this separation can be operated by frontal, zonal and displacement techniques like most other separation processes, and coined the term, `zone electrophoresis' for all types of separations in which discrete solute zones are formed irrespective of the method used to stabilize these zones.

Robert Brown (1773-1858), a Scottish botanist and a paleobotanist, demonstrated `Brownian movement' (1827) with pollen of *Clarkia pulchella* suspended in water under a microscope. He observed that minute particles (now known to be amyloplasts and spherosomes) ejected from the pollen grains, executing a continuous jittery motion, and he observed the same motion in particles of inorganic matter and hypothecated that the effect was life-related. It is a known fact that all particles (in nature) in solution as well as in their gaseous state exhibit random movement. Molecular diffusion and occasional contact due to Brownian movement result in the formation of new compounds which can be isolated in their pure state by different techniques like crystallization, filtration, chromatography and centrifugation. Electrophoresis is one among them.

The term `*electrophoresis*` is Greek in its origin, and is a combination of two words, `*electro*` meaning `electricity` and `*phoresis*` or `*phoreein*` meaning `*to bear*` or `*carrying*` or `*being carried*`. Literally it means `*to carry with electricity*` or `*transport by electricity*`. The basis of this method is that, under the influence of an applied electric field, the particles of a solution with a positive charge can move towards the negative pole, and those with a negative charge can move towards the positive pole. Although many compounds of biological origin can be separated by chromatography, we have to depend on weak natural forces. But, for proteins and amino acids and the like, this poses a serious problem. To overcome this difficulty, electrophoresis was developed. According to Dr. William H. Heidcamp, electrophoresis is a "*powerful and inexpensive molecular separation technique*" (Cell Biology Laboratory Manual). On the whole, the process aims to provide an accurate way of analyzing substances, which are difficult to separate using the conventional methods.

Definition

The process of electrophoresis is defined as: `the migration of suspended particles such as the macromolecules of protein, under the influence of an electric field` Electrophoresis is "an empirical technique used in the separation of charged molecules (positive and negative) like cells and proteins, according to their response in electric current".

How Separation Occurs in Electrophoresis?

Electrophoresis is a versatile tool in molecular biology. All different types of electrophoresis work on the same principle. The samples are normally placed on one end of a gel (a substance that consists of a molecular matrix of evenly sized and spaced openings) and an electric current is passed through the gel. The current moves the molecules in the sample through the gel. The rate at which the molecules move depends on both the size and the charge. The larger molecules will be slowed by the matrix while the small molecules will move more quickly. In general, the molecules with a charge will move rapidly while the uncharged molecules will move slowly.

Requirements for Electrophoresis

The basic requirements to carry out the process of electrophoresis are:

Sample – The sample should be feasible and stable to the conditions of electrophoresis.

Power Supply – A device which can provide the required amount of electric power to carry out the process is essential.

Medium – Basically an aqueous solution with a solid support matrix of varying texture and properties is essential.

Apparatus – This should be a non-conducting container to carry out the process.

Basic Principles of Electrophoresis

In general, this is a process of migration of charged molecules through solutions in an applied electric field. The technique is often classified according to the presence or absence of a solid supporting medium or matrix through which the charged molecules migrate in the system. The solution electrophoresis systems employ aqueous buffers in the absence of a solid support medium. These systems suffer from sample mixing due to diffusion of the charged molecules, with resultant loss of resolution during the application of the sample, separation and removal steps. Hence, these systems use varying densities of a non-ionic solute such as sucrose or glycerol to minimize the mixing of the materials due to diffusion. Even then, these have only limited application.

Most of the applications in biochemistry employ some type of zonal electrophoresis in which the aqueous ionic solution is carried in a solid support and the samples are applied as either spots or bands. Paper electrophoresis, cellulose acetate strip and cellulose nitrate strip and gel electrophoresis are the examples of zonal electrophoresis systems. All these are generally governed by one set of principles, such as: *Mobility (rate of migration) of a molecule= (applied voltage)(net charge on the molecule)/(friction of the molecule).* The mobility of a molecule increases with increased applied voltage and increased net charge on the molecule. Alternately, the mobility decreases with increased molecular friction or resistance to flow through the viscous medium, caused by molecular size and shape. The total actual movement of the molecules increases with increased time.

Most systems employ an equal and constant voltage on all of the cross-sectional areas of the paper strips, gels or solutions employed in the separation. These electric fields are defined in terms of volts per linear centimeter. Ohm`s law ***(V=IR)*** dictates that voltage ***(V)*** is a function of current ***(I)*** and resistance ***(R)***. The nature of the apparatus and buffer composition dictates the resistance in the system. Therefore, the current ***(mA)*** is often used to define the voltage requirements of separation. The resistance of the system is important because it will determine the amount of heat generated during the process. Since the electrophoretic mobility is also a function of temperature, heating of separation matrix must be controlled. If significant heating occurs during the process, it will be necessary to provide some means of cooling the apparatus so as to maintain a constant temperature. If the voltage or current applied to a system is constant throughout the separation, the mobilities of the molecules being resolved will reflect the other terms, namely, the net charge and frictional characteristics of the molecules in the sample.

The separation of larger molecules follows the general principles while other factors influence the general resolution of macromolecules. The friction experienced by the molecules during migration reflects both the molecular size and shape. The size may prove to be the most important determinant of mobility if the process of electrophoresis is carried out in a medium that offers significant barriers to the movement of macromolecules through it (as is the case with polyacrylamide and agarose). If the charge to mass ratio is approximately equal, the size becomes the sole determinant of mobility. The shape of the molecules is not very significant in small molecules, in which the bonds are free to rotate, size alone determines their friction. The macromolecules often have defined shapes with specific axial ratios (length to width ratios) and as such both the size and the shape influence the migration. The molecules with high axial ratios generally demonstrate lower mobility than more spherical molecules that have equal weight and charge. The macromolecules may also deviate from the general principles because of interaction with ions because of charge-dependent inter-molecular association.

Electrophoresis - Method

Electrophoresis is one of the highly efficient techniques for analysis but it is time-consuming, expensive and also needs technical skill. It is a both qualitative and quantitative analytical technique. It is also similar to other techniques like chromatography, but differs in terms of the types of samples analyzed, the method used for separation and the principle. Electrophoresis is the main analytical separation technique that is essential and frequently used leading method for the separation of the molecules in molecular biology and medicine (like the investigation of proteins and nucleic acids). It is a reasonably easy, inexpensive and a powerful technique used for the separation and characterization of proteins, nucleic acids and sub-cellular particles like viruses and other small organelles. It is a method of separation of charged molecules in an electric field so as to make them migrate towards opposite charged electrodes. The migration is due to charge on the molecules and the potential applied across the electrodes. These molecules migrate at different speed and to different lengths based on their charge, mass and shape. The technique is mainly based on the mobility of ions in an electric field. The separation of the molecules mainly depends upon the charge distribution of the molecules that are being separated, irrespective of the type of the apparatus and the medium through which the molecules migrate. The process can be one-dimensional (on one plane) or two-dimensional (on two planes). The former method is normally used for the separation of most proteins and nucleic acids, while the latter method is generally used for finger-printing. When conducted properly, this method can be accurate in resolving all the cellular proteins. The support medium can be formed into a gel within a tube or it can be layered into flat sheets. The tubes are normally used for one-dimensional separations while the sheets have a larger surface and are better used for two-dimensional separations. The separation of the molecules within a gel is determined by the relative size of the pores formed within the gel.

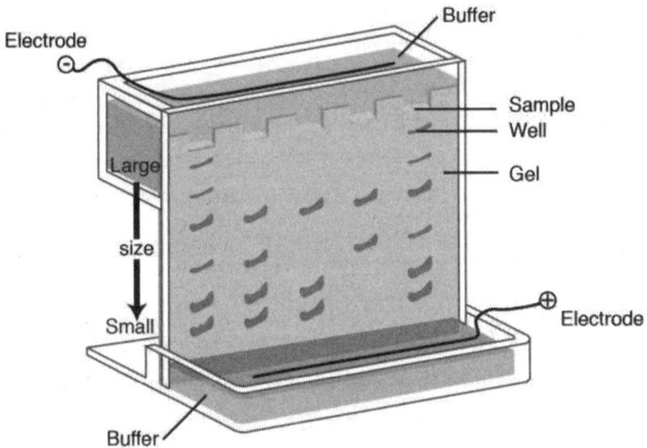

Electrophoresis –Method

In this process, the proteins are separated on the basis of the charge, which can be either positive (+) or negative (-) depending upon the hydrogen-ion-concentration (pH) of the buffer. Normally, the column of the gel is partitioned into 3 sections, the Separating gel or Running gel, the Stacking gel and the Sample gel. The sample gel can be eliminated by the introduction of the sample via a dense non-convective medium such as sucrose. The electrodes are attached to the ends of the column and the electric current is passed through the partitioned gels. The electrodes are arranged in such a way that the upper bath is the cathode (-ve) while the lower bath (at the ground level for safety reasons) is the anode (+ve). The ions have different migration rates depending on their total charge, size and shape and therefore they get separated. If the anions are allowed to flow toward the anode, then the system is referred to as an *anionic system*, while if in the opposite direction, the cations flow to the cathode it is called a *cationic system*.

The principle of electrophoresis is that the charged particles of a sample migrate in an applied electrical field, and if conducted in solution, the samples get separated according to their surface net charge density. The most frequent applications use gels (such as polyacrylamide and agarose) as a support medium. The presence of the matrix adds a sieving effect so that the particles can be characterized by their charge and the size (molecular weight) as well as their shape, secondary structure and those parameters which can influence electrophoretic properties either jointly or separately. Protein electrophoresis is normally performed in the presence of a charged detergent such as Sodium dodecyl sulphate (SDS) which helps to equalize the surface charge and allows for the determination of protein size on a single gel. The additives are not necessary for the nucleic acids which have a similar surface charge irrespective of their size.

Factors that affect Electrophoresis

This technique is one of the most important tools in many scientific labs (diagnostic labs, pharma labs, forensic labs etc.). The rate of migration of the particles during the process will depend on the following factors: the sample, the electric field, the medium and the buffer. The charge/mass ratio of the sample determines the electrophoretic mobility. The higher the charge, greater is the mobility and the charge depends on the pH of the medium. The bigger (size) molecules have a smaller mobility compared to the smaller molecules. The shape of the molecule also determines the speed, for example, the globular protein migrates faster than the fibrous protein. The mobility of the ion is the rate of migration under unit potential gradient, and an increase in the gradient increases the rate of migration. The medium which is inert can exert adsorption and molecular sieving effects of the particles, thus influencing its rate of migration. Adsorption is the retention of the component on the surface of the supporting medium. The rate and the resolution of separation can be reduced by adsorption. The media like polyacrylamide, agar, starch and sephadex have cross-linked structures and give rise to pores within the gel beads. The molecules which are larger than the pores are excluded from entering the gel beads as such they migrate faster (sephadex). The molecules like polyacrylamide, starch and agarose though they are larger, made to squeeze through the pores, and thus smaller molecules pass through the pores easily while the larger one are retarded. The buffers affect the mobility of the sample in different ways and the choice of buffer largely depends upon the type of the sample that is being subjected to electrophoresis. The common buffers are: formate, acetate, citrate, phosphate, EDTA, acetate, pyridine, Tris and barbitine etc. Ionic strength is a measure of the electrical environment of ions in a solution (usually between 0.05 and 0.1 M). With increasing ionic strength, the buffer ions carry larger load of the current while the sample ions carry a meager amount of the current. With decreasing ionic strength, the sample ions carry a larger share of the current leading to a faster separation. The pH not only determines the degree of ionization of organic compounds but also affects the rate of migration of these compounds. An increase in pH increases the ionization of organic acids while decrease in pH increases ionization of organic bases. e.g. amino acid is an ampholyte as it has both acidic and basic properties.

Physical Background of Electrophoresis

The macromolecules present in the buffer solution become charged, the sign and the magnitude of the total electrical charge depending on the pH of the medium. When this solution is transferred into an insulated channel like a glass tube and a voltage is applied at the ends of the channel, an electric field will be formed and direct current will pass through the solution. The field strength is the potential difference at the ends of the channel (or its section) with respect to its length (v/cm). Under the influence of the electrical field, the macromolecules migrate towards the cathode or the anode according to their net charge and the migration velocity being limited by the frictional force.

The essence of electrophoretic separation is differential velocity of the molecules of different shape and size. The original sample gradually gets separated into different zones of molecules with identical migration velocity which get distributed along the channel. At present, agarose and polyacrylamide gels (with a wider range of pore size) are exclusively used. By altering the pH of the buffer solution, it is possible to change the electric charge of the macro-molecules and the shape can be affected by the introduction of denaturing agents into the buffer. All these account for the extreme flexibility to the method. At the same time, this method is prone to some specific problems like broadening of the zones due to diffusion, evolution of heat in the liquid, distortion of the zones and deterioration of the resolution due to non-uniformity of gel temperatures. After separation, the zones should be fixed immediately to prevent further diffusion by removing the gel and placing it in acidified alcohol to enable the proteins and nucleic acids to precipitate at the same place. After fixation, the zones are stained by immersion of the gel in stain solution which binds firmly to protein or nucleic acid.

Now-a-days, instead of tube gels, slab gels are used. These are prepared by polymerization between two glass plates and these have the advantage of allowing several samples to be run simultaneously. The samples are applied to one side of the gel at equal intervals from each other. In the electric field, they undergo independent fractionation to give its own set of bands. Radio-active methods are also used to detect the separated zones besides staining. These include the registration of the bands on the film by auto-radiography or fluorography and other methods of measuring activity in liquid scintillation counters. The use of slabs saves time and space in multi-sample analysis. It also ensures a constant density and strength of the electric field as such identical conditions for the fractionation of different samples prevail. It is also easier to cool the thin slab gels than the tube gels and comparison of the samples for composition is easy and reliable. The use of different support for the liquid phase like cellulose acetate film, filter paper, thin layers of silica gel, sephadex are being used.

Instrumentation for Electrophoresis

The electrode apparatus consists of a high voltage supply, the electrodes (spirals made of platinum wire), electrode reservoirs, the buffer and a support (operating channel) for the buffer such as the filter paper, cellulose acetate strips, polyacrylamide gel or a capillary tube. Open capillary tubes are used for many samples while other supports are needed for biological samples like protein mixtures or DNA fragments. After the completion of separation, the support is usually stained to visualize the separated components. The resolution can be improved by the use of isoelectric focusing. The pH gradient is maintained by the supporting gel. As the protein migrates down the gel, it reaches a pH that is equal to its iso-electric point, and at this pH the protein remains neutral and no longer migrates (focused into a sharp band on the gel).

Electrophoresis – Apparatus

The electrophoretic apparatus developed by Tiselius consists of compartmented cells forming a U-tube connected to an anode and a cathode compartment. The protein solution is placed at the bottom of the U-tube and the tube is kept immersed in a water bath at 4^0C to minimize the convection current and the migration of the protein can be visualized by a *Schlieren* (shadow) optical system.

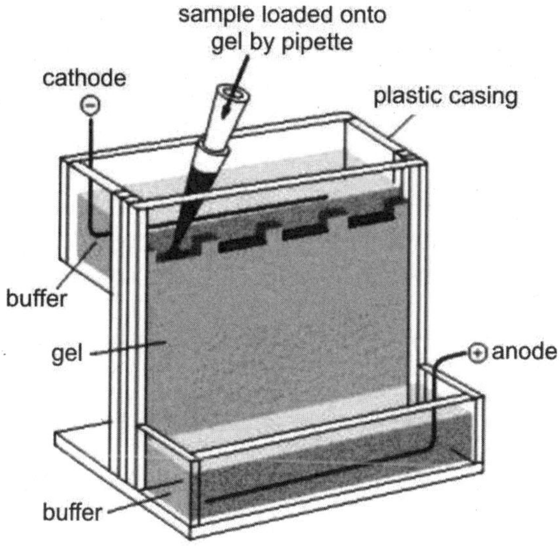

Electrophoresis – Apparatus

If we consider that the protein solution is composed of 3 compartments, A, B and C respectively and that these compartments carry the charges of the same sign (either +ve or –ve) but of a different magnitude. The solution will be homogeneous from its ascending boundary to the descending boundary before the electric current is applied. On the flow of the current, the three compartments will separate. If the magnitude of the electric charge is in the order A>B>C, the components will separate in such a way that the faster moving component A will be present in its pure form in the ascending boundary while C will be in the descending boundary of the tube. The component B will always be mixed with A or with C. The rate of migration of the protein is measured by observing the movement of the boundary as a function of time. During migration, the concentration gradients will be set up at the boundaries of the column. These can be measured by optical systems since most of the proteins are colourless. The migration of the coloured proteins like haemoglobin can be readily observed. This method is referred to as the `free boundary electrophoresis' or `free electrophoresis', meaning that it occurs in free solution.

An important modification of this technique includes the migration of proteins and other charged molecules in an electric field passing through a solution supported by inert material like moistened filter paper, starch gel, silica gel, cellulose sponges or glass powder. Such a technique is referred to as *zone electrophoresis* or *ionophoresis* which allows ready separation of components of differential mobility into zones. Later, these zones may be located by staining with appropriate dyes. The components may also be extracted from these zones. Thus, this method is a combination of electrophoresis and chromatography. This has proved much useful for the study of serum proteins as well as the cleavage products of proteins and nucleic acids.

THEORY OF ELECTROPHORESIS

All the factors that govern the movement of charged particles in an electric field fall into the following categories:

a. The characteristics of the ion, namely, its charge, its sign, its magnitude, its size, its shape and its tendency to dissociate and its amphoteric behavior.

b. The factors of the environment in which the ion is present, like the concentration of the electrolyte, ionic strength, dielectric properties, chemical properties, hydrogen-ion-concentration (pH), temperature, viscosity and the presence of non-polar molecules which may influence viscosity or dielectric properties or which may interact with the charged complexes.

c. The characteristics of the applied field, its intensity and distribution along the migratory path, and

d. The secondary interactions between the factors [(a) and (b)], either electrostatic or by Van der Waal`s forces which may further influence the migration.

In a system (salt-free and very dilute), the force exerted (F) on a charged particle by an applied electric field is equal to the field strength (X) and the net charge (Q).

$$\mathbf{F = XQ} \quad \ldots\ldots\ldots\ldots\ldots\ldots \quad \text{(eq. 1)}$$

A particle moving in a constant field (constant force) in a viscous medium, attains a constant velocity because the viscous retarding force (F`) opposing the movement increases in a linear fashion with acceleration. The viscous retarding force depends upon the geometry of the particle, and for a sphere is given by Stoke`s law:

$$\mathbf{F`/V = 6\pi rn} \quad \ldots\ldots\ldots\ldots\ldots\ldots \quad \text{(eq. 2)}$$

where, V = electrophoretic velocity, n = viscosity of the medium and r = radius of the particle. Thus, for an isolated spherical particle obeying Stoke`s law in the steady state, the two opposing forces are equal (F = F`) and we have from equations (1) and (2):

$$\mathbf{X\,Q = 6\pi\,r\,\eta\,V \text{ or } X/V = 6\pi\,r\,\eta/Q}$$

The electrophoretic mobility, u, is the velocity, V, measured under standard conditions i.e., per unit field, X, expressed as volts per cm or

$$\mathbf{u = V/X \text{ or } u = Q/6\pi\,r\,n}$$

Experimentally one cannot be restricted to ideal conditions, buffers or other salts are nearly always present and electrical interactions between charged ions can no longer be ignored as is permissible under the conditions of infinite dilution.

The electrical forces that are exerted between ions in aqueous solution e.g., Na^+ and Cl^- are of the same order as thermal forces and appreciably influence the distribution of other ions present. This results in a change in the net force that is applied by the electric field. This is the principal cause of decreased mobility with increasing ionic strength. The force applied remains XQ but a new retarding force is added, owing to the fact that ions of opposite charge, group about the migrating ions resulting in local ionic atmosphere of opposite charge. As a consequence, the fluid surrounding the charged particle moves in a direction opposite to that of migrating particle with an attendant decrease in mobility. The degree of effectiveness of various electrolytes in creating this retarding effect depends upon their concentration and the charge. It does not depend upon the specific chemical nature of the ion but primarily on its valence.

We have seen that the velocity, V, with which an ion migrates, is proportional to electric force acting on per unit charge. This force is in turn, proportional to the field strength. Field strength, X, is expressed as potential in volts divided by the distance between the electrodes in cm. In practice, geometrical considerations make it impracticable to measure this distance directly. A simple Ohm's law transformation permits to measure field strength, *X*, at any cross section, s, between the electrodes to be calculated from the expression.

$$\mathbf{X_s = i/q_s x k_s}$$

Where, i = current in amperes, q_s = area at cross section under consideration in square cm. k_s = specific conductivity in mhos/cm across cross section.

In order to carry out any type of electrophoretic separation, it is necessary that convection be controlled so that resulting mixing of the separated solutes does not take place and thus obscure or destroy the resolution. In the classical moving boundary method of Tiselius, convection control is affected by proper regulation of temperature gradient. In zone electrophoresis convection is prevented by the supporting medium, provided the medium is not overloaded with electrolyte.

It is obvious that greater the difference in the mobilities of any two ions ***A*** and ***B*** more easily and quickly they are separated; thus if u_a and u_b are the mobilities of components ***A*** and ***B*** respectively which migrate distance d_a and d_b and X be the field strength and t be the time of separation. Then the difference in mobilities is represented by the equation:

$$d_a - d_b = (u_a - u_b) \times X \times t$$

For weak electrolytes and amphoteric substances which are of considerable biological significance, the degree of separation depends also upon the degree of their dissociation.

During the course of electrophoresis, free diffusion occurs as a function of time, concentration and temperature. The diffusion, therefore, sometimes leads to limitation in the choice of field strength-time relationship chosen to effect a given separation. The rate of diffusion is roughly inversely proportional to the initial width of the sample section and imposes limits on the extent to which the width of sample section may be reduced in order to achieve equivalent separation of components with shorter migration distances. Thus, to obtain maximum resolution in lesser time, applied field strength and shorter migration channel, the starting zone must be of infinitesimally small size.

TYPES OF ELECTROPHORESIS

Electrophoresis is mainly of 3 types namely: microscopic electrophoresis, moving boundary or frontal electrophoresis and zone or zonal electrophoresis.

a. Microscopic Electrophoresis or Micro-electrophoresis

This technique is defined as *"the technique for observing the movement of microscopic particles in a fluid under the influence of a magnetic field"* or *"direct microscopic observation and measurement of the velocity of migration of ions or other charged bodies through a solution toward oppositely charged electrodes, also known as cytopherometry"* (McGraw-Hill Science & Technology Dictionary - 2003)

This is a method of studying electrophoresis of dispersed particles using optical microscopy. The electrophoresis is carried out in a closed medium with critical observations by focusing and adjusting the lens of the microscope. This is a variation of electrophoresis wherein an electric field is applied so that the particles in solution move in a horizontal direction on a glass slide. A microscope is used to focus one or several particles and their motion is observed as a function of time. The motilities are calculated by measuring the time a particle takes to move a prescribed distance. This method provides image of moving particles which is its unique advantage. This is carried out in a closed medium with critical observations by focusing and adjusting the lens of the microscope. This method provides an image of moving particles (like observations of RBCs, neutrophiles and bacteria) which is its unique advantage. The complexity of this method is associated with electro-osmosis generated by electric field influence on the double layers of the sample cell walls. In the usually used closed cells, this creates back flow (poiseuille type) leading to parabolic velocity profile. There are two stationary layers where fluid does not move and the position of these stationary layers regarding cell walls depends on the cell geometry. It is possible to focus the microscope on these stationary layers and observe particle`s motion not affected with electro-osmosis. The second complexity comes from the necessity of diluting the sample, if it was initially concentrated. The concentration must be sufficiently low for observing individual particles. The measurement in the low polar fluids presents additional problems. These systems have low electrical conductivity and low dielectric permittivity. The mobility is low and this requires high electric field, by factor 10 and higher. This technique is used to study and observe microscopic objects such as the colloidal particles, blood cells, protozoa or even protein molecules that adhere to tiny spheres resolving under the influence of an electric field. The study can be made in a tube of the desired solution that is kept for observation under a microscopic field.

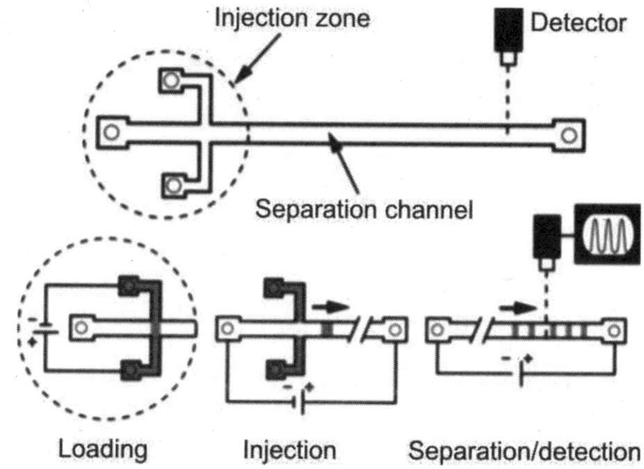

Micro-scale Electrophoresis – Schematic diagram

b. Moving Boundary Electrophoresis or Moving Zone Electrophoresis

This is also called as *Free Electrophoresis* or *Free Boundary Electrophoresis* or *Microscopic Boundary Electrophoresis.* According to McGraw-Hill Dictionary of Scientific & Technology Terms, this is defined as "*a U-tube variation of electrophoresis analysis that uses buffered solution so that all ions of a given species move at the same rate to maintain a sharp moving front*". This can also be defined as "*a free flow electrophoresis tracking available optical method for protein solution and buffer solution at the interface between a mobile in the electric field, usually for U-shape measurement tube*". This method is employed to measure the movement of the boundary of a mass particle in the buffer and can be used to study the electrical behaviour of molecules and ions in large quantities. Such a method has helped a lot in the identification of many proteins.

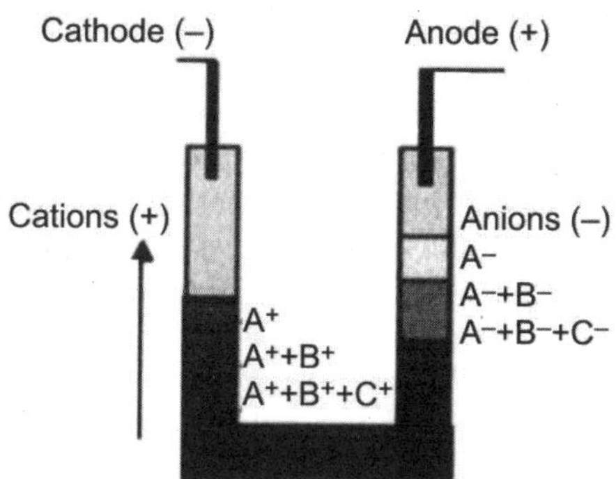

Moving Boundary Electrophoresis - Schematic Diagram

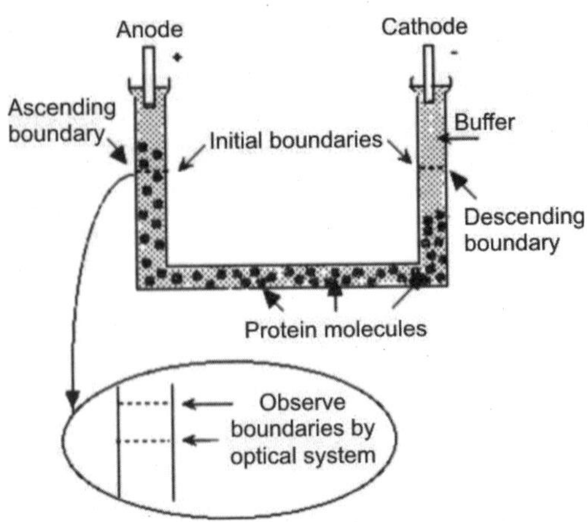

Moving Boundary Electrophoresis - Method

The technique was first performed by Arne Wilhelm Kaurin Tiselius (1937) in a fluid electrolyte. He was awarded the Nobel Prize (1948) for his work on the separation of colloids through electrophoresis, the motion of charged particles through a stationary liquid under the influence of an electric field. The apparatus consisted of a U-tube, the horizontal lower portion being filled with a mixture of the substance under examination dispersed in a suitable buffer. The two vertical limbs were filled solely with buffer and the cathode and the anode dipped into the buffer at the top of the each limb respectively. On the application of an electrical field, the boundary of the particles with positive charges moved towards negative pole on one end of the tube and *vice versa* and got easily separated. In this method, it was usually the advancing front of each component of the migrating sample that was observed. Hence, this is known as the moving boundary electrophoresis. The bands are best identified and measured using the Schleren optical system. This closely resembles a separation that would be obtained chromatographically by frontal analysis. This method has several disadvantages such as heat and convection current which drastically affect the results.

c. Zone Electrophoresis

This is also called as *Ionophoresis*. In this technique, separation occurs on a solid medium of inert material such as a strip of moistened filter paper. The molecules flowing through the solid supporting medium separate into discrete zones. Hence, the name zone electrophoresis. The individual constituents on separation occupy definite zones or bands. The filter paper was first introduced to hold the electrolyte as the anti-convection medium by Consden, Gordon and Martin (1946) in electrophoresis. This method uses only small sample sizes and the nature of the supporting medium (paper, cellulose acetate, starch-gel, agar-gel, polymethacrylamide, mixed gel) introduces additional factors influencing mobility. The rate of migration depends on the mobility of the particles and also on the electro-endosmotic current (in the case of carriers with polar properties), on the currents due to evaporation (caused by heat generated through the joule effect), and on the gradient of the electrical field. In practice, the mobility of the electrophoretic zones and their signs are ignored, and the zones are located by experience or by comparison with those given by a reference substance treated in the same way. After separation of the constituents, the position of colourless substances may be determined by treating the electropherogram with a reagent that will convert them to coloured or fluorescent derivatives. For quantitative purposes, the zone (spot) may be carefully separated, and the substance eluted with a suitable solvent, and then determined by a sufficiently sensitive method (spectrophotometric measurement) either directly or after a chemical reaction. In a different method, after conversion to a coloured derivative, the zone intensity can be measured with the help of a scanning densitometer. This technique has several advantages over the moving boundary method such as – the method is simple, inexpensive and helps separate individual components; the technique makes it possible to study smaller amounts of the material even at lower concentrations; the method permits 2-dimensional electrophoresis for high resolution; the technique allows the devices for continuous flow preparative works; the method helps in the detection and identification with different reagents and dyes; and the quantification can be done by densitometric scanning, autoradiography procedures.

For electrophoresis on a supporting medium, the apparatus consists of: a source of direct current, preferably of stabilized voltage. A parallel-piped chamber of glass or rigid plastic material with an air-tight lid to ensure the maintenance of an atmosphere of saturated humidity is required. Two insulated electric leads are sealed through the walls of the chamber, one at either end, each lead having an internal connector to which are attached the electrodes of platinum wire. The chamber should be fitted with suitable safety devices to ensure that the electrical supply is disconnected when the lid is removed. Two double troughs provided with a central length-wise partition are inserted into the chamber one at each end. Alternatively, the troughs may be integral parts of the chamber, and one platinum electrode is laid along the bottom of each outer trough compartment. The electrodes are connected through external insulated cables to an electric power source having an output of not less than 450 VDC at 150 mA. The power source should be provided with a means of indicating controlling the voltage and of indicating the current consumption. Additional circuitry may be incorporated to stabilize the voltage. A holder device is essential, and when paper or cellulose acetate is used, the carrier strips impregnated with the conducting liquid are stretched by an appropriate arrangement and the ends are immersed into the electrode troughs. In gel electrophoresis, an adherent on an even layer of gel is placed on glass and the electrical connections are attached at each end. For paper electrophoresis, a chamber about 50 cm long, 38 cm wide and 4.5 cm deep, with troughs about 37 cm long externally, 5 cm wide and 2 cm in depth internally, is suitable. To carry out electrophoresis on cellulose acetate strips, it is preferable to use a smaller chamber than for paper electrophoresis; one measuring about 25 cm by 24 cm with troughs of 10 cm by 23 cm. For gel electrophoresis, the inert carrier consists of a 1-2 mm thick layer of an agar or starch gel of suitable consistency and shaped as an elongated rectangle.

Glass powder, silica gel, agar, starch gel, cellulose sponges, cellulose acetate, starch blocks and polyacrylamide gel etc., are other substances used in place of the filter paper to give better results. A streak of the mixture to be analysed and separated is spotted horizontally on a clean strip of filter paper, and the strip is placed in a horizontal position with its two ends dipped in the buffer, which is kept in two containers at either end, one with a negative electrode and the other with a positive electrode. On the application of an electrical field, the individual constituents of the mixture with different surface charges move towards their respective poles. The spreading of the constituents depends on the amount of their surface charges. This results in the adsorption of different individual constituents at different places on the filter paper strip in the form of distinct bands or zones. These bands can be stained with dyes, separated and identified as in chromatography and analyzed qualitatively or quantitatively.

This technique is a combination of electrophoresis and chromatography, and is useful for the study of the serum proteins, cleavage products of proteins, nucleic acids and isozymes. This method has been improved by Tiselius, Svenson and Ingra Brattsten, Theodore Wieland and Fritz Turba and E. L. Durrum.

Specific Techniques of Electrophoresis Commonly Used in Biochemistry

There are numerous applications of electrophoresis. Routine protein electrophoresis performed in clinical laboratories is the oldest method and therefore the most frequently used method. With the advent of molecular diagnostics, several other electrophoresis methods have become very important, highly automated, and have several important applications.

Types of electrophoresis are: Routine electrophoresis, High resolution electrophoresis, Polyacrylamide gel electrophoresis, Capillary electrophoresis, Isoelectric focusing, Immunochemical electrophoresis, Two-dimensional electrophoresis and Pulsed field electrophoresis

The different types of electrophoresis generally used in life sciences are: Paper electrophoresis, Disc electrophoresis, Moving boundary/frontal electrophoresis, Column electrophoresis, Zonal electrophoresis, Capillary electrophoresis, Immuno-electrophoresis, and Iso-electric focusing.

A) Paper Electrophoresis (PE)

This is a commonly used method for the analysis and resolution of small molecules. This is not used to resolve macro-molecules like proteins because the adsorption and surface tension associated with paper electrophoresis usually alter or denature the proteins resulting in poor resolution. As a routine, 2 different methods are used to apply the samples to the paper.

a. Dry Application Method (of the sample)

In this method, a sample of solutes dissolved in distilled water or a volatile buffer is applied as a small spot or as a thin stripe on a penciled `origin line` marked on the paper. Required standards of known compounds are applied at other locations on the origin line. The line should be in the centre of the paper, if the electrophoretic migration is anticipated toward both the poles of the system. If migration is anticipated in only one direction, the origin line should be near one end of the paper. After the evaporation of the solvent with the sample, the paper is dampened with the buffer, by uniform spraying or by dipping and blotting the ends of the paper so that wetting from both the ends of the paper meets at the origin line simultaneously. This method has the advantage of allowing small initial sample spots and better resolution of similar mobile compounds. This is also disadvantageous because the dipping or spraying requires considerable skill to avoid spreading of the applied samples.

b. Wet Application Method (of the sample)

In this method, the samples dissolved in distilled water as concentrated solutions are applied to the paper dampened with buffer. This method is simpler to perform and usually yields larger spots and poorer resolution because of the diffusion of the sample.

After the sample is applied to damp paper, the paper is placed in the electrophoretic chamber so that both the ends are in contact with the reservoirs of the buffer at the electrodes. The paper must be positioned to allow maximum migration toward the correct electrode, and the origin line is not in the centre of the paper. The chamber is covered or closed to protect against electric shock, the required electric field is applied to the system.

The application of the electric field and the resultant resistance to the current flow in the buffered paper generates heat (such generation of greater amount of heat is the only difficulty with paper electrophoresis). This heat dries the paper which leads to more resistance to current flow causing greater resistance. If drying of the paper is prevented, the heat will change the flow of the current and resistance properties of the system which finally result in the distortion of the molecular migration.

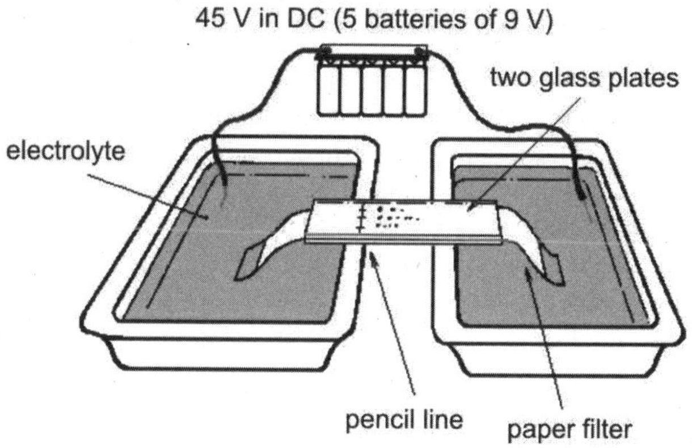

Apparatus for Paper Electrophoresis

The modern paper electrophoretic systems have been designed to compensate for potential heating problems because of these difficulties. A majority of the low voltage systems are portable and also can be operated conveniently in cold rooms or refrigerated chambers. Contrary to this, high voltage systems employ a cool flat-bed system which helps to dissipate the heat or operate in cool bath of an inert and non-polar solvent, such as Varsol (a petroleum distillate). This helps in the absorption of heat generated by the system, without mixing with either water, buffers or samples on the paper. After the elctrophoretic resolution of the samples on the paper (for the time and voltage used for optimal separation), the current is turned off, the paper is removed and dried followed by the determination of the presence and location of molecules of interest.

Disc Electrophoresis or Discontinuous electrophoresis (DE)

This method was developed by Ornstein and Davis for the analytical separation of protein mixtures. By this method, the proteins are concentrated into thin starting zones and are then separated by the combined action of electrophoresis and molecular sieving. According to the authors, discontinuous electrophoresis solves two problems during the separation of proteins in gels with small pores, namely, it prevents the precipitation of proteins during sample entry into the gel and it promotes the formation of well defined bands. The discontinuity is based on four parameters, namely, gel structure, pH value of the buffer, ionic strength of the buffer and nature of the ions in the gel and in the electrode buffer. The gel matrix is divided into two areas, namely, the resolving gel and the stacking gel. The proteins are separated according to the principle of isotachophoresis and from stacks in the order of their mobility (stacking effect). The individual zones are concentrated and the proteins migrate slowly and at constant speed towards the anode till it reaches the limit of the separation gel. The stacking and the resolving gels are only cast together just before electrophoresis because, when the complete gel is left standing for a long time, the ions diffuse towards one another. The proteins are in a homogeneous buffer medium and start to separate according to the principle of zone electrophoresis. This method affords high resolution and good band definition.

Moving boundary/Free boundary/Frontal electrophoresis

This method refers to electrophoresis in a free solution and was developed by Arne Tiselius (1937), the motion of charged particles through a stationary liquid under the influence of an electric field. He used a U-shaped tube filled with buffer solution and electrodes immersed at its ends. The applied sample could be any mixture of charged components (like a protein mixture). On the application of voltage, the compounds will migrate to the anode or cathode depending on their charges. The change in the refractive index at the boundary of the separated compounds is detected using Schlieren optics at both the ends of solution in the tube.

Column electrophoresis

This is any method of electrophoresis in which the support material is contained in a solution, and such a technique is usually used for the preparative scale separations by electro-focusing or isotachophoresis.

Zonal Electrophoresis

This refers to any electrophoretic technique in which the components are separated into zones or bands in a buffer, and stabilized in solid, porous or any other support medium (such as filter paper, agar gel, or polyacrylamide gel). Otherwise, it is also a technique that makes possible the separation of charged macromolecules and the characterization of each molecule in terms of its electrophoretic mobility.

Immuno-electrophoresis

This is a general name for a number of biochemical methods for the separation and characterization of proteins based on electrophoresis and reaction with the antibodies. All variants of this process require immunoglobulins (antibodies) reacting with the proteins to be separated or characterized. These methods were developed and used extensively during the second half of the 20th century. The types include 1-dimensional immunoelectrophoresis (*immunoelectrophoretic analysis* – classical method wherein proteins are separated by electrophoresis), 2-dimensional quantitative immunoelectrophoresis (*crossed immunoelectrophoresis* – the method in which the proteins are first separated during the first dimension), 1-dimensional quantitative immunoelectrophoresis (*rocket-immunoelectrophoresis* – used for quantization of human serum proteins), *fused rocket immunoelectrophoresis* (a modification of rocket immunoelectrophoresis used for the detailed measurement of proteins in fractions from protein separation experiments) and *affinity immunoelectrophoresis* (used for the estimation of binding constants – with lectins – or for characterization of proteins with specific features – like ligand bonding).

B) Capillary Electrophoresis (CE)

This is a new method of analytical electrophoresis that finds an increasing application in biochemical research. The material to be analyzed and the electrophoretic medium (an aqueous conducting liquid) are placed in long, capillary tube with a fine bore (50 to 100 cm long and 25 to 100 micrometer diameter on the inside). A very small sample (in the range of nanolitre) is placed at one end of the capillary before being subjected to electrophoresis under electric fields up to 20 to 30 kV. The analytes are separated and detected as they emerge from the other end of the capillary tube (any one of the methods used commonly on HPLC).

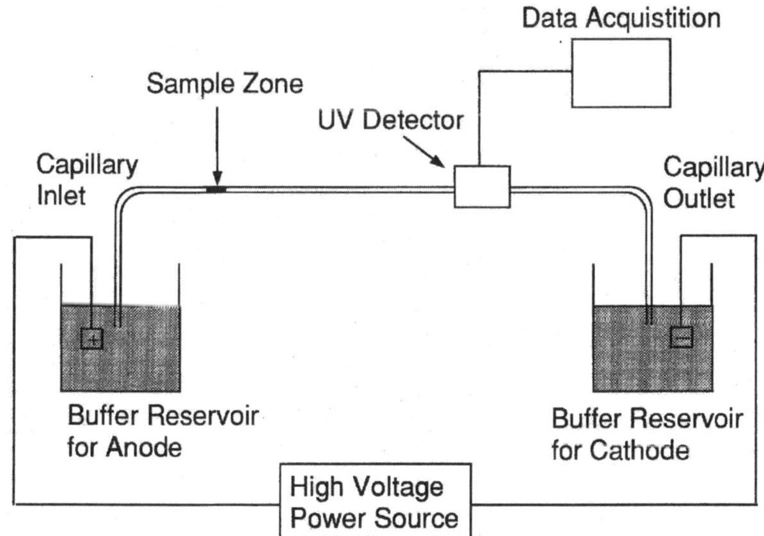

Capillary Electrophoresis - Schematic Diagram

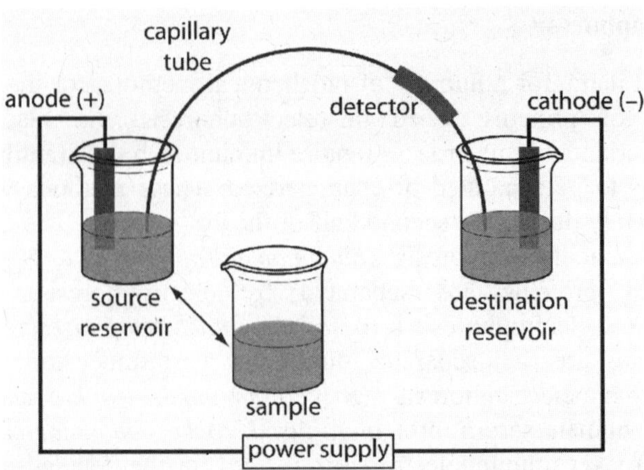

Capillary Electrophoresis – Arrangement of Apparatus

This method offers the advantages of extremely high resolution, high speed and high sensitivity for the analysis of extremely small samples (but this is not useful as a preparative method). This method has been proved to be useful for the separation of DNA molecules that differ in their size by as little as a single nucleotide. This method is the basis of separation of polynucleotides because of its high resolution in some of the newer designs of DNA sequencers. The method can also be adapted to the separation of uncharged molecules by the inclusion of charged micelles of a detergent such as SDS in the aqueous electrophoretic medium. If a mixture of solute molecules that partition between the aqueous medium and the hydrophobic interior of the micelles is introduced into a system, they can also be separated electrophoretically. This method is highly adaptable and the range of its applications and optimal methodology are still being explored.

The special types of CE include: *Capillary Sieving Electrophoresis* (CSE – an agent is included in the CE system which sieves the analytes or separates them based on their size), *Capillary Gel Electrophoresis* (CGE – a porous gel is placed in the capillary, like polyacrylamide gel employed in SDS-PAGE or to add to the running buffer a large polymer that can entangle with analytes and alter their rate of migration), *Micellar Electrokinetic Chromatography* (MEKC – also called as *Electrokinetic Chromatography*, in which a charged agent is placed in the running buffer that can interact with neutral substances and make them migrate with the electro-osmotic flow), *Capillary Isoelectric Focusing* (CIEF – used in chemical analysis and separation, and involves the production of a pH gradient across the capillary for the separation of zwitter ions), and *Affinity Capillary Electrophoresis* (ACE – used in the separation of chiral analytes through the use of binding agents like cyclodextrins or proteins. The method is also used in clinical and pharmaceutical assays and for the study of biological interactions).

C) Gel Electrophoresis (GE)

In this method, the molecules are separated in aqueous buffers supported within a polymeric gel matrix. These systems have several disadvantages such as: they can accommodate larger samples than mist paper systems as such they can be used for preparative scale electrophoresis of macro-molecules. Secondly, the character of the gel matrix can be altered at will to fit in a particular application (because the gel has an ability to enhance the friction that governs the electrophoretic mobility). The lower concentrations of the matrix material or a low degree of cross-linking of the monomers in gel systems (polymerized) allow them to be used largely as stabilizing or anti-convection devices with a relatively low frictional resistance to the migration of macro-molecules. Alternatively, higher concentrations (of matrix material) or a higher degree of cross-linking monomers are used to generate greater friction that results in molecular sieving (a situation in which the viscosity and the pore size largely define electrophoretic mobility and migration of the solutes). As such, the migration of macro-molecules in the system will be substantially determined by their molecular weights.

In electrophoretic systems many gel-like agents are used of which agarose (a polygalactose polymer) gels have proven highly successful when applied to very large macro-molecules like nucleic acids, lipo-proteins etc. The most useful and most versatile gels in gel electrophoretic separations are polyacrylamide gels because they readily resolve a wide range of proteins and nucleic acids. These are formed as a result of polymerization of a monomer, acrylamide and a cross-linker, N, N`-methylene-bis-acrylamide. Both these are stable by themselves or mixed in solution, but polymerize readily in the presence of a free –radical generating system. To induce the polymerization process, the scientists generally use either chemical or photochemical free-radical sources.

In the most commonly used chemical method, the free radical initiator, APS (ammonium per sulfate) is added along with a catalyst, N,N,N`,N`-tetra-methyl-ethylene-diamine (TEMED). In the presence of a monomer, cross-linker and an appropriate buffer, the two components generate the free radicals that are needed to induce polymerization. In a less widely used photo-chemical method, APS is replaced by a photo-sensitive compound (like riboflavin) that helps to generate free radicals when irradiated with ultra-violet light. Modifications can be made to produce a gel for a particular application. If larger pores are needed, the amount of monomer and/or the cross-linker can be decreased and/or increased in the polymerization solution respectively. The pore size required for a particular system will depend on the difference in size of the compounds to be resolved. Any methods used to produce these gels must be followed exactly each time they are cast, as reproducible separations require uniform gel-forming conditions. Many procedures and applications of polyacrylamide gels employ the gel in some sort of `slab format`. e.g. PAGE and Iso-electric Focusing. This technique may be Tube Gel Electrophoresis, or Slab Gel Electrophoresis (Vertical and Horizontal).

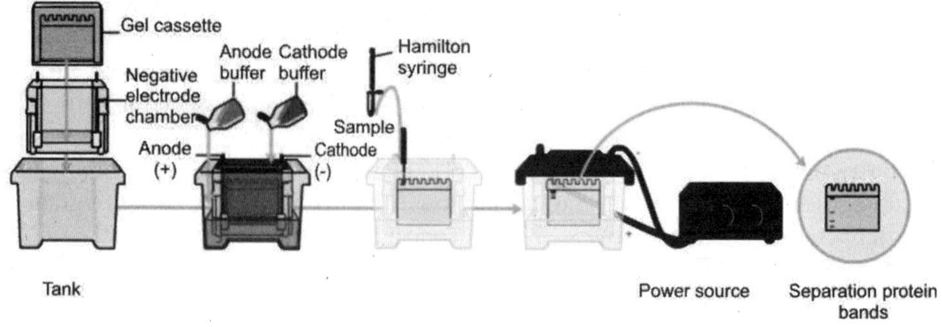

PAGE Technique

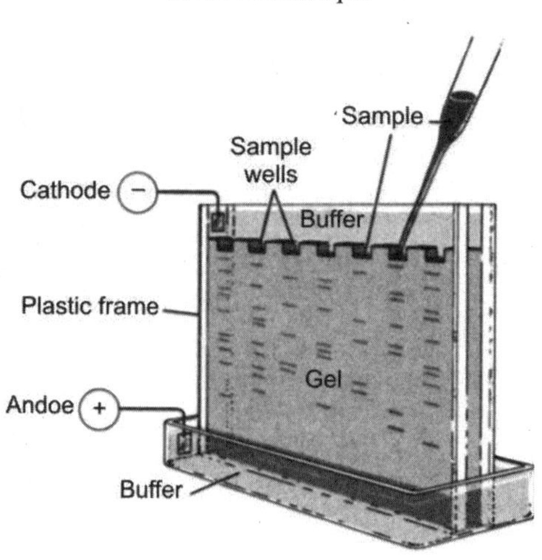

SDS Gel Electrophoresis

D) Slab Gel Electrophoresis

Electrophoresis is conventionally conducted on plates or slabs. To maintain the ionic buffer solution on the plate, some anti-convective medium or gel is necessary and thus the method is called slab-gel electrophoresis. The typical gel material is polyacrylamide or agarose. This may be either vertical or horizontal.

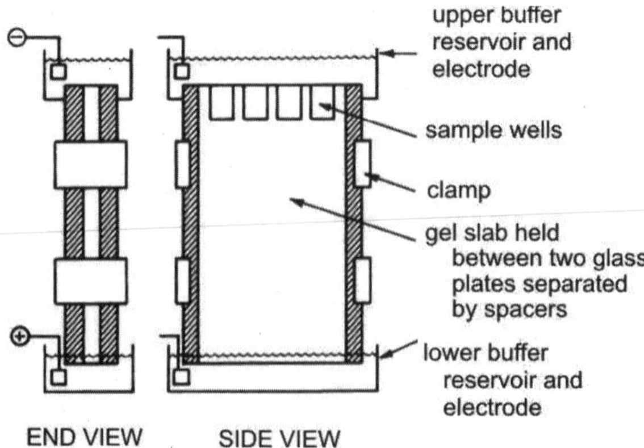

1. Vertical Slab Gel System

A gel electrophoresis apparatus must allow the researcher to maintain a uniform electric field across the gel, provide cooling to prevent thermal artifacts, and allow access to the gel for sample loading and monitoring the run. Two types of apparatus are in common use: vertical and horizontal. Vertical gel systems are further subdivided into slab gels and tube gels. In general, agarose gels are run in the horizontal format, while acrylamide gels are run vertically.

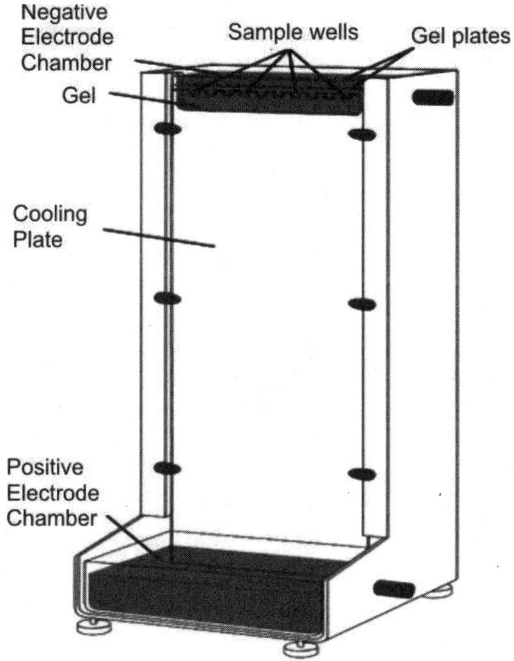

Vertical Gel Electrophoresis Apparatus

A typical vertical slab gel apparatus is used for sequencing and this system shows the components common to all vertical slab systems. The gel is cast between two glass plates, separated by spacers, typically less than 2 mm thick. The gel is mounted in the system so that the top is in contact with the negative electrode chamber, and the bottom is in contact with the positive electrode chamber. Unlike the horizontal system, the only connection between the buffer chambers is through the gel, and this allows precise and reproducible control of the voltage gradient. Because of the high resistance of the thin gel, the apparatus must have provisions for cooling. The front of the gel cassette is exposed to the air, while the back of the gel is held against a metal plate which dissipates the heat rapidly. In some systems, the upper buffer chamber extends almost to the bottom of gel, and the upper buffer is used for cooling. The relatively small amount of current carried through the gel means that buffer recirculation is generally not required (In Vertical Gel Electrophoresis apparatus, the gel is clamped into the apparatus so that the lower end is immersed in the lower buffer chamber, and the upper end forms one wall of the upper chamber. The gel provides the only electrical connection between the two buffer chambers. Cooling is provided by a metal heat sink behind the gel. This type of apparatus allows high voltages (>1500 ^{o}V) to be used. The gel is clamped into the apparatus so that the lower end is immersed in the lower buffer chamber, and the upper end forms one wall of the upper chamber. The gel provides the only electrical connection between the two buffer chambers. Cooling is provided by a metal heat sink behind the gel. This type of apparatus allows high voltages (>1500^{0}V) to be used).

In a standard mini gel apparatus, the gels – generally 10 cm or smaller – have become the standard for many applications because of their ease of preparation and handling and short run times. The gels are cast between the glass plates, but in the mini gel system, the cassettes are mounted on to a U-shaped frame so that the cassettes themselves form the sides of the negative electrode chamber. The assembly is placed in a tank of buffer which contains the positive electrode, and this means that the gels are effectively submerged in the buffer during the run, providing the optimal cooling. (In the vertical mini-gel apparatus, the upper (negative) electrode chamber is formed by the two gel cassettes, clamped to the "core". This assembly is submerged in the lower (positive) electrode buffer chamber, which provides cooling. As with the full size vertical system, the gels provide the only electrical connection between the two chambers).

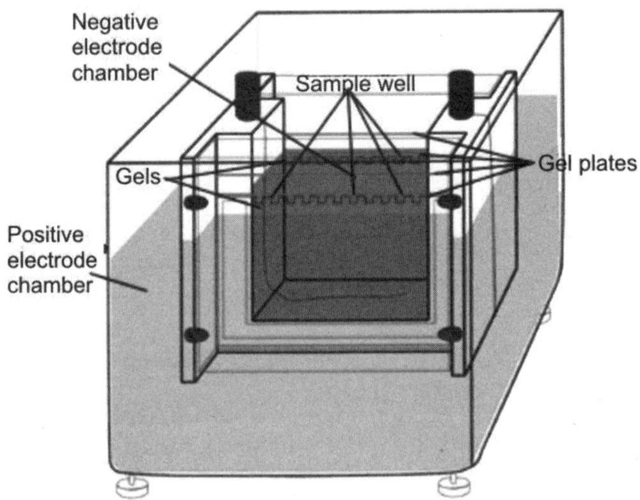

Vertical Mini-gel Apparatus

In general, the vertical slab gels are loaded through the top, under a layer of buffer and the gels are monitored during the run through the front glass plate. The fact that the body of the gel in these systems cannot be accessed until the end of the run can be an inconvenience. So, some sample recovery techniques used on horizontal gels are not available for the vertical gels. However, the resolution and reproducibility of the vertical polyacrylamide gels more than compensate for this.

The vertical gels are cast in a cassette made up of two glass plates separated by spacers which run along the sides of the plates. The bottom of the cassette is sealed by some temporary means (such as a tape, agarose or a gasket). The gel monomer solution is treated to initiate the process of polymerization, and poured into the cassette, and a comb is inserted into the top of the cassette to form the sample wells, and the gel is allowed sufficient time to polymerize. After polymerization, the bottom of the gel is unsealed and the cassette is mounted into the apparatus.

2. Horizontal Slab Gel System

In its simplest form, a horizontal gel apparatus consists of a box which is divided into two compartments by a platform in the middle. The gel is placed on this platform, and the buffer is added until the gel is fully submerged. The electrodes in each compartment supply the electric field. The resulting current flows through both the gel and the buffer over the gel, so the thickness of these must be controlled for fully reproducible results. Cooling is provided by the buffer which surrounds the gel, and this buffer is often recirculated to prevent the development of a pH gradient and also to aid in temperature control. Access to the gel is through the overlaying buffer. The samples are loaded through this buffer layer, and the apparatus has a clear lid to allow the run to be monitored.

One major limitation of the horizontal apparatus is that, since the two compartments are connected by a layer of buffer, it is not possible to use discontinuous buffer systems. Another limitation is that the gels are cast in trays which are not covered. Because atmospheric oxygen has full access to the upper surface of the gel, acrylamide will not polymerize in this system. Horizontal systems are primarily used to run agarose gels which are run in continuous buffer systems and are not affected by oxygen. For agarose gels, the simplicity and ease of use of the horizontal system make this system the best choice.

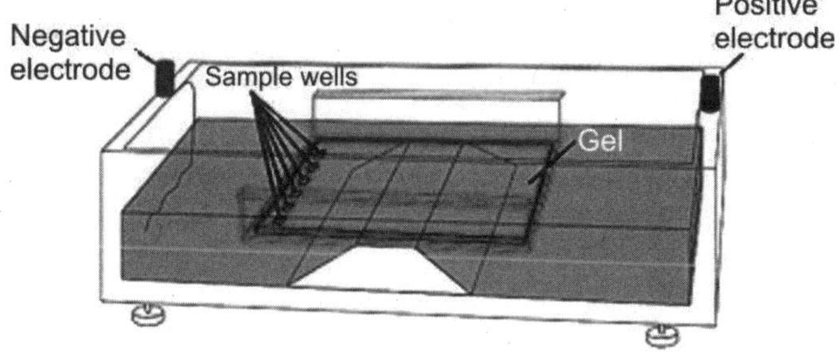

Horizontal Gel Electrophoresis Apparatus.

(The gel rests on a platform which divides the apparatus into two chambers. Note that the buffer level is higher than the surface of the gel, so the two buffer chambers are connected. Samples loaded into the wells will migrate toward the positive electrode).

E) SDS-PAGE (Sodium Dodecyl Sulfate – Polyacrylamide Gel Electrophoresis)

These systems are used to determine the number and size of protein chains or protein subunit chains in a protein preparation. In the beginning, the protein preparation is treated with an excess of soluble thiol (2-mercaptoethanol) and SDS. The thiol reduces all disulfide bonds (-S-S-) present within and/or between the peptide units, while the SDS (an ionic or denaturing detergent) binds to all regions of the proteins and disrupts most non-covalent inter-molecular protein interactions. These two result in total denaturation of the proteins in the sample, thus yielding unfolded, highly anionic polypeptide (-vely charged) chains. These chains are then resolved within a polyacrylamide gel saturated with SDS and the appropriate buffer carrying the current. The excess SDS in the gel helps to maintain the denatured state of the protein during separation. The SDS-coated polypeptides (one SDS molecule per two amino acids) create a situation in which the charge – to – mass ratio of all of the proteins in the sample is approximately the same. The intrinsic charge on the individual polypeptide chains is insignificant as compared with the negative charge imposed on them by the presence of the SDS. The friction of the molecules as they migrate through the matrix is now the major factor that influences the differences in their mobility. The friction of the molecules is governed by the pore size of the matrix: the larger polypeptides which experience greater friction will migrate more slowly than smaller polypeptides which will migrate more rapidly. Hence, SDS-PAGE allows the separation of proteins on the basis of size.

SDS-PAGE employs two buffer/polyacrylamide gel compositions in a single slab referred to as the *stacking gel* and the *running* or *resolving gel.* The samples of protein are first introduced into wells cast within the stacking gel after they are mixed with a viscous sample buffer (30% glycerol) containing SDS and thiols. After loading the samples, voltage is applied to the system (3-4 V/cm^2) between two separated pools of glycine buffer with a pH of 8.3. Glycine carries an average charge (-0.1 per molecule at pH 8.3 of running buffer) and almost no net negative charge (at pH 6.8 of stacking gel), while SDS-coated proteins carry a high negative charge that is essentially independent of the pH of the system. At a low pH in the stacking gel, the anions of glycine lose negative charge to display decreased mobility, while the ions of chloride contained in the stacking gel migrate ahead of proteins due to their small size (low friction) and full negative charge. The negatively charged proteins migrate into the stacking gel at a lower rate than that of chlorides, but faster than that of virtually uncharged glycine anions. As a result, the proteins accumulate ahead of the advancing glycine front and finally stack into concentrated, narrow bands at the interface between the stacking gel and the running gel. In the meanwhile, ions in the stacking gel readily migrate into the running gel.

As the advancing front enters the running gel, the high concentration of the buffer (pH 8.8) and acrylamide (of decreased pore size) in the gel trigger two events, namely, the high pH buffer imparts a greater negative charge to the glycine anions, whose migration gets retarded in the buffer (of pH 6.8) in the stacking gel. The glycine ions (because of their smaller size and increased anionic character) quickly overtake the proteins as they migrate through. In addition, the reduced pore size of the gel also imparts a significant frictional component to the mobility of each individual protein present. Due to the equal charge-to-mass ratio, all the proteins in the system will migrate through the gel on the basis of their size.

The relative mobility of each of the anionic polypeptide chains is a function of the logarithm of its molecular weight. The relative mobilities of the set of polypeptides of known molecular weight in the sample is determined and plotted as log molecular weight versus relative mobility. The standard curve thus produced from the analysis of the polypeptides of known molecular weights can then be used (along with the relative mobilities of the unknown polypeptides in the sample), to estimate their molecular weight.

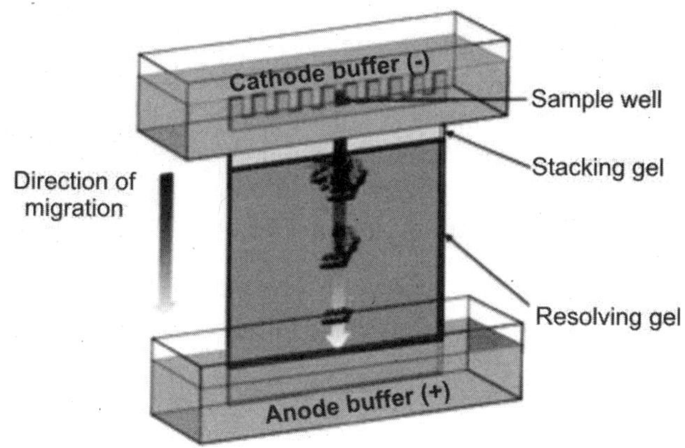

SDS-PAGE Apparatus

The proteins that have been separated through SDS-PAGE may be most often visualized by two methods. The first method involves saturation of the gel with a solution of acetic acid, methanol and water containing a dye called Coomassie Brilliant Blue R-250. The methanol and acetic acid fix the proteins within the gel matrix, while the dye binds to the proteins in the gel. After the gel is destained with the same aqueous acetic acid/methanol solution without the dye (the dye gets removed from the portions of the gel that do not contain the protein), the proteins on the gel become clearly visible as dark blue bands on the gel. This method is sufficiently sensitive to detect a protein band containing as little as 0.1 to 0.5 micrograms of a polypeptide. Silver staining, an alternative method, is sensitive to about 10 ng of protein contained in a single band on the gel. The method begins by gel saturation with a solution of silver nitrate, followed by the addition of a reducing agent to promote reduction of Ag^+ ions to metallic silver (Ag), which precipitate on the proteins in the gel whereby the protein bands appear black. The method is technically difficult than the first method, and as such is used only when extreme sensitivity is essential.

F) Iso-electric Focusing (IEF)

This is a technique by which the macromolecules (zwitter ions – substances with both acidic and basic groups like amino acids) are separated on the basis of their iso-electric points (pI, pH values at which they carry no net charge). Similar to SDS-PAGE, this technique can also be carried out in a slab format. A pH gradient is established in the gel with the help of ampholytes. The ampholytes are small polymers (about 5000 Da) with random distribution of weakly acidic and weakly basic functional groups (such as carboxyls, imidazoles, amines etc.). The gel containing the ampholytes is connected to an electrophoresis system with a dilute acid solution (H^+) in the anode chamber and a dilute basic solution (OH^-) in the cathode chamber. When voltage is applied the current flows due to migration of the charged ampholytes present in the gel. These migrate either towards either pole consistent with their charge distribution. Those with lower isoelectric points (those with more carboxyl groups and a net negative charge) migrate towards the anode, while those with high isoelectric points (those with more amine groups and a net positive charge) migrate towards the cathode. Finally, each ampholyte will reach a position in the gel that has a pH equal to its isoelectric point, and when this condition occurs, the ampholytes carry no net charge and will no longer migrate in the gel. After a sufficient period of electrophoresis, the population of ampholytes will act as local buffers to establish a stable pH gradient in the gel. This forms the basis for the separation of macromolecules that are present in (or subsequently introduced into) the system.

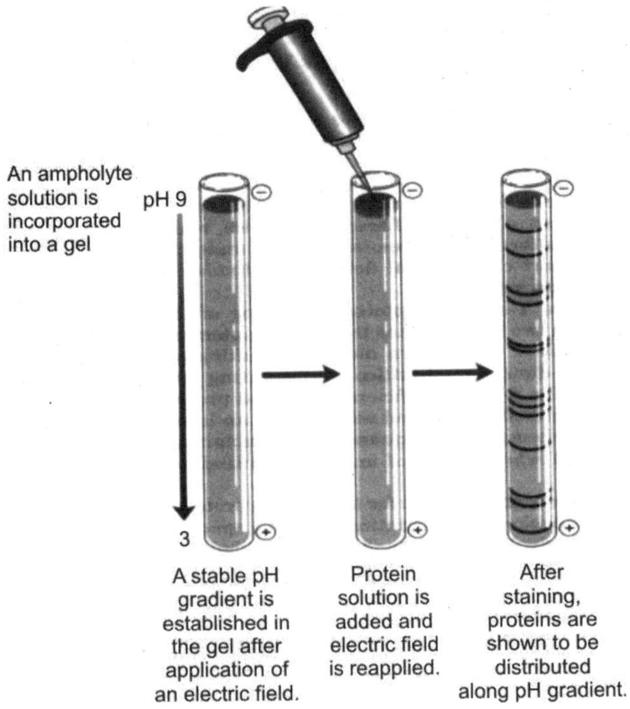

Iso-electric Focusing – Method

As a pH gradient is established in the gel by the ampholytes, the proteins also migrate toward their respective poles until they reach a pH in the gel at which they too carry no net negative charge (the pI of the protein). At this point, the attractive forces applied to the protein by the anode and the cathode are equal and as such the protein no longer migrates. The different proteins in the sample are focused to particular portions of the gel where the pH is equal to their pI values. When the ampholytes and the proteins reach a pH value in the gel equal to their pI and no longer migrate, the separation is said to be complete. When the migration stops, the current (I) also decreases. In this method, proper care must be taken not to impose a friction factor. In case, polyacrylamide gels are used the concentration of the gel should not exceed 8%. The technique can also be performed with other slab matrix materials like agarose or sucrose density gradient. One must also be aware that ampholytes are commercially available in a wide pH range for separation of a molecule of interest, for e.g. pH 3-5, pH 2-10 or pH 7-9 etc. In earlier times, this method was employed as an analytical technique in the slab format, but more recently, it is being performed on a large scale and non-denaturing format.

G) 2-D (2-Dimensional) Electrophoresis

This is a gel electrophoresis which is a high resolution technique for complex protein mixtures. In this method, two different types of separation are carried out on a single sample, out of which the first separation is based on iso-electric point (iso-electric focusing) while the second is according to size (SDS-PAGE).

The method begins with the application of a small band of sample to the top of a support (agarose or polyacrylamide gel with large pores) for use in isoelectric focusing. During the first separation, some proteins have been separated according to their pI values, but still many proteins with similar isoelectric points and overlapping bands remain as such a second separation becomes necessary. The first gel is turned on its side and placed at the top of a second support (polyacrylamide gel) for use in SDS-PAGE. During this process, a separation according to size occurs in which each band from the first separation has its own lane on the SDS-PAGE gel. This results in a series of peaks that are now separated in two dimensions across the gel (one on the basis of pI and the other on the basis of size). This technique makes it possible to resolve a much larger number of proteins than by either IEF or SDS-PAGE alone. At the end of separation, the protein bands can be detected either by straining with Coomassie Blue or Silver nitrate in the location and measurement of the bands. Another option is analysis by mass spectrometry. The interpretation and analysis of many protein bands in a single sample requires the use of computers to help image and catalogue the location of each band and to correlate the information with that obtained by other methods like mass spectrometry.

H) Agarose Gel Electrophoresis (AGE)

This is the principal technique used to determine the size of high-molecular-weight nucleic acids (DNA and RNA). Agarose is a long polymer of galactose and 3,6-anhydrogalactose linked through alpha-glycosidic bonds. Agarose can be readily isolated from sea-weeds. The polymers may contain up to 100 monomeric units, with an average molecular weight of around 10,000 Da. The gels are cast by dissolving the white agarose powder in an aqueous buffer containing EDTA and either Tris-acetate (TAE) or Tris-borate (TBE) as the buffering species. When the sample is heated to just below the boiling point, agarose powder dissolves in the buffer to form a clear solution. As the solution slowly cools to room temperature, hydrogen bonding within and between the polygalactose units in the solution will cause the formation of a rigid gel with a relatively uniform pore size. The induction of this reaction involves no chemical reaction unlike polymerization for polyacrylamide gels.

The pore size can be controlled by the percentage of the agarose dissolved in the solution. A high percentage gel (3% wt/wt) will have a smaller pore size than a lower percentage of gel (0.8% wt/wt). The agarose percentage to be cast in the gel will be determined by the size of the various molecules to be resolved during the process, the smaller the molecular weight of the molecules to be resolved, the higher percentage of agarose (with a small pore size) the gel should contain.

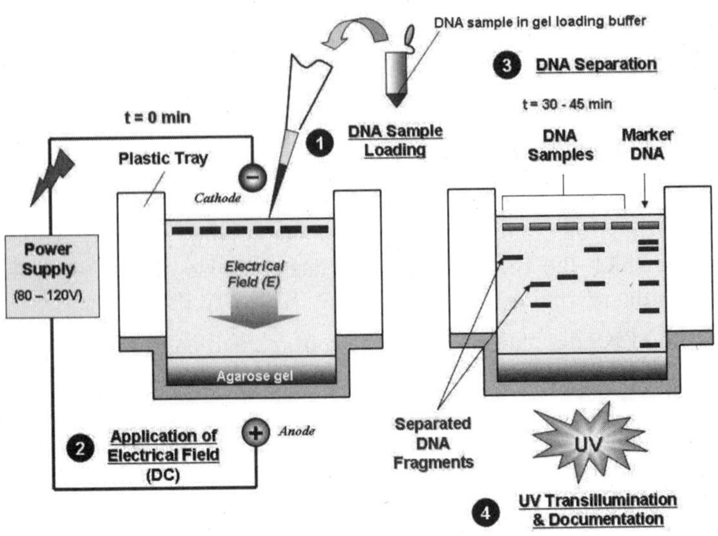

AGE – Principle

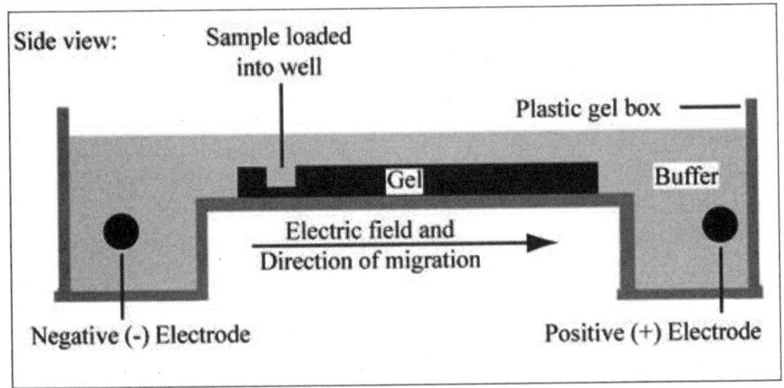

AGE – Side View

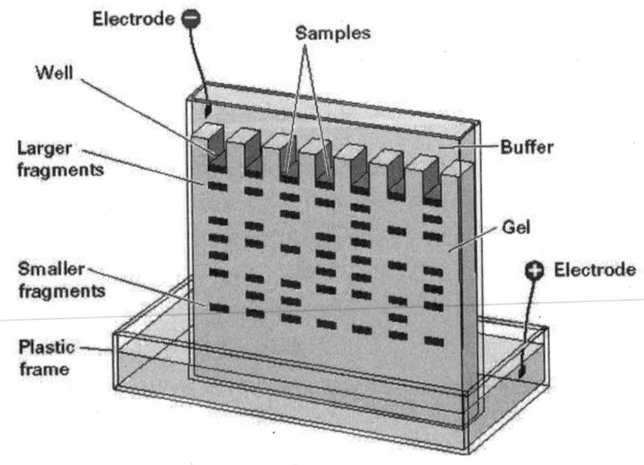

Agarose Gel Electrophoresis – Apparatus

The solution heated to dissolve the agarose, is momentarily cooled and poured into a slab mold fitted with a Teflon or plastic comb at one end. After the polymerization of the solution, the comb is removed to create wells into which the desired DNA or RNA samples will be applied. The gel is then transferred to a chamber and is covered with the same TAE or TBE buffer completely. The nucleic acid sample is mixed with a viscous buffer (30% glycerol) containing one or more tracking dyes (bromo-phenol blue dye will migrate at the same rate as a DNA molecule of about 500 base pairs, while xylene-cyanole dye will migrate at the same rate as a DNA molecule of about 4000 base pairs) that will be used to monitor the progress of the process.

The nucleic acid samples are loaded into the wells of the gel, along with a sample of DNA fragment of known molecular weight (equal number of base pairs) in one of the wells. The standards will be used following the separation to help in the determination of the size of the nucleic acid samples present in the unknown sample. As the DNA is negatively charged, the cathode (-ve electrode) should be connected to the side nearest to the wells in the gel, while the anode (+ve electrode) is connected to the opposite side of the apparatus. Approximately 4 to 6 V/cm of voltage is applied to the system, and the process is continued until the bromo-phenol blue dye front reaches the end of the gel. In this method, a stacking gel is not employed as in the SDS-PAGE. The nucleic acids focus very quickly at the buffer-gel interface before entering the matrix as they have a much greater frictional component in the gel than they do in the buffer inside the wells.

Nucleic acid visualization is possible by the use of fluorescent dyes like ethidium bromide which has an ability to intercalate between the bases of nucleic acid duplexes (DNA double helix). After the process, the gel is placed in a solution of TBE or TAE buffer containing ethidium bromide, which diffuses into the gel and gets associated with the nucleic acids. It is de-stained for a short period in buffer without the dye (so as to remove the dye from the areas on the gel that do not contain nucleic acids), and later exposed for a brief period to ultraviolet light (256 to 300 nm) to reveal the nucleic acid fragments as orange or pink fluorescent bands in the gel. The pattern of the beads is recorded by photographing the gel in a dark chamber while the gel is exposed to UV light from below. A filter is used to screen out UV light so that most of the light that exposes the film is from the fluorescence of the nucleic acid bands that contain tightly bound ethidium bromide.

The method of determination of the molecular weight of an unknown nucleic acid is the same as that of protein in SDS-PAGE. The log molecular weight of the nucleic acid samples of known size is plotted against their relative mobilities to produce a standard curve. From the relative mobility of an unknown nucleic acid fragment on the same gel, the molecular weight and the number of base pairs that the fragment contains can be readily determined (because a single base has an average molecular weight of approximately 320 Da, while a single base pair has an approximate average molecular weight of 640 Da).

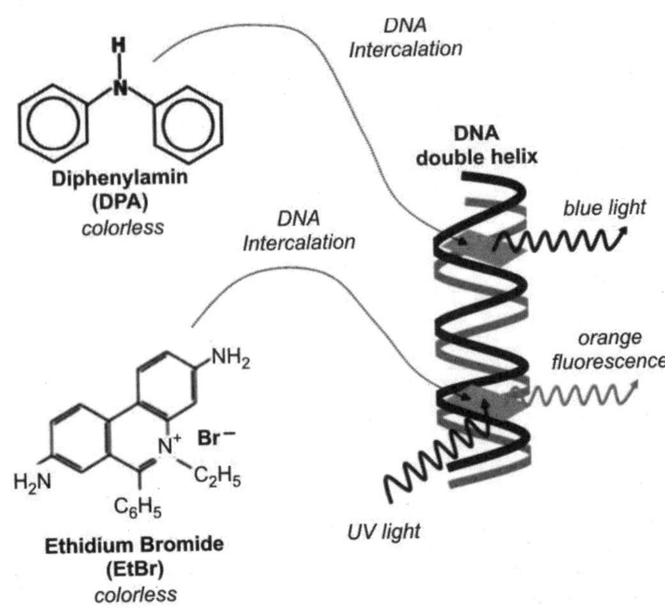

Intercalating Dyes for DNA Visualization

DETECTION METHODS – BLOTTING

Blotting is an approach for detection in gel electrophoresis where a portion of the analyte bands is transferred to a second support (like nitro-cellulose) where they react with a labeled agent. There are several types of blotting in practice.

1) Southern Blot

The most common method is a Southern Blot (named after its discoverer, Edwin Southern, a British Biologist), which is used to detect specific sequences of DNA by having the sequences that bind to an added known sequence of DNA that is labeled with a radio-active tag or with a label that can undergo chemi-illuminiscence.

2) Northern Blot

This was developed after the southern blot and is almost similar to it. This is used to detect the specific sequences of RNA by using a labeled DNA probe.

3) Western Blot

This method is used to detect specific proteins on an electrophoresis support. The proteins are first separated on a support electrophoretically and then blotted out on to a second support like nitro-cellulose or nylon. The second support is then treated with labeled antibodies that can specifically bind the proteins of interest. After the antibodies and proteins have been allowed to form the complexes, any extra antibodies are washed away and the remaining bound antibodies are detected through their labels, indicating whether there is any of the protein of interest is present. This technique is used to screen the blood for the HIV virus by looking for the presence of proteins from the virus in the samples.

Other Methods

Recently, the use of instrumental methods for band analysis on electrophoretic supports is becoming increasingly popular. Mass spectrometry is a popular method for determining the molecular mass of a protein in a particular band. This analysis is accomplished by removing a portion of the band from the gel (or sometimes the gel is directly looked into) and examining the band by *Matrix-assisted laser desorption/ ionization time – of – flight mass spectrometry* (MALDI-TOF MS). This method makes it possible to identify a particular analyte (like a protein) by its molecular mass even when there are many similar analytes in a sample.

ELECTROPHORESIS – APPLICATIONS

Electrophoresis is one of the most important techniques used by the molecular biologists and biochemists. The most important applications of electrophoresis are most evident in the health (medical) industry, molecular biology, microbiology, genetics and biochemistry, and include the analysis of antibiotics and vaccines. This has become a vital tool in many areas of science and society. The application of this process in studies of antibiotics dates back to the 1950s. With the application of this process, experts are not only able to synthesize new antibiotics but are also able to analyze which bacteria are resistant to antibiotics. One of the many important applications of the process is the analysis of vaccines. Many vaccines (such as influenza vaccine, hepatitis vaccine and polio vaccine) have been purified, processed and analyzed through this process. The other important application is the analysis of proteins and DNA. Through this process, specific DNA sequences can be analyzed, isolated and cloned. This process also allows the researchers to map and account for the differences in the genetic code of the species. The analysis of DNA also provides a reliable tool in forensic investigations and paternity tests. The amount of proteins in the blood and/or urine can be measured and compared to established normal values to indicate an infection or a disease.

The need of an effective method for the separation and analysis of mixed proteins has been uniquely met by the recent improvements in electrophoretic method. The ordinary principles of qualitative analysis are inadequate to deal with proteins, whose components are similar and whose characteristics are determined less by their composition rather than by their structure. Further, proteins may be easily modified and their distinctive properties may be destroyed by the chemical treatment necessary for their separation. A distinctive structural characteristic of the proteins is used in the separation, and the forces applied are extraordinarily harmless in the physical methods which are better fitted.

The applications of different types of electrophoresis are as follows: *Paper electrophoresis* is widely used in clinical medicine to analyze the proteins of blood and other body fluids. The clinical biochemists consider this technique as a convenient, inexpensive and dependable means of diagnosing the disease like myeloma. *Membrane foil (cellulose acetate) electrophoresis* is widely employed in clinical investigations of monoclonal gamopathies, immunofixation of monoclonal components, analysis of haemoglobin variants, separation of urinary proteins and identification of Bence Jones proteins. *Gel electrophoresis* is used for the separation and analysis of macromolecules like DNA, RNA or proteins and their fragments on the basis of their size sand charge. It is also used in clinical chemistry to separate the proteins by charge and/or size and also in biochemistry and molecular biology to separate a mixed population of DNA and RNA fragments by their length, and also to estimate the size of DNA and RNA fragments or to separate the proteins by their charge. This can also be used for the separation of nanoparticles. In molecular genetic diagnosis or genetic fingerprinting, this is used to analyze the PCR products. This method can also be effectively used in forensics, genetics and microbiology. *Starch gel electrophoresis* (horizontal and vertical) is used for polymorphic enzyme studies in plant specimens. *Agarose gel electrophoresis* was extensively used for investigating the DNA cleavage efficiency of small molecules and as a useful method to investigate various binding modes of small molecules to supercoiled DNA. This is used as a diagnostic tool to visualize the fragments of DNA. This is also a common forensic technique and is used in DNA fingerprinting. *Counter immunoelectrophoresis* (crossover electrophoresis) is used in the diagnosis of certain diseases in cattle. *Protein gel electrophoresis* is a simple way to separate proteins prior to downstream detection or analysis. *Cellulose acetate electrophoresis* is an important technique in clinical diagnostics. In addition to its usefulness in research techniques, agarose gel electrophoresis is a common forensic technique and is used in DNA fingerprinting. It is also highly effective in the separation of trace minerals.

RECENT ADVANCES IN ELECTROPHORESIS

Introduction

Electrophoresis is one of electro-kinetic effects brought about by the action of an electric field in producing a relative motion of a molecule (or a particle or a cell) and the liquid medium in which it is dissolved (or suspended or immersed), and which depends on a property that differs in the molecule (or particle or cell) compared with the medium. If the molecules are a part of a permeable (or porous) solid phase, the medium moves through the solid phase. If the molecules are in solution (or suspension), they move through the medium. The properties that produce these effects may be the electric charge, conductivity, dielectric constant, magnetic permeability or a combination of such properties.

The property in the case of the medium may be homogeneous (in which the steady motion is produced for different components in solution or suspension) or alternatively, it may be non-homogeneous (when the different components moving to different equilibrium positions). The electro-kinetic effects may be used to study the particle property of the molecule, or particle or cell or to separate mixtures on the basis of the property for analytical or preparative purposes. Some of these aspects have been exploited to only a limited extent, and include: dielectrophoresis, electro-magnetophoresis and combined electro-magnetophoresis and electrophoresis. Such methods are more promising for particle or cell separation than for molecules in solution.

The electrokinetic effect depends on the electric charge on the molecule (or a particle or a cell), and this may arise from the acidic or basic ionic groups in the molecule (or near the surface of the particle or cell) or by adsorption of small ions to the surfaces. On the application of the electric field, each charged object will migrate relative to the medium at a velocity determined by the electrophoretic mobility.

The important methods in electrophoresis that have been developed recently are as follows:

1. Non-linear Frictiophoresis or Non-linear Electrofrictiophoresis

This technique is the unidirectional drift of a particle in a medium caused by the periodic driving force with zero mean. This is possible due to the non-linear dependence of the friction-drag on the velocity of the particle. This was discovered theoretically and is also known as the non-linear electrofrictiophoresis. In the beginning, a periodic driving force with zero mean is able to entrain a particle into an oscillating movement without unidirectional drift, because the integral momentum provided to the particle by the force is zero. The unidirectional drift can be recognized if we take into account that the particle itself loses the momentum through transferring it further to the medium it moves in/at. If the friction is non-linear, it may so happen that the momentum loss during the movement in one direction does not equal to that in the opposite direction and these causes the unidirectional drift.

2. ZIFE (Zero Integrated Field Electrophoresis)

In the separation of the DNA fragments, zero mean periodic electric field is used in zero-integrated field electrophoresis (ZIFE), where the field time dependence is used, which allows separation of the long fragments in agarose gel, non-separable by standard constant field electrophoresis. The long DNA geometry and its manner of movement in a gel (known as reputation) do not allow applying directly. For FIGE, an interesting phenomenon, the so-called "band inversion" has been described, where manipulates DNA molecules with a higher molecular weight move faster than the smaller ones. Such a behaviour is a major drawback of this method and this might be one reason for the low acceptance of the field inversion technique. To circumvent this problem which can lead to misinterpretations when separating the unknown DNA molecules, Carle *et al.* (1986) proposed the use of pulse time ramps, which means that the pulse time is steadily increased or decreased during the run.

The preparation and manipulation of very large DNA molecules for PFGE requires more care than normally used for smaller molecules. The general protocol used is the preparation of DNA directly in solid agarose blocks (plugs or beads). The intact cells are encapsulated in agarose (plugs or beads) and are treated with different combinations of enzymes and detergent to remove the cell walls, membranes, RNA, proteins and other materials in order to obtain naked DNA. This is possible because the detergents and the enzymes can diffuse by Brownian motion through the agarose pores of the plug, while the large pieces of DNA cannot and remain sequestered inside the plug. In ZIFE, which is used for the 10–2000 kbp range, the product of the field strength and the pulse time is approximately the same for backward and forward pulses, although the actual field strengths and pulse times are different. This regime avoids band inversion effects in a pre-selected size window, but at the high end of the size range results in a fairly slow migration of the molecules. A similar pulse shape, but with a positive bias in the forward direction, is used in the size range above 2000 kbp. In general, for megabase-sized molecules, extremely short, high voltage spikes in the reverse direction are inserted into the l9ng pulses to avoid the trapping that is commonly observed for large DNA molecules at high field strengths.

3. Pulsed - field Gel Electrophoresis (PFGE)

This is a technique used for the separation of large DNA molecules by applying to a gel matrix with an electric field that periodically changes the direction. David C. Schwartz and Charles Cantor (1984) developed a variation on the standard protocol to separate very large molecules of DNA effectively by introducing an alternating voltage gradient to improve the resolution of larger molecules. This became known as PFGE, and the development of this method expanded the range of resolution for the DNA fragments by as much as two orders of magnitude. The procedure of this technique is relatively similar to a standard gel electrophoresis, except that instead of constantly running the voltage in one direction, the voltage is periodically switched among three directions – one that runs through the central axis of the gel and the remaining two that run at an angle of 60° either side. The pulse times are equal for each direction resulting in a net forward migration of the DNA. For extremely large molecules (up to around 2 Mb), the switching interval ramps can be used that increases the pulse time for each direction over the course of a number of hours – for example, increasing the pulse linearly from 10 seconds at zero hours to 60 seconds at 18 hours. Such a procedure takes longer than normal gel electrophoresis due to the size of the fragments being resolved and the fact that DNA does not move in a straight line through the gel. PFGE may be used for genotyping or genetic fingerprinting, and it is commonly considered a 'gold' standard in epidemiological studies of pathogenic organisms. The sub-typing has also made it easier to discriminate among the strains of *Listeria monocytogenes* and to link environmental or food isolates with the infections clinically.

Many PFGE techniques have been developed since 1983, based on an empirical approach to find the optimum electrode geometry, which coupled to the right choice of pulses that would give good separations. This is the characteristic of a field where the applications of techniques are important, but the basic principles are poorly understood. The separation of DNA molecules in OFAGE (orthogonal field alternation gel electrophoresis), FIGE (field inversion gel electrophoresis), TAFE (transverse alternating field electrophoresis), CHEF (clamped homogeneous electric field), PACE (programmable autonomously controlled electrode gel electrophoresis), PHOGE (pulsed homogeneous orthogonal field gel electrophoresis), ZIFE (zero integrated field electrophoresis, and ODPFGE (one-dimensional pulsed field gel electrophoresis) – all depend in some way on the size-dependent retardation/ reorientation of large molecules. At present, the electric field geometry involves the inversion of the direction of the field in one dimension and the change of the magnitude of the electric field, with the exception of ZIFE and ODPFGE. Recently, the pulsed field technique using one-dimensional electric field geometry has been developed and a variant of this method called ZIFE.

4. Isotachophoresis (ITP)

This is a technique in analytical chemistry which is used for the selective separation and concentration of ionic analytes. It is a form of electrophoresis in which the analytes are separated on the basis of their ionic mobility, a factor which refers to the fact that how fast an ion migrates through an electric field.

Conventionally, in ITP separations, a discontinuous buffer system is used. The sample is introduced between a zone of fast LE (leading electrolyte) and a zone of slow TE (terminating or trailing electrolyte). Generally, the LE and TE have a common counter-ion, but the co-ions (that have the charges with the same sign as the analytes of interest) are different. LE is defined by the co-ions with a high ionic mobility, while the TE is defined by co-ions with a low ionic mobility. The analytes of interest have an intermediate ionic mobility. The application of an electric potential results in a low electrical field in the LE and a high electrical field in the TE. The analyte ions situated in the TE zone will migrate faster than the surrounding TE co-ions, while the analyte ions situated in the LE will migrate at a slower rate, and the result is that the analytes are focussed at the LE/TE interface. ITP is a displacement method in which the focussing ions of a certain kind displace other ions. If present in sufficient amount, the analyte ions can displace all electrolyte ions, thus reaching a plateau concentration. The multiple analytes with different ionic mobilities will form multiple plateau zones. In plateau mode ITP separation, these can readily be recognized by stair-like profiles, while in peak mode ITP, the analytes will focus in sharp Gaussian-like peaks. The completion is characterized by a dynamic equilibrium in which all the co-ionic zones migrate with equal velocities. Hence, its name as ITP (*iso* = equal, *tachos* = speed and *phoresis* = migration). ITP is exactly equal to the steady-state-staking step in discontinuous electrophoresis.

5. Transient Isotachophoresis (tITP)

tITP is a popular form of ITP which alleviates the limitation of the conventional ITP that it has a limited separation capacity because of the overlap of the analyte zone. In this method, the analytes are first concentrated by ITP, and then can be baseline separated by zone electrophoresis. tITP is usually accomplished by dissolving a sample in the TE and sandwiching the sample/TE plug between the LE zones or vice versa; a sample/LE plug can also be sandwiched between the TE zones. In the first case, the analytes are focussed at the front TE/LE interface. In the meanwhile, the back of the TE plug comes dissolved in the LE because the faster LE ions overcome the TE ions. When all the TE ions are dissolved, the focussing ceases and the analytes are separated according to the principles of zone electrophoresis. Now-a-days, tITP is more widespread than the conventional ITP because it is easily implemented in capillary electrophoresis (CE) separations as a pre-concentration step, making CE more sensitive while profiting from its powerful separation capacities.

6. Multi-junction IEF System

The increased demand for faster and easy-to-use protein separation tools has accelerated the evolution of IEF towards in-solution separations. In this context, a multi-junction IEF system was developed to perform fast and gel-free IEF separations. The multi-junction IEF system utilizes a series of vessels with a capillary passing through each vessel. Part of the capillary in each vessel is replaced by a semipermeable membrane. The vessels contain buffer solutions with different pH values, so that a pH gradient is effectively established inside the capillary. The buffer solution in each vessel has an electrical contact with a voltage divider connected to a high-voltage power supply, which established electrical field along the capillary. When a sample (a mixture of peptides or proteins) is injected in the capillary, the presence of the electrical field and the pH gradient separates these molecules according to their isoelectric points. The multi-junction IEF system has been used to separate tryptic peptide mixtures for two-dimensional proteomics and blood plasma proteins from Alzheimer`s disease patients for biomarker discovery.

7. Dielectrophoresis (DEP)

In DEP, the force is exerted in a dielectric particle when it is subjected to a non-uniform electric field, and this force does not require the particle to be charged. In the presence of electric fields, all the particles exhibit dielectrophoretic activity. The strength of the force depends strongly on the medium and electrical properties, and also on the shape and size of the particles, as well as the frequency of the electric field (which at a specific frequency allows the separation or the orientation and the manipulation of the nano-particles and nano-wires). In addition, a study of the change in the dielectrophoretic force as a function of the frequency also allows the electrical properties of the particles to be elucidated. It was Herbert Pohl (1950s) as far back as the early 20th century, who described, named and first understood this phenomenon. But, recently, it has been revived recently due to its potential in the manipulation of micro-particles, nano-particles and cells.

This process occurs when a polarisable particle is suspended in a non-uniform electric field, which polarizes the particle, and the poles experience a force along the field lines, which can be either attractive or repulsive according to the orientation on the dipole. As the field is non-uniform, the pole experiencing the greatest electric field will dominate over the other, and the particle will move. The orientation of the dipole is dependent on the relative polarizability of the particle and the medium, in accordance with Maxwell-Wagner-Sillars polarization. The direction of the force is dependent on the field gradient rather than the direction of the field, DEP will occur in both AC (alternate current) as well as DC (direct current) electric fields; and the polarization (the direction of the force) will depend on the relative polarizabilities of the particle and the medium. If the particle moves in the direction of increasing electric field, the behaviour is referred to as the *positive DEP* (or pDEP), and if acting to move the particle away from the high field regions, it is referred to as the *negative DEP* (or nDEP). As the relative polarizabilities of the particle and the medium are dependent, this also allows the elimination of the electrophoretic motion of the particles due to an inherent particle charge. The electro-rotation and travelling-wave dielectrophoresis (TWDEP) are the phenomena associated with DEP, and these require signal generation equipment in order to create the required rotating or travelling electric fields. As a result of this complexity DEP has found less favour than conventional DEP among the researchers. This method allows the separation of the cells or the orientation and manipulation of nanoparticles and nano-wires. Further, a study of the change in DEP force as a function of frequency can also allow the electrical (or electrophysiological) properties of the particle to be elucidated. DEP can be used to manipulate, transport, separate and sort out different types of particles. It is also being applied in fields, such as medical diagnosis, drug discovery, cell therapeutics and particle filtration. This has also been used in conjunction with the semiconductor chip technology for the development of DEPArray Technology (Menarini Silicon Biosystems) for the simultaneous management of 1000s of cell in a microfluidic device. Single microelectrodes on the floor of a flow cell are managed by the CMIS chip to form 1000s of "Dielectrophoretic Cages", each capable of capturing and moving one single cell under control of a routing software. DEP has made it possible to characterize and manipulate the biological particles such as blood cells, stem cells, neurons, pancreatic β-cell, DNA, chromosomes, proteins and viruses. DEP has also been applied for the separation of both live and dead cells to study cell-cell interaction. DEP can also be used to detect apoptosis soon after drug induction, measuring the changes in electrophysiological properties.

8. Optical Dielectrophoresis

The use of photo-conductive materials (such as in lab-on-chip devices) allows for the localized inducement of dielectrophoretic forces through the application of light. One can project an image to induce forces in patterned illumination area, allowing for some complex manipulations. During the manipulation of the living cells, this technique provides a non-damaging alternative to optical tweezers, as the intensity of light is about 1000 times less.

9. Affinity Electrophoresis

This is an important technique that is used widely to separate and analyze the bio-molecules in biology and medicine, and both qualitative and quantitative information can be gained through this method. This phenomenon can be applied through different strategies that are based on the changes in the patterns of biological macromolecules that result from the interactions and processes of formation of the complexes. The fragments of nucleic acid can be characterized by their affinity to other molecules (transcriptional factor proteins). A shift in the mobility induced by a charged detergent helps in the identification of hydrophobic membrane proteins. The very first technique was cross electrophoresis developed by Nakamura et al (1959) which has been used in the detection of the enzyme-substrate complexes and also in the analysis of antigen-antibody reactions. A variety of these techniques have been developed to provide good qualitative results and have good separation abilities. The pioneering technique is affinophoresis that utilizes a charged, hydrophilic polymer that binds to the molecules with a suitable affinity (affinophore). In this method, only the molecules that bind to the affinophore can be induced to migrate by the application of an electric field. The development of the affinity probes (affinity ligands – the molecules that have affinities for specific molecules) occupies an important role in the growth of this technology at present. Different types of apparatus have also been developed for capillary affinity electrophoresis and microchip electrophoresis. This method is also used in a new technique where a PVDF (poly vinylidene fluoride) membrane is used as a molecular matrix (in the separations of large glycoproteins).

10. Difference Gel Electrophoresis (DIGE)

DIGE is a form of gel electrophoresis where up to three different protein samples can be labelled with size-matched, charge-matched, spectrally-resolvable fluorescent dyes (such as Cy3, Cy5, and Cy2) prior to two-dimensional gel electrophoresis. The three samples are mixed and loaded on to IEF (isoelectric focusing chromatography) for the first dimension and the strip is transferred to a SDS-PAGE. The gel is scanned with the excitation wavelength of each dye one after the other, after the gel electrophoresis, so that each sample can be seen separately (if the gel is scanned at the excitation wavelength of the Cy3 dye, one can see in the gel only the sample that was labelled with that particular dye). The technique is used to see the changes in protein abundance (between a sample of a healthy person and the sample of a person with disease), the post-translational modifications, truncations and any other modification that might change the size or the isoelectric point of the proteins. The binary shifts might be left to right (change in the isoelectric point), vertical (change in the size), or diagonal (change in both the size and isoelectric point). The reciprocal labelling is done to make sure the changes are not due to dye-dependent interactions. The advantages of this technique are: this overcomes the limitations in traditional 2D-electrophoresis that are due to inter-gel variation, which can be considerable even with the identical samples. As the proteins from the different sample types (for example, healthy/diseased, virulent/non-virulent) are run on the same gel, they can be directly compared. To carry out this, traditional 2D-electrophoresis requires large numbers of time consuming repeats.

11. Discontinuous Electrophoresis (colloquially Disc Electrophoresis)

This is a new method developed by Ornstein and Davis (Mt. Sinai Hospital, New York, N. Y - 1964), for fractionating the serum proteins and enzymes. This possesses a great sensitivity, speed and reproducibility and requires a sample as little as 3 micro-litres for a routine separation. The technique derives its name from its dependence on the discontinuities The pore size of the gel may be varied by simply changing the concentration of the inert polymer, polyacrylamide in the support medium where the separation occurs and also from the form where the separated fractions appear as a stack of flat discs in the gel column. Hence, this is a PAGE technique by which separations of as many as 20 distinct and concentrated protein fractions in the serum samples are possible. The discontinuity is based on four parameters, namely – structure of the gel – pH value of the buffer – ionic strength of the buffer – and nature of the ions in the gel and electrode buffer. The electrode buffer contains glycine (low net charge at pH 6.8 of stacking gel) and as such has a low mobility. The proteins move towards the anode slowly at a constant speed till they reach the limit of the separation gel. The mobility of the proteins depends on the size and charge, the pH rises to 9.5 and the net charge increases.

The separation is accomplished in a column of acrylamide gel consisting of 3 layers – the large pore upper gel which contains the sample – a large pore spacer gel in which the electrophoretic concentration and fraction alignment takes place – and the small pore lower gel in which the actual separation occurs. The remaining arrangement includes an upper and lower buffer reservoir, electrodes for each reservoir, and a constant power supply source. Electrophoresis is carried out in a vertical position, with the containers (gel) attached to the upper reservoir and the lower ends immersed in the buffer solution of the lower reservoir. The electrodes are arranged in such a way that the upper one is the cathode and the lower one is the anode. The sample ions migrate toward the small pore gel in the lower section of the gel column. The current is applied for a specific period of time, after which the gels are rimmed out of the containers and placed in special staining solutions for the location of the separated fractions.

12. Microfluidic Chip-based Electrophoresis

Microfluidic chip-based electrophoresis is a promising alternative to CE, since it has the potential to provide rapid protein analysis, straightforward integration with other microfluidic unit operations, while channel detection, nitrocellulose films, smaller simple sizes and lower fabrication costs.

Fluidic based cell-on-chip (CoC) approach – Microchip based biosensors show a promising future for monitoring cellular nanotoxicity as they allow rapid, real-time and multi-sample analysis creating a versatile, non-invasive tool that is able to provide quantitative information with respect to alteration in cellular function under various nanomaterials exposure. Most of these approaches for nanotoxicity assessment at present are based on multiple cell screening. By some experiments, it has already been elucidated that a chip-based biosensor capable of selective trapping of single cell dielectrophoretically in a micro-well and the same electrode can also be used for further electro-chemical study of the captured cell. Such electrodes are independently addressable for capturing the single cell in an individual micro-well, as well as study an independent cell on order. Such a CoC can be used for the nano-material based toxicity. The cell behaviour can also be monitored in terms of catacalomine release from the cell vesicles. The electro-chemical and optical measurement-based fluidic chip platform have been explored for cell analysis over the last decade which hold the potential to be used for the assessment of nanotoxicity.

ChIP (Chromatin Immunoprecipitation Assay) microfluidic systems include automated electrophoresis platforms and kits to facilitate the experimental procedure and analysis with fewer incidences of contamination errors. The automated systems are available with/without a processor or laptop, and kits can be chosen by DNA fragment size range. For example, Agilent DNA 1000 and 12000 kits are specific for fragments ranges from 25 to 1000 bp or 100 to 12000 bp, respectively. These kits are also supplied with the reagents and, consumables and microfluidic chips. The sizing and quantification of the nucleic acids and proteins are simplified with microfluidic–based automated electrophoresis systems that require minimal volumes of the sample, produce the results in less than an hour, and are achieved with just a push of a button. Any genomics lab's requirements will be met by evaluating the products dedicated to microfluidic ChIP technology. This technology offers a growing set of tools for manipulating small volumes of fluids to control the chemical, biological and physical processes that are relevant to sensing. The development of such tools based on lithographic techniques enable the integration with electronic and optical elements necessary for the construction of the operation sensor. The microfluidics has now reached the point where the individual and multiple integrated processes can be carried out for a wide variety of applications. The future developments will be concerned with bringing such technology into commercial devices operated by the non-specialist, and require standardization of the microfluidic components, and the improvement of real-world interfacing, allowing intuitive handling and operation by unskilled end-users. The commercialization of these systems will become more frequent whether in POC devices or in a 'chip-in-a-box' format. The devices for basic research will also remain highly important especially on the biological front as the researchers strive to produce the `human-on-a-chip`.

13. Iontophoresis

This is a process of transdermal drug delivery by use of a voltage gradient on the skin. The molecules are transported across *stratum corneum* by electrophoresis and electro-osmosis, and the electric field can also increase the permeability of the skin. These phenomena, either directly or indirectly, constitute the active transport of matter due to an applied electric current. The transport is measured in unit of chemical flux, commonly μmol/(cm^2/hour). This has experimental, therapeutic and diagnostic applications. This technique is useful in laboratory experiments in neuropharmacology. The transmitter molecules naturally pass signals between the neurons. The neurotransmitters and other chemical agents can be artificially administered very near living and naturally functioning neurons by micro-electrophoretic techniques, including microiontophoresis, the activity of which can be simultaneously recorded. This can be used to elucidate their pharmacological properties and natural roles.

Therapeutically, EMDA (electromotive drug administration) delivers a medicine or other chemical through the skin. It is an *injection without a needle*, and can be described as *non-invasive*. It is also different from the dermal patches which do not rely on an electric field. This method drives a charged substance, usually a medication or a bioactive agent, transdermally by repulsive electromotive force, through the skin. A small amount of electric currernt is applied to an iontophoretic chamber places on the skin, containing a charged active agent and its solvent vehicle. Another chamber (or a skin electrode) carries the return current. One or two chambers are filled with a solution containing an active ingredients and its solvent vehicle. The positively charged chamber (the anode) will repel a positively charged chemical species, whereas the negatively charged chamber (cathode) will repel a negatively charged species into the skin. This is used to treat some types of palmo-plantar hyperhidrosis. The tap water is often the chosen solution for mild and medium forms in this treatment. In very serious cases, a solution containing glycopyrronium bromide or glycopyrrolate (a cholinergic inhibitor) can be used. Diagnostically, *iontophoresis of acetylcholine* is used in research (to test the health of the endothelium – acetylcholine is positively charged and is placed in the anode chamber). Similarly, *pilocarpine iontophoresis* is used as a part of cystic fibrosis (to stimulate sweat secretion). Sometimes, *Reverse iontophoresis* is used for the removal of molecules from within the body for detection. Electroosmosis is currently used in device such as GlucoWatch, which allows for the blood glucose detection across the layers of the skin.

14. Temperature Gradient Gel Electrophoresis (TGGE) and Denaturing Gradient Gel Electrophoresis (DGGE)

TGGE and DGGE are two different forms of electrophoresis which use either a temperature or chemical gradient to denature the sample as it moves across an acrylamide gel. These two can be applied to nucleic acids (like DNA and RNA) and less commonly proteins. The former relies on temperature dependent changes in the structure to separate the nucleic acids. The latter, separates the genes of the same size on the basis of their different denaturing ability which is determined by their base pair sequence. The latter was the original technique while the former was a refinement of it. TGGE provides a "*sequence dependent, size independent method*" for the separation of DNA molecules. It separates the molecules and gives additional information about the melting behaviour and stability.

15. Free Flow Electrophoresis (FFE) or Carrier-Free Electrophoresis

FFE is a matrix-free electrophoretic separation technique. This is an analogous technique to CE, with a comparable resolution, that can be used for scientific questions (where semi-preparative and preparative amounts of the samples are required). This method is used to separate the samples quantitatively according to the differences in charge or isoelectric point. Because of its versatility, a wider range of protocols for the separation of the samples (like rare metal ions, protein isoforms, multi-protein complexes, peptides, organelles, cells, DNA origami, blood serum and nano particles) exist. The main advantage of this method is the fast and gentle separation of the samples dissolved in a liquid solvent without any need of a matrix (such as polyacrylamide in GE). This method also ensures a very high recovery rate since the analytes do not adhere to any carrier or matrix structure. This method allows a fast separation of preparative amounts of the samples with a very high resolution because of its continuous nature and high volume throughput. In addition, the separations can also be conducted under native or denaturing conditions.

This technique was developed by Kurt Hannig (1960s) at the Max-Planck Institute, Germany, and until 1980s, it was a standard technology for the separation of cells and organelles. It was even tested in space to minimize the sedimentation under zero gravity. It also became a standard method for cell sorting as flow cytometry. The further developments of this method also focussed on the separation of proteins and charged particles, and some research groups are presently working on the miniaturized versions of FFE systems (or micro-FFEs).

The technique can be described in brief as follows: The separation chamber consists of a back plate (consists of a cooled aluminium block, covered with a plastic-covered glass mirror) and a front plate (is now-a-days made of PMMA, while in earlier times, glass has been used). The distance between the two plates can be adjusted by spacers, and is usually between 0.1–0.5 mm. The front plate also contains the inlets for the separation buffers and the sample, the outlets for the fractionation tubes and the platinum wires as the electrodes. The operator is able to change the pH, salt concentrations or composition, by applying different buffers over the multiple buffer inlets, and therefore, the separation conditions over the width of the chamber. The separation buffers and the sample are applied by peristaltic pumps, to ensure a laminar flow. A high voltage electric field is applied perpendicular to the laminar flow. The analytes in the laminar flow can be separated by charge density and/or isoelectric point. This method can make use of different modes of electrophoresis (like isotachophoresis, isoelectric focussing or [interval] zone electrophoresis). At the end of separation cell, the separated sample is split up as the fractionation tubes and collected in microtiter plates. Later, the samples can be characterized by all the major techniques, such as HPLC, LC-MS, mass spectrometry (ESI/MALDI – depending on the protocol used) or electrophoresis (IEF/SDS-PAGE, 2D PAGE).

The standard applications include the high resolution separation of the protein complexes, membrane proteins, protein and antibody isoforms, cells, subcellular compartments (like organelles and ribosomes), and liposomes. It is possible to separate protein isoforms, that vary by less than 0.02 pH units, with a throughput of around 3 mg/h, by making use of very flat H gradients, generated by ampholytes. To increase the resolution or denature the samples, it is also possible to add additives to the buffers. Typically, the concentrations of urea up to 8M, 0.1–1.0% of detergents like CHAPS, CHAPSO, Digitonin-β-D-maltoside, Octyl-ß-D-glucoside, Triton-X-114 (IEF) and DTT up to 50 mM are tolerated.

16. New Technique to Separate Microsatellite Alleles

This is a new method to analyze and visualize SSR (simple sequence repeats: microsatellites which are tedious, time-consuming and expensive) data quickly and accurately to compare the ability of four other methods. The individual PCR reactions consisting of DNA from several dogwood (*Cornus florida* L.) cultivars and two SSR primer pairs were assembled for the analysis using agarose gel, polyacrylamide gel and QIAxcel System. Two separate PCR reactions consisting of the same component plus a fluorescent-labelled primer were set up for analyses using the CEQTM 8000 Genetic Analysis System and ABI 3130xl DNA Sequencer. These five methods were assessed for their advantages and disadvantages. Polyacrylamide gels had the highest resolution of all the alleles, while the agarose gels had the lowest resolution. But, with both the separation media, it was found difficult to score the size of the alleles. CE with the Genetic Analysis System and DNA Sequencer easily separated the products and determined the allelic size, but was more expensive than agarose or polyacrylamide gels. The QIAxcel System has lower resolution than CEQTM 8000 Genetic Analysis System and ABI 3130xl DNA Sequencer. But, QIAxcel System was rapid and cost effective compared to the two widely used sequencers, and also provided a computer generated gel image. The QIAxcel System using a 12-channel, sieving-gel cartridge is an affordable device for SSR assays used for mapping and population diversity analysis for researchers in small to intermediate labs.

Conclusion

Electrophoresis is currently the most widespread technique used for performing functional proteomics (the large-scale analysis of alterations in protein expression levels). Nevertheless, several limitations inherent to this technology have restricted the full potential of this protein differential display methodology for years. This has even led to the abandonment of 2D electrophoresis by several groups that switched to performing gel-free functional proteomics analyses based on liquid chromatography and mass spectrometry. Meanwhile, important recent advances in 2D electrophoresis, such as the introduction of fluorescent 2D difference gel electrophoresis and numerous protein pre-fractionation techniques, have thoroughly modernized 2D electrophoresis, making it again one of the preferred methods for the analysis of protein expression differences in many laboratories.

■■■

CHAPTER - 5

HERBARIUM TECHNIQUE

Herbarium – What it is?

In plant sciences, herbarium (plural: herbaria) is an anglicized term, *herbar* – that refers to a collection of dried, pressed and preserved plant specimens mounted on sheets and arranged in the sequence of an accepted classification which is readily available for reference to systemic botanists.

A herbarium is a collection of plant specimens preserved for long term study. These include many pressed and mounted plant specimens with flowers/fruits, and all are generally referred to as herbarium specimens. There are over 300 million specimens preserved for research in herbaria world-wide. The plants that are preserved serve as a ready reference record for checking up the description of new plants collected during the plant collections. A picture of the plant can be obtained with the help of the specimens kept in an international herbarium.

A herbarium is a repository, desired to ensure long-term preservation of scientific plant samples, generally referred to as specimens. The herbaria are analogous to plant libraries, with the specimens corresponding to the pages in books and most herbaria function under the aegis of an academic institution, a museum or a botanical garden. Most herbaria are housed in a room full of specially designed steel cases, herbarium cabinets. The interior of the cabinet is divided into 2 or 3 vertical sections and each section is subdivided into 6 - 10 compartments. The cabinets are lined up side-by-side in rows, with sufficient space for the doors to open between the rows.

According to Fosberg (1946), the herbaria provide all the information about the collected plants in different forms, such as recorded notes, photographs and hand-drawn illustrations etc. In majority of cases, the herbaria are often associated with botanic gardens. The term, herbarium, also refers to the building where the specimens are stored or to the scientific institute that not only stores the specimens but also conducts research on the specimens. Herbaria are usually associated with universities, museums or botanical gardens. The first herbarium is believed to have been established in Bologna, Italy (1570) by Luca Ghini. Now, there are around 4,000 herbaria in over 165 countries. A world catalog of public herbaria, *Index Herbariorum*, is published periodically by the International Association of Plant Taxonomists (IAPT) and is available on the web. Each herbarium in the index is assigned an official acronym or code that is used as a standard for referring to the institution and its specimen. The herbarium specimens are useful as references for plant identification and for the determination of plant locations and ranges, abundance, habitat and flowering and fruiting periods. They are used for studies in which the differences between plant species are evaluated and described (in monographs) or in which the species growing in a region are reported (in floras).

PLANT COLLECTION

The plant specimens are to be collected from different localities and at different times of the year. The plants of interest may be collected and mounted on the herbarium sheets. These plants are to be pressed on the spot or may be collected in a *vasculum* to be placed in the plant press after sometime. These specimens must include the maximum possible part of the plant. The herbs should invariably be collected with their roots. In case of bigger or larger plants, a twig of about 30 cm in length is to be collected. The collected specimens are to be tagged immediately and for each plant about 5-6 specimens are to be collected. The record of the locality and the place of collection is also to be noted.

The fresh plants or plant parts required for the herbarium are collected in three different ways. The most common method of collection is to collect the individual plant from the site of its occurrence and press it immediately on the site. This is done in a plant press which consists of a pair of rectangular wooden or metallic frames containing folded blotting paper sheets or newspapers in between the frames. Sometimes, the collected specimens are kept in air-tight vasculum (metallic container) which is large enough to accommodate the plants/branches/leaves, inflorescences/flowers and fruits without folding. Before the collection, the interior of the vasculum is to be lined with blotting paper/newsprint to avoid wilting. Overcrowding of the specimens is to be avoided and the vasculum is to be always partly filled. The vasculum (plural: vascula) is always painted white as it reflects most of the sun rays so that the collected plants do not get dried up within a short time. The specimens should be kept in a good condition at least for about 12 hours and as such the vasculum is kept in a cool place. The specimens should be pressed immediately. There is still another method of collection wherein the collected specimens are carried either in polythene bags or ruck sacks (bags strapped to the shoulders) to be pressed immediately. This method is useful for the plants collected in tropical rain forests, preferably during the rainy season when the humidity in air is more.

The specimen to be collected for the use in a herbarium should always be in flowering and fruiting stage. The herbs are to be collected completely including the underground parts of the plant (roots, rhizomes, tubers, flowers, fruits etc.). In case of shrubs and trees, the twigs or branches (bearing leaves, flowers and fruits) are collected as it is not possible to collect the complete plant. A minimum of 5 specimens are to be collected preferably in different stages of flower and fruit development (young inflorescence, flowers in bud condition, mature flowers ready to open up, open flowers, young fruits, fruits with fully developed seeds, dehiscing etc.). The specimens collected without flowers and fruits are of no use at all (as identification of the plant becomes difficult) and they are to be discarded. One has to visit the same locality several times a year to collect the specimens in different stages of development. It is also better to take neat and clear photographs of the specimen and its parts as well as the surroundings where the specimens are collected (preferably using a good SLR camera) for reference at a later date.

WHY PRESERVE PLANT SPECIMENS?

The herbarium specimens are a vital resource because they store a wealth of information. They are generally utilized by the staff and the students in the universities, colleges and other research institutions, by personnel from different government agencies and conservation agencies as well as the general public. The specimens are utilized for research in taxonomy which includes the preparation of monographs, revisions and floras, manuals which provide species description, help in identification and also documentation about the occurrences of the specimen. The collections help specialists to document the range of variation in individual species so that they can be accurately described. The specimens also provide the earliest documentation of the natural vegetation of a particular area and also document the range of indigenous species as well as the introduction and naturalization of alien species. The specimens are also the primary source of data of rare plants. They also help in the identification of unknown plants. The data recorded on the label is a resource to see a particular species in its natural habitat and the ecological information is useful for getting an idea of the type of habitat for a particular species. The type specimens are the original specimens that are the basis for descriptions of a new species, are the permanent record that has already been described. In case of doubt related to the identity of a plant during the research these can be reexamined to check the validity of the original identification.

PRESERVATION OF THE PLANT SPECIMENS

To retain the original form the plants collected in the field are spread flat on sheets and dried. The drying is done in a plant press between blotters or absorbent paper. The specimens are then mounted on the sheets of stiff white paper and are labeled with all essential data (date and place of collection, description of the plant, altitude and special habitat if any). The sheet is placed in a protective case which is disinfected as a precaution against insect attack. The plant is poisoned to avoid insect attack. Some plants are soft, bulky or not amenable to drying or mounting on sheets for which different methods may be used. The cones and fronds of gymnosperms may be stored in labeled boxes. The flowers or fruits may be stored in formaldehyde and smaller mosses and lichens are air-dried and packed in small paper envelopes. All the details where and when the plant is collected, the flower colour (as it may fade over time) and the name of the collector is usually included.

EQUIPMENT NEEDED FOR COLLECTION OF THE PLANT SPECIMENS

The field collection is the significant part of the taxonomical studies. It is a simple process and needs careful handling of the selected specimens. The indispensable and essential articles necessary for field exploration are:

a. Digger / Trowel and Pruning Knife or Collecting Pick

To procure the specimens without damage very sharp tools are required. Diggers are most suitable trowels necessary for obtaining (digging up) the underground parts of the plants like rhizomes, corms, bulbs, tubers as well as root system so that the complete specimen (small herbs) is collected. A digger of suitable size and shape may be chosen according to the type of the soil and depth of the sub-aerial parts. Very sharp pruning and cutting knives with heavy sheath and dandelion knife and hammer with chisel blade are also necessary for collection and cutting of branches of herbs, shrubs and trees. The pruning knife with a curved or hooked blade is used as a cutting instrument which has a pointed blade with a sharp edge and a handle. A digger or a trowel is a pointed tool used to dig in the dirt and plant more easily. These tools are to be brightly painted to avoid loss in the field during collection.

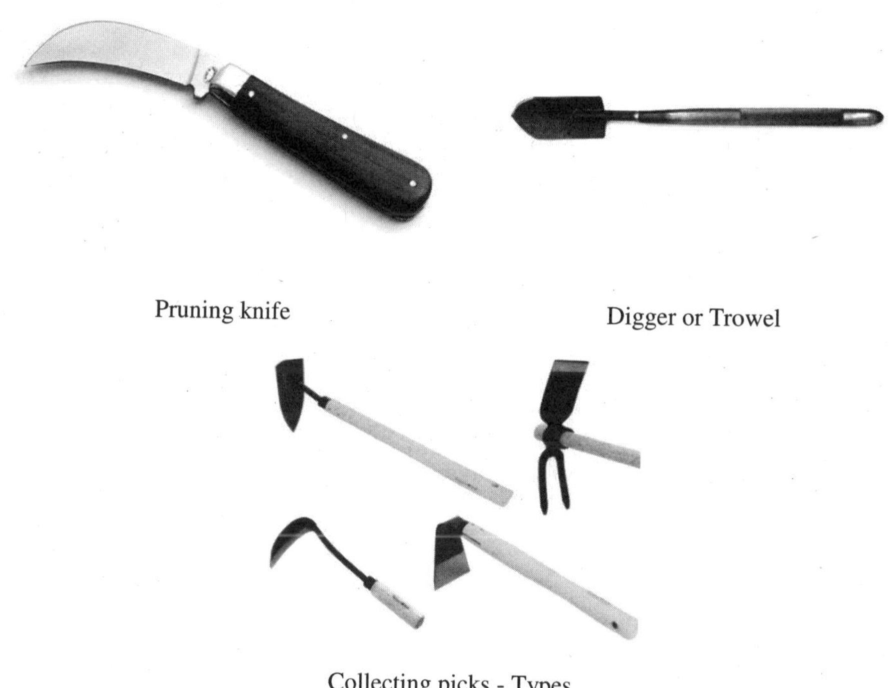

Pruning knife

Digger or Trowel

Collecting picks - Types

b. Machete and Secateur or Pruning Shears

Machete is a strong knife which is required for cutting the hard branches/twigs (flowering and fruiting shoots at the top and also the parasites along with the host stem – this is a large cleaver-like knife with a blade typically 32.5 to 45 cm long and approximately under 3 mm [0.12 in] thick. Secateur (a small pair of shears for pruning with a pair of pivoted handles and with a single cutting blade that closes against a flat surface) or pruning shears (also called as hand pruners or secateurs) are a type of scissors used for plants; shears with strong blades used for light pruning of woody plants, and to prune hard branches of trees and shrubs, sometimes up to 2 cm thick and an instrument for cutting the twigs. These are used in gardening, arboriculture, farming, flower arrangement and nature conservation and are necessary for cutting the woody and hard branches and other parts of the plant.

c. Sickle with Long Handle

A sharp bent sickle with a long handle is very useful for obtaining the twigs that are attached at a height on big trees. To prepare the sickle, small pieces cylindrical wooden rods about 60 cm are selected and each rod is fitted with a screw and a socket and joined together to make a long supporting sickle at the extremity. This helps to collect the twigs which are placed higher from the level of the ground and after the use, it can be unbolted. The screw and the joining system would facilitate easy carrying to the field of collection. Whenever desired, a long handle can also be used.

d. Vasculum and Polythene or Plastic Bags

A vasculum is a container generally used by botanists to keep field samples (collected plant material) viable by maintaining a cool, humid environment and these are typically flattened tin cylinders, carried horizontally on a strap so that the specimens lie flat and lined with moistened cloth. A metal vasculum is useful for temporary handling and to accommodate the specimens and to retain their freshness before pressing. It is always preferable to press the specimens following the collection. The prolonged delay in pressing is not favourable for the specimens as they lose their freshness after sometime. A standard vasculum is oval in shape, 45-60 cm in length, 20 cm in width and 25 cm in its depth. It is provided with a hinged door with a screwed fastener to keep it air-tight to avoid wilting of the specimens by moisture conservation. On either side, the vasculum is provided with a pair of buckles attached to a strap for easy carriage. The top of the vasculum has a handle for holding. Vasculum (in Latin, meaning a `small vessel') was popularized by William Withering (1770), is also the name of the newsletter of the Society of Herbarium Curators.

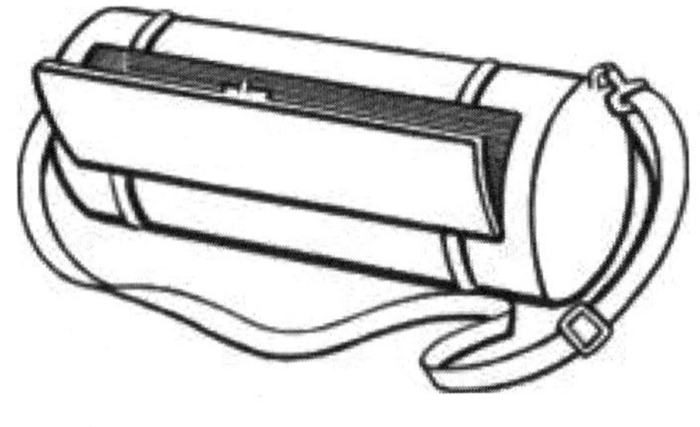

Vasculum - Line drawing

Vasculum - Image

A good number of clean polythene bags are needed as they make convenient containers to carry and store fresh plant material. After collection, the mouth of the bags should be tightly closed with a thread to avoid wilting/drying. Polythene bags or plastic bags are useful in plant collection as they are impervious to moisture and hold the plants fresh for a longer duration, and because of their light weight, they can be carried conveniently. The different specimens are put into separate bags and together they can be put either in the vasculum or light pack basket so as to avoid mixing up of different plants and their parts and also keep them isolated to avoid the difficulty of separation and selection.

e. Field Note Book

It is always essential to carry a bound, field note book while on plant collection. The complete data on collection such as the locality and the place of collection, the name of the collector, date and time of collection, name of the plant (if known), altitude, habit and habitat of the plant, the size of the plant, type of branching, nature of the root system, association with other plants, geography of the plant, colour of the flower, nature of the fruit and the seed(s), nature of the bark, appendages on vegetative organs if any, ecological condition of the locality, topography, type of the soil, the average population in the area per sq. metre and any other relevant information may be noted down in the note book. Each specimen is given a definite number in the note book which coordinates with the number given on the labels after mounting the specimen on to the herbarium sheet which facilitates identification and avoids confusion. Even duplicate specimens are given the number and this should be done on the paper containing the specimen and as far as possible use of slips or small pieces should be avoided because the chances of losing the slips are more. In general, serial number will be allotted to the specimens.

Note: Each page of the field note book is perforated at its base which can be easily separated from the page. The lower slip is further perforated into 5 parts, and each part is provided with a number that corresponds to the number of the page. The 5 pieces are tied to 5 separate specimens (parts) of the same plant collected from a particular locality. The specimen page of the field note book is given below:

Flora of .... (District)

Number: .... Date: ....

Name: .... Habit: ....

Local Name: .... Habitat: ....

Family: .... Flower – Colour: ....

Locality: .... Collected by: ....

Remarks (Ecological notes such as soil condition, altitude etc.) if any:
....

....

....

....

The number on the label should necessarily correspond to the number of the page (of the field notes) on which the details of plant was noted.

f. Plant Press and the Folders

Plant press along with folders (blotting sheets or newsprint) is required for pressing and quick drying of the specimens following collection. A best plant press has the ability to hold the plant material under a constant and firm pressure and helps to dry up the specimens to a stage of crispness. The specimens which contain the leaves, inflorescence, flowers, fruits and any other special modified parts of the plant are said to be good.

The plant press consists of top and bottom plates of equal size, 45-50 cm in length. The plates are wooden hard sheets placed at right angles to one another forming lattice work. The joints between the plates are firmly screwed. In between the plates, are the *folders, blotters* and *ventilators* (*corrugates*). The folders (moisture absorbing sheets such as newsprint) are of similar size as the press and serve the purpose of temporary files for the collected specimens. The blotters are the absorbent smooth-faced corrugated card boards of desired thickness. The assembled plant press is tightly bound together by strong and long straps of leather or canvas. The beauty of the herbarium sheet mainly depends on the care in the method of pressing. The plants should be placed in the folders in such a way that they appear natural as far as possible. If the specimens are larger, they are bent into a V, N, M or W shape which is maintained by paper slips inserted over the bent portion of the stem. Every care is to be taken to avoid breaking of the plant near the bending point. During the arrangement of the specimen on the folders, the leaves should not be crowded. The bulky organs are trimmed on their back and a cotton pad is slipped on the side for proper pressing. The folded and ugly leaves are to be removed and the corolla should be properly spread over. The uneven parts of the specimen are uniformly pressed with the help of cotton or blotting pads. Some parts like the leaves are to be turned back so that both the surfaces (upper and under surfaces) are visible in one and the same view.

There are 3 types of plant presses that are in common use. Commonly, a plant press consists of a pair of wooden planks (30 x 45 cm or 12 x 18 inches) made of shisham, teak or any other wood of a good quality. 4 strong bolts (20 cm long and 2 cm in diameter) are vertically fitted at the 4 corners of the lower plank. The upper plank (with 4 holes at the 4 corners equal to the size of the bolts of the lower plank) so that it allows the bolts to pass through the holes which are tightened with the help of the nuts. The pressure over the specimens is adjusted (either increased or decreased) by tightening and loosening the bolts as per requirement.

The second type of plant press consists of a pair of wooden or metal frames (the size of which is same as mentioned above) and the specimens are kept within the folded sheets of paper kept on the blotter and also above them. These are tied by means of leather straps or stout cords that pass over the frames. The straps are tightened in such a way that all parts of the specimen remain flat and also they are close to blotters which help in removal of the moisture from them. The specimens are kept within the folded sheets of blotters/newsprint and are arranged one above the other within the frames and strapped or screwed up tightly to remove the moisture from the fresh plant parts and to make them crisp/dry.

In case, the plant press is not available, the blotters containing the specimens may be kept under a box with a weight of about 20 kg and this may be gradually be increased up to 50 kg. If the plants are hydrophytic and fleshy, it is better to boil the specimens in water before transferring them into blotters which avoids blackening of the specimens.

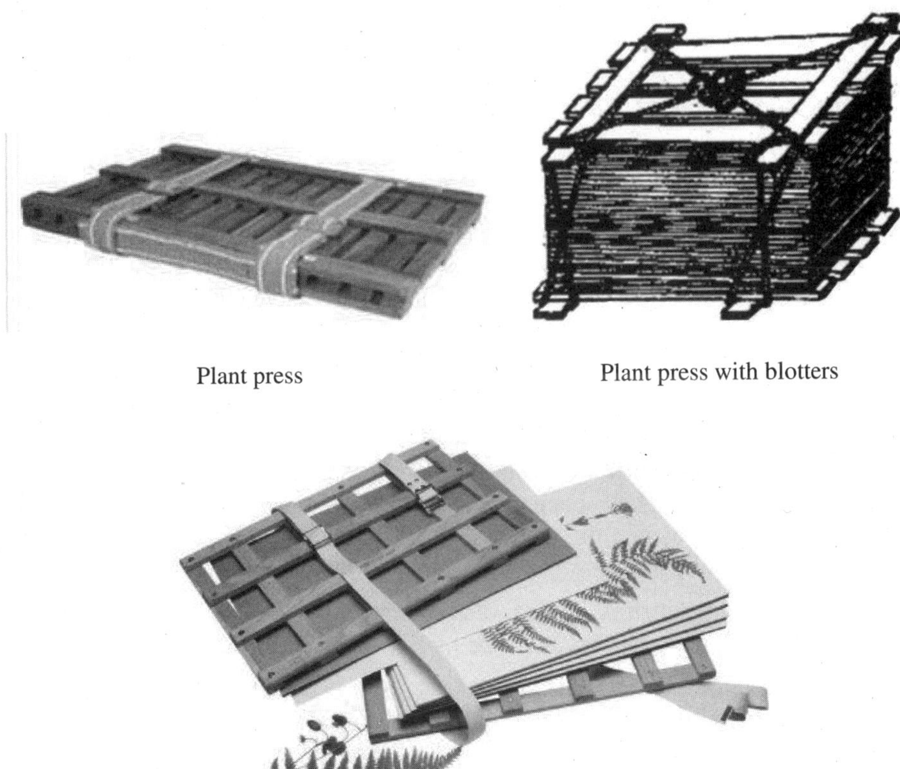

Plant press

Plant press with blotters

Plant press with specimens

g. Spare Blotters/Blotting Material

Several spare sheets of folded blotters/newsprint for use in the plant press should be kept readily available. The preference regarding the use of blotting material differs. Always the best quality blotters are selected which absorb a given volume of water/moisture in a shortest period. Such blotters not only absorb the moisture quickly from the plant that is pressed in the plant press but also get dried up more quickly for further use. The thickness is also very important and thicker and heavy blotters are least efficient. Blotting paper sheets and newsprint are most commonly used and those which are heavier than the newsprint are to be avoided as they fail in allowing proper drying of the plant specimens. The standard size of the folded blotter should be 30 x 45 cm. The newsprint is cut length-wise into 2 halves and each half sheet is folded and cut at one end not exceeding the length of the plant press. Every care should be taken while placing the specimen in the folded blotter that no portion of the blotter/specimen is outside the press. The blotters should be changed every 24 hours, dried and reused. The changing of the blotters depends upon nature of the specimen (with more or less amount of moisture within it) and the humidity in the surroundings. In case of fleshy parts of the plants and during the rainy season of the year, the specimens should be replaced at shorter intervals.

h. Forceps, Long Rope, Pencil

Pair of forceps (different sizes) is always essential for opening the flowers and study the floral envelopes and essential organs (stamens and carpels) of the flower for easy identification. Sometimes, during the collection, a long rope is needed to draw out and collect the aquatic specimens from ponds/lakes as also the parasites (along with the host stem) from the top branches of host trees. It is always better to use a pencil to record the notes rather than a pen. The pen is not suitable in case the papers in the field book get moistened.

METHOD OF PLANT PRESSING

The most important step in the preparation of the herbarium which needs careful attention is arranging the plant specimens within the pressing paper. Before pressing, the specimens should be carefully selected. The following points are to be taken note of during pressing the specimens.

1. If the specimen is an herb, the underground parts of the plant like root system, rhizome etc., are to be included.
2. As the sterile specimens (without flowers and fruits) are useless, one must make sure that the twigs are in flowering and/or fruiting stage.
3. The specimens should be free from insect pests, fungal or bacterial infection or any other pathological symptom.
4. All parts of the plant are to be arranged within the pressing paper. The art of pressing comes only from experience.

a. The maximum surface of the pressing paper is covered with the specimen to be pressed. Only one plant is to be placed on one sheet.

b. A single specimen should be pressed in a folded pressing paper. In case of a specimen with larger foliage, it may be cut into 2 or more pieces and each piece is arranged and pressed into 2 or more number of folded papers.

c. If the specimen is a small herb, many specimens may be pressed in one paper.

d. If the herbs are longer than 40 cm, they are to be folded *in toto* in the shape of V, N or M- before pressing. The points of the folds should be moderately macerated before bending the stem to avoid breaking (in plants with brittle stems).

e. To avoid overlapping of the plant parts, pruning is essential. Every care is to be taken to leave the basal portion of the pruned parts of the plant (such as petiole, pedicel etc.) attached to the plant to indicate their natural position. The care is to be taken to see that at least 2-3 leaves are left attached to the stem to account for the phyllotaxy (arrangement of the leaves on the stem).

f. Some of the leaves should be arranged in such a way that their under surface faces upward.

g. In case of large and pinnately compound leaves, it is necessary to excise all the leaves except one. If the leaf is still too large, the pinnae (leaflets) of one side may be removed leaving the terminal leaflet as it is (if present).The complete leaf (single) can also be split lengthwise so that only one half of the leaf is pressed.

h. Larger pinnately compound leaves are commonly split into half (lengthwise) and the remaining half is discarded during pressing.

i. If the herbs are exceptionally tall, they are pressed in 3 parts, namely, one-third from the base, the middle portion and the remaining one-third from the top indicating their position on the paper.

j. Sometimes, it is not possible to accommodate more than a single leaf and the inflorescence within a single herbariumn sheet. In such cases, the phyllotaxy is noted down on the paper. The complete petiole and portion of the stem from which it is produced is to be included.

k. In specimens possessing gamopetalous corolla, a few flowers should be pressed separately. Some of the flowers are split open and pressed to expose the essential organs (stamens and carpels) and also the nature of thalamus (hypo-, epi- or perigynous).

l. Before pressing, the roots and the underground portions of the plant should be thoroughly washed to remove the soil particles attached to them.

m. If the underground organs are fleshy (tubers etc.), they should be sliced open to facilitate quick drying. Paper padding or cotton wadding around such organs helps in quick drying.

n. The blotters/newsprint should be changed frequently to avoid discolouration and moulding depending upon the nature of the specimen and the humidity in the surrounding environment.

o. Some plants belonging to families Iridaceae and Commelinaceae etc., produce flowers with deliquescent perianth/corolla that quickly become semi-liquid and these often stick to the pressing paper and become so thin that they cannot be easily removed after drying. In such cases, a layer of absorbing cleaning tissue paper is put over and beneath the flower before pressing. The tissue can be easily peeled off when the specimen becomes dry.

p. Sometimes, the parts of the specimen need holding in a desired position which is not possible when more than a pair of hands is needed and the collector is alone. In such a case, for holding the parts, wet newspaper strips (2.5 – 10 cm) are used which can be removed after the specimen gets dried in the plant press.

q. If the portions of the plant to be pressed are bulky, the newspaper pads (of the same thickness as that of the bulky portion) are placed around them to provide a uniform pressure on the foliage and flowers. Some botanists use pieces of sponge-rubber (2 cm) but, the use of newspaper is cheaper. The use of cotton pads between the layers of the newspaper has been suggested by Fosberg (1939) which gives a very good result especially in case of fleshy and bulky plants.

r. Many aquatics which are filmy and filamentous are difficult to be arranged on the pressing paper if taken out of water. Lawrence (1951) has suggested a method to float unprinted pressing paper on the water surface till it is completely wet. The floating plants are then slid over the paper and the paper is lifted carefully in a sloping manner allowing all the water to drain out. Later, the wet paper along with the specimen is placed in the paper press to be pressed. When it is dry, the mucilage produced from the specimen may cause it to adhere to the paper and the complete sheet is mounted as it is without making any effort to remove the aquatic plant or its parts.

METHOD TO KEEP WET PLANT MATERIAL IN ADVERSE CONDTIONS

In tropical regions particularly during the rainy season when the atmosphere is more humid, the collectors feel it difficult to press the plant or its parts in the plant press. The specimens are liable to be spoiled and decomposed if there is no proper arrangement for drying. Two special techniques have been advocated by which the specimens may be pressed without being spoiled or decomposed and dried after sometime when the conditions are congenial.

According to the technique proposed by Schultes (1947), the specimens are pressed between the blotters in a plant press for 24 hours. Later, they are taken out and dipped into a solution of 40% formaldehyde (2 parts) and water (3 parts) for few seconds. After dipping, the specimens are transferred into pressing papers and all the sheets are piled up one above the other without blotters or corrugates in between. After a few hours, the entire bundle is then sealed inside cellophane or several layers of thick rubberized cloth bag. Finally, the specimens are removed and dried when convenient.

The specimens can be kept in wet condition even for a month, and if for any reason, the drying is delayed, they may be taken out and dipped again in commercial formaldehyde-water solution after a month. While working with this solution, to prevent cracking of the skin on the hands, rubber-gloves should be used. The plant specimens prepared by this method are of an inferior quality as they are prone to lose their natural colour and beauty.

An improvement over this method was proposed by Fosberg (1947) wherein Schultes` solution was replaced with 1 part of formaldehyde and 2 parts of 70%

alcohol, and this solution was applied with flat brush (5 cm) instead of dipping the specimen into the solution. This was found to be a better control from decomposition as the solution penetrates into the specimen. The substitution of 40-50% alcohol as a preservative was observed to be more satisfactory by Hodge (1947) and the use of hydroxyquinoline sulphate as a preservative was advocated by Moore (1950). Mercuric chloride (15 g) mixed with 70% alcohol (1 litre) has been found to serve as a good preservative. The specimens are dipped into this solution for a few seconds and then shifted to the pressing papers. The disadvantage is that the solution leaves a blackish mark on the pressing paper which cannot be avoided. But, the treated specimens neither get spoiled nor got decomposed, and are also well protected from insect damage.

METHODS OF DRYING THE PLANT SPECIMENS

After pressing the specimens in the press, the press is locked for about 24 hours. Later, the press is unlocked, blotters are removed and the papers are turned back after which the specimens are rearranged. When the parts have become flattened, it is easier to rearrange all the parts in the desired manner which results in a nice herbarium specimen. These papers are once again kept between fresh blotters and are locked again for the next 24-36 hours. Once again the blotters are changed and the specimens are locked for the next 48-72 hours. If the material is fleshy and succulent, it may take a much longer time to dry but mostly the specimens get dried up within a week. This method also has some disadvantages like: the drying requires a much longer time, blotters are to be constantly changed during the process (3-4 times) and the wet blotters are to be dried in the sun to be used once again. A larger number of blotters are needed for drying and a lesser number of specimens can be processed. The specimens are liable to be infected by fugal pathogens and larvae if present in the flowers may eat away some of the essential parts.

WHAT MAKES A GOOD SPECIMEN?

The plants should be pressed as soon as possible after their collection and before they wilt. The collected specimens should be pressed flat and are to be dried between the sheets of absorbent paper (newspapers, blotters etc.). The entire lot of the specimens can be strapped together in a plant press or they may be stacked carefully with a weight on them.

To be accurate in identification, the collector should need to know where the plant was collected and the distance from the town along what road etc., and whether the plant was growing in full bush or semi-urban area, pasture or in a garden. For the wild plants, what sort of location was it – a wet land, dry/scrub, forest, rocky slope or a river area, and the other plants around it. The altitude of the place of collection, rock-type, soil-type, and latitude and longitude are also to be noted. The appearance of the plant collected is to be noted, and was it a tree, shrub, prostrate shrub or a herb or a climber is to be noted. In addition, the colour of the flowers when fresh and the colour and the texture of the bark and whether it is a parasite and if so, the host on which it was observed is also to be noted. The collected samples of about 30 cm long are adequate and the sample should necessarily include buds, flowers and/or fruits as well as a piece of stem bearing typical leaves. If the shape and size of the leaves is highly variable on the plant, separate pieces from different parts of the plant are to be taken to represent the variation. For some species, the fruits/seeds are essential for a positive identification. For some groups like eucalypts, acacias, herbs, grasses, aquatic plants, ferns, algae, fungi, lichens and liverworts, the collectors require special information as they are difficult to be correctly identified. Good data is always essential for the collected plants as it helps in the identification of new species when they turn up. Such specimens are scientifically valuable and can be incorporated into the herbarium.

THE DATA TO BE RECORDED IN THE FIELD NOTE BOOKS

The herbarium sheets generally carry all the data from the field and also carry the original plant specimens. Hence, it is the ultimate reference point for the associated data and for all purposes. The herbarium specimens serve a variety of purposes in taxonomy, ecology and horticulture. The ecological data is very important in practical taxonomy in describing and naming the taxa. The ecological data associated with the specimen like its geographical location including the altitude or depth at which it has been collected; physiography or topographic position like top of the mountain, sea shore, lake, sand dune or alluvial plain; the substratum like parental rock, soil, bottom of the lake or upon the buildings; type of vegetation such as dominant locally, structure of the species or the community type; type of the environment like water, rock, trees, litter etc., also play a dominant role in providing names of the plants in taxonomy. All the information is also essential in determining the necessary requirements of living plants at all the stages from propagation in the nursery to shifting into the garden in horticulture.

The important criteria that limit the type of data to be collected in the field note book are:

The data must be simple enough to be understood by people without any ecological training; the data must be easily reducible to numbers for computerization which requires a strict classification of data, and simple logical terminology; and it must be easily interpretable in themselves i.e., by reference to the herbarium sheet alone because the sheets will be used by outsiders also. The terminology must be all-inclusive and all the descriptive terms should be needed to express all the data. The description should elaborate the plant`s peculiarities in the given notes.

The data recorded should not be misleading and it should be always from the general to specific. For example, the substratum should be as – rock and sandstone (rock is a general substratum and the sandstone is specific). Similarly, the immediate environment of the plant should be recorded as – topography is to be recorded as hills with a steep slope and location must be – say 50 km NE from .... and 0.6 km along an east bank from northern end. A soft pencil (preferably HB or F) should be used for recording and all the measurements must be metric.

The data to be recorded in the field note book (under 15 headings) must be as follows: always one page of the field note book should be used for a single collection. A single collection is generally defined as the material taken from one plant on a particular day of the calendar. Hence, a herbarium specimen, cuttings, and seeds, all from the same tree on a particular date (say, 14th February 2014) comprise a single collection. If a further specimen taken from the same tree a week later does not belong to the same collection, it must be recorded separately. But, in the case of very small plants (annual herbs) which do not provide enough material, several plants from a single location or population are to be regarded as a single collection. The transplants (whole plants) may also be grouped with a voucher from another plant as a single collection.

Registration Number (1)

The space after the heading in the field note book (Flora of .... (a place or a locality like State, District, City, Mandal or a Village), Department of Botany .... [Name of the College/University/Institution]) is intended for the registration number (1) of the specimen and this is to be added later when the field note book is registered but not to be entered during the field trip.

Collector Name and Collector Number (Coll.) [2]

Here, the name of the collector and his personal collecting number should be entered. The name must have all the initials, and with the full surname. This must be a single number series starting from 1 increasing throughout your collection. During writing the number on the tags attached to the specimens, it is always preceded with the full initials of the collector for example, PSN 5650 as the tag is the only means of tracing a specimen to its filed note book. One number is always assigned exactly to each collection. If one gets assistance from other collectors, their names are to be listed after the main collector`s number. The acknowledgement of assistance in collection is only a courtesy during now-a-days. e. g: T. Pullaiah 1982 (tag – TP 1982); T. Pullaiah 2873, D. A Moulali & P. S. P. Babu (tag – TP 2873).

Date (3)

The date always refers to the date of collection and the standard way of expressing it is: 14 FEB 1985 i.e., with the numbers for the day and the year and three letters, upper case, for the month.

State and District (4)

The state in which the collection is made should be indicated by one of the standard symbols like, AP, HP, UP, TN, KL, PO, MH, DL and KA etc. The district is indicated by the standard two letters (lower case) in the field note book Each state has been divided into a number of districts for planning the field work and simplifying the descriptions of the locality.

Locality (5)

The locality of the collection must be recorded precisely, preferably to the nearest $1/10^{th}$ km or at the worst the nearest 1 km, from, a precise and a well-known map point. If it is a remote locality, its relationship to a better-known place should be given. Now-a-days latitude and longitude gives correct location of the place where the plant is collected.

Topography (6)

The details of topography recorded in the field note book during plant collection give botanists, ecologists and horticulturists a general insight into the ecosystem to which the plant is adapted. To record topography adequately, three levels of details are necessary such as landscape, landform and the position on the landform.

The landscape is a description of the general topography of a region, and it can be determined from a map. It is always necessary to be entered in the field to place specific topography in context rather than being determined from a map. The details are represented as *coastline* (the land adjacent to the sea), *plain* (the level land of low altitude), *undulating land* (rolling land with slopes up to 8 degrees), *hills* (the slopes in range from 8 to 20 degrees), *mountains* (the higher and generally steeper than hills) and *plateau* (elevated plains).

The landform is a description of the more localized features of an area and it is expressed by type (a noun) and in most cases a qualifier (an adjective or a descriptive phrase). The qualifier can be magnitude or simply further description and the magnitude should be specified as often as possible such as the size for depressions and rises, degrees of slope, size of the river etc. Further descriptive qualifiers should be added as required like eroded slope, rocky outcrop or isolated hill etc. As the landform varies considerably and is not easy to categorize, the following terms with qualifiers should be used to give adequate coverage as per the needs, like *depressions* (valley, gully, depression or sinkhole), *rises* (peaks, ridges, hills, mine dumps, roadside embankments, railway embankments or channel embankments), *slopes* (cliffs, alluvial slopes – along with the degree of rise such as: up to 2 degrees [gentle], up to 50 degrees [moderate], up to 150 degrees [steep] or above 150 degrees [very steep], and descriptive qualifiers [steep rocky slope] can also be added), *artificial slopes* (roadside or railway cuttings), *flats* (desert, pavement, floodplain, river terrace), *waterways* (river, stream, creek, canal, spring or channel), *water bodies* (lakes, lagoons, swamps, marches, bogs, seepage areas or reservoirs), and *coastal areas* (bays, islands, tidal channels, estuaries, coastal cliffs and coastal hills or coastal dunes).

The position on the landform gives the specific location of the specimen and it is described as follows: 100 m up 200 m vertical cliff, the edge of the stream, upper steeper slope or the midpoint of steep spur.

Aspect (7)

The direction in which any piece of land is sloping gives valuable information on the micro-climatic conditions on that slope, especially in relation to the amount of the sunlight received. This should be specified to the nearest 45 degrees, such as N, NE, E, SE, S, SW, W and NW.

Substratum or Substrate (8)

Substratum or the substrate is that medium in which a plant is fixed and in most cases it will be soil, but it can also be rock, host, water or artificial environment.

In the case of the parasites, try to record the type of the host and the position of the parasite on the host (a root, stem, a branch) and also whether it is a complete or a partial parasite. The details related to host`s substrate are also to be added. In the case of epiphytes, try to record whether it is aerial or terrestrial and the type of the host and the position on the host along with the details of host`s substrate.

In the case of watery environment, try to record the nature, such as *inland-aquatic* (whether the submerged plants are fixed to the bottom of the stream or the lake and if so record the type of the surface to which they are attached), marine (whether the submerged plants are attached below the high water mark and if so try to record the type of the surface to which they are attached) or free floating (try to record whether the plant is at or below the surface, inland aquatic or marine and also record the rate of movement of water and the degree of salinity).

In the case of artificial environment, try to record the type of the substrate whether it is a brick wall, fence post, mine dump or concrete etc.

Vegetation (9)

In vegetation, mainly two variables are to be recorded namely, the structure and dominant vegetation.

The structure is determined from the life-form, height and the density of the tallest level (stratum) in the vegetation. These are: trees (5-10 m in height or 10-30 m in height and trees above 30 m in height), shrubs (0-2 m in height or shrubs 2-8 m in height), and herbs. The disturbed vegetation may be regenerating or secondary (logged, overgrazed, burned, cleared, inundated, diseased or droughty) or artificial (crop, planted forest, garden, lawn or weed community).

Every care is to be taken in selecting the dominant plants of the given area. The descriptive terms such as dry or wet sclerophyllous forest, rainforest, dry deciduous forest and scrub jungle etc, are used.

Latitude (Lat.) and Longitude (Long.) (10)

The locality is expressed in latitude and longitude coordinates and in descriptive terms in herbaria. This allows people not familiar with the region to place the locality on a small scale map and also allows direct comparison between the localities. The coordinates are to be given to the nearest second (in a GPS) or a minute. e. g: Lat.: 45^0 28` 41" S and Long.: $142^0$ 19` 32" E.

Altitude (11)

The altitude is given in meters and the precision should be to the nearest 10 m but in mountainous country it should be nearest 50 m while in the case of the coastal region, it must be nearest 1 m.

Habit (Description of the Plant and Notes) (12)

During description of the habit, the features of growth form and size which cannot be seen on the specimen or a distorted specimen are to be indicated.

The growth form of the plant is described as a tree, shrub, under-shrub, herb, fern, parasite, vine (liana or liane), grass, sedge, rush, rosette tree, rosette shrub, succulent shrub, mat plant or cushion plant.

The habit is described by a variety of qualifiers like thicket-forming and clonal; straight, crooked, slender and robust; open, dense and diffuse; many/few/long/short-stemmed; divaricate, drooping, erect, spreading, whorled and intricate stems/ branches/branchlets; much/few-branched, erect, leaning, decumbent, procumbent, prostrate, ascending and pendulous; weak, trailing and straggling; cylindrical, rounded, pyramidal, spreading and weeping crown; tufted; and shortly/extensively/ insiduously creeping, rhizome, rooting at nodes etc.

With reference to size, always record the height of the plant unless the whole plant is taken as a voucher. If one wants to record the breadth, diameter at breast height (dbh) or other measurements, "tall, broad and dbh" should be written. For example, a straight tree, long-stemmed, spreading crown, 25 m tall, 1 m in dbh; a fern, shortly creeping rhizome, 70 cm; pendulous woody parasite, 2 m; decumbent open shrub, 40 cm tall x 200 cm broad etc.

The section for notes is reserved mainly for the additional features which are not present on the specimen or lost during processing and preservation of the specimen, which applies mainly to note the type of the bark and colour of the flower etc. Anything noteworthy not covered in the categories mentioned should be mentioned here such as "female plant is DVL 1508" etc.

Type of the Material, Abundance and Duplicates (13)

With reference to the herbarium, the type of the material that makes up the collection, are ticked.

Sheet – the pressed plant specimen (one should always be collected),

Spirit – flowers etc., that have been put in 70% alcohol,

Fruit sep. – bulky fruits destined for carpological collection,

Image – photographs or digital images or line drawings, and if more than one image is to be included, write the name,

Dissect – floral card with the dissected floral parts as in orchids,

Cryptpkt – the specimens to be filed vertically in separate envelopes, usually mosses, liverworts and lichens,

Cryptbox – bulky cryptogam specimens that require protection in a box such as fruiting bodies of fungi, and

W`block – wood and bark samples.

With reference to the information about the material (living) collected to propagate, tick the relevant items, such as cuttings, plants, seeds, spores, bisexual, male or female flowers.

Abundance (dominant, common, frequent, occasional or rare) refers to the expression of the commonness of a species or population in the vegetation in which it is found. In the field note book, it represents the abundance of population which the specimen represents that is to be recorded.

Phenology refers to the collection of reproductive and vegetative state of the specimen collected. Tick the relevant state such as – buds, flowers, fruits, seeds, fertile parts, sterile parts and bark.

Name (14)

Normally, four levels are provided for recording the name and taxonomic position of the specimen namely, group, family, field name (botanical name as far is known) and determined name (to be determined by a botanist and is not to be recorded in the field).

Determined by (Det.) (15)

During the entry of the determined name (always not in the field) one should write the name (initials and surname) of the botanist who made the determination, followed by the determination.

Finally, in the case of a doubt, the more general terminology must be recorded rather than erroneous information in the field notes. In case of more than one collection at a given locality, the data is to be repeated from one page to the next, and the repetition must be indicated by leaving the category blank. Every page of the field note book is to be filled out as the blanks are interpreted as repetition. In case, if one is unable to record the data for a given category for some reason, "no data" (ND) is to be entered, though the category is relevant. In case, if a category is irrelevant in the given locality, "not applicable "(NA) is to be entered. Sometimes, the situation cannot be described by the standard terms, and in such a case, it can be described by some simpler and less general terms, and, if the situation is unique which can occur very rarely, it can be described in own terms in the "notes" category. These must always go from the general to specific.

MOUNTING OF THE DRY PLANT SPECIMENS

The mounting of the specimens is the process of attaching the dried and pressed plant specimen on a suitable hard standard paper sheet (herbarium sheet) on one side. Only one plant is to be placed on one sheet. The specimens are mounted on a herbarium sheet of the size, 11.5 x 16.5 inches or 29 x 42 cm. The sheet must be long lasting and durable and should be heavy enough to support the specimen attached to it. A 100% rag paper is the best though it is expensive (used by curators of the most of larger herbaria). A paper of lesser rag content which is cheaper and easily available can be used but it is not advisable as it easily bends under the weight of the specimen which is liable to be broken. Hence, thick drawing sheets are used. The mounting is generally done with the help of glue/paste (fevicol). Glue-coated (unstained) white paper strips are generally used for sticking. The unstained glue may be directly applied to the back of the specimen by placing it flat on the surface of a glue-coated glass plate (36 x 51 cm) and later shifting it on to the herbarium sheet carefully with the help of a forceps. The excess of the glue/paste can be removed by placing a pressing paper (unprinted newsprint or a blotter), gently pressing it over the specimen, taking off and rejecting it. The glue can also be applied at the back of the specimen with the help of a brush before transfer. Sometimes, the strips of the adhesive tape (it is best to avoid because the tape does not remain permanent) or ordinary thread or a string may also be used for mounting. The specimens are labeled by affixing the labels at the bottom of the sheet on the lower right hand side near the corner. The fruits and seeds after collection are placed in paper/polythene packets/envelopes, properly sealed and mounted at one corner of the herbarium sheet with pin. During drying, the specimens become very brittle, so careful handling is required to avoid crumpling and cracking of the parts. The use of quick drying liquid plastic as a good medium for mounting the specimens has been advocated by Andrew and Archer (1950) and Rollins (1955). During the preparation of this plastic, ethyl cellulose and resin are dissolved in a solvent (methyl alcohol and toluene) to give thick sticky syrup which is used for mounting. This liquid is sprayed on the lower surface of the dry specimen either with wash bottle (with nozzle) or squeeze bottle (with slim nose) or a spray gun (pressure-oil gun) and placed on the sheet which dries very quickly. With age, this may become yellow but it will never crack. This is a very quick method for mounting the specimens on to the sheets.

Mounting a specimen – Fern

Mounting – a Flowering plant

LABELING THE PLANT SPECIMENS

The herbarium sheet is given a final touch following mounting of the specimen by pasting the labels. The labels are very important and they are pasted with glue on the lower right hand corner of the herbarium sheet. Each label is 3-5 inches (7-12 cm and may vary by 1-2 cm on either side) long and must include all the information related to the growth pattern and the name. The top of the label should contain the name of the state, province or the country from where the collection is made, name of the plant family, local name of the plant, botanical name of the specimen (genus and species with authority), locality or area of collection, habitat, name of the plant collector, field number, date of collection, followed by an additional information about the habit, distribution, association with other plants, flower, branching, height, nature of bark, altitude, geographical information, soil characters etc. The labels can be hand-written (use of water-proof black is preferred and should be legible), type-written or printed.

Specimen Label

Scientific Name: ....

Name of the Family: ....

Common /Vernacular Name (in English): ....

Location of the Specimen:

State: ....   District: ....

Mandal: ....   Village: ....

Habitat:

Latitude:..........................   Longitude: ....

Physical characteristics of the Collection site: ....

Biological characteristics of the Collection site: ....

Description of the plant:

Herbaceous or Woody – Herb/Shrub/Tree – Annual/Biennial/Perennial – Erect/Vine

Plant - Height: ....   Flower – Colour: ....

Other important characters: ....

Collector`s Name: ....

Collection Number: ....   Collection Date: ....

FILLING OR ARRANGEMENT OF THE PLANT SPECIMENS

This refers to the arrangement of the specimens in the herbarium or placement of pressed specimens according to a classification system. The pressed specimens are to be handled with special care and are placed in tight steel cases provided with the pigeon holes. Alphabetical arrangement is done for genera and species in small herbaria. In larger herbaria, the specimens are arranged in a phylogenetic sequence, and are placed number-wise on the basis of their phylogenetic position to reveal the interrelationships between the species. Preparation of the index is an essential part and most common is an index containing the family names. Some herbaria use technicolour folders to depict physiographic areas, and the folder colour indicates the distribution of the plants.

SPECIAL CARE OF THE PLANT SPECIMENS

The specimens in a herbarium can be stored indefinitely by taking special precautions against exposure to moisture, insect attack, fungal attack etc. The specimens are to be poisoned by coating them with mercuric chloride solution. Sometimes, the pungent para-dichloro-benzene (PDB) can be placed in the racks to avoid insect attack. So also, insecticides like dichloro-diphenyl-trichloroethane (DDT), carbon-disulphide (care is to be taken as this is inflammable) and cyanide gas (care is to be taken as this is poisonous) are also used. It is always better to use those chemicals which can repel the insects rather than powerful insecticides. Excess pressure during handling may result in the fragmentation of the parts/specimen and if so, the fragmented parts are kept in a separate envelope and placed along with the herbarium sheet. Every care is to be taken to save the herbarium from fire and other natural calamities. The type specimens are to be stored in steel cabinets and in fire-proof buildings.

SAFE STORAGE OF THE SHEETS/SPECIMENS

These specimens are stored in wooden or steel cabinets/almirahs in a separate room (herbarium room) for safe deposit and future record. The almirahs are specially designed with 2-3 tiers of pigeon holes, each 48 cm deep, 33 cm wide and 21 cm high. The almirahs are air-tight and dust-proof and mainly protect the specimens from the insects. The specimen sheets are grouped (stacked) by the species and placed in a large light-weight folder with label on its bottom edge.

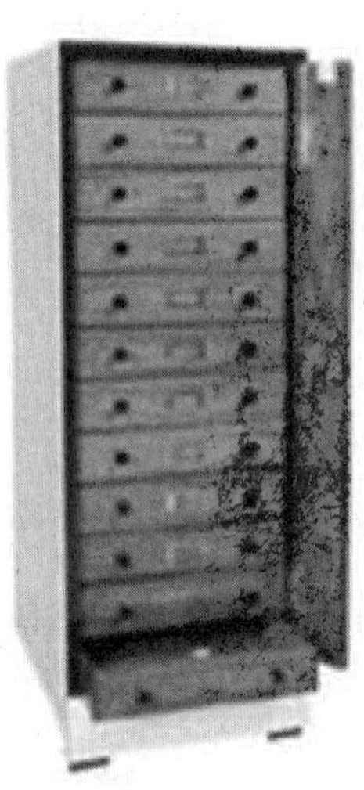

Herbarium Cabinet

Green Plant Herbarium – Safe storage in air-tight cabinet

Such groups are placed into larger heavier folders according to the genus. These are sorted by family according to a well accepted and standard international system of classification. Most of the herbaria in the world follow the system of Benthan and Hooker or Engler and Prantl. It is customary to arrange the genera and the species in an alphabetical manner while in some larger herbaria these are separated by the continent to which they are indigenous. Finally, these are placed into pigeon-holes in the cabinets. The modern herbaria maintain an electronic database of the collections. Some herbaria are virtual where the specimens have been digitized.

SAFE PRESERVATION OF THE SPECIMENS (PLANTS)

Because of the likelihood of being attacked by the insects (herbarium beetle, drugstore beetle and booklouse) and fungal pathogens, extreme care is taken in the preservation of the specimens in herbaria. The insects complete their life cycle within the dried specimens and eat away the plant parts damaging the specimens and the herbarium beetle is the worst of all which completes its life cycle within 2 -3 months. Hence, insect repellants and insecticides are used to store the specimens for a long time. The insect repellants (naphtha cakes/balls) and PDB (para-dichloro-benzene) prevent the attack by driving them back and also checking their advance towards the specimens and are neither lethal nor toxic to insects. According to Merrill (1948), 2 parts of naphthalene compounds mixed with 1 part of PDB (2-3 ounces of this mixture is kept in a muslin cloth bag and kept in the pigeon-hole almirah and are refilled once a year) will be more effective to prevent insect attack. The insecticides kill the insects either by coming in contact with the insects or digested by them. The common insecticides which are in use are: cyanide gas, PDB (para-dichloro-benzene), carbon-disulphide gas, DDT (dichloro-diphenyl-trichloroethane), mercuric chloride, mixture of ethylene chloride (3 parts) and carbon-tetrachloride (1 part) etc.

Cyanide gas is extremely poisonous and is most effective. The specimens are fumigated with this gas in an air-tight chamber. Cyanide gas is exclusively used at Kew. The specimens in an air-tight chamber are exposed to the vapours (at 43 – 60^0C) of PDB (more a repellant than an insecticide) for 3-4 days to kill different types of insects that cause damage. Carbon-disulphide is highly inflammable and volatile (10-15.5^0C) and the specimens are fumigated in an air-tight chamber for 2 or 3 days to kill the insects and their eggs. Every precaution is taken that the specimens don't catch on fire. The specimens can also be fumigated with a mixture of ethylene-dichloride and carbon-tetrachloride (3:1 by volume) in an air-tight chamber which is volatile at room temperature and also most effective. According to Howard (1947), dusting of the specimens with powdered DDT (25-50% or 100%) after taking out from the plant press is most effective. The dusting is to be repeated after 1-2 years. A saturated solution of mercuric chloride in 95% alcohol at room temperature (or a mixture of 2% mercuric chloride added to petroleum ether [1 part] to 95% alcohol [2 parts]) is a very strong poison for many insects that attack the specimens. 1 part of the stock solution is mixed with 1 part of alcohol and the dry specimens are dipped into it and are kept in the blotters for 24 hours before mounting. This is not advised for the mounted specimens as mercuric chloride can dissolve some of the plant pigments leaving a black residue on the sheets.

Some herbaria store their type specimens in a special cabinet or use coloured labels on them, but are often interspersed within the collections. These specimens are kept in cellophane covers and preserved separately in special cases with proper labeling. According to Lawrence (1951), if the herbarium is not fire-proof, the type specimens are to be removed and kept in a place where they are protected from all sorts of damage, including fire. During examination, the sheet is to be lifted with both the hands so that it will not bend to prevent breaking of the specimen due to its brittle nature. The sheet should not be turned over like the page of a book to prevent damage.

USE AND SIGNIFICANCE OF THE HERBARIUM

Herbaria are very significant for those people who are actively engaged in the taxonomic work and comprise the collections of dried, pressed and nicely preserved plant specimens. These dry specimens are used for the identification of fresh specimens, comparison between the floras or pollen morphological studies. Herbaria are very essential to study plant taxonomy, geographic distribution of plants and stabilization of nomenclature. The specimens in a herbarium may be useful to catalogue or identify the flora of an area. Similarly, a larger collection of specimens from a single area or locality can be used to write a field guide or a manual to help in the plant identification of that area. With a larger number of specimens, one can understand the variability of form and their natural distribution. Herbarium is also useful in preserving historical record of change in the vegetation over time. The specimens represent the only record of original distribution of plants in case of plants that have become extinct. This also helps to track the changes in the climate as well as human impact on plants. The determination of plants for cultivation is solely dependent on herbarium specimens as it is impossible to procure the living material in a laboratory from the natural fields or forests whenever desired. The herbaria also serve the purpose of a repository of viable seeds for rare species, and also prove very useful as sources of plant DNA to be used in taxonomy and molecular systematics. It is always necessary to preserve the dry specimens for reference. These specimens are available for reference whenever required.

ROLE OF HERBARIUM IN MODERN PLANT TAXONOMY

The herbarium techniques have been proved to be excellent material for pursuing many investigations with the modern aspects of plant taxonomy. Just a glance into the herbarium may acquaint a botanist with the flora of distant areas. The dry specimens help in correct identification of unknown plants and plant material that is continuously collected from various parts of the world. The national herbaria which work on a wider range of taxonomic and floristic projects are extremely valuable economically. The herbarium also serves as a major adjunct to population studies with the geographical limitation. The dry material is of much significance in carrying out important anatomical investigations. The herbarium is a wealth of pollen material and serves as a source of material for studies of exhaustive floras like pollen flora of Indo-gangetic plain on the basis of Duthie`s collection. With the material in the herbarium, the variations in time and space of plants can be studied which may throw some light on evolution of plants. The herbaria provide a place for the voucher specimens to be preserved for verification of the data for cytological investigation whenever required. Similarly, with the help of the herbarium the geographical variations can also be analyzed numerically.

■■■

CHAPTER - 6

CHROMATOGRAPHY

Introduction

The separation of chemical components is vital in any type of chemical analysis. When one is trying to identify any unknown substance, the sample must first be simplified as much as possible into its constituents. The unknown can then be characterized by individual identification of its parts. This does not imply that the separated chemical components are recovered after the separation and analyzed. Usually the analytes are irretrievable. The separated compounds are compared to the known standards. It is always important to have an idea of what compounds are being searched for in the first place.

Chromatography accurately describes what happens to ink on a wet paper? Literally, it means, `*colour writing*`. It is a misnomer (*kinetic partition* would be a better term) because it often does not involve colour, paper, ink or writing. It is actually a way of separating out a mixture of chemicals in the form of a gas or a liquid. With the ink and paper trick, we have a liquid (represented by ink) dissolved in water or any other solvent creeping over the surface of a solid (represented by the paper). The essential thing in this technique is that we have some mixture in one state of matter (a gas or a liquid) moving over the surface of something else in another state of matter (a liquid or a solid) that stays where it is.

Chromatography studies the separation of molecules on the basis of their differences in structure and/or composition. In general terms, this involves *in the movement in the preparation to be separated (test preparation) over a stationary support*`. The molecules interact differently with the stationary support which leads to the separation of similar molecules. The molecules that display tighter interaction with the support tend to move more slowly than those with weaker interaction, as such, different types of molecules can be separated from each other as they move over the support. The separation can be carried out using different types of support material which includes immobilized silica on glass plate (as in TLC), volatile gases (as in GC), paper (as in PC) and liquids which contain insoluble hydrophilic molecules (as in LC).

Primarily a separation technique, chromatography is mostly employed in analytical chemistry. It can also be used for preparative purposes, in particular, for the isolation of smaller amounts of materials with a high intrinsic value. It is a single step process wherein a mixture can be separated into its individual components and at the same time helps to provide a quantitative estimate of each constituent. The samples may be solid, liquid or gaseous and range in their complexity from a simple blend of two components to a multi-component mixture which differs widely. The analysis can be carried out on a simple and inexpensive thin layer plate or on a complex costly instrument.

Definition

`*Chromatography*` is derived from two Greek words, `*chroma*` meaning `colour` and `*graphein*` meaning `to write` (literally, *colour writing*, was used and named in the first decade of 20th century), was originally applied by Micheal Tswett, a Russian Botanist (1906), and is a method of analysing the materials involving the separation by selective absorption of the various compounds as identifiable bands. In general, it involves a test material being carried by a moving phase (liquid or gas) through a stationary phase (solid or liquid). Hence, this technique involves separation of components of a mixture (and is thus a form of purification) for their further use and their identification by colour. This is a process whereby a mixture of solutes may be resolved into its components by exploiting the differences (differential partitioning between the mobile and stationary phases). The principle behind chromatography is that the rate of migration depends upon the rate of interaction of the solute with the two phases, one being the mobile phase and the other stationary phase as the compounds travel through the supporting medium. The smaller the affinity of a molecule has for the stationary phase, the shorter the time spent in a column. Repeated acts of sorption and desorption that take place during the movement of the sample over the stationary phase determine their rates of migration providing the separation. This technique has become a potential and versatile tool for separating and analyzing a vast variety of chemical compounds in quantities from picograms (10^{-12} gram) to tons.

Chromatography has been variously defined as "*a broad range of physical methods used to separate and/or to analyze complex mixtures*". The components to be separated are distributed between the two phases, a stationary phase (bed) and a mobile phase (which percolates through the bed).

It is a "*process of separation that can be achieved by distributing the components of a mixture between two phases, a stationary phase and a mobile phase. The components held in the stationary phase are retained longer in the system than those that are distributed selectively in the mobile phase. As a consequence, the solutes are eluted from the system as local concentrations in the mobile phase in the order of their increasing distribution coefficients with respect to the stationary phase, ipso facto, a separation is achieved*".

It is also defined as: `a technique of separation of substances according to their partition coefficients below or their relative solubilities in two immiscible phases`.

Principle

The principle of chromatography is *capillary action*. The attraction of water to the paper (*adhesive force*) is larger than the attraction of the water to itself (*cohesive force*), and hence water moves up the paper. The moving substance is called the *mobile phase* and the substance that stays is the *stationary phase*. As the mobile phase moves, it separates into its components on the stationary phase, which can be later identified. The components of a moving liquid mixture spread out because they travel at different speeds over the stationary solid. As the liquid starts to move, some of its molecules (being energetic) are sucked towards the surface of the solid and stick there temporarily before being pulled back again into the liquid they came from. This exchange of molecules between the surface of the solid and the liquid is a type of gluing or adhesive effect called the *adsorption* (not absorption, where the molecules are permanently trapped inside). To be more effective, we need the components of the mobile phase to separate out as much as possible as they move past the stationary phase. So, the latter must be something with a larger surface area like a filter paper sheet, a solid made of finely divided particles, a liquid deposited on the surface of a solid or some other highly adsorbent material.

In this method, the separation of the components of a mixture is the function of their different affinities for a fixed or stationary phase, such as a solid or a liquid, and their differential solubility in a moving or a mobile phase, such as a liquid or a gas. The separation starts to occur when one component is held more firmly by the stationary phase than the other which tends to move on faster in the mobile phase. Thus, the underlying principle of chromatography is first to absorb the components of a mixture on an insoluble material and then to differentially remove or elute these components one by one with suitable liquid solvents.

Chromatography - Early History

The practice of chromatography is much older than the theory. A very old practice among the rural folk was to diagnose jaundice by soaking the rice grains in the urine of the patient wherein the rice grains turned yellow indicating that the patient is positive for the disease. This method is a forerunner of chromatographic isolation of uroporphyrin and the bile pigments by adsorption on starch or cellulose.

Chromatography has a long history and column chromatography as well as paper chromatography was probably employed empirically in antiquity. The basic idea underlying paper chromatography is quite old; Pliny, the Elder (*Gaius Plinius Secundus*: 23-79 AD), the Roman writer and the great savant of Imperial Rome, noticed that different liquid substances separated when they were absorbed by a sheet of papyrus.

The *Kapillaranalyse* (capillary analysis) developed by Friedlieb Ferdinand Runge (1855 – Friedlob occasionally misnamed as Friedrich), a German analytical chemist; Christian Friedrich Schonbein (1799-1868), a German-Swiss chemist; and his student, Friedrich Goppelsroeder (1860s) for the separation of coloured dyes and other chemicals on paper or cloth is an older cousin of paper chromatography. The credit for the first recorded work on column chromatography should go to Reed (1892) who published on the use of tubes of kaolin for the separation of potassium chromate from eosine and of ferric chloride from copper sulphate.

Mikhail Semenovich Tsvet (sometimes spelled as Tsvett, Tswett, Tswet, Zwet and Cvet; (1872 - 1920), an Italian-born, Swiss educated, Russian botanist, was the first to recognize and invent adsorption chromatography (in 1903-1906 – his last name *Tsvet* is Russian for both `colour` and `flowering`) as an efficient analytical method of separation. He used a simple type of liquid-adsorption column chromatography with calcium carbonate as an adsorbent and petrol ether/ethanol mixtures as eluent to separate chlorophylls and carotenoids during his research on plant pigments and generated many coloured bands (yellow, orange and green) on the adsorbent.

While working on plant pigments, Tsvet filled a tube with calcium carbonate and poured a mixture of pigments dissolved in petroleum ether into the tube and then poured a solvent through the tube. As they trickled through the tube, the various pigments adhered to the calcium carbonate. Each one of the pigments had a different degree of attraction to calcium carbonate and so the weakly attached pigments got displaced by the ones that were more strongly attracted. These pigments were forced downwards until they too adhered to calcium carbonate. This process caused the different pigments to separate at different levels within the tube, resulting in a series of coloured bands. For this reason, Tsvet called the process *chromatography* which means `*written in colour*` in Greek. By using this method, he was able to determine that the extract of leaves contained two chlorophylls, four xanthophylls and carotene.

The technique of chromatography has speeded up the development of modern biochemistry. It has made the separation, identification, quantitative estimation and the study of many of the substances of biological origin possible. The substances like pigments, sugars, amino acids, organic acids, phenolic acids etc., can be separated quantitatively and studied. A number of different types of chromatographic techniques have been developed during the 20th century. Some of them are: *Column chromatography*, *High performance liquid Chromatography* (HPLC), *Paper chromatography* (PC), *Thin layer chromatography* (TLC), *High pressure thin layer chromatography* (HPTLC), *Gas chromatography* (GC), *Ion exchange chromatography*, *Gel filtration chromatography* (GFC) and *Affinity chromatography*. Out of these, two common handy methods are PC and TLC.

CLASSIFICATION OF CHROMATOGRAPHIC METHODS

The principal methods of chromatography are represented by: *Frontal chromatography, Displacement chromatography* and *Elution chromatography.*

a. Frontal Chromatography (Adsorptive Filtration)

This is a procedure in which the sample (either a liquid or a gas) is fed continuously into the chromatographic bed, and no additional mobile phase is used. The sample components migrate through the column at different speeds and finally break out into a series of fronts and only the least retained component exits the column in a pure form and can be isolated while all others exit as mixed zones. The resulting chromatogram is referred to as a breakthrough curve (used in literature as *frontalgram*). The capacity of the column is an important parameter because it determines when the column is saturated with the sample components. Then, the mixture flows through the column with its original composition. The two classical applications of frontal chromatography are the purification of gases and solvents in addition to the purification of proteins initially used. It is also used as a tool for the determination of single-component adsorption isotherms. The first non-analytical use was introduced by James and Phillips (1954) who used this method to measure the adsorption isotherms of benzene and cyclohexane on charcoal.

b. Displacement Chromatography

This is a procedure in which the mobile phase contains a compound (referred to as the *displacer*) which is more strongly retained than the components or the sample under examination. The sample is fed into the system as a finite slug on the head of the column. The components are resolved into consecutive rectangular zones of highly concentrated pure substances rather than the solvent separated peaks. This is basically a preparative technique by which higher product concentration, higher purity and increased output may be obtained compared to other types. It was Arne Tiselius (1943) who first classified chromatography as frontal, elution and displacement types. The technique of displacement chromatography was redeveloped by Prof. Csaba Horvath (1981) by employing modern high pressure columns and equipment. Earlier, this was applied to preparative separations of amino acids and rare earths and for isotope separation. The advantages of this method include: the method is more effective than elution type for difficult separations; the method is simple in its operation as there are no gradients; the method results in the isolation of the product in dilute buffer wherein there will not be any salt contamination; the method is useful as the product can be recovered in a more concentrated form; the method results in sharper separations for higher yields; and the method is beneficial in that it operates on standard stationary phases and requires no special medium.

c. Elution Chromatography

This is a procedure in which the mobile phase is continuously passed through or along the chromatographic bed and the sample is fed into the system as a finite slug.

The chromatographic methods are classified according to the nature of the mobile phase (into gas or liquid chromatography), the nature of the stationary phase (either solid or liquid film on support) and the nature of the separation mechanism (as adsorption, partition, ion exchange or permeation chromatography). Elution chromatography is the most commonly used development procedure. Other optional procedures include frontal and displacement chromatography.

The methods of chromatography can be differentiated on the basis of the physical means of bringing the stationary and mobile phases into contact or according to the shape of the chromatographic bed, as: Column chromatography and Planar chromatography.

1. Column Chromatography

This is a separation technique in which the stationary phase lies within a tube. The particles of the solid stationary phase or the support coated with a liquid stationary phase may fill the whole inside volume of the tube (called the *packed volume*) or be concentrated on or along the inside tube wall leaving an open, unrestricted path for the mobile phase in the middle part of the tube (referred to as the *open-tubular column*).

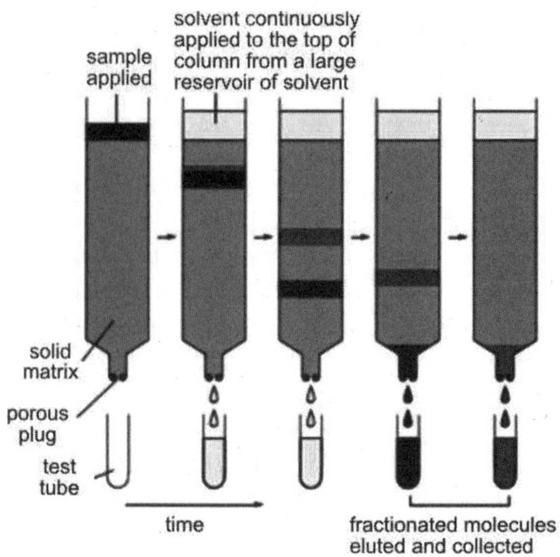

Column chromatography – Procedure

In this technique, the stationary phase is held in a narrow tube through which the mobile phase is forced either by pressure or gravity. Examples: Simple column chromatography (CC), high pressure liquid chromatography (HPLC) and gas chromatography (GC). Column chromatography can be further differentiated on the basis of the types of stationary and mobile phases and the kinds of equilibrium involved in the transfer of the solute between the two phases, such as liquid chromatography (LC) and gas chromatography (GC). This method is used to purify individual chemical compounds from the mixtures. This method is advantageous because it is relatively of low cost and the stationary phase used in the process is easily disposable which prevents cross-contamination and degradation due to recycling. The column is generally prepared either by dry method or wet method. In the former, the column is first filled with dry stationary phase (powder) followed by the addition of the mobile phase which is flushed through the column until it is completely wet. From this point it is never allowed to run dry. In the latter, slurry of the eluent is prepared with the stationary phase (powder). It is carefully poured into the column taking every care to avoid the air bubbles. A solution of the organic material is added on to the top of the stationary phase with the help of a pipette. This is covered over by sand or cotton or glass wool for protection. The eluent slowly passes through the column and helps the organic material to advance. Often an eluent reservoir (separating funnel) is put on the top of the column. During the process, the eluent is collected in a series of fractions.

2. Planar Chromatography or Open-bed Chromatography

This is a separation technique in which the stationary phase is present as or on a plane The plane can be a paper, serving as such or impregnated by a substance as the stationary bed (as in *paper chromatography*) or a layer of solid particles spread on a support like a glass plate (as in *thin layer chromatography*). Sometimes, planar chromatography is also termed as *open-bed chromatography*.

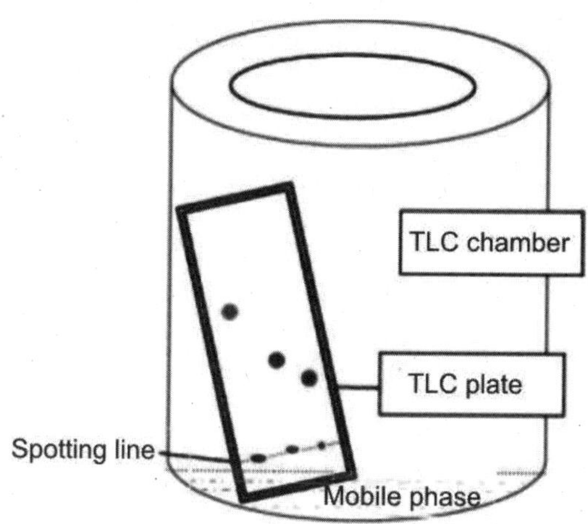

Planar chromatography – Apparatus

In this technique, the stationary phase is supported on a flat plate or in the fibres of a paper. The mobile phase moves through the stationary phase by capillary action or by gravity. Examples: Paper chromatography (PC) and thin layer chromatography (TLC). The separation occurs in space after a fixed time period rather than in time at a fixed location. The separated components appear as spots on the plane.

Classification on the basis of the Type or the Physical state of the Mobile phase

These methods are categorized into four types, gas-liquid chromatography (GLC), gas-solid chromatography (GSC), liquid-liquid chromatography (LLC) and liquid-solid chromatography (LSC). Sometimes, gas-liquid partition chromatography (GLPC) may also be found in literature.

A. Gas chromatography (GC)

This is always carried out in a column. The mobile phase is a gas.

B. Liquid Chromatography (LC)

This can be carried out either in a column or on a plane. The mobile phase is a liquid. Now-a-days, liquid chromatography utilizes very small particles and a relatively high inlet pressure and is often referred to as *high-performance* or *high-pressure liquid chromatography* (HPLC)

C. Supercritical-fluid Chromatography (SFC)

This is a technique in which the mobile phase is a fluid above and relatively close to its critical temperature and pressure. This method finds use in industry mainly for the separation of chiral molecules and uses the same columns as that of standard HPLC systems. This is now commonly used for achiral separation and purification in pharmaceutical industry.

Classification on the basis of Mechanism of Separation

These methods are classified on the basis how the component molecules of the mixture get separated. The methods under this category are: adsorption chromatography, partition chromatography, ion-exchange chromatography, affinity chromatography and exclusion chromatography.

1. Adsorption Chromatography

In this technique, the adsorption is based on the differences between the adsorption affinities of the sample components for the surface of an active solid.

2. Partition Chromatography

In this method, the separation is based on the differences between the solubility of the sample components in the stationary phase (as in gas chromatography) or on the differences between the solubility of the components in the mobile and stationary phase (as in liquid chromatography).

3. Ion-exchange Chromatography

In this method, separation is based mainly on the differences in the ion exchange affinities of the sample components. In the present day, this is carried out on small particle high efficiency columns and usually utilizes conductometric or spectroscopic detectors and is referred to as *ion chromatography* (IC).

4. Exclusion Chromatography

In this technique, separation is based mainly on exclusion effects such as the differences in the molecular size and/or shape or charge. The term *size-exclusion chromatography* is also used when the separation is based on the molecular size. The terms, *gel filtration* and *gel-permeation chromatography* (GPC) were used earlier to describe this process when the stationary phase is a swollen gel. The term, *ion-exclusion chromatography* is specifically used for the separation of ion in an aqueous phase.

5. Affinity Chromatography

This method characterizes the particular variant of chromatography in which the unique biological specificity of the analyte and ligand interaction is utilized for the separation.

SPECIAL TECHNIQUES OF CHROMATOGRAPHY

These techniques are represented by: *normal-phase chromatography, reversed-phase chromatography, isocratic analysis, gradient elution, stepwise elution, two-dimensional chromatography, isothermal chromatography, programmed-temperature chromatography* (also referred to as temperature programming), *programmed-flow chromatography* (also called as flow programming), *programmed-pressure chromatography* (also called as pressure programming), *reaction chromatography, pyrolysis-gas chromatography, post-column derivatization* etc.

HYPHENATED (HYBRID) TECHNIQUES AND ALTERNATIVE METHODS OF DETECTION IN CHROMATOGRAPHY

The hyphenated techniques have been developed by coupling of a separation technique and an on-line spectroscopic detection technology over the last two decades, and significantly have broadened their application in the analysis of biomaterials, especially the natural products. These techniques refer to a combination of chromatographic and spectral methods (one or more) to exploit the advantages of both. Hirschfield two decades ago introduced the term `hyphenation` to refer to the on-line combination of a separation technique and one or more spectroscopic detection techniques. Most often the other technique is some form of chromatography. Now-a-days, these are known as hyphenated or hybrid techniques. Recently, these methods have received ever-increasing attention as means to solve the complex analytical problems, and demonstrated both the quantitative and qualitative analysis of unknown compounds in complex natural products. These methods are useful in the analysis of crude extracts (pre-isolation analysis) or fractions from different natural sources, isolation and detection of natural products, chemo-taxonomic studies, chemical fingerprinting, quality control of herbal drugs and products, de-replication of natural products and metabolomic studies.

The hyphenated analytical methods are represented by GC-mass spectrometry, LC-mass spectrometry, Pyrolysis GC-mass spectrometry, LC-nuclear magnetic resonance spectrometry and others (Examples: CE-MS, GC-IR, ICP-MS, LC-IR or LC-FTIR, LC-PDA-MS, LC-MS-MS, LC-NMR-MS, LCPDA-NMR-MS, ESI-MS-MS, SPE-LC-MS and LVI-GC-MS etc.).

CHROMATOGRAPHY – BASIC PRINCIPLE

Each method consists of two phases, the stationary phase (column packing material, either a solid or a liquid) and the mobile phase (the solvent, usually a liquid or a gas). The stationary phase refers to the support or the bed that is fixed, while the mobile phase refers to the medium through which the mixture to be separated moves in a particular direction. The mixture gets separated by distributing its component molecules between the two phases. The distribution may take place on the basis of the structure, size, shape, charge or molecular weight of the molecules. The technique can be either analytical (makes use of only a small amount of a mixture to determine the components) or preparative (a process of purification where the components of a mixture are separated and used for further studies) based on the main objective for which it has been followed.

TYPES OF CHROMATOGRAPHY

Chromatography is mainly of two types namely, Adsorption chromatography and Partition chromatography.

1. ADSORPTION CHROMATOGRAPHY

The addition of a gas, liquid or a solid dissolved in a liquid to the surface of a solid body is known as `adsorption`. In this method, the unknown substances are separated and identified making use of adsorption. Many solids like charcoal, alumina and other metallic acids, metallic salts, silica gel and special sugars are used as the adsorbing agents. The preparation of the adsorbent is very important as even small impurities greatly influence the process by blocking the site. One of the main impurities is water which can be removed by heating, and the process is called as the `activation`. `Deactivation` is a reverse process in which controlled amount of water is added. The addition of compounds to the adsorbent to reduce its activity in gas chromatography is called `poisoning`. Another important point to be remembered is the physical state of the adsorbent, The particle size is to be usually small to allow the equilibrium to be set off rapidly and to minimise the diffusion effect, but must not be so small that impedes the flow of the solvent. The coarse particles are removed by `sieving`. Another point to be remembered is that the adsorbents with very high surface area should not be used as they act as catalysts and may bring about chemical changes.

In adsorption chromatography, the stationary phase is a finely divided adsorbent such as alumina or silica gel and the mobile phase can be a gas or more commonly a liquid. Acetone, Petroleum ether, Butanol, Acetic acid, Water, Chloroform, Benzene, Phenol, Ammonia and hydrates of organic and inorganic solvents are used as resolving fluids. Usually two to four mixed in a definite proportion are used.

Adsorption chromatography was the earliest method used in the history of chromatography. It is also known as Column Chromatography, as solid columns of inert material are used for the adsorption of different substances. The mixture to be separated is taken in the form of a solution, which is then poured down the column. Different components travel down the column at unequal rates because they are adsorbed by the material of the column to a different extent.

Essential Requirements

The essential requirements of the method include: a column, an adsorbent and the solvent. The column is usually a glass tube with means for supporting the packed adsorbent and openings for the admission and collection of the mobile phase of the solvent. The adsorbent should be inert and insoluble in the solvent to be used. Sometimes, inert substances such as the diatomaceous ooze are mixed with relatively active adsorbent to improve the flow. Some commonly used adsorbents are: Alumina, Calcium Carbonate (for plant pigments), Calcium Phosphate (for proteins), Celite or Super cell (for proteins), Cellulose (for organic compounds), Glass (for high molecular weight organic compounds), Kaolin (for porphyrins, carbohydrates, vitamins – A, D and E), Magnesium oxide, Magnesium silicate (for steroids and carbohydrate derivatives), Silicic acid (for hydrocarbons, amino acids and carbohydrate derivatives), and Sucrose (for chlorophyll). The solvents should be pure and stable. Further, the recovery of the dissolved substances from the solvent should be easy. Acetone or organic compounds of the type are frequently used in the chromatographic separation of the plant pigments.

2. PARTITION CHROMATOGRAPHY

Partition chromatography involves the partition between two liquids rather than adsorption by a solid from a liquid. Here, the stationary phase is a liquid which is held on an inert porous supporting liquid. The mobile phase may be a gas or a liquid.

Principle

Partition chromatography is based on the distribution of a mixture between two immiscible liquids. Paper is most widely used as a supporting medium for this technique. The filter papers manufactured by Whatmann, Schleicher, Schull and D`Arches are some of the common ones used for this purpose. Whatmann no. 1 is the standard slow-running paper which is ideally suitable. Sometimes, even thicker papers, Whatmann no. 3 and no. 4 are also used. The papers which are used for special purpose include Whatmann – CM 82, Whatmann – DE 81 and Whatmann – ET 81. The solvents used are mostly immiscible with water or partially miscible. Phenol, Butamol or Butanol mixtures and a wide variety of organic acids and alcohols are used as solvents in this method.

Method

The method is carried out on sheets of filter paper and columns of thin layers of powdered cellulose, moist silica gel or diatomaceous earth (called kieselguhr and celite). The porous medium acts solely as a support for water which acts as a partitioning agent. Developed by Martin and Synge (1934), this technique has revolutionized the separation or detection of the reaction products and the determination and identification of compounds of biological origin. Consden, Gorden and Martin (1944) described paper chromatography in detail where separations were done mainly by partition.

3. PLANAR CHROMATOGRAPHY

This is a separation technique in which the stationary phase is present on a plane, which can be a paper used as such or impregnated by a substance as the stationary bed (PC) or a layer of solid particles spread on a support like a glass plate (TLC). The different compounds in the mixture travel different distances according to how strongly they interact with the stationary phase as compared to the mobile phase. Later, the components may be identified.

This method includes paper chromatography (PC), thin layer chromatography (TLC) and electro chromatography (electrophoresis – EC) all of which make use of a stationary phase (a sheet or a flat sheet of paper or any other material such as metal, glass or a plastic plate coated with a suitable adsorbent with the help of a binder). The mobile phase moves through the stationary phase by capillary action assisted by gravity or an electric potential. Earlier, planar chromatography included 2-dimensional chromatography, and it is called planar because the stationary support consists of the plane surface (of a paper or a smooth glass plate).

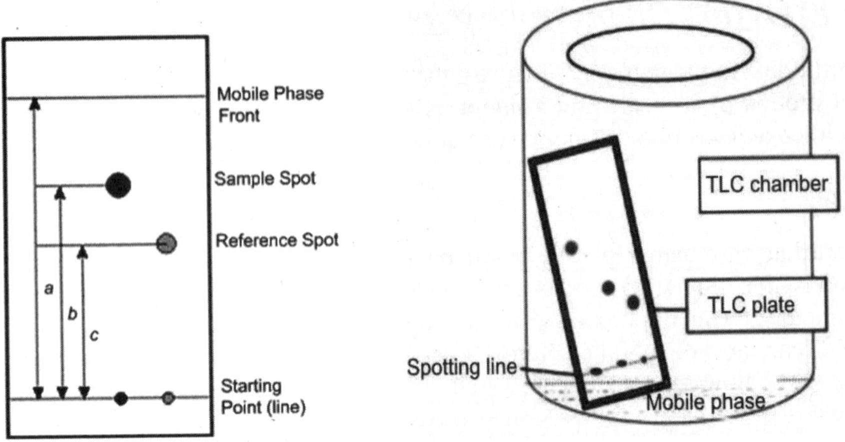

Chromatogram in Planar Chromatography Planar chromatography (PC) – Apparatus

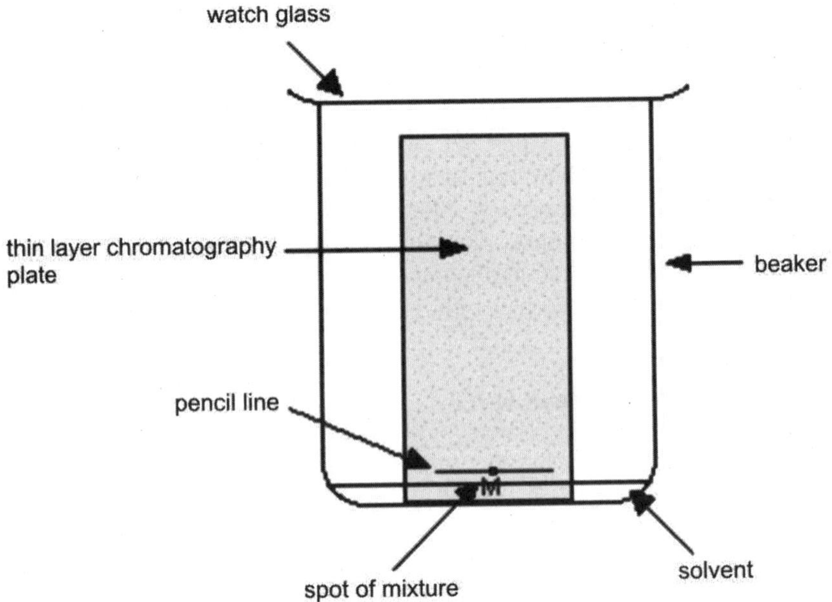

Thin Layer Chromatography – Apparatus

PAPER CHROMATOGRAPHY (PC)

This is an analytical method used for the separation and identification of mixtures especially the pigments and can also be used in experiments of secondary or primary colours. This is still a powerful teaching tool in the laboratory, and is useful for separating the complex mixtures of similar compounds like amino acids. 2-way PC also referred to as 2-dimensional chromatography involves the use of two solvents and rotation of paper 90^0 in between.

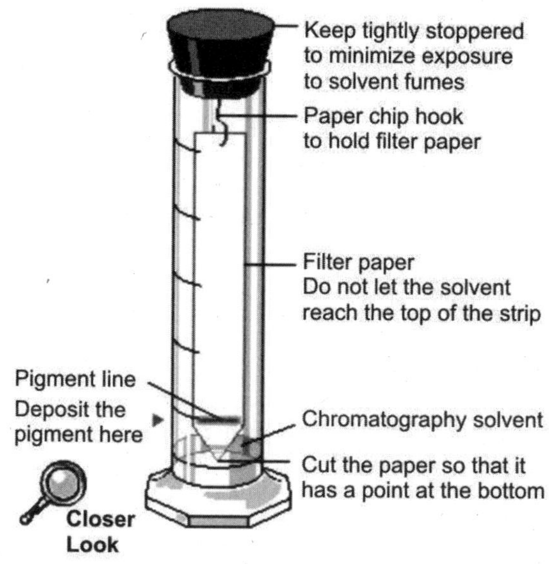

Paper chromatography – method

GENERAL PROCEDURE – Paper Chromatography (PC)

Paper chromatography is a simple analytical method for separation of substances. This is one of the easiest methods of chromatography, and is a method of planar chromatography (stationary phase is in form of a plane). This follows the basic principle of chromatography, which states that substances or components are distributed in between the stationary phase and the mobile phase. It is an analytical technique, where only a small amount of a sample is used for separating and identifying its components. Like any other method this has two phases, the stationary phase and the mobile phase. The porous paper serves as the stationary phase, and a solvent (either water or ethanol) serves as the mobile phase. Paper can be a blotting paper or a filter paper or any special paper. It is always better to use Whatmann No.1 filter paper or chromatography paper, since they are pure and have uniform fibers. The cellulose of the filter paper acts an inert support. The materials and equipment required for performing this procedure are easily available, and a simple paper chromatography experiment can be carried out at home.

The sample is a mixture, in which the components are to be separated and studied. The sample should be in liquid form for paper chromatography. In schools and colleges, various ink types are used as sample for paper chromatography practical. The sample can also be a complex mixture such as blood, polluted water, plant pigments, amino acids etc.

The developing chamber is a container or glassware that can be sealed properly. Sealing is important in order to create an atmosphere saturated with solvent vapour inside the chamber. The shape and size of a chamber differs with reference to a particular experiment.

The chromatography paper or the stationary phase is cut based on the shape and size of the developing chamber. Usually, the paper shape is in the form of a rectangle, for example, 10-15 cm length and 4-5 cm width. The paper size should be such that it can be suspended without touching the sides of the developing chamber. The solvent or the mobile phase can be water, alcohol or a mixture of solvents. It should be added into the developing chamber up to a certain level in the bottom. The other requirements needed are pencil, ruler, tape, scissors, capillary tube, calculator, notepad, etc.

The procedure for paper chromatography is quite simple as compared to other methods of chromatography. In order to achieve a good separation, one has to follow the following procedure carefully.

1. The chromatography paper is cut into rectangular strips and a line is marked (with a pencil to avoid contamination) on the paper at about 2-3 cm from the bottom. The paper with its corresponding sample is properly labelled to avoid confusion. The hands are to be kept clean before handling chromatography paper.
2. The sample is placed (loaded) as a spot on the starting line with the help of a capillary tube or ink filler. Proper care should always be taken to avoid spilling while loading the sample on to the paper.
3. The chromatography paper (loaded) is placed in the developing chamber, containing the solvent or the mobile phase. While placing the paper, every care is taken that the solvent level doesn't reach the starting line or the sample spots. The paper is suspended without touching the sides of the chamber; which will otherwise lead to a poor separation. The chamber is sealed properly (as it should be air-tight).
4. By capillary action, the solvent rises up the paper or the stationary phase and dissolves the sample. The components of the sample move along with the solvent in an upward direction. The speed of movement depends on the attraction of the solvent molecules to the paper and the differential absorption of the solute components in the solvent. The more the attraction or affinity a component has, the slower it moves up and vice-versa. Thus, different components cover different distances within the same time.
5. When the solvent has reached near the top level of chromatography paper, it is removed, and the level is marked with a pencil. This level or height is called the "solvent front". The different spots of varied colors are examined as each spot represents a specific component of the sample. If the spots are not distinct to locate, the paper should be viewed under ultra-violet light, ninhydrin or iodine vapors. The spots are carefully circled with a pencil.
6. Retention Factor (Rf) is `*the ratio of the distance travelled by the substance to the distance travelled by the solvent*`. Rf value is always between 0 and 1 and has no unit. The Rf value of unknown compounds is compared with the Rf value of known compounds for identification.

The distances of the solvent front and also the distances travelled by the components (taking the center point of the spots) are measured. The retention factors of the components are calculated by using the relation, Rf = distance travelled by the substance/distance travelled by the solvent. Since the distance covered by the components varies, the resulting Rf values will also vary. The Rf values of the unknown components are compared and matched with Rf values of the known compounds and the substances present in the particular sample are identified.

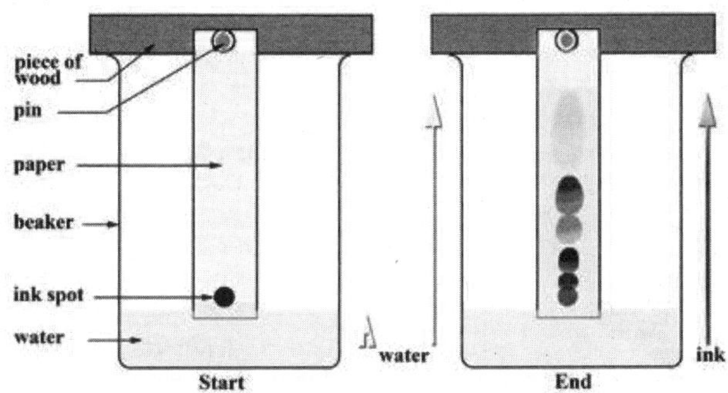

Simple PC – Laboratory setup

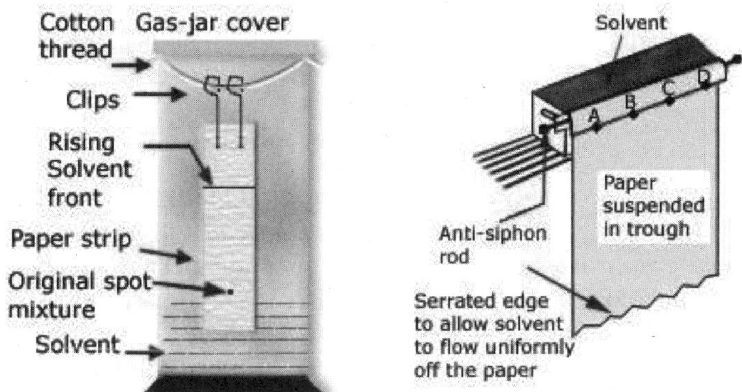

2-way PC – Ascending and Descending

Paper chromatography can be an ascending chromatography or a descending chromatography, based on the direction of running a chromatogram. The time required for running a chromatogram varies from one hour to several hours. In general, it is used to separate most coloured compounds and is widely used in artificial and natural pigment research, in clinical research, hospitals, and manufacturing industries and in forensic science studies. The disadvantage of paper chromatography is that it is not a reliable method for separating complex mixtures.

This is a useful technique because it is relatively quick and requires small quantities of the material for resolution. The substances are distributed between a stationary phase and a mobile phase. The former is usually a piece of high quality filter paper (laboratory filter paper, chromatography paper and white paper towel are good) while the latter is a developing solution (water, ethyl alcohol, isopropyl alcohol, acetone are the best) which travels up carrying the samples with it. The components of the sample mixture will separate readily according to how strongly they adsorb on the stationary phase versus how readily they dissolve in the mobile phase. If water is used as a solvent, the less polar the colour, the lower it will rise on the paper.

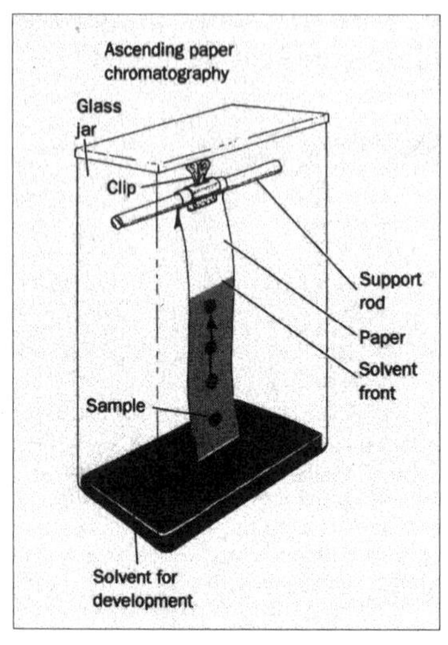

Ascending PC – Equipment

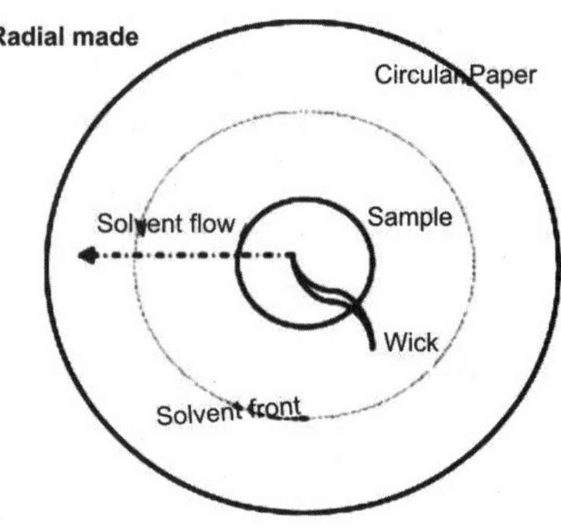

Circular or Radial chromatography

Thin Layer Chromatography (TLC)

This is also referred to as *open column chromatography* (by some workers) and also as *drop chromatography, strip chromatography, spread layer chromatography* and *surface chromatography* (by others). TLC not only combines the advantages of paper chromatography and column chromatography, but in certain aspects, it is superior to either of the two. TLC appears to have been first developed and successfully utilized by Schraiber (1939) while she was working with Izmailov at Kharkov Chemistry and Pharmacy Research Institute who used the techniques for pharmaceutical analysis. She not only invented TLC (1939) but also was the first to use fluorescence as the separation indicator or the detection system. Unfortunately, her work does not seem to have been taken into account and the technique has been later rediscovered by Kirchner (1951).

This is a technique commonly used to separate mixtures, and is performed on a sheet of glass, plastic or aluminum foil coated with a thin layer of adsorbent material, usually silica gel, aluminum oxide or cellulose (stationary phase). After the sample has been applied on the plate, a solvent or solvent mixture (mobile phase) is drawn up the plate via capillary action. As different analytes ascend the plate at different rates, separation is achieved. TLC involves the use of a particular sorbent on an inert sheet of glass, plastic or metal. The solvent is allowed to travel up the plate with the sample spotted on the sorbent just above the solvent. Depending on the nature of the sorbent, the separation can either be partition chromatography or adsorption chromatography (cellulose, silica gel, alumina or cellulose on a flat, inert substrate).

The major advantage is the disposable nature of the plates. The samples do not have to undergo the extensive clean-up steps. The other advantage is the ability to detect a wider range of compounds in a cheaper way using the very reactive agents or indicators like iodine vapour and sulphuric acid. Non-destructive detection like fluorescent indicators in the plates and examination under a ultra-violet lamp also enable scrapping off of the purified samples and analysis by other techniques. This method can be used to monitor the progress of a reaction, identify the compounds present in a given mixture, and also to determine the purity of a mixture. The applications include the analysis of fatty acids, detection of pesticides or insecticides in food and water, analysis of dye composition of fibres in forensics, assay of radio-chemical purity and identification of the constituents of medicinal plants.

Preparation of the Plate

The plates are commercially available. They are prepared by mixing the adsorbent (silica gel) with a small amount of an inert binder (calcium sulfate and water) and the mixture is spread (as a thick slurry) on an unreactive carrier sheet (usually glass, thick aluminum foil or plastic sheet). The plate is dried and activated by heating in an oven for 30 minutes at 110°C. For analytical purpose, the thickness is around 0.1 - 0.25 mm and for preparative TLC around 0.5-2.0 mm.

Technique of TLC

A small spot (containing the sample) is applied to the plate (at the centre), about 1.5 cm from the bottom. The solvent is allowed to completely evaporate off to achieve a better separation. In case of a non-volatile solvent, the plate is dried in a vacuum chamber.

The required amount of an appropriate solvent (elutant) is taken in a suitable transparent container (separation chamber – either a boiling test tube, a collection jar or a specimen jar) to a depth of less than 1 cm. A filter paper strip is inserted into the chamber so that its bottom touches the solvent, and the paper lies on the wall of the chamber and reaches the top of the container. The chamber is closed with a glass plate (lid) and is left for a few minutes to allow the solvent vapour ascend the filter paper and saturate the chamber. (No saturation results in a poor or no separation).

The TLC plate is placed in the chamber so that the spot(s) of the sample do not touch the elutant and the lid is closed. The solvent moves up the plate by capillary action, meets the sample (mixture) and carries it up the plate (elutes). When the solvent front reaches no higher than the top of the TLC plate in the chamber, the plate should be removed and dried.

Process of Separation

Different compounds in the sample travel at different rates. This is due to differences in their attraction to the stationary phase, and because of differences in their solubility in the solvent. The separation of compounds is also based on the competition of the solute and mobile phase for binding places on the stationary phase. It is believed that strong solvents push the compounds up the plate while the weaker solvents barely move them. The order of the strength or weakness depends on the coating of the TLC plate. Hence, the solvents are to be carefully selected.

Analysis

As the separated chemicals are colourless, many methods are used to visualize the spots. Vapours of iodine are used as general unspecific colour reagent. The TLC plate can be dipped into or the specific colour reagents are sprayed on to the TLC plate such as Potassium permanganate (oxidation), iodine or bromine. The chromatograms can be transferred to a PVDF membrane and then subjected to further analysis like mass spectrometry (Far-Eastern Blotting). When the spots are visible, the R_f value (retardation factor) of each spot can be determined by dividing the distance travelled by the product by the distance travelled by the solvent front using the initial spot site as the reference. These values depend on the solvent used and the type of the TLC plate (not the physical constants).

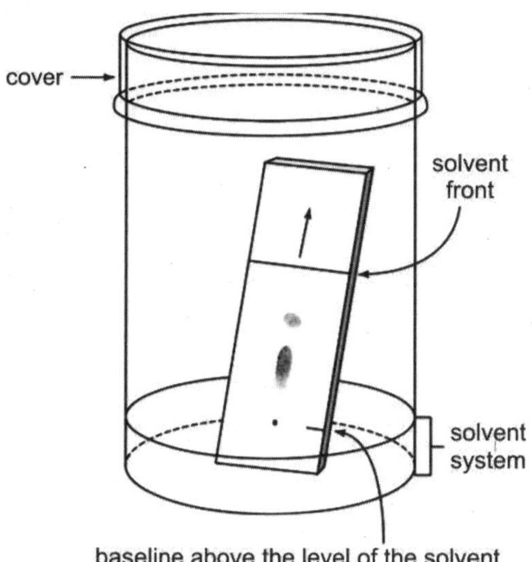

TLC – Diagram

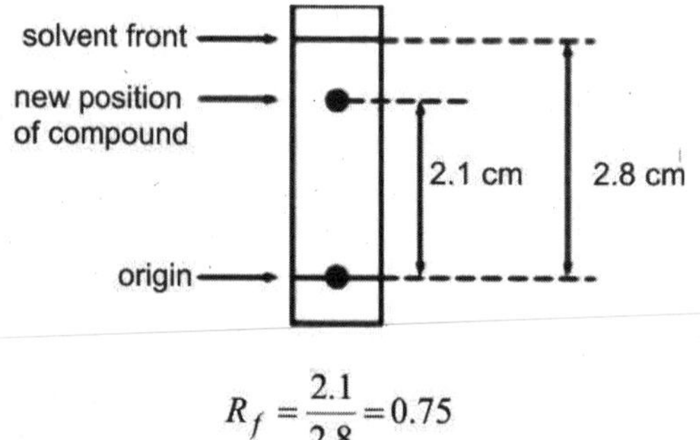

$$R_f = \frac{2.1}{2.8} = 0.75$$

TLC – R_f value calculation

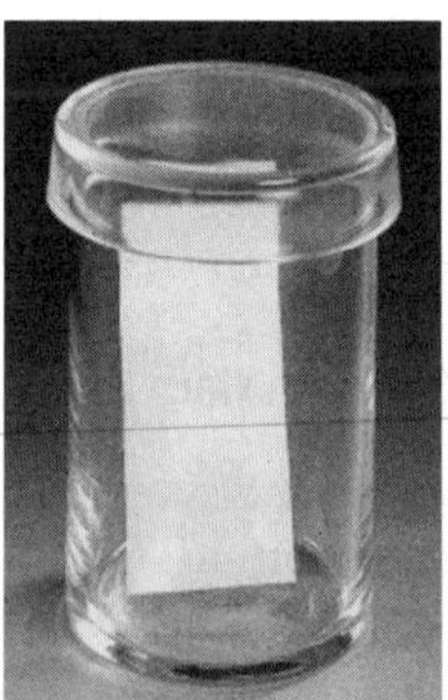

TLC chamber

Advantages and Disadvantages of TLC

TLC is very simple and inexpensive to use. The material needed for TLC is little too. The best solvents used in TLC can be applied to other methods like HPLC. More than one compound can be separated on the plate. The solvents can be easily changed and several different solvents can be used for the desired results. It is also very easy to check the purity of a compound under UV light. Most of the compounds can be identified by comparison of literature values of R_f. The conditions can be easily be modified to increase the optimization for the resolution of the required component.

The main disadvantage is that the plates do not have long stationary phases and as such the length of separation is very limited. The detection limit is a lot higher. This operates on an open system and is controlled by humidity and temperature which influence the nature of the chromatogram. The problems if any can be solved or avoided easily. The size of the sample spot should not be larger than 1-2 mm in its diameter. If the spot is larger, it causes the overlapping which influences the resolution of different components. Uneven advance of the mobile phase results in inaccurate R_f values. To avoid streaking, the sample solution is diluted. The sample should be always above the level of the solvent.

Superiority of TLC over PC

The advantages and superiority of thin layer chromatography over paper chromatography are as follows:

1. As the supporting medium or the adsorbent is inorganic, concentrated sulphuric acid spray followed by heating may be used to locate the substances on the chromatogram by charring and rendering visible any spots of organic compounds.
2. The amino acid mixtures which require 18 hours for separation on paper require as little as 3 hours using cellulose thin layer chromatography.
3. The advantage also lies in the choice of the adsorbents which can allow the separation of substances not possible on paper.

After localization of the substances, the selected spots may be easily scrapped off and the compounds can be isolated by extraction, and can be quantified.

High Pressure Liquid Chromatography (HPLC) or High Performance Liquid Chromatography

HPLC is a development over column chromatography which is used to separate a mixture of compounds to identify, quantify and purify the individual components of the mixture, like separation and quantification of drugs like steroids in samples of urine and the levels of vitamin-D in samples of serum. Long back it was realized that the use of particles of a smaller size (3 or 5 or 10 micro-meter) with a very narrow size distribution would greatly improve the resolution if the flow rate and column dimension could be adjusted to minimize the band broadening. The pumps were developed to easily handle the required chemicals and pressures. Traditional column chromatography is described as `*normal*` where non-polar solvent and the polar surface are used. If the system uses polar solvents like water, methanol or acetonitrile and a non-polar surface, it is referred to as `*reverse-phase*`. The latter is the method of choice for larger non-volatile bio-molecules. Recently, `*universal detectors*` have been emerged (evaporative light-scattering).

In this technique, the sample is forced by a liquid at high pressure (the mobile phase) through a column which is packed with a stationary phase composed of irregularly or spherically shaped particles (a porous monolithic layer or a porous membrane). This method is divided into two sub-classes namely, normal phase liquid chromatography (NPLC) and reversed phase liquid chromatography (RPLC). In the former method, the stationary phase (silica) is more polar than the mobile phase (toluene) while in the latter method; mobile phase (mixture of water and methanol) is more polar than the stationary phase (octadecylsilyl – C_{18}).

GENERAL PROCEDURE – High-Performance Liquid Chromatography (HPLC)

HPLC is a powerful chromatographic tool in analysis of a mixture of compounds in analytical chemistry and biochemistry which helps identify, quantify and purify the individual components of the mixture.

This technique utilizes different types of stationary phases contained in columns; a *pump* which moves the mobile phase as well as the sample components through it; and a *detector* which is capable of providing characteristic retention times for the components and area counts which reflect the amount of each analyte passing through the detector. The detector also provides the information related to the analyte like UV spectroscopic data. The retention time varies depending on the strength of its interaction with the stationary phase, the composition and the rate of flow of the mobile phase as well as the dimensions of the column. Generally, this utilizes small sized columns (250 mm or shorter and 4.6 mm internal diameter or smaller, packed with smaller particles) and higher pressures of the mobile phase compared to ordinary methods of liquid chromatography. In this method, a pump provides the higher pressure that is required to move the mobile phase along with the sample components through the densely packed column. The particles are capable of providing a better separation on the columns of shorter length when compared to ordinary method of liquid chromatography.

The sample is introduced in a discrete small volume into the mobile phase that percolates through the column. The components move through the column at different velocities and the velocity depends on their chemical nature, on the nature of the column and also on the composition of the mobile phase. The retention time is at which time the analyte emerges (elutes) from the column and this is measured under particular condition (characteristic of the given analyte). The use of packing material of smaller particle size requires higher operational pressure (also called the back pressure) which improves the degree of separation (chromatographic resolution). Any miscible combination of water with different organic solvents like acetonitrile and methanol are the common mobile phases used in the technique. Sometimes, water-free mobile phases are also used. The aqueous component may contain buffers, acids like formic acid, phosphoric acid or trifluoro-acetic acid, or salts which help in the resolution. The composition of the mobile phase is either kept constant (isocratic elution mode) or may be varied (gradient elution mode) during the analysis, and the former is effective in the separation. In the latter, the composition of the mobile phase varies from low to high eluting strength and this is reflected by the retention time of the analyte (the analyte with high eluting strength producing the faster elution). The partitioning is similar to that in liquid-liquid extraction, but is continuous but not step-wise. The choice of the components of the mobile phase, the additives like salts or acids, and gradient conditions depend on the nature of the column and the nature of sample components. In order to find out the method which gives the best separation, a series of trial runs are performed with the sample.

HPLC, an ideal complementary technique to GLC, since its inception in the late 1960's, has made significant practical impact in pharmaceutical, clinical, forensic, environmental and industrial research and development analyses. Preparative HPLC is important in the isolation and purification of various compounds. There are two main classes of column: "normal" and "reversed" phase.

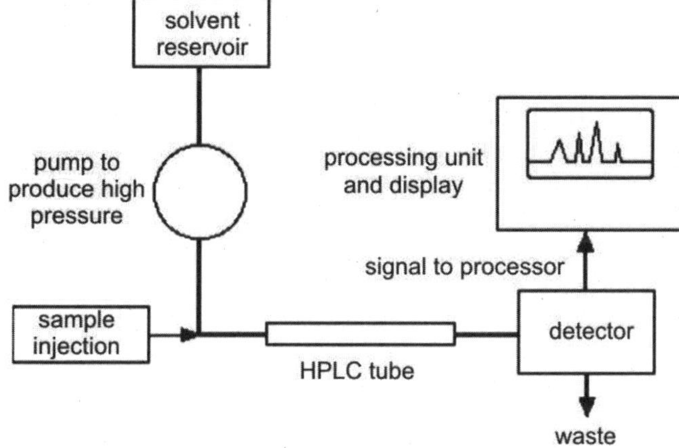

HPLC – Flow sheet

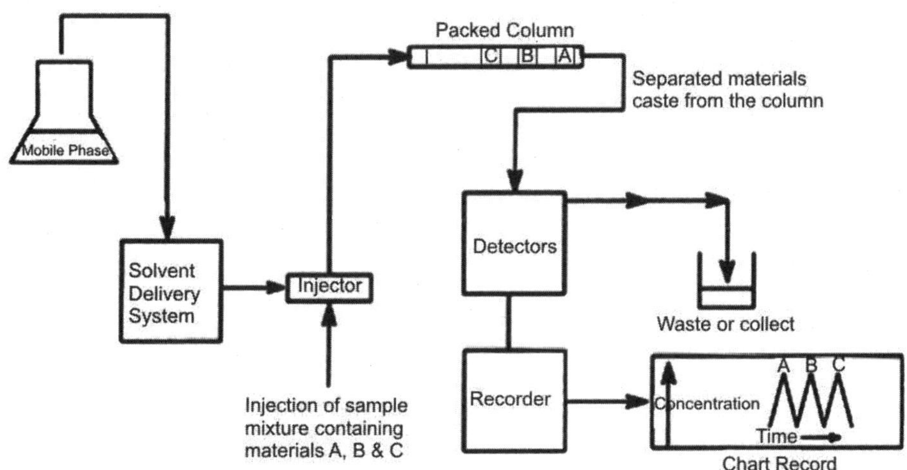

HPLC – Schematic diagram

Normal phase columns are usually packed with silica gel; and work in the partition/adsorption mode in the same manner as in conventional chromatography. Reversed phase chromatography, the most common form of HPLC, is a type of partition chromatography. The reversed phase columns are very non-polar and are packed with chemically bonded octadecylsilyl coated silica. Other popular bonded phase columns have octasilyl, cyanopropyl, or phenylsilyl packings. The eluent used with reversed phase columns is relatively polar (MeOH/H_2O). In this method, the more polar components of a mixture elute first, and increasing the polarity of the solvent increases the retention time of a particular component, which is the reverse of normal adsorption chromatography. HPLC is especially suited to the analysis of compounds not readily assayed by GLC (thermo-labile compounds can be analysed by HPLC at ambient temperatures, as well as highly polar or non-volatile compounds).

Identification of the compounds

During chromatographic analysis, the compounds can be identified by inspection under ultra-violet light as some of the compounds, for example, gibberellins and phenolic acids fluoresce under ultra-violet light. They can be detected by spraying certain reagents and dyes, for example, spots of amino acids turn deep violet on spraying with Ninhydrin. They can be identified by elution and examination of their physical properties such as pH, optical rotation, electrical conductivity, refractive index etc. If the substance is radio-active, it can be detected by using Geiger-Muller Counter, and if the substance is biologically active, it can be identified by bio-assay etc.

Resolution Front (*R_f*)

In chromatography, the migration of a particular solute is measured in terms of its R_f value. The position of a spot in a chromatogram is recorded as its R_f value. It is often a diagnostic feature of a compound. The R_f value for a component is defined as: `*as the ratio of the distance travelled by the component to that travelled by the solvent front, both measured from the loading point of the mixture*`. It is also defined as: `*the measure of the velocity of migration and as the quotient of the distance of the substance from the starting point divided by the distance of the solvent from the same point*`.

R_f = Distance from origin run by the compound/Distance from origin run by the solvent or

R_f = Distance travelled by the solute/Distance travelled by the solvent.

R_f value is only of a limited value in the diagnosis of a compound. The distances are calculated from the start i.e., the centre of the original spot. The R_f value for a given compound varies with the type of the solvent used, temperature, saturation of the chamber and the concentration of the sample. Similar R_f values do not necessarily mean the compounds are identical as different compounds may have identical R_f values in a given solvent system. The value of R_f is always one (unity) or less than one (unity). It is a physical constant. The published results may differ from worker to worker due to different methods adopted. Hence, while giving R_f values, the details of the technique used and temperature must be specified. It is always preferable to chromatograph known standards along with the unknown mixture so that R_f values can be easily compared and the substances be identified, and because the standard values are more reliable and reproducible.

Forces that operate

The movement of the solute molecules on the chromatogram depends on the net result of a number of forces operating in the system. They may be, propelling forces or retarding forces.

Propelling Forces

The propelling forces include the `Capillary force` and the `Solubility force` of the solvent. The Whatmann filter paper is made of numerous fibrils which are placed very close to each other, forming a net work of capillaries. The solute rises through these capillaries as a result of capillary force. The smaller the bore of the capillaries, the greater the height to which the solute rises.

The solubility force refers to the capacity of the solvent to dissolve the solute in it. The greater the solubility of the solute in the solvent, the greater is the height to which it rises.

Retarding Forces

The retarding forces include the `Gravitation force` and the `Partition force`. In ascending chromatography, the solute has to move against gravitational force acting from below which tends to retard the movement of the solute in the upward direction. In descending chromatography, the gravitational force does not figure as a retarding force, but it assists in the propulsion of the solute.

The partition force refers to the force between the liquid and the liquid molecules. The interspaces between the fibrils on Whatmann filter paper are occupied by the solvent molecules. So, the solute molecules for their movement have to displace them. If the spaces are not filled with the solvent molecules, the movement of the solute through them would be more rapid.

Hence, the movement of the solute on the filter paper in either direction, i.e., ascending or descending, is the net result of interaction between various forces of propulsion and retardation. The movement of the solute molecules is exhibited only when the propelling forcers exceed the retarding forces in their magnitude.

CHROMATOGRAPHY - APPLICATIONS

The technique of chromatography has many advantages over other techniques of separation in analytical studies. The technique is very sensitive and reliable provided the method is carried out carefully without any contamination. The main advantage of this method is that the complex mixture can be separated accurately by using only a few micrograms of the sample. The second advantage is that separation takes lesser time as compared to other methods. Also, the setting up of the experiment is simple and easy. The only disadvantage of this method is that it is very sensitive and as such improper set up or contamination even in nanograms will give a different result. The applications are not only manifold, but also are essential in today`s industry. For example, pesticides (DDT in the ground water) and insecticides (PCBs-poychlorinated biphenyls) are removed by this process in chemical industries. It is also used to separate toxins from the drinking water and also to monitor air quality, as a major testing tool. In pharmaceutical industry, this is used to prepare large amounts of pure materials that are required in making drugs, and it is also used the check the presence of any contamination in the manufactured drugs. The method is also crucial to analyze the exact chiral compound so that correct medicines can be manufactured. In food industry where proper maintenance is necessary to ensure quality of the food, this technique is used to separate additives, vitamins, preservatives, proteins and amino acids. It is also used in the detection of drugs in urine, and the separation of chemical traces in the case of fire in houses or buildings. It is also a very popular method in forensic science for the purposes of investigation in cases of crime.

Chromatography is normally used in many industries and laboratories. Chromatography is used to analyze, identify, purify and also quantify the substances in a mixture and the only disadvantage of the technique is that it requires time and skill. This is a very special technique of separation as complex mixtures can be separated with a great precision (for example: drugs, plastics, flavouring material, foods, pesticides, tissue extracts, fuels, air samples and water samples). Even more identical proteins that may vary only by a single amino acid can be separated. If the right adsorbent material, carrier fluid and operating conditions are employed, this method helps to purify any soluble or volatile substance of a mixture. Chromatographic methods are used extensively in forensics, from analyzing the body fluids for the presence of illicit drugs, to fiber analysis and blood analysis from the crime scene, and also at the airports to detect the residues from the explosives. The process does not alter the molecular structure of the compounds and as such it will provide a way to obtain pure chemicals from various sources. Chemists use this technique in the laboratory even to track the progress of a reaction. By the observation of the sample spots on the plate, they can easily find out when the products start to form and when the reactants have been used up (i.e., when the reaction is complete). Chemists and biologists also use this method to identify the compounds present in plants. The method works well on very large and very small scales and it is used both by the scientists who study micrograms of a substance in the laboratory as well as by the industrial chemists who try to separate tons of material. The method is also used to determine whether sweets contain any prohibited dyes within them. In pharmaceutical industry, this is used to check whether the product is synthesized from the reactants or not and in petrochemical industry, it is used to check the purity of the fuel and its additive property.

Liquid chromatography is mainly employed for the separation of highly polar and large molecular weight materials with a very low volatility like non-volatile liquids and solids. This technique can be used to easily separate peptides, polypeptides, proteins and larger biopolymers. This method can also be used in trace analysis (drugs and drug metabolites present in the blood) such as determination of tetra-hydro-cannabinol carboxylic acid in urine in persons who have recently smoked marijuana, and determination of tricyclic antidepressant drugs in the serum of blood.

Compared with liquid chromatography, GC has different applications. It is used for the separation of volatile materials. This is also useful in the separation of amino acids, steroids and fatty acids with a high molecular weight (free fatty acids from whole milk and separation of a sample of lime oil). GC can handle multi-component mixture such as essential oils, hydrocarbons and solvents more easily. The most common hydrocarbon analysis carried out is that of gasoline. Another important application is in studies of pollution (for separation and analysis of potential carcinogenic aromatic hydrocarbons in the environment), forensic work and general trace analysis. This technique is ideal for food and beverage assays and tests.

CHROMATOGRAPHY – APPLICATIONS IN BIOLOGY AND CHEMISTRY

In biology and chemistry, it is most often necessary to separate the components of a mixture which are very similar and are difficult to separate by chemical or physical methods. Chromatography and electrophoresis are the two powerful modern methods utilised for such a purpose.

1. Ion exchange chromatography has been widely used for the separation of amino acids and peptides, and also organic weak cations or anions from inorganic salts. It can also be used in the preparation of deionised water in the laboratory, but not non-ionic contaminants.
2. Paper chromatography can best be used in the separation and identification of amino acids in a mixture.
3. Thin layer chromatography can be used to separate the constituents of a mixture of amino acids, of neutral lipids and phospholipids which can also be estimated in a shorter time than in paper chromatography.
4. Gel chromatography is highly applicable for the separation of carbohydrates, proteins, peptides, amino acids, vitamins, steroids, neuropeptides, hormones and drugs. The separation is achieved very quickly. So, this is indispensable in research laboratories.
5. Gas liquid chromatography (GLC) is used to separate substances as vapours in an inert gas like nitrogen, argon or helium over liquids (e.g.: esters) which can be identified by physical or chemical means or flame ionisation.
6. The hyphenated techniques like LC-MS and LC-NMR are extremely powerful and informative tools for screening the crude extracts of plants and the former technique allows the analysis of small non-polar compounds to large polar constituents like oligosaccharides, proteins, and tannins present in natural product extracts.

CHROMATOGRAPHY - USES IN PHARMACOGNOSY

PC is one of the most common types of chromatography used for separating and identifying coloured mixtures like plant pigments. TLC is the standard analytical chromatographic method in many organic and pharmaceutical laboratories which has not yet been completely displaced by other methods due to its simplicity and high speed. Column chromatography is generally used for preparative applications from micrograms up to kilogram. GC is used to separate Gaseous mixtures. Ion exchange chromatography is used for the separation of almost any type of charged molecule including proteins, nucleotides and amino acids. Hydrophobic interaction chromatography (HIC) is normally used for the separation of immobilized hydrophobic groups (in amino acids) which are usually buried away on the inside of the protein. Gel filtration chromatography helps separation of proteins on the basis of their shape and size. Affinity chromatography is the most powerful technique available which can potentially allow a one-step purification of the target molecule and is used for the purification of any protein provided that a specific ligand is available such as recombinant proteins or lectin affinity chromatography where lectins are used for the separation of components within the sample. HPLC, a form of column chromatography is used to separate, identify and quantify compounds. Normal phase HPLC (NP-HPLC) separates the analytes on the basis of their polarity. Size exclusion chromatography (SEC), also known as gel-filtration chromatography is a low resolution chromatography which helps the separation of particles on the basis of their size. This method is often reserved as the final step of purification and is also useful for the determination of tertiary or quaternary structure of purified proteins. This method is widely used to determine the molecular weight of polysaccharides and is the official technique that has been suggested for the molecular weight comparison of different commercially available low molecular weight heparins by European Pharmacopoeia. Ion exchange chromatography is widely used in the purification of water, separation of proteins (IEC) and separation of carbohydrates and oligosaccharides (high pH anion exchange chromatography). Bio-affinity chromatography mainly works on the principle of the ability of biologically active substances to form stable, specific and reversible complexes due to the formation of a bio-specific bond by concerted action of many forces (Van der Waals interaction, electro-static interaction, dipole-dipole interaction, hydrophobic interaction and the hydrogen bond) in the complementary binding site.

CHROMATOGRAPHY – APPLICATIONS FOR THE ANALYSIS OF HERBAL DRUGS

Liquid chromatography has its important application in pharmaceutical analysis, forensic chemistry, forensic toxicology, toxicological analysis, food chemistry and analysis of herbal drugs. This method is used for the separation of secondary metabolites and metal ions in the herbal drug mixtures.

Gas chromatography is useful in analysis of herbal drugs, forensic chemistry, forensic toxicology, toxicological analysis and food chemistry. This is a common method used for the separation of volatiles and secondary metabolites during the analysis of herbal drugs.

Thin layer chromatography finds its use in analysis of herbal drugs and food chemistry. This is a simpler and a rapid method for checking the purity of herbal drugs (secondary metabolites and volatiles) as well as detecting the pesticide or insecticide residues in the crude drugs.

Paper chromatography has its main application in the analysis of herbal drugs. This is the most common type of chromatography used for the separation of amino acids and anions, RNA fingerprinting, separation of histamines and antibiotics.

HPLC methods are useful for the qualitative and quantitative determination of artificial colourants (rhodamine-B), aflatoxins, coumarins, genotoxic anthroquinones, isoflavones (phytoestrogens) in soy beans and soy products; isoflavones in red clover; pyrrolizidine alkaloids, caffeine, catechines (in green tea); and opium and other alkaloids in many plants and mother tinctures. This method is widely used in food chemistry and pharmaceutical industry and also finds its use in clinical analysis of herbal drugs (biopharmaceutics, cocaine in urine, alcohol in blood etc.); for the screening, determination and measuring of narcotics (opium alkaloids, cocaine and morphine); for health consequences of acute and chronic use of marijuana, and for several other applications to forensic problems in forensic chemistry and forensic toxicology.

CHROMATOGRAPHY – LATEST AND ADVANCED TECHNIQUES

(The Role of Chromatography in the Modern Laboratory)

The chromatographic techniques are an important element of majority of organic testing labs whether it is a liquid phase or gaseous phase, but, they are only a part of a much bigger picture. Chromatography is a complex process; even then the fundamental principles that drive the method still remain the same. Basically, it is the separation of a complex mixture of components (a blend of hydrocarbons in a fuel sample or a combination of pharmaceuticals in a waste water effluent) into separate entities, the main aim of the process is to separate out the components before they can be detected by a detector.

Evolution of Chromatography

The science of chromatography over the past 100+ years has improved to enable fast and reliable separations to be undertaken. Though there are several forms of this technique, the two most common within the environmental labs are liquid chromatography (often referred to as LC or HPLC [high performance liquid chromatography] and gas chromatography. LC (HPLC) is a direct evolution of the ink spot test. The samples are introduced into the system via high precision injectors which are capable of injecting sub-microliter to several milliliter volumes. one can have an access to a wider range of packed columns instead of a modified filter paper, that contain ultra-purified silica or polymers which have had their surfaces modified to alter their interaction characteristics. The columns vary in their size depending on the application in hand, but in analytical applications, 1–5mm in diameter and 100mm to 250mm is common. The flow is under direct control by high pressure pumps which are capable of running at 100s if not 1000s of bar pressure, though the solvent is used to enable the separation of the components. The entire system is controlled by the computer with additional variables like low rate, temperature and composition of the solvent are under the control of the researcher/analyst. Hence, separation of a limitless range of substances (compounds) is possible. The examples are organic acids, phenolic compounds, vitamins, fluorinated surfactants (PFOs) and drugs of abuse.

Gas chromatography (referred to as GC operates in the gas phase), is a technique where in the separating components that separate must be volatile and thermally stable (unlikely to break down), and follows the same principles of LC, but the mobile phase to drive the separations is a combination of heat and gas flows (instead of a liquid solvent). Similar to HPLC, the liquid samples are introduced in to the GC via a high precision injector (the volumes are 1–5 micorlitres), and the injection port is heated (to 200–300°C), so that the liquid injected is quickly vapourised and becomes a gas. The sample (as vapour) is swept into the analytical column by a flow of high purity gas (the mobile phase is referred to as a carrier gas). The columns are much thinner (1 mm in diameter) and much longer (between 10 to 60m long) than for LC application. These are made of drawn glass tubing coated with special polymers not to break under extreme temperature changes in general use). The flow is under the direct control by the high pressure pumps which run at 100s of bar pressure. The entire system is controlled by a computer with additional variables (like flow rate and temperature under the control of the analyst). The efficiency of the GC column is much greater than that of an LC column, and the peaks in GC are much sharper and many components can be separated with a given time frame. The examples include: separation of petroleum hydrocarbons, PCBs (polychlorinated biphenyls), pesticides (like DDT and Lindane) and many more.

Chromatography – Improvements – Multi-dimensional Analysis

The present day technology enables multi-dimensional separations to take place using two columns simultaneously. The technique is referred to as GC×GC or LC×LC depending on the application and enables greater separation when compared to the standard single column approach.

In multidimensional approach, the two analytical columns of different stationary phases are connected together prior to the detector. Whilst they are in series, the flow of the mobile phase (or carrier gas) does not travel linearly to the detector through them, but it is more of a modulated approach. In GC×GC, the second analytical column is shorter than the main primary column. The first column separates on boiling point providing the initial separation of the components prior to entry into the modulator. Later, after the transfer to the second column it separates on the basis of the polarity and the components are detected. The combination of the columns gives a significant increase in the separation of the components, on the standard x-axis versus y-axis and a third z-axis.

The methods based on the partition are effective on the separation and identification of small molecules, such as the amino acids, carbohydrates and fatty acids. The affinity chromatography like ion-exchange chromatography is more effective in the separation of the macromolecules like nucleic acids, and proteins. Paper chromatography is used in the protein separation, as well as in the studies related to the synthesis of protein. Gas liquid chromatography is used in the separation of alcohol, ether, lipid and amino groups as well as the observation of enzymatic interactions. Molecular-sieve chromatography is used especially for the molecular weight determinations of proteins. Agarose-gel chromatography is employed for the purification of RNA, DNA particles and viruses. Gas chromatography is used for gases, and mixtures of volatile liquids and solid material. Liquid chromatography is used for thermally, unstable and non-volatile samples. The main purpose of the application of chromatography as a method of quantitative analysis is to achieve a satisfactory separation with in a suitable time interval. Various methods have been developed and some of them include:

1. Dye-Ligand Chromatography

Reactive dyes are known to bind proteins with high affinity, but with varying binding capacity and specificity. The chemical composition of most of these reactive dyes is comprised of azo dyes, anthraquinone, or phthalocyanine, linked to a reactive group, often a mono- or dichlorotriazine ring. This unique binding of various proteins to the dyes is still enigmatic and is a multifarious display of electrostatic, hydrophobic hydrogen bonding. The development of this technique was based on the demonstration of the ability of many enzymes to bind purine nucleotides for Cibacron Blue F3GA dye. Cibacron Blue 3-GA can bind using other mechanisms, such as ionic forces (most dominant method), hydrophobic interactions, or exclusion-diffusion. Synthetic dyes (also called as - *biomimetic dye-ligands*) can act as alternative ligands for proteins and can be used as part of a protein purification protocol for chromatography. The ligands that interact with the targeted proteins are used to capture these proteins as the sample is passed through the chromatography column. The ligands used in the chromatographic columns, such as antibodies, lectins and nucleotide co-factors, are unstable, expensive and have a low capacity. The dyes used as ligands are based on a chromophore, typically either azo, anthraquinone or phthalocyanine, and they also have a reactive part, typically a chlorotriazine ring. The chromophore gives the colour to the dye and should have sulfonic groups to allow it to be soluble in water. The reactive part allows the dye to bind to the matrix.

The planar ring structure with negatively charged groups is analogous to the structure of NAD. The analogy has been proved by the binding of Cibacron Blue F3GA dye to adenine and ribose binding sites of NAD (the dye behaves as an analogue of ADP-ribose). The binding capacity of such adsorbents is 10–20 fold stronger that of the affinity of other adsorbents. Under appropriate conditions of pH, the elution with high-ionic strength, and using ion-exchange property of the adsorbent, the adsorbed proteins are separated from the column. The process depends on a variety of factors, including dye concentration, buffer conditions, flow rate, and column geometry. The different steps that are involved are (i) Purification of the dye, (ii) Immobilizing the dye onto the chromatography matrix, and (iii) running the chromatogram.

The synthetic dyes are advantageous as they are inexpensive, can be easily immobilized on to the matrix of chromatography column, are resistant to chemical or enzymatic degradation, and they have a capacity far higher than other ligands. Hence, this method is good for large scale protein purification. But, the dyes are not as specific as other types of ligands, because of their potential usage. Such a low specificity is an advantage as it could reduce the need for the use of multiple ligands.

Dye ligand chromatography is rapidly arising as a suitable substitute to specific affinity chromatographies owing to their easy accessibility and economical, and effortless immobilization. Despite being synthetic in nature, the triazine-based dyes have a high affinity for many proteins and enzymes. In the immobilized form, the dye has rapidly been identified and used as an affinity ligand in chromatographic isolation of a variety of proteins including dehydrogenases, kinases, serum albumin, interferon, several plasma proteins, and a host of other proteins owing to their ability to mimic the configuration of substrates, cofactors, or binding agents, thereby leading to high recovery and many a time purity in tandem, both achieved in a single chromatographic step.

2. Hydrophobic Interaction Chromatography (HIC)

HIC (salting out) is one of the most widely used methods for the separation and purification of proteins in their native state. It is also useful in the isolation of protein complexes and also in studying the protein folding and unfolding. This technique separates the molecules based on their hydro-phobicity, and is a useful technique for protein purification while maintaining the biological activity due to the use of conditions and matrices that operate under less denaturing conditions.

In this method, the proteins containing both hydrophilic and hydrophobic regions are applied to an HIC column under high salt buffer conditions. In this technique, the adsorbents prepared as column material for the ligand-binding in affinity chromatography are used. The ligands used in HIC media have alkyl or aryl groups (alkyl, Butyl, Octyl, Ether, Isopropyl etc.). This technique is based on hydrophobic interactions between the side-chains bound to the chromatography matrix. The general considerations for HIC include the factors, such as ligand, matrix, salt concentration, pH and temperature. In recent years, HIC has emerged as a powerful technique for the purification of the biological compounds on a laboratory scale as well as at the industrial scale. This combines the non-denaturing characteristics of salt precipitation and the precision of chromatography to obtain high resolution along with good recovery. It also preserves the biological activity of the proteins due to a weaker interaction than other chromatographic techniques.

There are three major theories that help to explain the mechanism behind HIC. They are: the *salting out theory* - the *thermodynamic theory* and the *surface tension, Van der Waal`s theory*.

The salting out theory states that the hydrophobic amino acids form protected hydrophobic areas upon folding in aqueous solution, while the hydrophilic amino acids allow the proteins to bond with the hydrogen molecules in water, as a result of which the probability that the protein will dissolve in water will increase if enough hydrophilic areas are present in the surface of the protein molecules. Upon the addition of anti-chaotropic salts, such as ammonium sulphate and sodium sulphate in the solution, some of the water molecules will interact with the salt ions instead of the charged part of the protein. When the protein – protein interactions in the solution become stronger than the solvent – solute interactions, the protein molecules react freely with one another, allowing them to aggregate and finally precipitate out of the solution. So, the addition of salt in the solution reduces the solubility of different proteins to varying extents.

The thermodynamic theory states that the interaction between the hydrophobic molecules is an entropy-driven process based on the second law of thermodynamics. The hydrophobic interaction between two or more non-polar molecules is a polar solvent solution is a spontaneous process that is governed by a change in entropy.

Such interactions can be altered by controlling the temperature or by the modification of the polarity of the solvent.

The surface tension theory suggests that the hydrophobic interaction in HIC is dependent on Van der Waals forces between the proteins and the immobilized ligands since these forces increase as the ordered structure of water increases in the presence of salting out salts.

3. Pseudo-affinity Chromatography

The term "*pseudo-biospecific affinity ligands*" (introduced by Vijayalakshmi in 1989), encompasses simple, small, and chemically defined molecules either of biological origin, such as single amino acids, or of non-biological origin, such as triazine dyes from the textile industry or metal chelates. Extensive investigation has been carried out to elucidate the mechanisms involved in *pseudo-ligand affinity chromatography* using textile dyes, and, empirically, it has been attributed to the chemical and steric structures of dye and protein. Possibly, a variety of interactions especially ionic and/or hydrophobic influences with a varying share in the binding and differ from protein to protein and from dye to dye. Some compounds (anthraquinone dyes and azo dyes) can be used as ligands in order to target the proteins, because of their affinity, especially for the enzymes such as dehydrogenases, kinases, transferases and reductases. The dyes that are used to mimic the ligands, but they do not display high specificity, and therefore, the dyes are able to capture a variety of proteins. A study by Tulsani and co-workers looked at this technique to separate three enzymes, from the muscles of rabbit, namely, lactate dehydrogenase, pyruvate kinase and aldolase. In this method, cross-linked guar and cross-linked pectin were used to immobilize different traizine dyes. In this method, guar and pectin were first washed in deionized water, then suspended in the fresh deionized water. The dye was then added to the gel, stirred for 5 minutes. The salt was added to this mixture, and stirred further for about 30 minutes. Sodium hydroxide was added to the mixture and stirred for 16–18 hours, and the mixture was finally washed in deionized water. The dye-bound gel was placed into 2 ml columns. The crude protein sample was loaded on to the top of the column when the phosphate buffer had passed through the column. When the sample had passed through, the column was washed with the buffer, to remove the proteins that were unbound. This was verified by measuring the absorbance of the buffer that leaves the column at 280nm with the help of a spectrophotometer. Once clean, the bound protein was released by using the same buffer, but with added salt, NAD^+ or sodium pyruvate. Of the three enzymes, it was found that the dyes 1014 and 1015 retained LDH, and 1015 bound to cross-linked pectin also retained pyruvate kinase, while aldolase was not retained by any of the 10 dyes tested in this experiment.

Similarly, a study by Urbas and co-workers looked at optimizing the methods used to remove serum albumin and IgG from the plasma. They not only used pseudo-affinity chromatography to selectively remove the serum albumin, but combined this with affinity chromatography so that both serum albumin and IgG could be removed at the same time. Though this method was found to be efficient, the Mimetic Blue SA A6XL dye used in pseudo-affinity chromatography to remove the serum albumin was not highly specific. The most known type of this type of chromatography is IMAC (Immobilized Metal Affinity Chromatography).

The three classes of ligands, namely, the amino acids, triazine dyes and metal chelates recognize the molecules of protein through almost similar mechanisms. All the immobilized ligands act as electron acceptors. The complementary donor groups in the proteins are the NH-containing, SH-containing or OH-containing lateral groups of amino acids like histidine, cysteine, tryptophan, tyrosine and serine. Among the three classes, the single amino acids, like histidine and tryptophan have an inherent advantage over the other two classes, by virtue of their biological (non-toxic) and biocompatible nature (the purification of plasminogen and fibronectin using immobilized lysine and arginine has been reported – Suamari et al., [1976] and Vuento and Vaheri [1979]). The aromatic amino acids present a more versatile potential as ligands than the aliphatic amino acids. Tryptophan immobilized on to polysaccharide matrices was shown to be capable of separating interferon and cellulases from *Trichoderma viride*. The heterocyclic amino acid histidine has proved to be a more versatile immobilized ligand with an enormous potential in affinity chromatography purification of a plethora of proteins.

4. Immobilized Metal Affinity Chromatography (IMAC)

Also called as MCAC (*Metal-Chelate Affinity Chromatography*) was first demonstrated successfully by Porath and collaborators in 1975 for human serum proteins. IMAC of proteins is a HPLC technique. It has the ability to differentiate a single histidine residue on the surface of a protein, and it can bind proteins with dissociation constants of 10^{-5} to 10^{-7}. This is specialized variant of affinity chromatography where the proteins or peptides are separated according to their affinity for metal ions that have been immobilized by chelation to an insoluble matrix. At pH values around neutral, histidine, tryptophan and cysteine form complexes with the chelated metal ions (like Zn^{2+}, Cu^{2+}, Cd^{2+}, Hg^{2+}, Co^{2+}, Ni^{2+} and Fe^{2+}) which can then be eluted by reducing the pH, increasing the mobile phase ionic strength, or adding EDTA (ethylene diamine tetra acetic acid – 0.05M) to the mobile phase. It has wide application in molecular biology for the rapid purification of the recombinant proteins, as poly-histidine fusions and for the membrane proteins and protein aggregates where detergents or high ionic strength buffers are required.

5. Perfusion Chromatography (PC)

In 1974, Fred Regnier, a chemistry professor, Purdue University, was the first to use HPLC to successfully isolate the proteins and peptides, and he has brought another innovation to the field of chromatography. Regnier and Noubar Afeyan, a bio-processing engineer, and his collaborator, Purdue University, have created an HPLC system, called Perfusion Chromatography (PC), which separates the bio-molecules using a new type of material in its stationary phase. They also found that the materials helped in better separations under the conditions of high flow rates. When the flow rate of the mobile phase is increased, the separation efficiency decreases, causing a more dilute and a less pure material to come out of the column. But, this was not the case with PC stationary phases. The two scientists have established a new company, called *Perseptive Biosystems Inc.* in Cambridge, Massachusetts.

6. Supercritical Fluid Chromatography (SFC)

This is a separation technique (sometimes called as the *Convergence Chromatography* - CC) in which the mobile phase is a fluid above and relatively close to its critical temperature and pressure. This method utilizes carbon dioxide as the mobile phase and hence the entire chromatographic flow path must be pressurized. The supercritical phase is a state in which the liquid and gas properties combine.

7. Expanded Bed Adsorption (EBA) Chromatography

This chromatographic separation captures a target protein from a crude feed stream when it passes through a chromatographic column system containing adsorbent beads. The unprocessed crude compound can be treated directly in the chromatographic column, avoiding clarification and pre-treatment steps using this technique.

8. Nanoflow Liquid Chromatography (NLC)

This is a new technique created NASA (National Aeronautics Space Administration, USA) which allows to scan the extremely small meteorite samples for the key ingredients of life, and help the researchers and scientists unravel the mystery of life's origin on Earth. This can also be used to examine the extra-terrestrial materials small in scale like inter-planetary dust particles, micrometeorites, and cometary particles according to NASA. Dr. Michael Callahan (2014) and his team of researchers from NASA`s Goddard Space Flight Centre, Greenbelt, Maryland, USA, Md, have found amino acids, used to make the proteins, after analyzing the carbon-rich meteorites. This was an interesting finding, called Nanoflow Liquid Chromatography. Callahan and his team also found the evidence of the components that are responsible for creating DNA and other molecules that are a key to the building blocks of life, after examining the meteorite. Callahan and his team used a nanoflow liquid chromatography instrument to extract the samples from a meteorite, sorted the molecules for the samples, and added them to a spectrometer which analyzed them molecule by molecule to detect the potential amino acids, the key components of a protein. With this new technique, humans may understand and know where life comes from in the coming years, but substantial enough evidence is needed.

9. Micellar Liquid Chromatography (MLC)

MLC is a form of reversed phase liquid chromatography that uses aqueous micellar solutions as the mobile phase. Armstrong and Henry (1980) were the first to introduce the use of micelles in HPLC. This method is used mainly to enhance retention and selectivity of different solutes that would otherwise be inseparable or poorly resolved. MLC has been used in different applications which include the separation of mixtures of charged and neutral solutes, the direct injection of serum and other physiological fluids, the analysis of pharmaceutical compounds, the separation of enantiomers, the analysis of inorganic organo-metallics etc.

The main drawback of this technique is the reduced efficiency that is caused by the micelles. Despite its poor efficiency, this is a better choice than ion-exchange liquid chromatography or ion-pairing liquid chromatography for the separation of charged molecules and mixtures of charged and neutral species.

The micelles are composed of surfactant (or detergent) monomers with a hydrophobic moiety (or tail) on one end, and a hydrophilic moiety (or head group) on the other. The polar head group may be anionic, cationic, zwitterionic or non-ionic. When the concentration of a surfactant in solution reaches its CMC (critical micelle concentration), it forms micelles (which are aggregates of the monomers). The CMC is different for each surfactant, as is the number of monomers which make up the micelle, termed as the aggregation number (AN).

The applications of this method include: the foremost advantage is its ability to directly inject physiological fluids (such as plasma, serum and urine). Martinez *et al.* found MLC to be highly useful in analyzing the class of drugs called β-antagonists (also called as beta-blockers) in urine samples. With this method, direct injection is often possible, with retention times of less than 15 minutes for the separation of up to 9 beta-antagonists. Another application is the analysis of desferrioxamine (DFO – very common drug for the removal of excess of iron in patients with chronic and acute levels) in serum. It was found that reverse phase HPLC was a better and more sensitive technique in this case, despite the time savings in direct injection. MLC mimics, but enhances the selectivity by ion-pairing reagents for the separation of active ingredients in pharmaceutical drugs. Hydrophilic drugs are retained by MLC due to solubilisation into the micelles. By using MLC, drugs in cold medications like acetaminophen, phenylpropanolamine HCL, tipepidine hibenzate and chlorpheniramine maleate have been successfully been separated, in addition to basic drugs like narcotics (like codeine and morphine) have also been separated using MLC. Another application is the separation and analysis of inorganic compounds, mostly simple ions. In future, the use of MLC appears to be extremely advantageous in the area of physiological fluids, pharmaceuticals and even inorganic ions. This technique has been proven to be superior over ion-pairing and ion-exchange for many reasons.

10. Electrochromatography

This is a chemical separation technique in analytical chemistry, biochemistry and molecular biology and is used to resolve and separate mostly larger biomolecules (like proteins). It is a combination of gel filtration chromatography (size exclusion chromatography) and gel electrophoresis. These mechanisms essentially operate in superposition along the length of a gel filtration column to which an axial electrical field gradient has been added. The molecules are separated by their size due to the gel filtration mechanism and by electrophoretic mobility due to the mechanism of gel filtration. In addition, there are secondary chromatographic solute retention mechanisms.

11. Capillary Electrochromatography (CEC)

CEC is a chromatographic technique in which the mobile phase is driven through the chromatographic bed by electro-osmosis. This is a combination of two analytical techniques, namely HPLC and CEC. The main aim of this method is to separate the analytes on the basis of their mass-to-charge ratio by passing a high voltage across the ends of a capillary tube, filled with the analyte. HPLC separates the analytes by passing them under high pressure through a column filled with the stationary phase. The interactions between the analytes and the stationary phase and the mobile phase lead to the separation of the analytes. In CEC, capillaries (packed with HPLC stationary phase) are subjected to a high voltage. Their separation is achieved by electrophoretic migration of the solutes and differential partitioning.

The advantages of CEC are as follows: Avoiding the use of pressure to introduce the mobile phase into the column, results in a number of advantages.

First advantage is, the pressure driven flow rate across a column depends directly on the square of the particle diameter and inversely on the length of the column. This restricts the length of the column and the size of the particle (the size of the particle is seldom less than 3 micrometer) and the length of the column is restricted to 25 cm. Electrically driven flow rate is independent of the length of the column and the size.

A second advantage of using the electro-osmosis to pass the mobile phase into the column is the plug-hole flow velocity profile of EOF, which reduces the solute dispersion in the column, increasing the column efficiency.

12. Weak Affinity Chromatography (WAC)

WAC is an affinity chromatography technique for affinity screening in drug development, and is an affinity-based LC that separates the chemical compounds on the basis of their different weak affinities to an immobilized target. The higher affinity a compound has towards the target, the longer it remains in the separation unit, and this will be expressed as a longer retention time. The affinity measure and ranking of an affinity can be achieved by processing the obtained retention times of the analyzed compounds. This technology is demonstrated against a number of different protein targets, – such as proteases, kinases, chaperones, and PPI (protein-protein interaction) targets. This technique has been more effective than the established methods of fragment based screening.

13. High Performance Liquid Chromatography (HPLC)

HPLC earlier referred to as the high pressure liquid chromatography, is a form of column chromatography, and an analytical technique used to separate, identify and quantify each component in a mixture. HPLC has been used for the manufacturing (of pharmaceutical and biological products), legal (detection of the performance enhancement drugs in urine), research (separation of the components of a complex biological sample, or of similar synthetic chemicals from each other), and medical (detecting vitamin D levels in the serum of blood) purposes.

The two important types of HPLC are: uHPLC (*Ultra High Performance Liquid Chromatography*) and FPLC (*Fast Protein Liquid Chromatography*).

a. Ultra High Performance Liquid Chromatography (uHPLC)

This is also known as UPLC. The standard HPLC method uses column particles with sizes from 3 to 5 μm and pressures of around 400 bar. uHPLC uses specially designed columns with particles down to 1.7 μm, and at pressures in excess of 1000 bar. The main advantage of this method is speed. The system is faster, more sensitive and relies on smaller volumes of organic solvents than the standard HPLC, and results in the ability to run more samples in less time. The columns age or degrade quicker if the systems are run at typical pressures greater than 800 bar. The new technology is being developed at present for uHPLC units to use the column particles with 1μm size, and pressure potentials up to 6800 bar.

b. Fast Protein Liquid Chromatography (FPLC)

FPLC is similar to HPLC and is used to separate or purify the proteins and other bio-molecules from complex mixtures. The main difference between the two is the standard working pressure. FPLC columns can only be used up to maximum pressures of 3–5 MPa. If the pressure of the HPLC can be limited, nearly every FPLC column may also be used in HPLC system.

Conclusion

In the beginning, the techniques of chromatography were used to separate the substances on the basis of their colour as was the case with plant pigments. But, with time, its application area was extended considerably and nowadays, the technique is accepted as an extremely sensitive and effective separation method. One of the useful separation and determination methods is the column chromatography, a protein purification method especially based on one of the characteristic features of the proteins. In addition, these methods are also used to control the purity of a protein. Similarly, HPLC technique because of its higher sensitivity, rapid turnover rate, and its use as a quantitative method can be used to purify the amino acids, proteins, nucleic acids, hydrocarbons, carbohydrates, drugs, antibiotics and steroids. These methods are valuable tools for biochemists, besides their application in the clinical laboratories. For example, PC is used to determine different sugars and amino acids in the body fluids associated with hereditary metabolic disorders. GC is being used to measure the steroids, barbiturates and lipids. Some of these techniques are also used in the separation of vitamins and proteins.

Hy-phe-nated Techniques or Hybrid Techniques

Hy-phe-nated HPLC (High-performance Liquid Chromatography) Methods

A hyphenated (or hybrid) technique is a combination (or coupling) of two different analytical techniques with the help of proper interface, to exploit the advantage of both. Mostly chromatographic techniques are combined with spectroscopic techniques, and these range from the combination of separation-separation, separation-identification and identification-identification techniques. It was Tomas Hirsch Feld (1980) who adapted the term "*hyphenation*" first to describe a possible combination of two or more instrumental analytical methods in a single run. The objective of coup-0lin is to obtain an information-rich detection for both identification and quantification compared to a single analytical technique, now-a-days referred to as a "*hyphenated technique*". The recent advances in the application of these techniques are useful in the pre-isolation analyses of the crude extracts from different natural resources, isolation and on-line detection of the natural products, chemo-taxonomic studies, chemical fingerprinting, the quality control of the herbal products, dereplication of natural products and metabolomic studies.

These hybrid techniques are mainly of two types, namely, double hyphenated and triple hyphenated techniques. The double hyphenated techniques are represented by: LC-MS (Liquid Chromatography-Mass Spectrometry); LC-NMR (Liquid Chromatography-Nuclear Magnetic Resonance); LC-IR (Liquid Chromatography-Infra Red); CE-MS (Capillary Electrophoresis-Mass Spectrometry); GC-IR (Gas Chromatography-Infra Red); GC-MS (Gas Chromatography-Mass Spectrometry); and GC-FTIR (Gas Chromatography-Fourier Transformed Infra Red Spectroscopy).

The triple hyphenated techniques are represented by: LC-API-MS (Liquid Chromatography-Atmospheric Pressure Ionization-Mass Spectrometry); ESI-MS-MS (Electro-spray Ionization-Tandem Mass Spectrometry); LVI-GC-MS (Large Volume Injection-Gas Chromatography- Mass Spectrometry); LVI-ESI-MS (Large Volume Injection-Electro-spray Ionization-Mass Spectrometry); LC-NMR-MS (Liquid Chromatography-Nuclear Magnetic Resonance-Mass Spectrometry); LC-UV-PDA (Liquid Chromatography-Ultra-violet – Photodiode Array Detection); LC-PDA-MS (Liquid Chromatography-Photodiode Array Detection-Mass Spectrometry); LC-PDA-NMR-MS (Liquid Chromatography-Photodiode Array Detection-Nuclear Magnetic Resonance-Mass Spectrometry); and SPE-LC-MS-MS (Solid Phase Extraction-Liquid Chromatography-Mass Spectrometry).

Nature represents an extraordinary reservoir of novel molecules and there is currently a resurgence of interest in natural products as a possible source of new lead compounds for the introduction into therapeutical screening programmes. To discover new bioactive natural products, the dereplication of crude extracts performed prior to the isolation work is of crucial importance for avoiding the tedious isolation of known constituents. In this respect, the chemical screening strategies based on hyphenated technique are more extensively used.

The advantages of these methods are: to solve the complex analytical problems – a fast and accurate analysis, a shorter analysis time, a higher degree of automation, a higher sample throughput, a better reproducibility, a reduction of contamination because it is a closed system, separation of quantification at the same time, an enhanced combined selectivity and there a high degree of information, and it provides an excellent separation efficiency as well as the acquisition of on-line complementary spectroscopic data on an LC or GC peak of interest within a complex mixture.

■■■

CHAPTER - 7

BOTANICAL MICRO-TECHNIQUE

Introduction

Micro-technique is the body of knowledge and the skills used in the preparation of plant or animal materials for the study under microscope. The development of our knowledge of the structure of the cell has been correlated to a large extent with the development of methods of micro-technique, and it is a significant fact that even Robert Hooke did not discover the cell until he prepared thin sections of plant tissues.

The advances made in micro-technique in plant and animal sciences (botany and zoology) are closely related to each other in such a way that the discoveries made by the zoologists in making the preparations were adopted by botanists, while the botanical methods were successfully utilized by the zoologists. The microscopists (neither botanists nor zoologists) have played a very appreciable part in the development of the methods of making the microscopical preparations; as such we are much indebted to the greatest extent to the microscopists. The most important contribution of the microscopists to micro-technique between 1860 and 1875 was the introduction of staining (of plant tissues). Although the attempts at staining were first made by microscopists on botanical material (or by botanists), the development to the technique of staining is almost wholly due to the zoologists. The development of the embedding methods is also due to zoologists. The fixing solutions are a comparatively recent development. The term fixation came into use in the early eighties (in the decade of 1880 – 1890). Between 1875 and 1880, the most noticeable progress in botanical micro-technique was the gradual adoption of the methods of staining. During the late seventies, other methods used by the zoologists were introduced into botanical micro-technique, namely, clearing before mounting, dehydration, embedding etc.

The preparation of permanent slides from microtome sections of materials collected consists of the following processes: *Selecting* the desired plants or parts of plants and if necessary dividing them into suitable pieces; the *Killing* and *Preservation* of the contents of cells and of cell structures in a condition approximating that in the living plant; *Embedding* in a matrix if necessary, in order to support the tissues for sectioning; *Sectioning* of tissues into very thin slices; and *Staining* the slices and covering with cover slip (cover glass) to a permanent slide.

In this chapter the methods of making microscopical preparations have been treated from the botanical standpoint, and the history of microscope and the development of other methods of culture have not been considered.

KILLING AND FIXATION

The killing and fixing of the material to be made permanent preparations are two important and distinct processes, but both are usually obtained by a single fluid which in turn is commonly a mixture of different chemical reagents. The term, *killing* means the sudden and permanent termination of the life processes. It does not apply solely to an entire organism, but to the individual cells of which the organism or tissues are composed. Any reagent used for killing must, therefore, reach every cell in the organism or a piece of detached tissue if it is to do its work completely and effectively. Killing invariably precedes fixation since the reagents which do the killing penetrate the tissues faster than those upon which the responsibility for fixation rests. For example, alcohol is primarily a killing agent.

Fixation is a process difficult to describe in easy terms, and is commonly said to be the preservation of all cellular and structural elements in as nearly the natural living condition as possible. A good fixative is one that changes the cell chemistry the least, and preserves the cell structure the best (Schiller, 1930). The fixation is an entirely empirical process, the success or failure of which is best judged by the quality of its usefulness when the complete preparation is examined. Many technicians do not need exact fixation, and they are satisfied with approximation. For example, a cytologist needs that the chromosomes be spread out to facilitate counting, the chromosomes are darkly stained and that the cytoplasm stains lightly or not at all. Accurate preservation of the internal structure and of the cytoplasm is not necessary for them. The most common explanation is that the fixation is required in order that the obscure or entirely invisible structures when the cells are observed in the living condition may be seen more distinctly, and in order that the soft structures may be hardened sufficiently for further treatment.

Earlier plant morphologists were interested more in definite cell regions than in cellular contents. They fixed the material so that they could stain things in the living cells which enabled them to make their precise observations. The fixation meant merely the preservation of structure, and this has been carried over into cytology. The final aim in fixing the protoplasm is to make invisible or barely visible structures easily observable.

Great care should be exercised in handling the tissues or organisms before placing them in the killing fluid. Very small organisms (filamentous, colonial and unicellular algae and filamentous algae, *Euglena*, small bryophytes) may be placed entire in the fluid. But, with larger organisms, dissection or reduction to smaller portions is required. The pressure of any sort is to be avoided and the work should be completed very quickly (after removal from the plant or from the aquatic habitat, the material should be immediately placed in the killing fluid without any lapse of time). The smaller the portion of tissue and the larger the surface exposed to the direct action of the killing fluid, the better the killing and fixation. The weaker fixatives penetrate with difficulty. The poor penetration of the fixative may result in the over-fixation of the outer layers while the innermost layers are under-fixed. The remedy in such cases is to use fixatives of stronger penetrating power. For example, acetic-alcohol and picro-aceto-formalin combinations penetrate rapidly and easily, while mixtures containing chromic and osmic acids penetrate poorly. In general, the organs or pieces of tissue should never exceed 5 mm for optimum results.

Fixation Images

Nearly all the fixing fluids in common use give the acid fixation image. The very few fluids giving the basic image have been compounded, and the most useful ones have been developed during the past few years. The variations do occur in each image according to the material. If the fixing fluid contains two or more substances, the resulting fixation image is determined by the component which penetrates more rapidly. The different reagents penetrate at different rates, and also according to their respective concentrations. For majority of purposes, a fluid that gives an acid image will be the most satisfactory while the basic fluids are usually employed for studies on cytoplasmic structure and inclusions like mitochondria.

The various *mixtures that give acid fixation images* are generally grouped according to whether the main ingredient is formalin, glacial acetic acid, chromic acid or different combinations of these three either alone or with the addition of other reagents. For example, the *acetic acid-alcohol mixtures* are represented by Carnoy-LeBrun`s fluid, Gilson`s fluid and Petrunkevitsch`s fluid; while *formalin-acetic acid-alcohol mixtures* are exemplified by Formalin-aceto-alcohol (FAA) and Formalin-propiono-alcohol (FPA).

The most commonly used killing and fixing fluids are: Absolute ethyl alcohol (fair if immediate results are desired), and 95% Ethyl alcohol (fairly general preservative) which should not be used at low temperatures, Chloroform (as a constituent of Carnoy`s fluid and never used alone), Formalin (best to use the pure solution used alone), Glacial acetic acid (as a constituent of fixing fluids), Nitric acid (fixative of little use), Picric acid (as a saturated solution), Osmic acid (Osmium tetroxide – highly expensive), Iodine (excellent as a weak solution of iodine and potassium iodide), Potassium bichromate (decidedly useful) etc.

Common Fixatives (FAA and FPA)

Formalin-Aceto-Alcohol (more familiar as FAA) is the standard preservative of botanical micro-technique which is used more extensively than any other fixative. The standard composition of FAA is: 90 ml of 70% ethyl alcohol, 5 ml of glacial acetic acid and 5 ml of formalin. Formalin Propiono - Alcohol (familiar as FPA) can be prepared by the addition of 25 ml of concentrated Propionic acid, and 25 ml of Formalin (approximately 37% of formaldehyde solution) to 450 ml of 50% Industrial Methylated Spirit. Propionic acid appears to give more adequate fixation than acetic acid. The proportion and the time of fixation and the manner of washing are identical as with FAA.

For hard woody materials, it is advisable to decrease the amount of acetic acid and to increase the amount of formalin since the latter penetrates more slowly than the former. This can be used with almost any plant material intended for anatomical or morphological study, and is not suitable for chromosome studies. The material can be left in it indefinitely with appreciable damage, and as such it is an ideal fluid to take on long collection trips. The minimum time of fixation is 18 hours. The killing fluid is washed out in two changes of 50% ethyl alcohol, while the woody material should be washed for two days in running water and softened for 3 to 6 weeks in a 50% aqueous solution of hydrofluoric acid, if necessary.

Characteristics of a Fixative

A good fixing agent should fulfil many criteria. They are: the fixative must kill the specimen quickly, but some chemical fixatives are toxic and are also harmful to a person`s health. The fixative must preserve the structure of the specimen without introducing deformation or other artefact. The structure should be sufficiently stable to withstand the dehydration and mounting. The fixative must enter the specimen well to react with all parts. The specimen must be sufficiently small so that the fixative can enter all the parts very easily. Some specimens contain air bubbles which prevent the fixative to reach all the parts, and in such a case, it may be necessary to apply a vacuum to remove the air.

Types of Fixatives

The chemical fixatives can be categorized into 4 groups, namely, alcohol and acetic acid, aldehydes, oxidation agents and tanning agents. The combination of alcohol and acetic acid denatures proteins, and the alcohol may also remove some lipids. This fixative is always preferred as it is less toxic than others. Aldehydes such as formaldehyde react with amino groups in the specimen. The oxidation agents react with lipids. The tanning agents react with proteins and with amino groups. The choice of the fixing agent must be carefully matched with the specimen, as some fixatives like alcohol may result in the shrinking of the specimen and therefore introduce artefacts. In some cases, it may be necessary to gradually increase the concentration of the fixative in order to prevent the formation of artefacts, which depends of the type of specimen.

Pure alcohol (ethanol) is suitable for fixing and storing plant specimens. Though the alcohol has a tendency to shrink cytoplasm, it does not affect the cell walls. It also hardens the plant material making it easier to cut with a microtome. Acetate (acetic acid) compensates the shrinking effect of alcohol, and the correct alcohol - acetate ratio should be fine tuned experimentally. The process of fixing should take place for about 24 hours.

After fixing, two more steps are necessary i.e., the fixative that is in excess has to be removed by washing the specimen and it has to be dehydrated. Several fixatives are water-based and the water has to be removed before mounting them in non-water based mounting medium. Dehydration is not necessary when mounting in glycerine gelatine. Dehydration is commonly done by placing the specimen in successively higher concentrations of ethanol. Later, the specimen is transferred into a solvent which is compatible to the mounting medium. Some mounting media require the specimen to be submerged in xylene which is toxic. Other media like euparal are able to directly accept the specimen from the alcohol. The clouding of the slide is an indication that there was still some water in the specimen.

Storage of Material in the Fixative

The collected material is stored in wide mouthed jar with a glass-stopper containing the fixative like FA (a mixture of formalin and alcohol) or FAA (a mixture of formalin, acetic acid and alcohol). Fixation is the preservation of cells and tissues in natural condition as far as possible. There are several fixatives in use but the best is that fixative which changes the chemistry (of cells and tissues) to minimum and preserves the cells and tissues to the best. It is also essential that the fixative is capable of penetration of all the cells and tissues as quickly as possible so that they are fixed without any further change.

The air captured should be expelled with the help of two jars (with 2-holed corks), and a vacuum pump connected to a water tap. The material and the fixative are taken in the first jar, while the second jar works as a trap between the first jar and the vacuum pump, and thus prevents the entry of water into the jar with the material and fixative. The water tap is opened and the pinch cock is allowed to open at short intervals to change the pressure to help in the sinking of the material in the fixative. Permanent sinking denotes that the air has been completely expelled.

Why the Material is to be fixed?

The plant material is mainly fixed to carry out anatomical or cytological studies. A proper fixative is to be selected for a particular material for study. The fixatives intended for cytological studies are of two types, namely, acid fixatives and basic fixatives. The fixatives of the former category (acid fixatives) help to preserve chromatin, nucleoli, spindle fibres and help dissolve the nucleoplasm and mitochondria, while the cytoplasm is fixed as spongioplasm. The latter fixatives (basic fixatives) help to preserve nucleoli, nucleoplasm and mitochondria and help to dissolve the chromatin and spindle fibres, while the cytoplasm is fixed as hyaloplasm along with vacuoles.

The fixatives that give acid fixation image are: Dioxan fixative, Chamberlain`s chrom-osmo-acetic, Gilson`s fluid, weak chrom-acetic, FPA (formalin-propionic-aclohol), Formalin-aceto-alcohol (FAA), strong chrom-acetic, Sass modified Bouin`s fluid, Carnoy`s fluids (of 2 types), Taylor`s modified Benda`s fluid, Petrunkevitsch`s fluid, Worcester`s fluid, Taylor`s chrom-osmo-acetic for smears, Flemming`s type medium: Chrom-osmo-acetic mixture, Carnoy-Le Brun`s fluid, Medium chrom-acetic and Radolph`s modified Navashin`s fluid (CRAF – solutiuon A and solution B). Similarly, the fixatives that give basic fixation image are: Zirkle-Frliki fluid, Marengo`s fixative and Zirkle`s fixative. The fixatives that are commonly used are FPA and FAA.

Common Fixing Fluids – Nature and Composition

a. Zenker`s fixative (Zenker`s solution or Zenker`s fluid)

This is a good routine and rapid fixative for animal tissues and is employed to prepare specimens of animal or vegetable tissues for microscopic study. It provides excellent morphological preservation and fixation of nuclear chromatin, connective tissue fibers and some cytoplasmic features but does not preserve delicate cytoplasmic organelles like mitochondria. The fixative contains mercuric chloride (corrosive sublimate), potassium dichromate, sodium sulphate, water and acetic acid. The fixatives containing mercuric chloride or potassium dichromate are toxic. Mercuric chloride can be replaced with the same weight of less toxic zinc chloride but it may not give the same quality of fixation as the original mixture.

The stock solution is usually made with 50 g of mercuric chloride, 25 g of potassium dichromate, 10 g of sodium sulphate (decahydrate), 50 g of glacial acetic acid and 950 ml of distilled water to complete 1000 ml (sometimes, before use, 5 ml of glacial acetic acid is added to 100 ml of the solution, if it has not been added to the stock solution). This fixative is stable for many years. If glacial acetic acid is replaced by 5 ml of formalin (37-40% formaldehyde), the resulting solution is called *Helly`s fixative* (also sometimes called formal-Zenker). This is stable for only a few hours because the formaldehyde and dichromate components react, producing formic acid and chromic ions, the orange solution becomes greenish. The fixation time is a few hours (4 – 24 hours). After fixation, the tissues should be washed overnight in running tap water before processing. The sections are to be treated with iodine thiosulphate sequence to remove mercury pigment. Because mercuric chloride corrodes metal, metal objects (including stainless steel) must be avoided. Dry paper towels serve as an effective barrier and they must be disposed safely.

b. *Bouin`s fixative* (Bouin`s solution or Bouin`s fluid)

This is a compound fixative used in histology and is composed of picric acid, acetic acid and formaldehyde. It is especially good for gastro-intestinal tract biopsies because it allows crisper and better nuclear staining than 10% neutral-buffered formalin. It is not a good fixative for electron microscopy, but it is good when the tissue structure with a soft and delicate texture must be preserved. Formalin-fixed tissue is normally mordanted with Bouin`s solution for better staining results in the trichrome stains. As the acetic acid lyses RBCs and dissolves small iron and calcium deposits in the tissue, a variant in which the acetic acid is replaced with formic acid can be used for both fixation and decalcification of the tissue.

Under the name, Bouin`s fluid, this is widely used for marine invertebrates. It is prepared as follows: 75 ml of Picric acid (saturated aqueous solution), 25 ml of formalin (40% aqueous solution of formaldehyde), and 5 ml of glacial acetic acid in 1000 ml of distilled water. The fixation time is about 6 (8 – 16) hours. After fixation, the tissue should be transferred to 70% alcohol. The variations are: *Gendre solution* (this an alcoholic version of Bouin`s solution which is saturated with picric acid instead of an aqueous saturated solution of picric acid and is useful when glycogen and other carbohydrates must be preserved within the tissue) and *Hollande solution* (this is a version of Bouin`s solution which contains copper acetate which stabilizes RBC membranes and the granules of eosinophils and endocrine cells so that there is less lysis than in regular Bouin`s solution).

c. *Carnoy`s solution* (Carnoy`s fluid or Carnoy`s fixative)

This is a very rapid fixative which has been used frequently for fixation when rapidly processed paraffin sections are required. For a thin small piece of tissue, only 30 – 60 minutes is required and dehydration takes place simultaneously. For normal tissues, the fixation time is about 30 minutes to 2 hours. After fixation, the tissue is transferred to absolute alcohol for storage. It is a good nuclear fixative but produces considerable shrinkage and cytological distortion. This is composed of 60% ethanol, 30% chloroform and 10% glacial acetic acid. It is composed of 1 g of ferric chloride dissolved in 24 ml of absolute alcohol, 12 ml of chloroform and 4 ml of glacial acetic acid. This is the same proportion by volume as the previous formula, with the addition of ferric chloride. This is used as a fixative for pap smear samples, as an agent for both the nuclear and mitochondrial DNA in different tissues and as an agent to preserve mucus, useful for tissue preparation before staining with periodic acid-Schiff base. There is another fixative often ascribed to Carnoy, which is a 3:1 mixture of ethanol and acetic acid, which is more correctly referred to as *Clarke`s fluid*.

d. *Champy`s fluid* (Champy`s fixing fluid)

The composition of the fixative is: 1 g of chromium trioxide, 3 g of potassium dichromate, 1 g of osmium tetroxide in 250 ml of distilled water (as a stock solution). Chromium trioxide and potassium dichromate are dissolved in 150 ml of distilled water. Similarly, osmium tetroxide is dissolved in the remaining 100 ml of distilled water and both the solutions are combined to form a stock solution. The solution is stable for some time and is suitable for use when cytological preservation is necessary. The pieces must be fairly small and 1-2 mm thick, never more than 3 mm. The time for fixation is from 12 - 24 hours. After fixation, the tissue is washed in running tap water overnight to remove the traces od chromium and osmium compounds.

e. Helly`s fluid

This is a good routine fixative which gives excellent results for cytoplasmic preservation, although it is not used much as it was in the past due to its mercuric chloride content. Helly`s base stock solution is identical to that of Zenker`s base stock solution. This is used as a microanatomic fixative for cytoplasmic granules, and nuclear staining and has some disadvantages as Zenker`s fixative. The composition of the stock solution is: 49.2 g of mercuric chloride (sometimes 50 g), 20 g of potassium dichromate (sometimes 25 g), 10 g of sodium sulphate and 960 ml (or 1000 ml) of distilled water. The working solution has the composition of: 48 ml of stock Helly`s base mixed with 2.5 ml of concentrated formalin (or the stock solution is mixed with 5 ml of formalin [40% aqueous solution of formaldehyde], which is made just before use. The fixation time is a few hours for thin pieces of tissue while 3 mm thick slices will require fixation overnight (12 – 24 hours). After fixation, the tissue is washed well with running tap water to remove the excess dichromate, and the sections are treated with iodine thiosulphate sequenced to remove the mercury pigment. The tissue is transferred to 90% ethanol (several changes) followed by absolute ethanol. The use of all metal objects is to be avoided as mercury corrodes the metal. Dry paper towels can be used and they must be disposed safely.

f. *Heidenhainm`s SuSa solution* (SuSa fixative or Sublimate-salt-formalin fixative)

The name SuSa is a contraction of two German words – *sublimat* and *saure* and sublimat refers to an older name for mercuric chloride (corrosive sublimate), because of its ability to corrode mucosa when used in medicine in the past and also its ability to sublime. *Saure* refers to trichloroaceticacid, a strong protein precipitant. It is a rapid fixative and gives good morphological preservation. This fixative is generally recommended for biopsies, and the period of fixation required is from 3 to 12 (sometimes 24) hours. The composition of the fixative is: 20 ml of formalin stock solution (40%), 4.5 g of mercuric chloride (corrosive sublimate), 0.5 g of sodium chloride (salt), 2 g of trichloroacetic acid, 4 ml of glacial acetic acid and 80 ml of distilled water. Mercuric chloride, sodium chloride and trichloroacetic acid in distilled water make up for solution-A, while glacial acetic acid and formaldehyde make up for solution-B. These two solutions are mixed together to form the stock solution.

The advantages of this fixative are: it penetrates rapidly, and produces good and even fixation with minimum shrinkage and hardening. It allows brilliant staining results with sharp nuclear detail. After fixation, the tissue is transferred directly to 95% or absolute alcohol. The main disadvantage is the slices should not exceed 1 cm in thickness, RBCs are poorly preserved and some cytoplasmic granules get dissolved.

How the Material is fixed?

The first step is to fix the material for wax embedding. This kills and preserves the cells so that no further changes take place. The material is to be fixed as soon as possible following collection. Hence, it is always preferable to carry the fixative to the plant than *vice versa*. Normally 28 ml glass bottles with rubber, lined metal screw caps half filled with the fixative are an ideal way of carrying the fixative into the field during collection. Each bottle should be provided with a blank label attached to note the name of the collected material. Each bottle should be used for the collection only one type of material (a stem, root, leaf, flower, fruit or a seed). The specimens should not be more than 5 mm thick and no greater than 10 mm at the widest point. Hard materials (roots and woody stems) should not exceed 5 mm^2. The specimens can be kept stored safely in FAA or FPA fixative for months. If the specimens float in the fixative, a water pump is used to extract the air. The specimens should be left in the fixative for at least 24 hours after they sink to the bottom of the fixative before dehydrating. Several types of botanical fixatives are available, but the most common all purpose fixatives are Formal Acetic Alcohol (FAA) and Formal Propionic Alcohol (FPA).

EMBEDDING THE MATERIAL IN PARAFFIN

The biological samples often need to be solidified to allow fine sectioning. Thin slices of the tissue improve the access of dyes, probes, and antibodies and reduce the overlay of different cell layers in the z direction. For light microscopy, paraffin wax is the most frequently used hard matrix for cutting the sections.

Principle

As paraffin is immiscible with water, the main constituent of the tissue, samples need to be dehydrated by progressively more concentrated ethanol baths. This is followed by a clearing agent, usually xylene, to remove the ethanol. Finally, the molten paraffin wax infiltrates the sample and replaces the xylene.

Thickness of the section

The paraffin sections for light microscopy are typically 5 µm thick. Paraffin wax does not provide a sufficiently hard matrix for cutting thinner slices needed for electron microscopy, typically 80 – 100 nm thick. So, epoxy or acrylic resins especially for immune-histochemistry are used.

Standard Protocol for Formalin-Fixed Paraffin Embedded Tissue

1. The tissues are *fixed* with 10% formalin or other fixatives for 24 – 48 hours at room temperature. Always make sure that you have enough fixative to cover the whole tissues. The fixative volume should be 5 – 10 times of the tissue volume.
2. The fixed tissues are trimmed into appropriate size and placed in embedding cassettes. .
3. The tissues are processed for paraffin embedding schedule as follows (for about a total of 16 hours)
    a. 70% ethanol, two changes, 1 hour each
    b. 80% ethanol, one change, 1 hour
    c. 95% ethanol, one change, 1 hour
    d. 100% ethanol, three changes, 1.5 hour each
    e. Xylene or xylene substitute (like Clear Rite 3), three changes, 1.5 hour each
    f. Paraffin wax (58 – 60°C), two changes, 2 hours each
    g. Embedding the tissues into paraffin blocks
4. The blocks are trimmed as necessary and cut at 3 – 10 µm (5 µm is commonly used).
5. The paraffin ribbon is placed in water bath at about 40 – 45°C.
6. The sections are mounted on to slides.
7. The sections are allowed to air dry for about 30 minutes and then baked in 45 – 50°C oven overnight. The baking temperature should not be allowed to higher than 50°C for sections thicker than 25 µm. Otherwise, the sections may crack, especially 25 – 50 µm thick sections, and result in sections falling off slides during the process of staining.
8. The sections are de-paraffinized in 2- 3 changes of xylene, 10 minutes each. It is always better to use three changes of xylene for sections thicker than 25 µm.
9. The sections are hydrated in 2 changes of 100% ethanol for 3 minutes each, 95% and 80% ethanol for 1 minute each. Finally they are rinsed in distilled water.

STAINS AND STAINING METHODS

The sections cut from the embedded blocks on a microtome will be ready for attachment to the slides and staining. There are many number of staining techniques, but the more general and widely used are staining with Safranin & Alcian Blue and Safranin & Fast Green. In olden days, the technicians placed much reliance upon the natural (textile) dyes, but very few of the natural and practically none of the technical dyes were in use. All the dyes now used in biological techniques are synthetic chemical compounds made from the substances found in coal tar.

Types of Dyes

1. Natural Dyes

Only three natural dyes (not yet been manufactured synthetically) are still used in botanical techniques and all are important cytological stains. One of them is cochineal and its derivatives, is of insect origin, while the others have their origin from plants belonging to Caesalpiniaceae.

Brazilin (dye obtained from different trees called *Brazilwood*, principally from *Caesalpinia crista* and *C. echinata*) is closely related chemically to haematoxylin but is neither so active nor so strong a stain. It has recently come into extensive use as a stain for smears. It is ordinarily used as a 0.5% solution in 70% alcohol which is allowed to ripen for about a week and kept in containers well stoppered and away from light and air. It reacts only after mordanting with ferric ammonium sulphate.

Haematoxylin is derived from *Logwood, Haematoxylon campechianum* is one of the most important of all stains. The solution itself has little or no affinity for tissues unless iron or aluminium is present in the latter. The stain is made up in combination or in conjunction with many metallic salts, mainly those of iron (in the ferric form), aluminium and copper. The colour effects vary with the character of the medium in which it is dissolved and according to after treatment. The colour is red in the presence of acids while it is blue in the presence of alkalis. The chromosomes and the pyrenoids are stained black while the cytoplasm of sporogenous cells is stained grey. When the sections are exposed to the action of fumes of ammonia, they turn blue.

In common use, as used in *Heidenhain`s iron haematoxylin*, the solution of the dye is a solution of 0.5% solution in distilled water. It is more satisfactory to prepare a 10% solution in absolute ethyl alcohol added to dilute a portion with distilled water to the 0.5% strength. A stable solution may be prepared by adding 5 cc of a 10% absolute alcohol solution to 100 cc of methyl cellosolve, 50 cc of distilled water and 50 cc of tap water that contain calcium compounds in a solution. If on shaking, the solution does not acquire wine-red colour, a pinch of sodium bicarbonate is added and shaken vigorously. Ripening occurs immediately and the solution can be used at once. The solution does not spoil at high temperatures and also retains its staining capacity for far longer periods.

Delafield`s haematoxylin is still being made according to the original formula in which the dye and mordant are combined. To 400 cc of a saturated aqueous solution of ammonium aluminium sulphate, drop-by-drop solution of 4 g haematoxylin crystals in 25 cc of 95% ethyl alcohol is added. The mixture is exposed to light and air for four days. Later, 10 cc of C.P Glycerine and 100 cc of methyl alcohol are added and it is allowed to stand for at least two months exposed to the air until the colour is sufficiently dark. It is dangerous and unsatisfactory to follow the recommendation to ripen the solution with hydrogen peroxide. The ripe undiluted solution is a powerful stain and the stain is differentiated in either water or 70% ethyl alcohol acidified with hydrochloric acid.

Harris haematoxylin resembles Delafield`s in its colour but is made up in a different matter. It is a superb *in toto* stain and the formula is

Haematoxylin crystals	5 g
Aluminium ammonium sulphate	3 g
50% Ethyl alcohol	1000 cc

The dye and the alum are dissolved with the aid of heat followed by the addition of 6 g of mercuric oxide (only the red powder) and boiled for 30 minutes. It is filtered and the brought up again to the original volume with 50% alcohol. The solution is acidified in the proportion of 1 drop of hydrochloric acid to each 400 cc of solution. *Delafield`s haematoxylin* will sometime stain chromatin well when the regular chromatin stains fail, and it also brings out cellulose walls more sharply than most of the other stains. The sporogenous cells are unusually well defined by both the stains.

Mayer`s haemalum* is recommended for the nuclei of filamentous algae and fungi, having little or no effect upon the cell walls and plastids. *Ehrlich`s haematoxylin has been extensively used with safranin on woody tissues. *Haematein* is also considered to be superior to haematoxylin on the ground that, it is easy to prepare, easy to use, saves time, and gives good results.

Cochineal is a yellowish-red powder obtained by grinding the dead bodies of the female cochineal insect and extracting the colouring matter. Cochineal and its derivatives are some of the most important bulk stains available. They are all progressive stains and cochineal itself is not used by botanists to any great extent.

Carmin is bright red in colour and is obtained by adding alum to cochineal. It is a complicate mixture of which the essential colouring matter is carminic acid. Both are easy to use and give astonishingly beautiful effects that are amazing. The cytologists have a considerable appreciation of carmin and extensively use in the *iron-acetocarmin smear technique.* Carmin is only slightly soluble in water at a neutral reaction and hence the solution must be either acidic (customary) or alkaline. Carmic acid is the active principle of cochineal and carmin, is a fairly strong dibasic acid. It is best known to botanists as the principal ingredient of *Mayer`s carmalum.*

Indigocarmin is indirectly a natural dye and is included in natural dyes because it is the sodium salt of indigosulphonic acid and has acid properties. It is bluish in colour, but when mixed with picric acid (to make *picro-indigocarmin* stain used in staining algae) it becomes green. It is 1.68% in water and 0.01% in 95% alcohol and is soluble at 25°C.

2. Coal - Tar Dyes

The classification of coal-tar dyes at present is based on the similarity in their chemical structure and six groups (Conn – 1936) have been recognized. They are: the *nitro dyes* (*Picric acid* – is a general cytoplasmic stain used as a differentiator for haematoxylin, safranin and the violets in alcoholic solutions); the *azo dyes* (*Orange-G* – is the most useful general cytoplasmic counter stain), *Bismarck brown* (also referred to as *Bismarck brown Y* – is a permanent stain used to brilliantly colour mucin and cellulose cell walls) and *Sudan IV* (also referred to as *Scharlach R* or *Fat ponceau R* – is a specific fat stain, which also stains lecithin, resins, latex, wax and cuticle while the chloroplasts are stained dull red); the *anthraquinone dyes* (*Alizarin Red S* – is used to differentiate chromatin and also for staining the chromosomes); the *quinone-imide group* of dyes (*Indamins* [none in common use], *Thaizins* [*Methylene blue* - is the most valuable bacteriological stain used by zoologists and rarely by botanists and a good stain for yeasts] and *Thionin* (is a stain of little use in botanical microtechnique but useful with animal tissues because of its meta-chromatic properties, as it imparts different colours to different cellular structures, and is specific for mucin which is stained red], *Oxazins* [none in common use], and *Indulins* [*Nigrosdin* – also called as *Nigrosin* or *Indulin black* – this cannot be satisfactorily used on plant material embedded in paraffin, but Chlorophyta or Cyanophyta can be mounted in glycerine with this stain]); the *phenyl-methane group* of dyes (*Diphenyl-methanes* [none commonly used]; *Diamino-triphenyl methanes* [*Fast green* - also referred to as *Fast Green FSF* – is a stain far superior to light green SF which does not fade at all, and is generally blue to bluish-green and rarely bright green on the stems and leaves of aquatics and most of the gymnosperms], *Light green* (also referred to as *Light Green SF* - is an excellent cytoplasmic stain used for cellulose cell walls, achromatic figure and is an exclusive stain for some filamentous algae, and fades in a shorter time) and *Malachite green* (also referred to as *Emerald Green* or *Light Green E* – is used to reveal clearly all histological elements like cell walls, endodermis, phloem, cytoplasm, nuclei and chloroplasts when used alone); *Triamino-triphenyl methanes* [*Acid fuschin* (also called *Rosanilin* – is used to stain the cortex, pith, parenchyma and cellulose cell walls, and is an excellent stain for the differentiation of zoosporangia and paraphyses in Laminariales, and sometimes also used as a mitochondrial strain), *Crystal violet* (also called as *Gentian violet* – is one of the most useful stains in cytology, the cytoplasm of the buds and root tips frequently get over stained and these are transferred to 2% aqueous acetic acid for about 30 minutes for easy chromosomal count) and *Aniline blue* (also called as *Aniline Blue WS*, *Cotton blue*, *Water blue* or *China blue* – is very valuable counter stain on plant tissues, stains cellulose walls and the achromatic figure, excellent stain for filamentous Chlorophyta)]; *Hydroxy-triphenyl methanes* [none generally used]); and the *Xanthene group* of dyes (*Pyronins* [none used in botanical techniques], *Rhodamines* [none used]; and *Fluorane derivatives* [*Eosin* (also referred to as *Bluish Eosin* – is an excellent counter stain for the gelatinous sheaths of plants like *Nostoc* and in cytology) and *Erythrosine*]). Other substances that act as stains include *iodine* and *osmic acid* which act as colour indicators of a micro-chemical nature. The nomenclature of these dyes is in great confusion because practically no system was used in assigning names to them. It is always good to follow the recommendations of the Commission on Biological Stains to acceptable names.

PROCEDURE OF STAINING

The staining methods are of two types, namely, *progressive staining* and *regressive staining*. In the former method, the tissue is left in the stain until it has reached the required depth of colour; while in the latter, the tissue is intentionally over-stained and the decolourised (differentiated) to the required depth of colour. The beginner should always use the method of progressive staining as it is a little difficult for him/her to judge when the required degree of differentiation has been reached. When he/she has reached some degree of success with this method, then he/she can try the method of regressive staining.

a. Single Staining

The sections are placed in the stain and left there until they are stained to the required depth (progressive staining), and this can be ascertained by placing the watch glass on the stage of the compound microscope and examining with the low power objective of the microscope. Normally, the sections take only two to three minutes, and the excess stain if any must afterwards be removed by placing the section in the solvent used for the stain, as the solvent will not remove the stain from the tissues. If a section is over-stained, or if the method of regressive staining is used, the degree of staining can be lessened by placing the section in acid alcohol. This is called *differentiation* because it differentiates the extent to which the various parts of the tissues are stained. The process of differentiation should be watched under the low power of the microscope as when staining, and when it is complete, the tissue should be washed in 70% alcohol.

b. Double Staining

The sections are stained with one stain at a time, the excess being removed as in single staining procedure before placing the section in the second stain. It may be necessary to differentiate and even to dehydrate before adding the second stain (such as when using Safranin and Light Green in Clove oil for the stems of plants).

The specific stains are mainly used to stain definite structures or substances like: Aniline blue, erythrosine, fast green and light green to stain achromatic figure; aniline blue, Delafield`s haematoxylin, fast green and light green to stain cellulose cell wall; crystal violet and safranin to stain lignified cell wall; safranin to stain suberised cell wall; aniline blue, erythrosine, fast green and light green to stain cytoplasm; crystal violet, erythrosine and safranin to stain cutinised cell wall; aniline blue to stain callose; safranin to stain chitin and proteins; crystal violet to stain mitochondria; crystal violet and iron haematoxylin to stain plastids; crystal violet, haematoxylin and safranin to stain nucleus; and haematoxylin and safranin to stain chromosomes.

Among the single stains, safranin or fast green are the best and are used alone to stain filamentous algae, fungi, sections of bryophytes, spores of pteridophytes, pollen grains of gymnosperms etc. Aniline blue or safranin is also suitable for algae. The common method of staining is as follows: the collected material is kept in a watch glass, and a few drops of stain are added so that the material is immersed in the stain. The material is allowed to remain for a few minutes to allow it to take the stain. The time of immersion varies with the type of the material. Later, excess stain is washed off in water and the washing is repeated till the stain stops coming out of it. In some cases, the excess stain is removed by acid water or acid alcohol if the water alone fails to do so. The stained material is mounted on slides for observation and study under the microscope.

More commonly, two or more stains (in combinations) are employed wherever tissue differentiation is necessary. In general, a combination of acidic and basic dyes of contrasting colours is used which permits the distinction of woody tissues from non-woody tissues. The combinations generally recommended are: haematoxylin and safranin; safranin and fast green; safranin and aniline blue; safranin and crystal violet and crystal violet and erythrosine.

DEHYDRATION OF THE MATERIAL

When once the plant material has been fixed, the next stage is to slowly dehydrate the material to slowly remove all the water contained in it. In order not to destroy the delicate tissues of the material, dehydration is done gradually in a series of alcohol/water solutions of increasing strength. Owing to the fact that water is immiscible with the oil used for clearing on the one hand and with the solvent of the mounting medium (usually xylene) on the other, it is always essential to see that all traces of water should be removed from the tissue. Such a removal of water is usually effected by means of ethyl alcohol and this process is known as dehydration. If the sections were placed directly into absolute alcohol, the cells would lose their water quickly and they would shrink as such their shape is altered. This alteration in shape can be avoided by adopting the following method:

The sections are placed in ethyl alcohol solutions of gradually increasing concentration for one or two minutes. The concentrations that are used generally are: 30%, 50%, 70%, 90% and absolute alcohol. It is always better to start with the concentration next above that in which the sections were last placed (for example, if the stain was aqueous, it is better to begin with 30% alcohol, but if it was alcoholic, one should begin with 70% alcohol). To ensure complete dehydration, finally transfer the sections to a second watch glass of absolute alcohol. Breathing on the absolute alcohol in the watch glass and in the bottle should be avoided; otherwise, it will no longer be absolute. It is always an advantage to cover the watch glass with a second one inverted over it. In place of absolute alcohol, *cellosolve* (ethylene glycol mono-ethyl ether) can also be used for dehydration of the sections. The cellosolve mixes with water, alcohol, clove oil and xylene and there is no need to use varying concentrations, as it does not cause shrinkage or alteration in the shape of the cells; nor is it necessary to clear after using it. The cellosolve dehydrates rapidly and because of this it is not recommended for the animal tissues as it may result in distortion. The sections are placed in cellosolve for about a minute and then mounted directly.

According to Johansen (1940), Tertiary Butyl Alcohol/Isopropyl alcohol schedule is as follows: The material is placed for 2 hours in 50% alcohol, followed by immersion in 70% alcohol overnight. Later, the material is transferred to 85% alcohol for 1 hour, followed by immersion in 95% alcohol for 1 hour. Then, the material is transferred to and kept in 100% alcohol for 1 hour, pure TBA for 1 hour, transferred to fresh solution of pure TBA for 1 more hour and finally in a fresh solution of pure TBA for 1 more hour. For harder tissues to get dehydrated, the period of time may be extended accordingly.

CLEARING THE SECTIONS

If alcohol has been used as the dehydrating agent, it must be removed. This process is known as clearing and it renders the tissues transparent. The best results are obtained by using clove oil for plant tissues for permanent preparations. Xylene can also be used but it has a tendency to cause the shrinkage. If clove oil is used, it may remove safranin from the tissues stained with it and counter staining with light green may displace safranin. In such a case, it is necessary to re-stain with safranin after going down through the alcohols as far as 70%, and then completing the dehydration. It is best to use *lacto-phenol* as the clearing agent for delicate algae and fungi.

DEWAXING THE SECTIONS

The slides have to be dewaxed first and traditionally this is done by immersing into two changes of xylene, not less than 10 minutes each. But, a new and safer alternative to xylene is the use of Hystoclear (in the same way as that of xylene).

MOUNTING THE MATERIAL ON TO THE SLIDE

The sections must be mounted in a suitable medium of about the same refractive index as the crown glass (1.5) for final examination under the microscope as well as for preservation.

Temporary mounts are prepared with glycerine (50%. Aq. – RI equal to about 1.39 – 1.34), while for permanent mounts, the best mounting medium for general use is *Canada balsam* dissolved in xylene (RI is equal to about 1.524). Sometimes, *euparal* (RI is equal to about 1.4) can also be used, even if the dehydration is incomplete.

During mounting, a drop of Canada balsam is placed on the centre of a clean microscopic slide, and the section is transferred to the balsam with camel`s-hair brush. It is covered with a clean microscopic cover glass (cover slip) against the finger and levering it down with a mounted needle, which will prevent the entrance of air bubbles. There should be no air bubbles in the balsam, and if any bubbles appear, they can generally be removed by gently warming the slide over a very small flame (of a spirit lamp). A white film if any in the balsam indicates incomplete dehydration. The slide is labelled and left to dry. The cover slip may also be ringed but this is not essential for ordinary purposes.

ADHERING SECTIONS TO THE SLIDES

The cut sections of the material should be adhered to slides using an adhesive. Traditionally many of these formulae use dilute formaldehyde solution to float and flatten the section on the slide (these are to be used in a highly ventilated area, preferably near an extractor fan).

1. Use of Haupt`s Gelatin Fixative

To 5 g of Glycerine jelly, 45 ml of water is added and the mixture is warmed and stirred well until the jelly has dissolved, and then allowed to cool. A drop of this jelly is smeared as a thin film on the slide and the sections are allowed to float on 2% formalin (2.5 ml of commercial formaldehyde in 97.5 ml of water). The slide is warmed gently until the section flattens (overheating is avoided as it will melt the wax and ruin the section). The slide is blotted carefully with a damp filter paper, and dried at 37^0C as long as possible (minimum 30 minutes).

2. Use of Land`s Gum Fixative

To 1 gram of Gum Arabic and 1 gram of Potassium Dichromate, 98 ml of water is added. The gum is mixed with a little of cold water to make it a paste, and Potassium dichromate is dissolved in the remaining warm water. The mixture is mixed well and warmed and stirred until it gets completely dissolved. It is allowed to cool. The prepared adhesive is smeared over a slide as a thin film and the above method (for adhesion of the section) is followed using water instead of formalin.

Alternatively, in this method, the sections are made to float on warm water (about 45^0C) until they flatten out. These are picked up directly onto a slide which has been thinly smeared with the adhesive. The slide is placed into the water under the section and lifted out horizontally. The section on the slide is centred; the excess water is dabbed away with filter paper and left to dry at about 35^0C to 45^0C overnight. Blotting and drying is carried out as mentioned above. It is always necessary to keep the sections dust-free while they are drying.

Care to be taken during mounting

The following care must be taken during mounting of a section on to a slide: The object should be mounted in the centre of the slide. No air bubbles should enter the medium while mounting which results in drying up of the medium and the spoilage of the preparation. Use a small quantity of the mounting medium as per the necessity so that it does not flow on to the slide. The preparation should be clean. The labels are to be pasted uniformly on one side of the prepared slide.

RINGING THE SLIDE/SEALING THE COVER SLIP

The slide is placed on a turntable and fixed in position by the springs provided. A brush is dipped in the ringing cement and the turntable is revolved rapidly by applying the finger to the milled edge. The brush is applied to the edge of the cover slip, covering it with a thin layer of the cement. This is allowed to dry and the process is repeated until a sufficiently thick ring is obtained. The temporary preparations (semi-permanent preparations) can be sealed with Canada balsam, gum, dammar, nail polish etc.

SLIDE LABEL - PREPARATION

During labelling the slide, one must state the name of the organism from which the tissue was obtained, the name of the object or the tissue, and, if it is a section, whether it is L.S. (longitudinal section) or T.S. (transverse section) or R. L. S. (radial longitudinal section) or T. L. S. (tangential longitudinal section). It is also an advantage to add the name of the fixing agent and the stain used for future reference, followed by the name of the person who prepared it with date. The algal and fungal material which is delicate can be mounted in lacto-phenol (in this case, it is always necessary to ring the slide (with the help of a ringing table).

FREE HAND SECTIONS OR SECTIONING BY HAND

The safety razor blade can be used to produce the sections which are thin enough for study under magnifications of 100 – 400xs. During sectioning, each cut section will slide onto the blade of the knife lubricated by 50% alcohol solution. The section can be examined temporarily in water (a 50% glycerine solution is needed if the section is meant to be kept for several months). The sections can be transverse, longitudinal or tangential.

Due to the relatively solid nature of the plant specimens, they are more amenable to simple hand sectioning. This is due to the fact that plant organs, especially stems and leaves are firm enough that hand sections can be obtained readily. It is also realized that one`s own hand sections are better than the prepared slides. In order to study the structural organization of the plant body, sections have to be made so that enough light can be transmitted through the specimen to resolve cell structures under the microscope. A free hand section is the simplest method of preparing specimens for viewing under the microscope. This method also allows examining the specimen in a few minutes. It is also suitable for a variety of plant materials, like soft, herbaceous stems and small woody twigs. The fixation is generally not required for a temporary preparation. Patience, experience and perhaps inherent skill are the chief requirements for this technique.

FREE HAND SECTIONING - PROCEDURE

A new double edge razor blade is selected, and the one edge of the razor blade is covered with masking tape to minimize the risk of cutting oneself. The blade is rinsed with warm tap water to remove the traces of grease from the blade surface if necessary.

The plant material is held firmly, and the material should be held against the side of the first finger of the left hand (or right hand) by means of the thumb. The first finger should be kept as straight as possible while the thumb is kept well below the surface of the material out of the way of the razor`s edge. Always note that it is not that easy to cut your own finger. So relax.

The razor is flooded with water which will reduce the friction during cutting as sections can float on to the surface of the blade. The razor blade is taken in the right hand (or left hand) and placed on the first finger of the left hand (or the right hand) more or less at a right angle to the specimen.

The razor is drawn across the top of the material in such a way to give the material a drawing cut (about 45° in the horizontal direction). This results in less friction as the blade passes through the specimen. Several sections are cut at a time, and the sections will definitely vary in their thickness, and there will be some to be used among the thick sections.

The sections are transferred to water, always by the use of a brush, rather than a forceps or a needle. The thinnest sections (the more transparent ones) are selected and transferred on to a clean glass slide and stained and observed under the microscope.

Additional support can be used to facilitate hand sectioning for delicate and hard to hold specimens like thin leaves and tiny roots. The pieces of the tissue can be inserted into a small piece of pith material (collected from fresh plants) or a small piece of Styrofoam support or slices of tuberous roots of carrot or the slices of stem tuber of potato. When once the tissue (material) is firmly in place, the technique of hand sectioning of the support material is applied, along with a section of the leaf or the root is cut.

The longitudinal sections are also difficult to be obtained by hand without any supporting material as small pieces of stem and root (needles of gymnosperms) are difficult to hold with one`s finger. But, it is possible to hold the tissue firmly for free hand sections by cutting a V-shaped notch or a small circular hole (at the centre) of the even diameter into the pith support and inserting the material into it for sectioning.

An important note: Special care should be taken for cross (transverse) sections during sectioning to see that the material is not cut obliquely. As long as the sections are not obliquely cut, even the thick sections are usable. During this process, a number of sections should be cut at the same time without worrying about the thickness of the sections at the time. A number of sections can be cut without moving the material or the thumb, by slightly and progressively increasing the pressure with the razor blade on the first finger, and simultaneously exerting increasing pressure on to the specimen by the thumb. It is always best to start cutting with the razor blade at the surface of the specimen rather than against the side of the material. It is usually not necessary that a section should be complete, as long as it includes a portion of the tissues from the centre to the outer edge (to the periphery or circumference) of the specimen in the case of root and/or stem.

EPIDERMAL PEELING - PREPARATION

The epidermis of most leaves can be readily removed by the method of scraping. First place a square of the leaf on the slide (keep the leaf along with a few drops of sodium hypochlorite), and then hold the leaf with a cork and scrape gently with a razor blade. Finally, trim off the thin part of the leaf, transfer to dish with tap water and brush off loose cells, and mount it on a slide. In some cases, scraping of the leaf is unnecessary. In such a case, peeling is done by folding the leaf to break the surface, and either stripping directly, pulling one part of the leaf downwards relative to the other, or by holding as thin a layer of the surface as possible between forceps and peeling it back.

a. Clearing the Section

Following the study of sections, observation is made with cleared sections. The sections can be cleared by transferring them from a dish of 50% alcohol to a dish of water with a fine brush or forceps. They are placed in a petri dish containing undiluted sodium hypochlorite (common household bleach) for 5 minutes for cell contents to dissolve. The sections are thoroughly washed with water (by using a fine forceps instead of a brush [not to allow the dissolution of the fine bristles of the brush].

b. Staining the Section

To enhance the details within the sections, there are several staining combinations methods in practice. The sections of the freshly cut material should be washed gently, to remove the cell debris (not to obscure the details once the section has been stained).

[NOTE: The details of Microtome and Microtomy are given in a separate chapter].

■■■

CHAPTER - 8

MICROSCOPY

Introduction

Microscopy is a technical field of *using microscopes to view the samples and objects that cannot be seen with the naked* (unaided) *eye* (the objects that are not within the resolution of the normal eye). This can be studied under three well known branches, namely, optical, electron and scanning probe microscopy.

Microscopy – Definition

The word, `microscopy' is derived from two Ancient Greek roots, *micros* meaning `small` and *skopeo* or *skopein* meaning `to view` or `to see` or `to look` (to view small objects) with an instrument used to see the objects that are too small for the naked eye. The science of investigating small objects using such an instrument is called *microscopy. Microscopic* means invisible to the eye unless aided by a microscope.

Types of Microscopes

The microscopes can be separated into several different classes as follows:

One type of grouping is based on what interacts with the sample to generate the image? That is, light or photons (optical microscopes), electrons (electron microscopes) or a probe (scanning probe microscopes).

Alternatively, the microscopes can be classed on whether they analyze the sample via a scanning point (confocal optical microscopes, scanning electron microscopes and scanning probe microscopes), or analyze the sample all at once (wide field optical microscopes and transmission electron microscopes).

1. Optical (Light) Microscope

The most common type of microscope and the first invented microscope is the optical microscope. This instrument contains one or more lenses producing an enlarged image of a sample placed in the focal plane. These have refractive glass and occasionally of plastic or quartz, to focus light into the eye or another light detector. The mirror-based optical microscopes operate in the same manner. Typical magnification of a light microscope (assuming visible light) is up to 1250x with a theoretical resolution limit of around 0.250 micrometer or 250 nanometers.

The LM creates a magnified, detailed image of seemingly invisible objects/specimens, based on principles of transmission, absorption, diffraction, and refraction of light waves. The different types of microscopes produce images of objects employing different strategies. In all instances (bright field, phase contrast and fluorescence), production of a clear and informative image is dependent on the magnification of the object, its contrast with respect to its internal/external surroundings and the ability to resolve the structural details.

In the LM, the objects are enlarged/magnified with a convex lens that bends light rays by refraction. The diverging rays from the points within the object (object points) are made to converge behind the convex lens and cross over each other to form image points (a focused image). The distance of the object from the lens divided into the distance of the focused image from the lens determines the magnification. In CM, there are usually two magnifying systems, one defined by the objective and the other defined by the eyepiece. Another important property of a lens is its focal length, which is defined by the distance from the lens at which parallel rays of light are focused. The visibility of the magnified object depends on contrast and resolution. In general, the contrast or differences in light intensity between an object and its background/surroundings render the object distinct. For colourless specimens, as is the case for most of the biological material, contrast is achieved in different ways. The object itself or the selected portions of the object may be stained, thus reducing the amplitude of certain light waves passing through the stained areas. This usually requires the fixation (killing) and staining of cells. Such stained specimens are generally observed using bright-field microscopy. On the other hand, several kinds of specially developed microscope systems may be used that can enhance the contrast of live specimens.

These systems include: oblique illumination, dark field, phase contrast, polarized light, differential interference contrast (*Nomarki*), reflection interference, fluorescence and video microscopy.

Special techniques (scanning confocal microscopy, vertico SM) may exceed this magnification but the resolution is diffraction limited. The use of shorter wavelengths of light (such as ultra-violet) is one way to improve the spatial resolution of the optical microscope (as are devices such as the near-field scanning optical microscope). A recent optical technique, *Sarfus*, increases the sensitivity of standard optical microscope to a point it becomes possible to directly visualize nanometric films (down to 0.3 nm) and isolated nano-objects (down to 2 nm diameter) and this is based on the use of non-reflecting substrates for cross-polarized reflected light microscopy. UV light enables the resolution of microscopic features as well as to image samples that are transparent to the eye. Near infra-red light can be used to visualize circuitry embedded in bonded silicon devices, as silicon is transparent in this region of wavelengths. In fluorescence microscopy, many wavelengths of light ranging from the UV to the visible can be used to cause samples to fluoresce to allow viewing by the eye or with the use of specifically sensitive cameras. Phase contrast microscopy is a technique in which small phase shifts in the light passing through a transparent specimen is converted into amplitude or contrast changes in the image. The use of this method does not require staining to view the slide, and this method made it possible to study the cell cycle in living cells. The traditional light microscope has more recently evolved into the digital microscope. Instead of directly viewing the object through the eye pieces, a type of sensor similar to those used in a digital camera is used to obtain an image, which is then displayed on a the monitor of a computer. The sensors use CMOS or charge-coupled device (CCD) technology depending on the application.

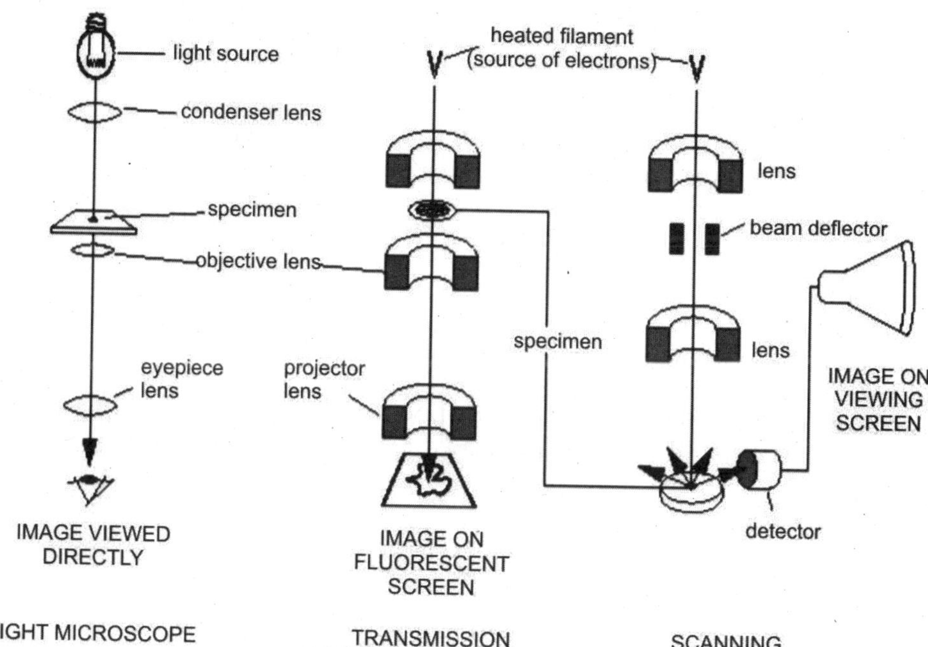

Comparison between LM, TEM and SEM

2. Electron Microscope

The electron microscope has 3 major variants, namely,

a. ***Scanning Electron Microscope (SEM),*** which generally looks at the surface of bulk objects by scanning the surface with a fine electron beam, and

b. ***Transmission Electron Microscope (TEM),*** which passes the electrons through the sample, analogous to basic optical microscopy. This requires careful sample preparation, since electrons are scattered so strongly by most materials. This is a device that allows people to see the objects that could normally not to be seen by the naked (unaided) eye.

c. ***Scanning Probe Microscope*** is of many types, namely: *Atomic Force Microscopy* (AFM), *Ballistic Electron Emission Microscopy* (BEEM), *Electro--static Force Microscope* (EFM), *Electro-chemical Scanning Tunneling Microscope* (ESTM), *Force Modulation Microscopy* (FMM), *Kelvin Probe Force Microscopy* (KPFM), *Magnetic Force Microscopy* (MFM), *Magnetic Resonance Force Microscopy* (MRFM), *Near-field*

Scanning Optical Microscopy (SNOM or *Scanning Near-field Optical Microscopy*), *Piezo Force Microscopy* (PFM), *Photon Scanning Tunneling Microscopy* (PSTM), *Photo-Thermal Micro-spectroscopy/ Microscopy* (PTMS), *Scanning Atom Probe* (SAP), *Scanning Capacitance Microscopy* (SCM), *Scanning Electro-Chemical Microscopy* (SECM), *Scanning Gate Microscopy* (SGM), *Scanning Ion-Conductance Microscopy* (SICM), *Spin Polarized Scanning Tunneling Microscopy* (SPSM), *Scanning Thermal Microscopy* (SThM), *Scanning Tunneling Microscopy* (STM), *Scanning Voltage Microscopy* (SVM), *Scanning Hall Probe Microscopy* (SHPM), *Scanning SQUID Microscope* (SSM), and *Recurrence Tracking Microscope* (RTM).

OPTICAL OR LIGHT MICROSCOPY

This method is the oldest, simplest and most widely used form of microscopy. This involves passing visible light transmitted through or reflected from the sample through a single or multiple lenses to allow a magnified view of the sample. The image can be detected either directly by the eye, or imaged on a photographic plate or can be captured digitally. The basic light microscope is made of the single lens with its attachments or the system of lenses and imaging equipment along with the appropriate lighting equipment, sample stage and support. The recent development is the digital microscope that uses as CCD camera to focus on the sample of interest and the image can be viewed on a computer screen, as such the lenses are not necessary.

The living cells in particular lack sufficient contrast to be studied as the internal structures are colourless and transparent. The commonest way to increase the contrast is to stain the different structures with selective dyes which often involve killing and fixing the sample. These techniques make use of the differences in the refractive index of cell structures.

In order to improve the specimen contrast or to highlight certain structures in sample, special techniques must be used to increase the contrast or label a specimen. They are as follows:

Transmission Electron Microscopy (TEM), *Scanning Electron Microscopy* (SEM), *Bright Field Microscopy, Dark Field Microscopy, Phase Contrast Microscopy* or *Phase Contrast Imaging, Differential Interference Contrast Microscopy* (DIC), *Interference Reflection Microscopy* (IRM) or *Reflected Interference Contrast* (RIC), *Fluorescence Microscopy, Confocal Microscopy, Single Plane Illumination Microscopy or Light Sheet Fluorescence Microscopy, Serial Time-encoded Amplified Microscopy* (STEAM), *X-Ray Microscopy, Scanning Probe Microscopy, Ultrasonic Force Microscopy (UFM), Ultra-violet Microscopy, Infra-red Microscopy, Digital Holographic Microscopy (DHM), Virtual Microscopy (Digital Pathology),* and *Laser Microscopy.*

1. Microscope – the Instrument

A compound microscope is composed of two elements, namely, a primary magnifying lens and a secondary lens system. The light is caused to pass through an object and then is focused by the primary and secondary lens,. If the light beam is replaced by an electron beam, the microscope becomes a transmission electron microscope. If the light is bounced off the object instead of passing through, it becomes a dissecting microscope. If the electrons are bounced off the object in a scanned pattern, it becomes a scanning electron microscope. The function of any microscope is to enhance the resolution. The microscope is used to create an enlarged view of an object such that we can observe the details not otherwise possible with the human unaided eye. Because of the enlargement, the resolution is often confused with the magnification, which refers to the size of an image. In general, the greater the magnification, the greater the resolution, but this is not always true. There are several practical limitations of lens design which can result in an increased magnification without increased resolution. So, to change the resolution, a different lens is often only the answer. The reason for the dichotomy between magnification and resolution is the ability of the human eye to see two objects. It is always necessary that the two objects be about 0.1 mm apart when held 10" from the face in order to detect them as two objects by us. If they are closer than 0.1 mm, we will perceive them as a single object.

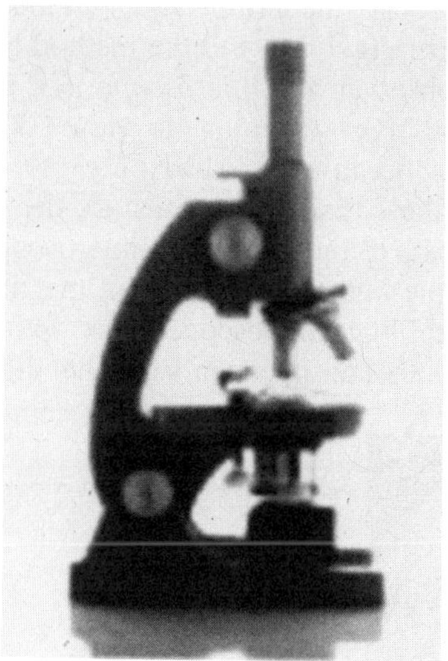

Compound Microscope

2. Why use an Optical Microscope?

The big advantage of light microscope is the ability to observe living cells. It is possible to observe a wide range of biological activity (like the uptake of food, cell division and movement). It is also possible to use *in vivo* staining techniques to observe the uptake of coloured pigments by the cells. This is not possible in real time use of electron microscopes because the specimen has to be fixed and it gets completely dehydrated during the process of sample preparation. The low cost of the optical microscopes makes them useful in a wide range of different areas, like education, the medical sector etc. In general, both the microscopes have different areas of application and do not complement each other.

3. Optical Microscopes – Different Types

The two most common optical microscopes are Compound microscope and Stereo microscope. The stereo microscope is commonly used to observe larger, opaque specimens as it fails to magnify as much as compound microscope (around 40x – 70x maximum), but gives a truly stereoscopic view of the specimen. This is because, the image delivered to each eye is slightly different. The stereo microscope does not necessarily require elaborate sample preparation. The compound microscope magnifies up to about 1000x, and the specimen has to be sufficiently thin and bright for the microscope light to pass through it. The specimen is generally mounted on a glass slide. The microscope is not capable of producing a 3-dimensional view, even if it possesses two eyepieces. This is because each one of the eyes receives the same image from the objective. The light beam is simply split into two.

THE COMPOUND LIGHT MICROSCOPE

a. How the Compound Microscope Works?

In a compound light microscope, the term, *compound* means that the microscope has more than one lens, the term, *light* refers to the method by which light transmits the image to the eye of the observer and *microscope* is a combination of two words, *micro* meaning `small` and *scope* meaning `to view`. The early microscopes were called simple microscopes because they had only one lens, and these work like magnifying glasses and these had limitations to the amount of magnification. Janssens added a second lens to the compound microscope to magnify the image of the primary lens. The simple light microscope can magnify an object to about 266x while the modern compound light microscope under optimal conditions can magnify an object from 1000x to 2000x the specimen`s original diameter.

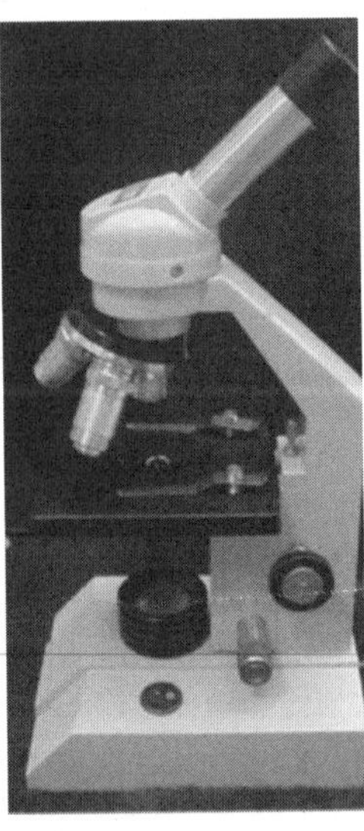

CMS

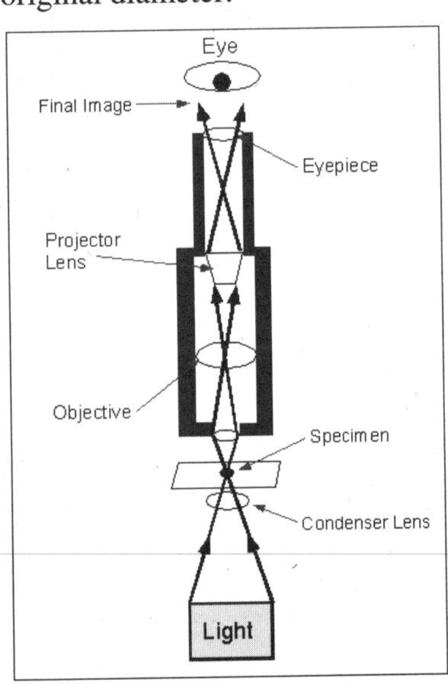

The path of light that travels through CMS

b. Parts of a Compound Light Microscope

The parts of a compound light microscope are: eyepiece, objectives, fine adjustment knob, power switch, stage, diaphragm and base, body tube, revolving nosepiece, stage clips, stage stop, coarse adjustment knob, aperture, arm and the light source.

ELECTRON MICROSCOPE

The electron microscope was co-invented by Heinrich Rohrer and Gerd Binnig along with Max Knoll and Ernst Ruska, in 1931 for which Ruska was awarded half of the Nobel Prize in 1986 (while the other half was shared between Rohrer and Binnig for inventing STM). In this, the electrons are speeded up in a vacuum until their wavelength is extremely short (only one hundred-thousandth that of white light). Beams of these fast moving electrons are focused on a cell sample and are absorbed or scattered by the parts of the cell so as to form an image on an electron-sensitive photographic plate.

Electron microscopes if pushed to the limit, can make it possible to view objects as small as the diameter of an atom. Most of these used to study the biological material can see down to about 10 angstroms, and it does allow the researchers to distinguish individual molecules of biological importance. It can magnify the objects up to one million times. All these microscopes suffer from a serious drawback, as no living specimen can survive under the high vacuum, they cannot show the ever-changing movements that characterize a living cell.

1. Electron Microscopes – Different Types

Mainly, there are two different types of electron microscopes, namely Scanning Electron Microscope (SEM) and Transmission Electron Microscope (TEM). In the transmission electron microscope, an electron beam is passed through an extremely thin section of the specimen and one will get a 2-dimensional cross section of the specimen. But scanning electron microscope is used to visualize the surface structure of the specimen, providing a 3- dimensional impression.

2. Light Microscope and Electron Microscope - Comparison

The light microscope is cheaper compared to an electron microscope (cost-wise).

The operation of the light microscope is also cheaper (it operates by any simple light source) compared to that of an electron microscope (is expensive to produce the electron beam).

Light microscope is small and easily portable (because of its smaller size) than an electron microscope (because of its larger size) which requires a special room to install.

The sample preparation for light microscope is simple and easier to operate than the electron microscope in which the sample preparation is a lengthy and complex procedure.

During preparation of the material for observation under light microscope, it rarely gets distorted while under an electron microscope, the material for observation gets easily distorted during the preparation.

Vacuum is not required for a light microscope, while vacuum is very essential for an electron microscope.

Under light microscope, the sample retains its natural colour while under an electron microscope, all the images appear only in black and white.

The light microscope magnifies the samples only up to 2000 times while the electron microscope can magnify the sample over 500,000 times.

The living or the dead material (sample) can be viewed under the light microscope while dead material (sample must be fixed in plastic and can be viewed in a vacuum) can be viewed under the electron microscope.

The relevant stains are often needed to make the cells clearly visible under a light microscope. But, the electron beam can easily damage the specimens during preparation as such they must be stained with relevant electron-dense chemicals (generally heavy metals like osmium, lead or gold) to enhance clarity.

3. Electron and Optical Microscopes - Basic Differences

Both these microscopes are technical devices which are used to visualize structures that are too small to be seen with the unaided eye. Both have relevant areas of applications in biological and material sciences. But, the method of visualization is different. Electron microscope use electrons and not photons (light rays) for visualization and compared to Light microscope, Electron microscope is a very recent invention.

4. Electron Microscope – Advantages over Light Microscope

The main advantage is that EMS has a higher resolution and is of a higher magnification (up to 2 million times). LMS can show a magnification only up to 1000 – 2000 times, which is a physical limit imposed by the wavelength of light. It is also possible to view the 3-dimensional shape of an object through an EMS (SEM). In SEM, due to the nature of electrons, the microscope has a greater depth of field compared to LMS. The higher resolution may also give the human eye the subjective impression of a higher depth of field.

5. Electron Microscope – Disadvantages over Light Microscope

Electron microscope is extremely expensive. The preparation of the sample is complex and much more elaborate. It is often necessary to coat the specimen with a thin layer of metal (like gold) and the metal is able to reflect the electrons. The sample must be completely dry which makes it impossible to observe living specimens. It is not possible to observe moving specimens (the specimens are dead). It is not possible to observe colour. The electrons do not possess colour and the image is only black/white. Sometimes, the image is coloured artificially to give a better visual impression. It always requires more training and experience in identifying the artifacts that may have been introduced during the sample preparation process. The energy of the electron beam is very high, and the sample is therefore exposed to very high radiation as such it is not able to live. The requirements of space are very high and they need a complete room for installation. Similarly the cost of maintenance is also high.

CYTOPHOTOMETRY OR MICROFLUOROMETRY

The first known use of cytophotometry was in 1952. The term is derived from three Greek words, *cyto* meaning `cell`, *phos* meaning `light` and *metron* meaning `measure`.

a. Definition

Cyto-photometry is defined as: `*A method in quantitative cyto-chemistry used to determine the chemical composition of cells in a histological preparation on the basis of the absorption of light by cells*`.

b. Cytophotometer

The cyto-photometer is `*an instrument for examining cells by determining the intensity or wavelengths of light transmitted through them*`.

The photometry of cells has been referred to by many names like: *micro-absorption photometry*, *micro-photometry*, *micro-spectrography*, *micro-colorimetry* etc, but recently, the term, cytophotometry has been widely accepted (Moses, 1952). In modern quantitative cytophotmetry, all light regions have been used: UV, visible and infra-red light and recently even x-rays and electron beams have also been tried for absorption measurements. In cytophotmetry, a beam of monochromatic light is passed through the histological preparation, the diameter of the beam being commensurate with the diameter of the cell or the cell structure. The concentration of the substance examined in the cell is found in accordance with the Bouguer-Lambert-Beer Law: $I = l_o e^{kel}$, where I is the intensity of the light after passage through the cell, l_o is the intensity of the light incident on the cell, k is the specific monochromatic absorption coefficient of the substance under study calculated per unit of its concentration at a given wavelength of light, and l is the length of the path traversed by the light in cell (in practical terms, the thickness of the histological preparation). After determining the concentration of the substance within the cell and measuring the volume of the cell, it is possible to calculate the total amount of the substance in the cell.

Cytophotmetry was developed by Torbjorn Caspersson, the Swedish histologist (1936). Dr. Walter Sandritter stimulated by the early studies of Caspersson devoted himself to cytophotmetry at a time when the photometer had to be built by hand before quantitative studies could be carried out. He along with his students was responsible for the origin of scanning, fluorescence and flow cytometers. He was successful in developing quantitative cytochemistry into standardized procedures for the identification of nucleic acids, proteins and other substances. He was responsible for bridging the gap between the nuclear DNA content that was measured cytophotmetrically, and the chromosomne analysis and also described the proliferative behaviour of the cell on the basis fo DNA values in the cell cycle. He also elucidated the problem of the DNA cytophotmetric distinction between reactive processes and neoplastic changes and correlated the concept of the tumour DNA stem line with the biology of tumour growth. He also made a quantitatively defined portrait of the individual cell by scanning cytophotometry, and studied euchromatin and heterochromatin as functional structures of the nucleus.

The sensitivity of the method is on the order of 10-12g. The accuracy of cytophotmetry is reduced by measurement errors resulting from the unequal distribution of the substance in the cell, this error can be avoided through cytophotmetry at two different emission wavelengths. UV cytophotometry helps it possible to determine the quantity of nucleotides, nucleic acids and proteins in unstained preparations according to their natural absorption of UV rays. Cytophotmetry is generally used in the visible region of the spectrum; it also makes use of the natural colour of individual substances or more often, the artificial staining of preparations with special histo-chemical dyes, which combine with the chemical components of the cell in determined quantities. The majority of the dyes aid in determining nucleic acids and proteins and their individual reactive groups in the cell and in ascertaining the activity of many enzymes.

PHASE CONTRAST MICROSCOPY (PCM)

PCM is a microscope that uses the change or phase when light is transmitted or reflected from a transparent or reflective specimen to form an image, used to study the living cells. PCM is a microscope that translates the differences in phase of the light transmitted through or reflected by the object into differences of intensity in the image (also called the phase microscope or phase difference microscope). A PCM splits a beam of light into two types of light, direct and refracted (reflected) and brings them together to form an image of the specimen. Where the lights are `in-phase`, the image is brighter and where the lights are `out of the phase`, the image is darker, and by amplifying these differences in the light, it enhances the contrast.

PCM is a special contrast enhancing imaging method for transmitted light microscopes and the easiest and most common way of imaging biological specimens. This was invented by First (Frederick) Zernike in 1934, and introduced into microscopic practice by August Kohler and Loos in 1941 and very soon it revolutionized biological and medical research. The main objective of this invention is to convert phase differences into amplitude variations that can easily be detected. PCM can be used to produce high contrast images of transparent specimens (living epithelial cells).

The method is specific in that it involves the separation of the direct zeroth order light from the diffracted light at the rear focal plane of the objective. To achieve this, Kohler placed a ring annulus in position directly under the lower lens of the condenser at the front focal plane of the condenser, conjugate to the objective rear focal plane. The hollow cone of light from the annulus passes through the specimen undeviated, and arrives at the rear focal plane of the objective in the shape of a ring of light, and the faint light diffracted by the specimen is spread over the entire rear focal plane of the objective.

The accessories required for a phase contrast are a sub-stage phase contrast condenser equipped with annuli and a set of phase contrast objectives, each of which has a phase plate installed. The condenser usually has a bright-field position with an aperture diaphragm and a rotating turret of annuli (each objective of different magnification requires an annulus of increasing diameter as the magnification of the objective increases). Each one of the objectives has a darkened ring on its back lens, and they can be used for ordinary bright-field transmitted light work with only a slight reduction in image quality. The outfit supplied by the manufacturer usually includes a green filter and a phase telescope. The latter is used to enable the observer to alight the condenser annulus to superimpose it onto the ring of the phase plate. A set of centering screws in the condenser allow manipulation of the annulus to align it while observing the back focal plane of the objective with the phase telescope (sometimes through a Bertrand lens).

Generally, use of an ordinary (bright field) transmitted light microscope produces images of very thin and transparent objects (such as living cells or biological tissues) which suffer from lacking contrast, which makes it very difficult to recognize and distinguish delicate sample structures or image details. This is because; light absorption of such objects is too low to provide sufficient amplitude changes of the transmitted light (bright dark contrast). The individual sample structures (like cell nucleus, cytoplasm and organelles) show little density differences leading to small differences of the refractive indices and to different optical path lengths. If light waves pass these structures, they experience a certain phase shift that corresponds to the respective optical path lengths. The phase contrast method is intended to convert such phase shifts into amplitude differences that are detectable by the human eye. PCM was very successful and ultimately gained widespread application which resulted in Noble Prize to Zernike (1953). PCM allowed scientists to see the details in unstained and/or living objects with a clarity and resolution never before achieved.

a. Principle

PCM is an optical illumination technique that changes the brightness passing through a transparent specimen in the image. The phase shifts are invisible to the naked eye. PCM is important in biology as it reveals biological structures that are not visible by the use of a simpler bright field microscope. As light travels through a medium, depending on the properties of the medium, light interacting with the medium causes the amplitude and the phase to change. It is always necessary to combine the light passing through the sample with a reference in order to make phase variations observable.

b. Setting up of a Phase Contrast Microscope

The microscope is set up, and the specimen is focused with the 10x objective. The cells of the cheek lining are readily available test material. Then, configure the microscope for Kohler illumination using the zero position (bright field) of the condenser. This is to assure the proper alignment of the objective, condenser and field diaphragm. After proper alignment, the condenser aperture diaphragm is opened up and the turret is swung into the 10 position (this automatically opens up the diaphragm). The green filter is placed in the light path and one of the eyepieces are removed. The phase telescope is inserted and while observing the back of the objective, the annulus centering screws are used to centre the annulus to the ring of the plate. A pinhole eyepiece or a Bertrand lens, if available, will allow a view of the back focal plane of the objective. Centering the annulus is often easier to do if the specimen is temporarily removed from the path of light. After alignment of the phase ring with the annulus, the eyepiece is reinserted and the specimen is placed back into its proper place on the microscope stage in the optical path. The same procedure is repeated for each objective, to make sure that the turret is rotated so that the appropriate phase annulus is positioned to match the objective magnification. Sometimes an individual push-in, centering and the annuli are provided by the manufacturer that can be inserted into the lower part of the common Abbe condenser. All such simple devices work well with the 10x, 20x and 40x phase objectives, but the condenser can receive only one at a time.

c. PCM – Applications and Limitations

PCM continues to be a widely used and an important tool, particularly for one studying living and/or unstained material (cells and tissues in culture). Currently this is also used simultaneously with reflected light fluorescence to reveal the areas of a specimen that do not fluoresce. These methods are useful particularly with specimens that are thin and scattered in the field of view. The technique is widely applied in biological and medical research, especially throughout the fields of cytology and histology. The method is utilized to examine living cells, tissues and microbes that are transparent under bright field illumination. This enables internal cellular components like membranes, nuclei, mitochondria, spindles, mitotic apparatus, chromosomes, golgi bodies and cytoplasmic granules from both the plant and animal cells and tissues to be readily visualized. The technique is widely employed in the diagnosis of tumour cells and the growth, dynamics and behaviour of a wide variety of living cells in cultures. Other areas in biology that benefit from this technique are hematology, virology, bacteriology, paleontology and marine biology. The industrial and chemical applications include mineralogy, crystallography, and polymer morphology investigations.

PCM has its own limitations such as: The images are surrounded usually by halos around the outlines of the details. These are optical artifacts which obscure the boundaries of the details. The annuli do limit the working numerical aperture of the optical system to a certain degree and thus reduce the resolution. PCM does not work well with thick specimens because of shifts in phase which occur from areas slightly below or above the plane that is in focus and confuse the image and distort the details of the image. The images appear gray if white light is used and green if a green filter is used.

FLUORESCENCE MICROSCOPY (FM)

1. Definition

A FM is `*an optical microscope that uses fluorescence and phosphorescence instead of/in addition to reflection and absorption to study the properties of organic or inorganic substances*`. The FM also refers to `any microscope that uses fluorescence to generate an image whether it is a more simple setup (an epifluorescence microscope) or of a more complicated design (a confocal microscope) which uses optical sectioning to get a better resolution of the fluorescent image`. This is an excellent tool for studying the material which can be made to fluoresce, either in its natural form (primary or autofluorescence) or when treated with chemicals capable of fluorescing (secondary fluorescence). This is now one of the fastest growing areas of investigation using the microscope. This has gained popularity ever since Coons (1941) developed methods to conjugate proteins to fluorochromes, and this is the most frequently employed technique of light microscopy in biomedical research, in particular neuro-biology, today. It also provides a window into the physiology at sub-cellular levels of resolution (allows direct visualization of working of physiological processes at a systems level in a living cell or tissue), of living cells. It also enables the study of diverse processes like location and association of proteins, motility and other phenomena like ion transport and metabolism.

2. Basic Principle

The specimen is illuminated with light of specific wavelength(s) which is absorbed by the fluorophores causing them to emit light of longer wavelengths (of a different colour than the absorbed light). The illuminated light is separated from the much weaker emitted fluorescence through a spectral mission filter. The components of FM consists of a light source (xenon arc lamp/mercury vapour lamp are common, and the more advanced forms are high power LEDs and lasers), the excitation filter, the dichroic mirror (dichroic beam-splitter) and the emission filter. The filters as well as the dichroic mirrors are chosen to match the spectral excitation and emission characteristics of the fluorophore used to label the specimen. In this way, the distribution of a single fluorophore (colour) is imaged at a time. The multi-colour images of several types of fluorophores must be composed by combining several single colour images.

Most FMs in use are epi-fluoroscence microscopes, where the excitation of the fluorophore and the detection of the fluorescence are done through the same light path (through the objective), and these are widely used in biology and are the basis for more advanced microscope designs (confocal microscope and the total internal reflection fluorescence microscope - TIRF). Most of the FMs that are used in the life sciences are of the epifluorescence design, wherein the light of the excitation wavelength is focused on the specimen through the objective lens. The fluorescence emitted by the specimen is focused to the detector by the same objective that is used for the excitation which will have a very high numerical aperture for greatest sensitivity. As most of the excitation light is transmitted through the specimen, only reflected excitatory light reaches the objective together with the emitted light and therefore this method gives a high signal to noise ratio. An additional barrier filter between the objective and the detector can filter out the remaining excitation light from fluorescent light.

3. Sources of light

FM requires intense, near-monochromatic, illumination which some widespread light sources (halogen lamps) cannot provide. The light sources used are of four main types, namely, xenon arc lamps and/or mercury vapour lamps provided with an excitation filter (along with LEDS with a dichroic excitation filter are commonly used for wide-field epifluroscence microscopes) are, lasers (mostly used for complex techniques like confocal microscopy and TIRF), super-continuum sources and high power LEDs.

4. Preparation of the Sample

The most suitable sample for FM must be fluorescent. Several methods are in practice to create a fluorescent sample. The main techniques are labeling with fluorescent stains (in the case of biological samples) or the expression of fluorescent protein. Alternatively, autofluorescence or primary fluorescence (intrinsic property of the sample to fluoresce) can also be used (but, this is often a problem). In biology, FM is a powerful tool as it allows the specific and sensitive staining of a specimen in order to detect the protein distribution or the distribution of other molecules of interest. As such, there is a diverse range of fluorescent staining techniques of biological samples and important of such methods are biological fluorescent stains, immunofluorescence and fluorescent proteins.

5. Biological Fluorescent Stains

M. Haitinger (1933) was the first to stain histological specimens with fluorescent dyes. Many stains fluoresce and selectively stain various tissue components. The fluorescence that results from this method of staining is called `secondary fluorescence'. Many stains have been designed for a range of biological samples, and some of these are small molecules which are intrinsically fluorescent that can bind a biological molecule of interest. Important examples of these are the nucleic acid stains like DAPI and Hoechst (excited by the UV wavelength light) and DRAQ5 and DRAQ7 (optimally excited by red light) all of which bind the minor groove of DNA and label the cell nuclei. Other stains are the drugs or toxins which bind specific cellular structures having been the derivatives with a fluorescent reporter (exemplified by phalloidin, used to stain actin fibers in mammalian cells). Other fluorescent molecules called fluorophores (like fluorescein, Alexa Fluors, DayLite 488) can be chemically linked to a different molecule which binds the target of interest within the sample).

6. Immunofluorescence

Albert H. Coons and N. H. Kaplan was the first to attach a fluorescent dye to an antibody which was subsequently used to localize its respective antigen in a tissue section. The power of this technique is the chromosomal localization of genes with a particular defect. This is a technique that uses the highly specific binding of an antibody to its antigen in order to label specific proteins/other molecules within the cell. In this method, the sample is treated with a primary antibody specific to the molecule of interest. A fluorophore can be directly conjugated to the primary antibody, or alternatively, a secondary antibody, conjugated to a fluorophore, which binds specifically to the first antibody can also be used (such as a primary antibody raised in a mouse which recognizes tubulin combined with a secondary anti-mouse antibody dervied with a fluorophore could be used to label microtubules in the cell).

7. Fluorescent Proteins (FP)

The modern techniques that are available for modifying DNA allows to genetically modify the proteins to carry a fluorescent protein reporter. This allows in biological samples to directly make a fluorescent protein of interest. The location of such a protein can then be traced, including the living cells.

8. FM – Limitations

The important limitation is that the fluorophores lose their ability to fluoresce due to photo-bleaching, which occurs as the fluorescent molecules accumulate chemical damage from the electrons excited during fluorescence. This can severely limit the time over which a sample can be observed by FM. Photo-bleaching can be reduced by many techniques (by the use of more robust fluorophores, by minimizing the illumination or by using photo-protective scavenger chemicals). FM with fluorescent reporter proteins enables the analysis of living cells, but the cells are susceptible to photo-toxicity, in particular with light of shorter wavelength. Further, the molecules have a tendency to degenerate reactive chemicals under illumination which increases photo-toxic effect. FM only allows the observation of the specific structures which have been labeled for fluorescence (the observation of a tissue sample prepared with a fluorescent DNA stain by FM only reveals the DNA organization within the cells and reveals nothing else about the morphology of the cell).

9. FM - Basics

The basic task of FM is to permit excitation of light to irradiate the specimen and to separate the much weaker re-radiating fluorescent light from the bright excitation light, as such only the emission light reaches the eye or the detector. The resulting areas shine against a darker background with sufficient contrast to permit detection. The darker the background, the more efficient the instrument is. These microscopes contain special filters and employ a special method of illumination to produce the images of fluorescent light emitted from excited molecules in a specimen. The filters are designed to isolate and manipulate two distinct sets of excitation and fluorescent wavelengths. A band of shorter excitation wavelengths from the illuminator and filters is directed to the specimen, while a band of longer fluorescence wavelengths emitted from the specimen forms an image of the specimen in the image plane. One must be able to select proper fluorochorme, filters and illuminator for a given application to perform this technique and evaluate the quality of fluorescence signals.

In FM, the sample one wants to study is itself the light source and this method is used to study specimens which can be made to autofluoresce. The FM is based on the phenomenon that certain material emits energy detectable as visible light when irradiated with the light of a specific wavelength. The sample can be fluorescing in its natural form like chlorophyll and some minerals, or treated with fluorescing chemicals.

The basic task is to let the excitation light radiate the specimen and then sort out the much weaker emitted light to make up the image. The microscope has a first filter that only lets through radiation with the desired wavelength that matches the fluorescing material. This radiation collides with the atoms in the specimen and electrons are excited to a higher energy level. When they relax to a lower level, they emit light. To become visible, the emitted light is separated from the much brighter excitation light in a second filter. The fact is that the emitted light of a lower energy and a longer wavelength is used. The fluorescing areas can be observed and they shine out against a dark background with high contrast.

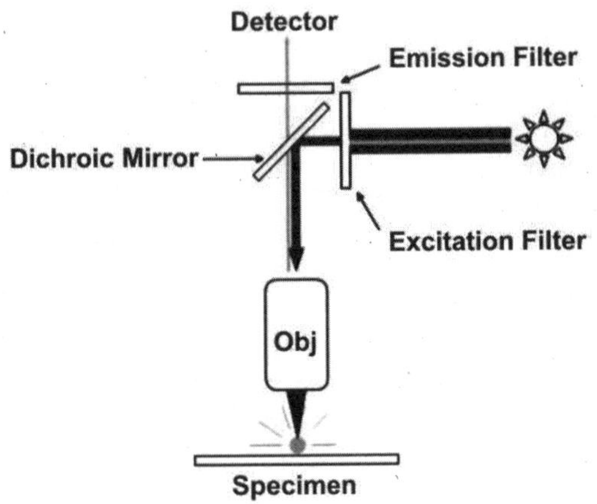

FM – Light Path

10. FM -Advantages and Disadvantages

FM is advantageous because of the following reasons: The technique is highly sensitive and highly selective too because one can label specific structures with appropriate probes. The method is highly versatile because of the availability of selective probes for different structures that are specific (such as antibodies to label the target proteins, specific probes for nucleic acids, selective probes for a variety of cellular components and fluorescent proteins for live studies). The techniques are capable of potentially providing superb contrast because the objects are self-luminous against a dark background. The methods have an ability to separate absorption spectrum from emitted spectrum by virtue of Stoke`s shift (when the electrons go from the excited state to the ground state, there is a loss of vibrational energy, as a result of which the emission spectrum is shifted to longer wavelengths than the excitation spectrum. The wavelength varies inversely to radiation energy, and the difference between the absorption [excitation] maximum and emission maximum is known as Stoke`s law or Stoke`s shift). The techniques are advantageous in that an extremely small number of fluorescent molecules (as few as 50 molecules per cubic micro-meter) can be easily detected.

The disadvantages of this technique are: *blurring* (which can be minimized by the use of optical sectioning or deconvolution), *bleaching* (can be avoided by the use of minimal illumination or by the use of anti-fade in the specimen), *bleed through* (can be overcome by choosing lasers and filters with great care or by the use of sequential image collection) and *autofluorescence* (can be reduced during specimen preparation or excluded with special imaging).

11. FM - Applications

FM is used extensively to study the intracellular distribution, dynamics and molecular mechanisms of a large variety of macromolecules and metabolites. The most important application of FM is in the determination of the intracellular distribution of macromolecules in formed structures such as membranes, cytoskeletal filaments and chromatin. It is also useful in the study of intracellular dynamics of macromolecules associated with binding dissociation processes and diffusion (fluorescence recovery after photo-bleaching or FRAP). The technique is also used in the study of nearest neighbours, interaction states and reaction mechanisms by FRET (fluorescence energy transfer) and by fluorescence correlation microscopy. In addition, this is also useful in the study of dynamics of single tagged molecules or assemblies *in vivo* using an extremely sensitive technique called TIRF (total internal reflection fluorescence) microscopy. FM also finds use in the determination of intracellular ion concentrations and changes in the concentrations for several ion species, including H^+, Na^+, K^+, Cl^- and many other metals. The technique is also used in organelle marking experiments using dyes that label specific organelles and cytoskeletal proteins; study of cell viability and the effects of factors that influence the rate of apoptosis in cells using a combination of dyes that are permeant and impermeant to the plasma membrane; and the examination of cell functions such as endocytosis, exocytosis, signal transduction, and the generation of transmembrane potentials using fluorescent dyes.

FM has many advantages compared to other optical techniques in that the use of fluorochromes had made it possible to identify cells and sub-microscopic cellular components with a high degree of specificity. The FM can reveal the presence of fluorescing material with exquisite sensitivity in that an extremely small number of fluorescing molecules, as few as 50 molecules per cubic micron can be detected. These techniques can be applied to organic material, former living material or to living material with the use of *in vitro* or *in vivo* fluorchormes (stains similar to the better known tissue stains which attach themselves to visible or sub-visible organic matter). These can also be applied to inorganic material (investigation of contaminated or semi-conductor waters) to monitor the rapidly changing physiological ion concentrations such as calcium, magnesium etc, and pH values in the living cells. The autofluorescence has been found useful in plant studies, coal petrography, sedimentary rock petrology, and in the semiconductor industry. But this is of minimal use in the study of animal tissues or pathogens. FM is often used for imaging the structural components of small specimens, like cells; for conducting viability studies on cell populations (whether they are alive or dead); for imaging the genetic material within a cell (such as DNA or RNA); and for viewing specific cells, within a larger population with techniques such as FISH.

FM is a rapidly expanding technique, both in the medical and biological sciences. This has made it possible to identify the cells and the cellular components with a high degree of specificity. Certain antibodies and diseased conditions or impurities in the inorganic material can be studied with the FM. Fluorescence microscopy is mostly used for biological samples (for medical diagnosis). It is currently investigated in which form fluorescence microscopy can be used for the detection of cancer. It is also useful for imaging living cells, particularly with multi-photon FM, it is possible, to watch cell division without severely influencing the cells, whereas cells often cannot tolerate single-photon excitation for extended time intervals.

12. Fluorescent Dyes – Types and Uses

The fluorescence of a dye is caused by its atoms that absorb the energy from other sources and produce the light. It is the light emitting quality along with their chemical makeup that distinguishes from other paints and dyes that make a fluorescent dye useful. Other characters in different types of these dyes include their ability to dissolve in water, possessing high detectability when mixed with other paints, dyes and chemicals. They can be easily seen with the naked eye and contain low toxicity compared to other dyes and paints, are stable in a normal aquatic environment, and possess high colour strength, superb fastness and good stability. The dyes range from colour types for coding and identification to chemical combinations that react differently depending upon the chemical dye contact.

The dyes come in multiple shades and have different light emitting properties. The dyes in the shades of green, orange, pink, violet, blue, red, purple, gray, rose, mustard and a variety of other shades are produced by Gene Link. Similarly, another type of dye is made by Tracerline, but not in specific colours.

The dyes have many applications most of which are industrial. These are used as penetrating liquid to find cracks in the materials and for painting plastics, synthetic resins and as an ingredient in printing ink and in paint ball products. In medical industry, they are used in molecular primers and probes in genetic testing, categorization and identification. The dyes also find use for detecting leaks in air-conditioning, gasoline and diesel engines, power-steering devices and hydraulic devices. The dyes can also trace limestone formations in land formations.

SCANNING ELECTRON MICROSCOPY (SEM)

SEMs can magnify the objects at upward of 300,000 times the size of the object where as the optical microscopes tend to have a magnification power of a few 100 times. SEMs also have tremendous depth of field compared to traditional microscopes, providing an almost 3-D image to analyze, as compared to the flat image an optical microscope produces. The microscope can also look past the surface of an object and tell the information about its composition.

a. SEM - How it works?

The SEM scans a focused electron beam over a surface to create an image. The electrons in the beam interact with the sample, producing various signals that can be used to obtain the information about the surface topography and composition.

b. Main Components

The basic and the main components of the SEM include: the source of electrons (the electron gun), thermionic guns, field emission guns, electromagnetic lenses, scanning coil, sample chamber, electron detectors and vacuum chamber and computer to display the images.

The electrons are generally produced at the top of the column, and accelerated downwards and passed through a combination of lenses and apertures to produce a fine beam of electrons which hits the surface of the sample. The sample is mounted on a stage that is contained in the chamber area and unless the microscope is designed to operate at low vacuums, both the column and the chamber are evacuated by a combination of pumps. The level of the vacuum will normally depend on the design of the microscope. The position of the electron beam on the sample is controlled by scan coils which are situated above the objective lens. These coils allow the beam to be scanned over the surface of the sample. This beam scanning (*rastering*) as the name of the microscope suggests, enables information about a defined area on the sample to be collected. As a result of such interaction of the electron beam with the sample, a number of signals are produced can then be detected by appropriate detectors.

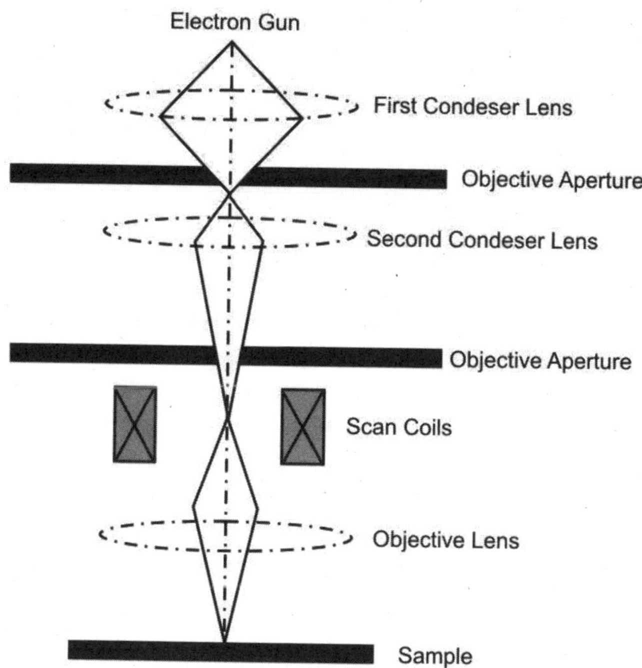

SEM – Schematic diagram of electron path

The SEM produces the images by scanning the sample with a high energy beam of electrons. As the electrons interact with the sample, they produce secondary electrons, back-scattered electrons and characteristic x-rays. These signals are collected by one or more detectors to form images which are then displayed on the computer screen. When the beam hits the surface of the sample, it actually penetrates the sample into a depth of a few microns, depending on the accelerating voltage and the density of the sample. Many signals, like secondary electrons and x-rays are produced as a result of this interaction inside the sample. The maximum resolution obtained depends on multiple factors like the electron spot size and interaction volume of the electron beam with the sample. Many SEMs have been known to achieve resolution below 1 nm. The modern full size SEMs provide resolution between 1-20 nm while the desktop systems can provide a resolution of 30 nm or more.

c. Source of Electrons (Electron Gun)

The electron gun produces the steady stream of electrons necessary for SEM to operate. These are typically one of the two types, thermionic gun and field emission gun. The former are the most common which apply thermal energy to a filament to coax electrons away from the gun and toward the specimen under examination. The latter create a strong electrical field to pull the electrons away from the atoms they are associated with. The guns are located either at the very top or at the very bottom of an SEM and fire a beam of electrons at the object under examination. The electrons don`t go where they need to, but to the next component of the SEM.

The electrons are produced at the source by thermionic heating, and are then accelerated to a voltage between 1-40 KV and are condensed into a narrow beam which is used for imaging and analysis. The 3 commonly used sources of electrons are: tungsten (W) filament, solid state crystal (CeB_6 – Cerium hexaboride or LaB_6 – Lanthanum hexaboride), and the FEG (field emission electron gun). The tungsten hairpin electron filament consists of an inverted V-shaped tungsten wire, about 100 micrometer long which is heated resistively to produce the electrons (the most basic type of electron source). Cerium hexaboride or Lanthanum hexaboride crystal source is a thermionic emission gun and is the most common high brightness source. This offers about 5-10 times the brightness and a much longer lifetime than tungsten wire. FEG is a wire of tungsten with a very sharp tip of less than 100 nm that uses filed electron emission to produce the electron beam. The small tip radius is advantageous in that it significantly improves emission and focusing ability of the microscope.

d. Lenses

Like any optical microscope, SEM uses lenses to produce a clear and detailed image. The lenses are not made of glass, but made of magnets capable of bending the path of electrons. By doing so, they focus and control the electron beam ensuring that the electrons end up precisely where they need to. A series of condenser lenses focus the electron beam as it moves from the source down the column, and the narrower the beam, the smaller the spot when contacting the surface (spot-size).

e. Scanning Coil

The scanning coils are used to deflect the beam in the x and y – axes after the beam is focused so that it scans in a raster fashion over the surface of the sample.

f. Sample Chamber

The sample chamber is the place where the specimens are placed. This must be very sturdy and insulated from vibration because the specimen must be kept extremely still for the microscope to produce clear images. The microscope is so sensitive to vibrations that it is often installed on the ground floor of a building. The chamber also manipulates the specimen, placing it at different angles and moving it so that there is no need to constantly remount it to take different images. The samples are mounted and placed into a chamber that is evacuated. This chamber can include a translation stage, tilt and rotation devices, feed throughs to the outside, temperature stages, optical cameras and a variety of other devices to assist in imaging the sample.

g. Detector

The detectors are like the eyes of the microscope which detect the various ways that the electron beam interacts with the sample. The detectors are capable of producing the most detailed images of the surface of the object and help to let know the composition of a substance. The detector collects the electrons coming off the sample. Typically 2 types of electrons are used for imaging, SE (secondary electrons) and BSE (back-scattered electrons).

h. Secondary Electron Detector

These electrons (SE) are low energy electrons produced when the electrons are ejected from the k-orbitals of the sample atoms by the imaging beam. The most popular detector is the Everhart-Thornley detector which consists of a Faraday cage which accelerates the electrons towards a scintillator. This in turn produces a current which is directed towards a photomultiplier and the amplified signal is read on the monitor.

i. Back-scattered Electron Detector

These electrons (BSE) are higher energy electrons that are elastically backscattered by the atoms of the sample. The atoms with higher atomic numbers backscatter more efficiently and as such this detector can give compositional information about the sample. These can be either scintillators or semi-conductors. Having a semi-conductor is advantageous that it can be split into sections, can be switched on and off to control the contrast and directionality which gives valuable topographical information about the sample.

j. Vacuum Chamber

The microscope requires vacuum to operate, without which the electron beam generated by the electron gun would encounter constant interference from air particles in the atmosphere. These particles not only block the path of the electron beam, but they would also be knocked out of the air and on to the specimen, which would distort the surface of the specimen.

k. SEM – Applications

SEM is widely used in a number of industries and laboratories to investigate the microstructure and chemistry of a range of both organic and inorganic materials.

SEMs also have some drawbacks like cost. Even the cheapest costs 10s of 1000s of rupees. The bulky and the complex instruments require considerable expertise to operate.

TRANSMISSION ELECTRON MICROSCOPY (TEM)

This is a microscopy technique in which a beam of electrons is transmitted through an ultra-thin specimen, interacting with the specimen as it passes through. From the interaction of the electrons transmitted through the specimen, an image is formed, magnified and focused onto an imaging device, like a fluorescent screen, or on a photographic film, or to be detected by a sensor like CCD camera. These microscopes are capable of imaging at a significantly higher resolution than a light microscope owing to the small de Broglie wavelength of electrons. This enables to examine finer detail – even as small as a single column of atoms, 1000s of times smaller than the smallest resolvable object in a light microscope.

1. Main Components

TEM has four main components – an electron optical column, a vacuum system, the necessary electronics (lenses for focusing and deflecting the beam and the high voltage generator for the electron source) and control software. A modern TEM typically comprises an operating console surmounted by a vertical column and containing the vacuum system, and control panels conveniently placed for the operator. The microscope may be fully enclosed to reduce interference from environmental sources, and is operated remotely.

This includes elements analogous to those of a light microscope. The light source of LMs is replaced by an electron gun which is built into the column. The glass lenses are replaced by electromagnetic lenses. Unlike glass lenses, the focal length (power) of magnetic lenses can be changed by changing the current through the lens coil. The ocular (eye piece) is replaced by a fluorescent screen and/or a digital camera. The electron beam emerges from the electron gun, and passes through a thin specimen, transmitting the electrons which are collected, focused and projected on top the viewing device at the bottom of the column. The entire path from the gun to the camera must be under vacuum.

The recent development of a commercial aberration–corrected TEM has enabled major advances in the capability of TEM. Without correction, the resolution is limited primarily by spherical aberration, resulting in a general blurring of the image, but also in a phenomenon called `delocalization', in which the periodic structures appear to extend beyond their actual physical boundaries. The ability to correct the spherical aberration leaves the reduction or correction of the effects of chromatic aberrations as the next major challenge in improving the performance of TEM. The chromatic aberration correctors have been successfully incorporated into the TEM platform, by their design and operation are substantially more complex than spherical aberration correctors.

ETEM (environmental transmission electron microscopy) uses a specially designed vacuum system to allow observing the specimens in a range of conditions approaching more natural environments, with gas pressures in the sample chamber as high as a few percent of atmospheric pressure. This is important for observing the interactions between the sample and the environment. This technique relies on pressure-limiting apertures and differential vacuum pumping to permit less restrictive vacuum conditions in the viscinity of the sample while maintaining high vacuum in the rest of the electron column.

2. TEM - Applications

TEM is designed specifically for the high contrast imaging needs in life sciences and soft matter research as magnifications beyond 1,000,000xs are attainable. TEM is a major analysis method in a range of scientific fields in physical and biological sciences. This also finds its application in cancer research, virology, materials science as well as pollution, nanotechnology and semi-conductor research. The other uses allow to observe modulations in chemical identity, crystal orientation, electronic structure and sample induced electron phase shift as well as the regular absorption based imaging.

3. TEM – Limitations

TEM technique has a number of drawbacks such as: many materials require extensive sample preparation to produce a sample thin enough to be electron transparent, which makes the analysis a relatively time consuming process with a low throughput of samples. The structure of the sample may also be changed during the process of preparation. The field of view is relatively small, raising the possibility that the region analyzed may not be characteristic of the whole sample. The sample may also be damaged by the electron beam particularly in the case of biological materials.

4. SEM and TEM – Differences

The most important differences between a SEM and TEM are: In TEM, the broad static beam is used while the SEM beam is focused to a fine point and scans line by line over the surface of the sample in a rectangular raster pattern. The accelerating voltages are much lower than in TEM, because it is no longer necessary to penetrate the specimen, and in a SEM they range from 50 to 30,000 volts. In TEM, the specimen need not be thin, greatly simplifying specimen preparation than in SEM. In SEM, the specimen itself emits secondary electrons. Some of the primary electrons are reflected BSE (back-scattered electrons) and these can also cause the emission of secondary electrons as they travel through the sample and exit the sample surface. If the sample is thin, the SEM may be operated in STEM mode with a detector located below the sample to collect transmitted electrons.

ELECTRON MICROSCOPY (EM)

a. What Happens in the Specimen During Electron Bombardment in EM?

Most of the specimens are not adversely affected by the electron bombardment as long as the beam conditions are controlled judiciously. Any one of the following may happen when the electrons impinge on the specimen.

Some of the electrons are absorbed as a function of the thickness and composition of the specimen and these cause *amplitude* (mass thickness) *contrast* in the image.

Other electrons are scattered over small angles depending on the composition and structure of the specimen and cause *phase contrast* in the image.

The electrons get scattered in crystalline specimens in distinct directions that are a function of the crystal structure and cause *diffraction contrast* in the image.

Some of the impinging electrons are deflected through the larger angles or get reflected (*backscattered*) by sample nuclei.

The impinging electrons can knock electrons from the sample atoms which escape as low energy *secondary electrons.*

The impinging electrons may also cause the specimen atoms to emit x-rays (*characteristic x-rays*) whose energy and wavelength are related to the specimen`s elemental composition.

The impinging electrons also cause the specimen to emit photons (light) and this is called *cathodoluminiscence.*

The transmitted beam electrons can be counted and sorted by an energy loss spectrometer according to the amount of energy they have lost in interaction with the specimen. The loss of energy carries the information about the elemental, chemical and electronic states of the sample atoms.

b. Electron Microscopy - Applications

EM finds a much wider use in different branches of science and some of the important applications are as follows:

In *structural biology*, 3-D techniques and electron tomography and single particle analysis allow the researchers to derive important information regarding protein domain arrangement, and in some cases, to trace the individual polypeptide chains. The combination of EM reconstruction with x-ray crystallography and NMR spectroscopy enables even greater structural details by fitting atomic scale structural models in the EM density maps.

In *cellular biology*, the high resolution cryo-microscopy avoids the alterations caused by the conventional preparation techniques to allow imaging of cell membrane structures and sub-cellular morphology in fully hydrated conditions.

In *tissue biology*, the ability of an EM to provide high resolution ultra-structural imaging over large areas and volumes of tissues or cells is invaluable in discerning critical relationships among components of biological systems across large differences in the spatial scale.

In *bio-material sciences*, the properties of the biomaterial and nano-particles are highly dependent on the structural characteristics that are readily observed using EM.

In *failure analysis*, the dual beam`s ability to cross section sub-surface defects and quickly prepare site-specific thin section sample for high resolution imaging in SEM/TEM allows the engineers to determine the root causes of manufacturing defects.

In *forensics*, EM is used to analyze criminal evidence such as gunshot residues, clothing fibers, handwriting samples and soil.

In *other automated particle analysis*, both natural and man-made inorganic particles including soil, coal, cement, fly ash and air-borne dust can be analyzed to provide a more detailed understanding of the impact of waste and pollution on the environment and health.

In *research*, EM is being applied successfully in the pursuit of a deeper understanding of the structure-property-function relationships in a wider range of materials, and processes like next generation fuel cell and solar cell technologies, catalyst activity and chemical selectivity, energy-efficient-solid-state lighting, and lighter, stronger and safer materials.

In *3-D nano-characterization*, STEM, TEM and DB tomography afford 3-D visualization at the nano-scale. Analytical techniques like EBSD (electron backscatter diffraction), EDSX (x-ray micro-analysis), and EFTEM (energy-filtered TEM) can also be extended to 3-dimensions, giving a world of new information relating structures to properties.

In *in-situ nano-processes*, the EM becomes a laboratory in a chamber with ESEM and ETFM techniques allowing dynamic control over temperature, pressure and gas type for *in situ* nano-process investigations. The researchers can visualize and correlate the structure, property and function of materials undergoing chemical and physical processes like catalysis, oxidation, reduction, polymerization, deformation and thermally induced phase transformations.

In *3-D nano-prototyping*, nano-prototyping is a fast, simple way to design, fabricate, and test small-scale structures and devices using an electron beam or a focused ion beam. Site-specific milling, lithography, or chemical vapour deposition can all be carried out at the nano scale to deliver high quality 3-D nano-proto-typed structures.

■■■

CHAPTER - 9

SMEAR AND SQUASH TECHNIQUE

Introduction

A cell is described as *a mass of protoplasm enclosed within a membrane* (cell or plasma membrane) containing a nucleus which is bounded by the nuclear membrane. The portion of cell between the plasma membrane and the nuclear membrane is known as the *cytoplasm*. Within the cytoplasm a variety of organelles are identified and these are specialized structures with individual functions and consists of the living material of the cell. *Cytology* is a scientific discipline that *deals with the study of cells, their physiological and structural properties*. Knowing the components of cells and how cells undergo divisions is the main aim of cytology. This also helps to understand various chromosomal mutations and genetic rearrangements. The gross structural changes in chromosomes can be evaluated by cytological procedures, which involve staining the chromosomes and evaluating their morphology *at metaphase of mitosis and meiosis*.

The counting of the chromosomes is the most definite way to differentiate ploidy of an organism, and in this technique, *smears* and *squashes* of the tissue are prepared which are then observed under the microscope for studies on chromosome counting, morphological differences and karyotyping. This technique is tedious and requires a longer time for analyzing. The sample taken should be of small quantity for a better analysis. The *meiotic chromosome smears* can be prepared using *pollen mother cells* isolated from immature anthers dissected from flower buds. Alternately, the *mitotic chromosomes* can be counted from the *root tip* or an *axillary bud* or *shoot tip squash* preparations. The *best stains* used to enhance the contrast of the microscopic images are: Propidium iodide, Hoechst stain, Aceto-carmine, Aceto-orcein, DAPI (4, 6-diamidino-2-phenylindole) and Acridine orange.

Numerous techniques can be used to prepare the tissue for microscopical examination, and the method selected being governed by a number of factors. These include the structures or inclusions to be studied, the amount and the nature of tissue to be examined, whether the specimen is fresh or preserved and the urgency of the investigation.

The fresh specimens may be examined as *teased* and *squash preparations*, *smears*, *touch preparations* or *frozen sections*. The cellular contents of small pieces of tissue not exceeding 1 mm in diameter can be examined by *squash* preparation (placing the tissue in the center of a microscopic slide and forcibly applying a cover glass). Staining can be carried out if necessary by making use of capillary attraction – a drop of a vital stain placed at the junction of the cover glass and the slide is drawn into contact with the tissue which absorbs it. The microscopic examination of the cellular material spread lightly over a slide in the form of a smear, is a technique which has wide application in histopathology. The method of preparation differs according to the nature of the material to be examined, but a as a general rule, the smears are made either by spreading the selected portion of the specimen over the surface of the slide with a platinum loop or alternatively, by making an opposition smear with the aid of a second slide. Smears may be examined either as fresh preparations in a similar manner to that described for teased preparations, or by using a supra-vital staining technique in conjunction with a warm stage. The *impression smears* (also known as the *touch preparations*) are prepared by bringing into contact the surface of a clean slide with that of a freshly cut piece of tissue microscopically by phase contrast or after applying vital stains. Alternatively, the impression smear can be fixed and stained according to methods generally adopted. The sections of about 10 – 15 μm thickness can be cut from fresh tissue frozen on a microtome with the aid of carbon dioxide or electro-thermal coupling units. The sections are then transferred from the microtome knife to a dish containing an isotonic solution from which they may be either attached to slides prior to staining or carried through the staining solutions by means of a glass rod.

Smear and Squash

The word *smear* comes from Middle English word, *smere*, from Old English word, *smerian*, meaning `to anoint or tub with grease, oil, and so on`, from Proto-Germanic word, *smerthan*, related to Old Norse word, *smjor* meaning `fat`, akin to Old High German word, *smero* meaning `grease` and probably to Old English word, *smiur* meaning `marrow`. The word was first used before 12th century.

Similarly, the word, *squash* comes from Middle English word, *squachen* and from Old French word, *esquassare* or *escasser*, meaning `to crush` or `to shatter` or `to destroy` or `to break`, and from Vulgar Latin word, *exquassare* and from Latin word, *ex-+quassare*, meaning to `to shatter`. Literally in English, it means `to press into a flat mass` or `to flatten out under pressure or impact` or `to press with force into a small space`. The first known use of the word was in 1590. The medical definition is `*a bit of tissue crushed between a slide and a cover glass and stained in situ especially in cytological study of chromosomes*`.

The different methods of making squash preparations for studying chromosomes involve the application of many chemical and physical agents for the pre-treatment and fixation of cells in division. A good technique should permit the chromosomes to become well spread during *squashing*, and at the same time, the fixation should bring out the details of chromosome morphology as distinctly as possible (for karyotype studies and chromosome measurements). It is also much more difficult to make good preparations from all species.

The techniques used in the preparation of a slide (the root tip or the anther) seem to be frustrating but it is always necessary to develop one`s own methods that serve the intended purpose for a longer time. The cells in the process of continuous division (in seedlings) are generally located in the apical meristem of the shoot and the root. These two areas of the seedling are very easy to dissect out of the freshly germinated healthy material. The easiest method of checking the chromosome number and the stage of division is by collecting the root tips.

The smear is a quick preparation which is mainly useful for studies in cytology where in the loose cells are uniformly and evenly spread out on a clean microscopic glass slide without causing any injury (contents, nuclei and chromosomes). The cells require no cementing adhesive.

The cells can also be killed and carefully fixed without any distortion. The slide after preparation is dehydrated, stained and made permanent (root tips and anthers).

SQUASH TECHNIQUE FOR ROOT TIPS - PROTOCOL

The general procedure of the squash or the smear technique (for root tips) involves 6 steps. They are: the germination of seeds or the bulbs to obtain young root tips; pre-treatment of the tips (apical meristems) to arrest the cells at metaphase to enhance the observation of chromosomes; fixation to preserve the material; hydrolysis to soften the cell walls and make the root tip malleable; maceration, staining and preparation of the slide; and observation and documentation.

a. Germination and Pre-treatment

The root tips are obtained from the seeds which are allowed to germinate either on wet blotting paper discs or sand free from soil and debris or moistened vermiculite (from the market). Before the use, the sand is thoroughly washed and the seeds are sown. The seeds of *Medicago falcata* (methi) or *Vicia faba* have been found to be more suitable because they germinate soon and because their chromosomes are bigger. The seeds of *Vicia faba* are to be soaked in warm water for 3-4 hours before planting to speed up germination. In the case of onion, the bulbs are grown at least 4 days before the experiment. The darkness always promotes the growth of the roots. The rapid root growth of onion (*Allium cepa*) is stimulated by allowing the bulbs to stand for a few hours in a beaker/gas jar containing 1 cm of water. The base of the bulb is placed in such a way that it is always in contact with water. The beaker is kept in a dark place for 3-4 days. When the roots are 2-5 cm long, the terminal 1 cm of the root is taken off and placed in ice water overnight at 4°C (pre-treatment to arrest the cell division) and fixed in Carnoy`s solution for a day.

b. Fixation and Storage

The roots are transferred to Carnoy`s fixative at 4°C overnight for preservation. The roots are stored in 70% ethanol at 4°C.

c. Hydrolysis

3-5 root tips are placed in a clean vial/watch glass/petri dish containing 1N HCl solution at 60°C for 8-10 minutes. The length of hydrolysis and the concentration HCl will vary for different species. Hydrolysis softens the cell walls and makes the root tip malleable.

d. Staining and the Preparation of the Slide

The root tip is selected and blotted on a paper towel/blotting paper. It is placed on a clean microscope slide and checked (the root tip should be slightly pointed and should appear opaque under the dissecting microscope). A drop of aceto-carmine (prepared) stain is placed on the tip and the smallest bit of the root tip (about 1-2 mm) is dissected. It is chopped into half, if possible, as the smaller (thinner) the section, the better the chance of getting a single layer of cells for observation. A clean micro-cover glass is placed on the slide at an angle, and slowly lowering it on to the stain will prevent bubbles from forming. Tap on the area of the root tip with the blunt end of a dissecting needle or the eraser end of a pencil by holding one edge of the cover slip down with a paper towel/blotting paper. Enough pressure is applied to squash the root tip under the cover slip but not to break it. The slide is slightly heated over the flame of an alcohol burner/spirit lamp and the cover slip is once again tapped to get an even flatter preparation. At the end, the root tip should be barely visible on the slide (best preparation).

e. Observation and Documentation

The slide is placed under the microscope at 10X power to locate the cells in division. The apical meristem of the root is generally located slightly behind the root tip, protected by a cell layer that sloughs off as the root pushes itself through the soil during growth. The cells should be small and cubical with a well-defined nucleus. At 10X power, the cells are scanned (whether in mitosis or not) and the objective is shifted to 40X to check the exact stage required (prophase, metaphase, anaphase or telophase of mitosis). Once the cell with required stage (preferably metaphase) has been located, prepare to draw the cell and record the observations. The slide is finally sealed with nail polish/paint/molten wax.

f. Drawing/Reproduction of the Microscopic Image

Drawing is done using the camera lucida. Only the outline of the chromosome structure is necessary. The drawing should be carried out on an un-ruled 8 x 10 inch sheet of white paper with a pencil (HB black pencil). The labeling should include the magnification (at which the cell has been observed), name of the plant (genus and the species), the chromosome number (2X or diploid) along with two types of chromosomes (somatic and sex chromosomes). The name of the person and the date of preparation (date and time of collection of the root tip) should also be appended on the sheet

g. Using Camera Lucida

At present, many methods are available for the reproduction of microscopic images such as freehand line drawings, micro-projection, photography, digitization etc. But, in the past, drawing an image was the only way for recording a microscopic image. Only few users of the microscope possess the requisite talent to make accurate freehand line drawings of the specimens observed under the instrument. But, these have the difficulty of judging the size and proportion of the individual parts of the specimen during drawing.

Hence, a drawing aide, the *camera lucida*, was devised to bring the microscope image and the drawing paper and the pencil in the same plane. This is done by the combination of a mirror and prism with an adjustable set of screens for reducing the strength of the light beams form the microscope and paper surface to produce equality. The drawing procedure involves sketching the outline, and the location of the centromere (of the chromosome). The main advantage of camera lucida over other methods of recording is that all the chromosomes need not be on the same plane as one can focus up and down a small amount to obtain all of the details. One can find it helpful to keep both eyes open while drawing the chromosomes. If one can find it difficult to see both the image and the end of the pencil, cover the drawing with one hand, check the location of the ends of the arms and resume the drawing.

PREPARATION OF ONION ROOT TIP SQUASH

Allium cepa (onion) is most useful for root tip squash for studying different stages of mitosis.

1. Root Tip Preparation

A large onion bulb with undamaged stem disc is selected. The stem is gently rubbed under running tap to remove the old and withered roots and the soil particles that are adhered to it. The bulb is placed on a glass jar of suitable size filled with water so that the stem disc is completely immersed in water (the water is changed two times a day). The adventitious roots begin to grow from the base of the stem disc in darkness after 3-4 days The bulbs fail to sprout if the temperatures are too warm [during summer]. The root tips should be collected at noon when the mitotic activity is at its peak (from 1.00 – 2.00 PM and 11.00 PM to midnight). They are excised when they have grown 2-3 cm long. The distal 1-2 cm tips are cut and fixed in a vial with freshly prepared fixative (acetic alcohol). The contents are shaken and the used fixative is replaced with equal volume of fresh fixative which is allowed to stand at room temperature for 24-48 hours. The fixed roots can be stored for 2-3 months at a lower temperature without impairment of the quality. If they are not to be used immediately, they can be safely stored in a refrigerator. Later, the fixative is taken off and 70% alcohol is added and the vial is shaken. The alcohol is replaced twice. The root tips are now ready for making the squash. This method does not require a long period of fixation. The material can be fixed for 24 hours in Carnoy`s fixative and stained by warming it in aceto-lacmoid + HCl stain for 5 minutes. HCl is important as it helps in the easy penetration of the stain into the tissues.

2. Preparation of the Reagents

a. Acetic Alcohol Fixative

This is prepared by mixing 1 part of glacial acid and 3 parts of absolute alcohol. This is to be prepared just before use.

b. Aceto-orcein Stain (1% solution in 45% acetic acid)

1 gram of orcein is taken in a flask and dissolved in about 55 cm^3 of boiling water added to it. About 45 cm^3 of glacial acetic acid is added to the flask and mixed

thoroughly. The mixture is filtered and stored in a glass-stoppered dark bottle in a refrigerator.

c. Aceto-lacmoid stain

This is prepared by mixing 0.2 g of lacmoid, 3.3 g of orcein and 100 cm^3 of glacial acetic acid. 2 drops of 1M hydrochloric acid is added to a watch-glassfull of the stain just before use.

d. Acetocarmine Stain

This is a basic dye prepared from the insect *Coccus cacti*. 10 g of carmine (powder) is dissolved in 1 L of 45% glacial acetic acid and refluxed for 24 hours. It is cooled rapidly and is filtered into a dark bottle and stored at 40°C (the solution can be stored for a long time). Staining can be intensified by the addition of ferric chloride. Hence, 5 mL of a 10% ferric chloride solution per 100 ml of per cent of acetocarmine is added.

1% staining solution:

Carmine	: 1 g
Glacial Acetic Acid	: 100 ml
Ferric chloride ($FeCl_2.6H_2O$ – 10%)	: 5 ml (Optional)
Total volume	: 100 ml

Note: *The stain stains clothes.*

e. Carnoy`s Fluid - Fixative

This is a very rapid fixative and has been used very frequently for fixation when rapidly processed paraffin sections are required. It is a chloroform-containing fixative which penetrates the tissues extremely rapidly and can fix small tissue pieces in minutes rather than the hours required for other fixatives. Only 30 to 60 minutes (sometimes to 2 hours) is required for a 1-2 mm piece of tissue and dehydration takes place at the same time. Delicate tissues can be damaged when transferred from aqueous solutions to this fixative, due to the extreme hydrophobicity of chloroform and the resultant rapid dehydration. This fixative has been used traditionally for cytological structures. It is a good nuclear fixative but produces considerable shrinkage as well as cytological distortion. Glycogen is preserved well as with most alcoholic fixatives. It is a fixative agent for both the nuclear and mitochondrial DNA in various tissues. This is used for fixing the root tips, anthers etc. This is the fixative usually meant when *Carnoy`s fluid* is specified, and there is another fixative often ascribed to *Carnoy`s fluid*, which is a 3:1 mixture of ethanol and acetic acid (but this is more correctly referred to as *Clarke`s fluid*). After the fixation, the tissue is transferred to 95% or absolute alcohol.

Carnoy`s fluid is of 2 types. The first one (also called as the *Farmer`s fluid*) is prepared by adding 15 ml of absolute ethyl alcohol to 5.0 ml of glacial acetic acid. The other fluid is prepared by the addition of 30 ml of absolute ethyl alcohol to 5.0

ml of glacial acetic acid and 15 ml of chloroform respectively. The composition is as follows:

Ethanol – Absolute	: 60 mL (Prepare just before use)
Acetic acid, Glacial	: 10 mL
Chloroform	: 30 mL

60 ml of ethanol is poured into a suitable container. To this is added 30 ml of chloroform. 10 ml of glacial acetic acid is added to it to give a total volume of 100 ml. All the three solutions are mixed well. The tissue s placed into the fixative for 1 - 3 hours. The fixed tissues are processed immediately or transferred to 80 to 95% of absolute alcohol.

Note: Small tissue pieces are fixed approximately for 1 hour, washed several times in absolute ethanol, infiltrated and embedded immediately.

f. Leuco-basic fuchsin (Coleman, 1940) Stain

The stain is prepared by dissolving 1.0 g of basic fuchsin in 200 ml of boiling glass distilled water and it is shaken well and cooled at 50^0C. It is filtered and to it is added 30 ml of N/1 hydrochloric acid. 3.0 g of Potassium metabisulphite is also added, shaken well and kept in a dark and cool place for about 24 hours. 0.5 g of Norit or activated charcoal is added to it and shaken well for about a minute, followed by quick filtration with the help of a coarse filter paper. The colourless stain is kept in a dark and cool place for future use.

g. Euparal – Mounting Medium

Euparal (yupa-ral) is a semi-synthetic microscopic mounting medium used in microscopy. It is slightly yellowish in colour, flows well and cures after a few days, and after curing, it becomes very hard but not brittle, keeping its elasticity. It also adheres strongly to glass and has a low refractive index of 1.483 (higher than Canada balsam). Euparal compared to other mounting media, has a significant advantage. The specimens can be directly transferred from alcohol into euparal while other mounting media like Canada balsam require the specimen to be transferred to xylene prior to embedding. It is also a clearing agent. This was first described and originally formulated and invented by G. Gilson (1904) and is liked for its ease of use and stability. This is used extensively in the mounting of entomological (insect) and histological (tissue) specimens, and is particularly suitable for the permanent mounting of cytological preparations. The objects like cells are stained with carmine or other stains and slides are passed through dehydration grades and finally mounted in a drop of euparal.

Gilson discovered that a combination of eucalyptol and paraldehyde was able to substitute for the alcohols without harming the pigmentation, and thus gave the mixture containing *EUcalyptol* (a solvent of euparal – dominant portion of *Eucalyptus globulus* oil, used as an insecticide) and *PARALdehyde* (a preservative and a solvent), the name EUPARAL. Now, it is exclusively manufactured by ANSCO Laboratories, Manchester, England.

The advantages of euparal are: it is possible to directly transfer the specimens stored in 70% alcohol into euparal for permanent mounting and it is not necessary to completely dehydrate the object by placing it into absolute alcohol; euparal has a low refractive index which can be an advantage to observe certain structures; euparal can be coloured green by adding some copper salt (*Euparal vert*) and this can further increase the contrast of the specimens stained with haematoxylin; euparal possesses reducing properties and as such prevents oxidation of some dyes, such as haematoxylin; and euparal possesses good fluidity, and does not pull strings and can be handled easily. Other advantages are: unlike other non-water-based mounting media, euparal does not use the harmful solvent xylene; it cures relatively quickly; and on the less serious side – it smells nicely (it should not be inhaled, as it is an irritant and flammable). The disadvantages of euparal are: when embedded in euparal, the acid sensitive dyes do not keep well; it does have the tendency to shrink a bit, and this can introduce air bubbles; and as it is flammable and an irritant, any contact with the eyes and the skin must be avoided.

h. Dioxan - Fixative

This fixative is prepared by adding 50 ml of dioxan to 6.0 ml of formalin, 5.0 ml of glacial acetic acid and 50 ml of distilled water. This fixative is generally used for unicellular and filamentous algae and fungi. The fumes are harmful and hence care should be taken during the preparation of the liquid fixative.

i. Radolph`s Fluid - Fluid for Killing and Fixing

Many fluids have been proposed form time to time by many botanists for killing. Of all these, Radolph`s fluid (CRAF or Craf – principal reagents are Chromium trioxide-glacial Acetic acid and Formalin) is a modification of Navishin`s fluid and is somewhat weaker, but, more convenient to prepare and is found to be superior as it gives satisfactory results (for plants in general and specially meant for root tips). It is used for the same purpose, which is to fix the chromosomes. The fixation time is 12 to 24 hours and after completion of fixation, the tissue is washed well in running tap water to remove all traces of chromium trioxide. It consists of two components, *solution A* and *solution B*. The composition of solution A is: 10 g of Chromium trioxide, 70 ml of glacial Acetic acid and 930 ml of Formalin; and the composition of solution B is: 300 ml of strong Formalin and 700 ml of Distilled water. Normally, both the solutions (A and B) are mixed in equal quantity just before use.

For critical work on certain subjects, different formulae are used. The acetic acid can be varied from 0.7 to 5% glacial acetic acid by volume. Craf V is suitable for the woody tissues (50ml of 1% chromic acid) while Craf III (30 ml of 1% chromic acid) is the most generally used formula. After several days, the colour of the solution will become olive or green and when this happens, the formaldehyde has reduced the chromic acid to chromous acid. Finally, it is washed in 70% ethanol three times, and transferred to 95% ethanol before proceeding to paraffin.

j. Killing and Fixation Fluid (Johansen – 1940)

This fluid is prepared by adding 10 ml of 10% solution of aqueous acetic acid to 1.0 ml of 10% aqueous chromic acid and a solution of 2.0% osmic acid in 7.5 ml of 2.0%

chromic acid and finally to 41.5 ml of distilled water. 1.0% saponin is finally added to the mixture.

k. Gentian violet or Methyl violet-2B

Gentian violet or Methyl violet-2B is a basic stain and its solubility in water is 2.93% while in ethyl alcohol it is 15.21%. 1.0% aqueous solution is boiled, filtered and cooled. There are 2 main shades of the stain indicated by R (reddish shade) and B (bluish shade). The latter is generally preferred to the former as it stains chromosomes and other tissues bright and clear.

l. Belling`s Iron-Aceto-Carmine

The Belling`s formula for the preparation of iron-aceto-carmine has been modified by Johansen (1940). The stain can be prepared as follows: 110 ml of distilled water is added to 90 ml of glacial acetic acid, and the solution is heated to its boiling point. At this point, 1 gm of carmine dye is added and the mixture is cooled to ice temperature and decanted. After decantation, a few drops of aqueous ferric acetate is added until a wine-red colour appears. Every care is taken that more than needed amount of ferric acetate is not added because carmine will get precipitated if added in excess.

3. Procedure for Squash

A clean glass slide is taken and near one edge of it the root tip is placed. The terminal portion of the root tip (about 2-3 mm) is excised and the proximal part of the root is discarded. A drop of aceto-carmine stain is placed in the middle of the slide and the cut tip is transferred into the stain with the help of a forceps. The root tip is teased into a number of pieces (length-wise) with a spear needle and a scalpel. With the help of a masher, the root tip tissue is crushed into a thorough pulp so that no large pieces are seen on the slide. A drop of stain is added if necessary but excess is to be avoided. The excess stain sticking to the slide is drained off. A clean cover glass is placed over the cell suspension and it is gently tapped with the eraser at the end of a pencil or the wooden end of a dissecting needle or with finger (thumb) to spread the material. The cover slip should not be moved during this process (the displacement of the cover slip will mar the preparation. Too much pressure is to be avoided as it may break the cover glass. The slide is placed over the steam bath or a hot plate until the stain at the edges of the cover glass begins to dry off. After drying, flatten the cells further by applying firm pressure from above holding the slide between the folds of tissue paper sheet. This expels the excess stain from under the cover glass and also helps spreading the chromosome content in an even layer. The slide can be used for observation for about 30 minutes. It can be used for a longer time (2-3 days) if it is sealed by nail polish/nail varnish temporarily. Upon storage, the stain gets intensified. The use of correct light is important during observation of the material and green light is most beneficial. A good squash preparation can be examined with an oil-immersion lens. The observations are recorded along with drawings (with the help of camera lucida).

4. Cleaning the Slide

A solution is prepared by dissolving 20 g of potassium dichromate in 100 ml of distilled water. 100 ml of concentrated sulphuric acid is added to it slowly. The mixture is cooled and stored for use. The fresh and unused glass slides are immersed in the solution for considerably a long time. The cleaning procedure is continued until the solution turns darker in colour to be replaced by a fresh solution. The slides are removed from the solution and thoroughly washed in running tap water. Later, the slides are immersed in strong alcoholic solution containing ammonia for some time. Finally, the slides are washed again in water and cleaned with a clean and dry muslin cloth for use.

CHROMOSOME SQUASH TECHNIQUE

1. Staining method

The fixative is drained off and the roots are placed in 1% acetocarmine solution for 1 to 3 hours. It is heated until the acetocarmine begins to boil. The root cap is cut off with a razor blade and the meristematic tissue is seized out with a lancet needle. A drop of acetocarmine or 45% acetic acid is added. A double edged razor blade is placed to one side and a cover slip is added. The cover slip is gently tapped with the needle end of a probe, and the razor blade is slided out and it is heated to a point just below boiling (steam will be seen forming beneath the slide), then, it is squashed quickly with the thumb or forefinger between two layers of filter paper. Care must be observed not to move the cover slip at this point.

2. Iron containing version of the stain

A solution of 45% acetic acid (45 ml of glacial acetic acid and 55 ml of distilled water) is heated to boiling. *Caution is always taken to wear personal protective equipment when in the lab with concentrated acids* (or any other time). *Always add the acid slowly into water and at any time do not add water into acid.* 0.5 g of carmine (powder) is added and heating is continued for 15 – 20 minutes while stirring. The resulting solution is cooled rapidly and filtered to remove any precipitate into a dark coloured bottle to be stored. A solution of 45 ml of glacial acetic acid and 55 ml of distilled water and 5 g of ferric oxide is made separately and this is slowly added (drop-wise) to 50 ml of the carmine solution until a precipitate starts to appear. 50 ml of carmine soloution is promptly added to the titrate mixture, and is filtered to remove any precipitate and stored for use.

3. Permanent Smear Preparation-Root Tip (Hillary`s Method)

The fixed root tip is thoroughly washed in water and stained with leuco-basic fuchsin for more than an hour`s time. It is dehydrated by changing it through dioxan (3 changes repeatedly). The dioxan fixative (adopted by Mc Whorter F. P and Weier E: 1936). The root tip is transferred on to a clean glass slide containing Canada balsam (1 drop) diluted with dioxin. A clean cover slip is put over the material and pressed gently with the tip of a finger so that the cells will spread underneath the slip. This method can also be successfully used for preparing the anther smears.

PREPARATION OF THE ANTHER SMEAR

The anthers are normally taken out from the flower buds at least 15 days before the pollen release. In some plants they are fixed during the mid day when meiotic divisions are good (*Tradescantia*). The anther smears are prepared by any one of the following methods.

1. Anther Smear – Killing, Fixing and Preparation of a Permanent Mount

A petri dish (16 cm in diameter) is taken and 2 glass rods are placed parallel to each other in it. The killing fluid is poured into it so as to immerse the glass rods. The slide containing the smear is inverted and placed over the glass rods in a horizontal manner in such a way that the smeared surface of the slide is in direct contact with the killing fluid. The slide is turned upside down for about 10 minutes and covered with another petri dish and left for sometime in the same position. The slide is placed in a dish (with the smeared surface facing above) and washed in slow running water for about 15 minutes. Then, the smear is immersed in a freshly prepared aqueous solution of 1.0% methyl violet-2B for about 15 minutes (the boiled solution should not be used). The smear is rinsed in water for about half-an hour. Later, this is differentiated in 70% and 95% alcohol respectively (each of the grade containing 0.5% of picric acid) for 10 seconds each. After differentiation, the slide is dipped in 95% alcohol containing 3 drops of ammonia for about 15 seconds, followed by dehydration in absolute alcohol for 10 seconds. Finally, the slide is once again differentiated in clove oil in a staining jar, from 10 -15 seconds moving back and forth. The slide is washed in xylol (with a few drops of absolute alcohol in it), transferred to pure xylol (for about 2 hours) and mounted in Canada balsam for observation under a microscope.

2. Preparation of Anther Smear (Method -1)

The fixed material (anther) is placed in a drop of aceto-carmine on a clean glass slide and the fluid is withdrawn after 5 minutes with the help of a blotting paper. A fresh drop of the stain is added to the material and the microsporocytes (pollen mother cells) are squeezed out with the help of a scalpel or by applying light pressure. The pieces of the anther wall are removed from the drop of the stain. A clean cover slip is put over the material and any amount of excess stain is removed with the help of a blotting paper. The margin of the cover slip is sealed with molten wax and the slide is left for few days to allow the stain to be penetrated into the material for observation under the microscope.

3. Preparation of Anther Smear (Method -2)

The smeared slide (containing the anther) is placed on 2 glass rods kept parallel to each other in a dish containing Carnoy`s fluid (fixative) for a few seconds facing the smeared surface downwards. Later, the slide is turned upside down for about 10 seconds. A drop of aceto-carmine stain is put on the smear and a clean cover glass is placed over it. Finally, the edges of the cover glass are sealed either with molten wax or Canada balsam diluted with xylol.

4. Preparation of Anther Smear (Method -3)

The anthers are placed in a container with 1 part of glacial acetic acid and 3 parts of absolute alcohol from half-a-day to a few weeks to improve the staining ability of their chromosomes. Later, the anther is taken out and transferred to a clean glass slide with a drop of aceto-carmine on it. The microsporocytes are squeezed out with the help of a scalpel of by the application of light pressure, the cell wall material of the anther is removed from the stain and a neat cover glass is placed over it. The slide is heated on a flame for one or two seconds and this is repeated 5-7 times to allow the cells to spread and flatten. Under no circumstances, the stain is allowed to boil. The slide is placed horizontally in a petri dish containing 10% solution of aqueous acetic acid so as to allow the cover glass to slightly rise above the material. The cover slip is carefully removed with a forceps, and both the cover glass and the slide are transferred carefully to a jar with equal parts of acetic acid and absolute alcohol for about 5 minutes. Both the cover slip and the slide are allowed to pass through the mixtures of 1 part of acetic acid and 3 parts of absolute alcohol; 1 part of acetic acid and 9 parts of absolute alcohol; and equal parts of absolute alcohol and xylol for 1 minute each in each mixture respectively. A thin drop of Canada balsam diluted with xylol is added to the smear and same cover glass is replaced over the smear. The material is finally observed under the microscope.

Squash and smear cytology has a better pick up rate than paraffin embedding and should be the preferred method when only one or few bits of tissue is available.

■■■

CHAPTER - 10

SPECTROPHOTOMETRY

Introduction

Spectrophotometry is one of the most widely used techniques in modern laboratories. This is a process using electro-magnetic radiation (light) to measure an unknown analyte concentration. Spectrophotometry is used to measure a wide range of analytes from cells to nucleic acids and proteins. Spectrophotometry is a scientific method based on the absorption of light by a substance and takes advantage of the two laws of light absorption, Lambert`s law and Beer`s law. Spectrophotometry (in chemistry) is the quantitative measurement of the reflection or transmission properties of a material as a function of wavelength. The more general term is `electromagnetic spectroscopy`. Spectrophotometry deals with visible light, near-ultraviolet and near-infrared, but does not cover time-resolved spectroscopic techniques. This method involves the use of a device called spectrophotometer, a photometer that measures the intensity as a function of the light source wavelength. The important features of the device are spectral bandwidth and linear range of absorption or reflectance measurement. The device is used for the measurement of transmittance or reflectance of solutions, transparent or opaque solids (polished glass) or gases. They can also be used to measure the diffusitivity on light ranges around 200-2500 nm with the help of different controls and calibrations. The calibrations are needed on the machine using standards that vary in type depending on the wavelength of the photometric determination. These devices can be used in physics, materials science, chemistry, biochemistry and molecular biology. They can also be used in industries including semi-conductors, laser and optical manufacture, printing and forensic examination and in chemical laboratories. It helps to determine the nature and the exact amount of the substances present in the target through the calculations of observed wavelengths.

History and Spectrophotometry

Europe during the Industrial Revolution provided many tools needed for science (mostly physics and chemistry) to solve crime as well as to the needs of the economic development. The UV spectrophotometer, the work-horse of the forensic chemistry laboratory in early and mid-20th century depended on Johann Ritter`s discovery of UV radiation (1804). Johann Wilhelm Ritter (1776-1810), a German chemist, physicist and philosopher, after the discovery of heat rays (infra-red radiation by William Herschel - 1800), looked for an opposite (cooling) radiation at the other end (beyond violet) of the visible spectrum. He noticed after a series of attempts that silver chloride was transformed faster from white to black when it was placed at the dark region of the Sun`s spectrum, close to its violet end. The chemical rays found by him were afterwards called ultraviolet radiation (1801). This was used to identify organic chemicals. The first working model was the UV spectrograph invented by Walter Noel Hartley (1877), London. He collaborated with Adam Hilger to produce a commercial instrument (1907) which proved to be cumbersome for use. The first working spectrophotometer was invented by Arthur C. Hardy working at MIT (Massachusetts Institute of Technology, USA) in the early 1920s. He used cesium photocells to detect the spectra. Though this was commercialized together with the General Electric Company (1930), the first "workhorse" spectrophotometer was the Beckman DU, invented by Arnold Orville Beckman (1940).

The modern spectrophotometer has taken decades to be invented. French scientists had drawn the preliminary sketches for a spectrophotometer in the 1890s but the technology of that time was insufficient to build one. The modern spectrophotometer was finally perfected in 1939-1940 by an American chemist, Arnold O. Beckman (1900-2004) and his colleagues, led by project leader Howard H. Cary, at National Technologies Laboratories, the company he started in 1935. This was the company's greatest discovery. The early spectrophotometer was invented before World War II and was used during the war by the government to check for food poisoning and for experiments to make atomic bombs. The invention of the spectrophotometer has impacted a wide range of fields from physics to chemistry and has led to the discovery of unknown properties of light. Before 1940, the chemical analysis process was a long venture taking weeks to complete with only 25 percent accuracy. In 1940, when the Beckman DU Spectrophotometer (which incorporated the amplifier from pH meter) was invented and introduced after making many improvements, it simplified the process greatly, requiring only a few minutes for analysis. The instrument set the standard in chemical analysis. Sometimes referred to as the "Model T of laboratory instruments," the Beckman DU spectrophotometer was a major contributor to the development of biochemistry for 30-plus years. "*The spectrophotometer improved biological assays from a process that took weeks and achieved 25 percent accuracy to one that took minutes and achieved 99.9 percent precision. Although new models were introduced through the years, many of these original instruments are still in use today.*" (www.scripps.edu-2004)

In the beginning there were performance issues with the spectrophotometer. These problems led to changes in design. The model-B spectrophotometer used a quartz prism instead of a glass prism which improved the UV capabilities of the device. Model-C soon followed with changes that raised wavelength resolution in the UV and three subsequent Model-C spectrophotometers were made. In 1941, the Model-D, also known as the Model-DU, was produced with a hydrogen lamp and other improvements. This design remained essentially unchanged from 1941 to 1976 when it was discontinued. By the time production on the Model-DU was halted in 1976, over 30,000 DU and DU-2 models had been sold. This instrument was used in clinics, industrial laboratories and in chemistry and biochemistry.

In 1981, Cecil Instruments produced a spectrophotometer that was microprocessor controlled. This automated the device and improved the speed. This spectrophotometer was more reliable than others made in this era. From 1984 to

1985, development was made in double beam versions of the instrument which developed into the Series 4000 model. With the 1990s came the addition of external software that provided PC control and onscreen displays of the spectra. Today, the development of the spectrophotometer continues and its applications range from science and medicine to crime scene investigation and law enforcement.

SPECTROPHOTOMETER – DEFINITION

The instrument used in spectrophotmetry, a spectrophotometer, can be variously defined. Spectro-photometer is an instrument for *measuring the intensity of light of a definite wavelength transmitted by a substance or a solution,* thus providing a measurement of the amount of material in the solution absorbing light.

SPECTROPHOTOMETER - COMPONENTS

The spectrophotometer is an apparatus for measuring the light intensity in a part of the spectrum as transmitted or emitted by particular substances. A spectrophotometer is composed of 6 major components, namely,

1. White Light Source

The device has a source of light with a continuous spectrum in either the visible or ultra-violet region of the electromagnetic spectrum. The light source is usually a tungsten lamp that procures a constant white light (representing all different colours) output. The bulb is fairly cheap and can be run continuously.

2. Diffraction Grating

This is a special plate with 100s of parallel grooved lines which act to separate the white light into the visible light spectrum (identical with that of a rainbow). More the lines, the smaller is the resolution of the wavelength.

3. Monochromator

This selects light within a narrow band of wavelengths. When once the light has been diffracted, a digital or mechanical knob is used to turn the diffraction grating and aim the desired wavelength at the exit slit.

4. Exit Slit

This is a small hole that allows only a small amount of the diffracted light to leave the instrument. The dimension of the exit slit is called the *band pass* (2-20 nm). If the grating is set at 510 nm, then the device with a band pass of 20 nm would allow wavelengths of light between 500 and 520 nm through the exit slit.

5. Sample Cell Holder

The selected light (coming out of the exit slit) passes through a pathlength/sample contained in a sample holder (sample compartment), usually a glass cuvette. The light directed through the cuvette either reacts with the coloured molecules and is absorbed or it passes through the solution and is transmitted. The transmitted light is sent to the detector. The sample cells should be matched and the matched sample cells (one for the reagent blank and the second for all the standards and samples) have the same optical properties and produce the same absorbance.

6. Light Detector

The transmitted light hits a photomultiplier tube which converts the light energy into electrical energy which results in a display.

SPECTROPHOTOMETER – DESIGN OF THE DEVICE

The spectrophotometers are mainly of 2 types, namely, *single beam* and *double beam*. The double beam spectrophotometer compares the light intensity between two light paths, one containing a reference sample and other path containing the test sample; while a single beam spectrophotometer measures the relative light intensity of the beam before and after a test sample is inserted into the device. The comparison measurements from the single beam devices have large dynamic range and are optically simple and more compact than those from the double beam devices which are easier and more stable. The single beam devices are optically simpler and more compact and are sometimes built on to telescopes or microscopes.

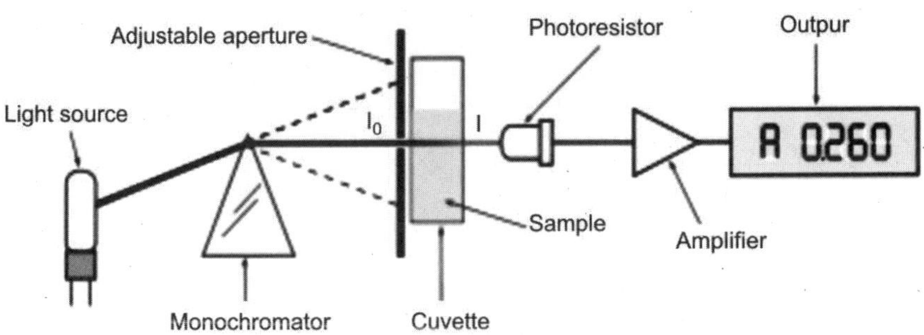

Single beam Spectrophotometer

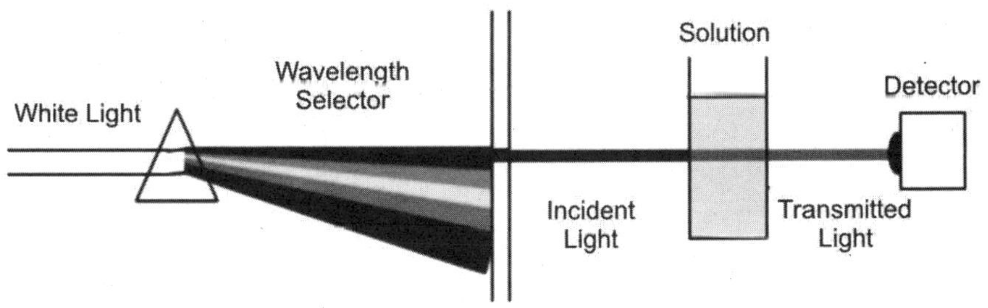

The difference between the incident and transmitted light indicates the absorbance

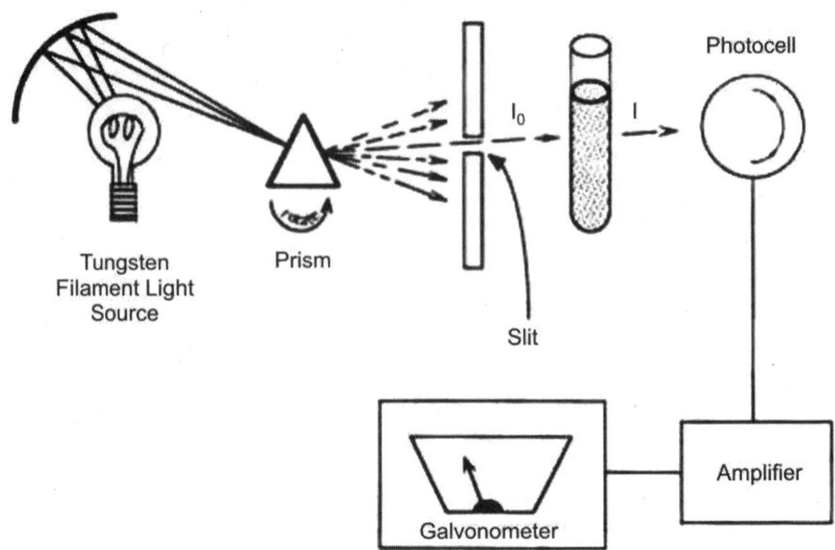

Design of a Spectrophotometer

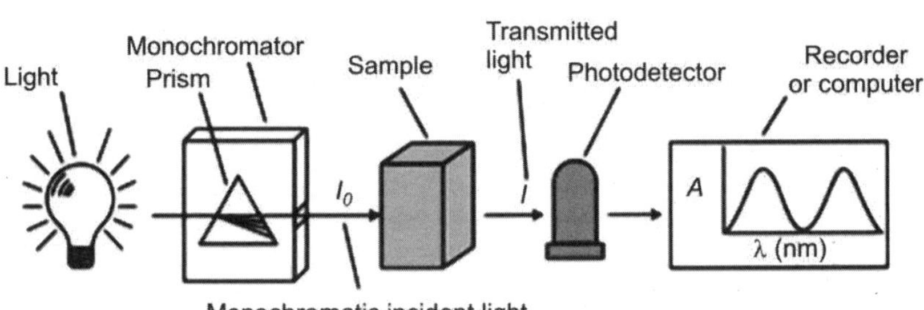

Spectrophotometer – Passage of light

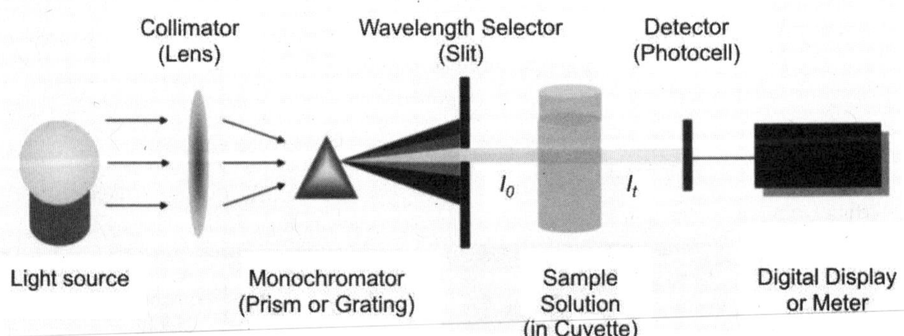

Basic structure of Spectrophotometer

Earlier spectrophotometers used a monochromator containing a diffraction grating (movable or fixed) to produce the analytical spectrum. If a photomultiplier tube or photodiode (single detector) is used, the grating can be scanned stepwise so that it can measure the light intensity at each wavelength. If charge coupled detectors (CCDs) or photodiode arrays (PDAs) can also be used, the grating is fixed and the intensity of each wavelength of light is measured by a different detector in the array of detectors. Fourier transform technique is used to acquire the spectral information in the most modern mid-infrared spectrophotometers (Fourier transform infrared spectroscopy). In a modern spectrophotometer, the sequence of events is as follows: The light source is imaged upon the sample, and a fraction of light is transmitted or reflected from the sample. The light from the sample is imaged upon the entrance slit of the monochromator, which separates the wavelength of light and focuses each of them on to the photodetector sequentially. The older spectrophotometers are to be calibrated by zeroing. The absorbency of a reference substance is set as a baseline value, and the absorbencies of other substances are recorded relative to the initial zeroed substance. The device then displays percent absorbency (the amount of light absorbed relative to the initial substance). The device quantitatively compares the fraction of light that passes through a reference solution and a test solution when making transmission experiments, and it compares quantitatively the fraction of light that reflects from the reference and test samples for reflectance measurements. The light from the source lamp is passed through a monochromator which diffracts the light into a rainbow of wavelengths and outputs the narrow bandwidths of the diffracted spectrum. Discrete frequencies are transmitted through the test sample and the photo-flux density (watts/meter square) of the transmitted or reflected light is measured with a photodiode or charge coupled device or other light sensor. The transmittance or reflectance value for each wavelength of the test sample is then compared with the transmission or reflectance values from the reference sample. The instrument comes in varied shapes and sizes, and looks like a bar code reader, like a fax machine or like a weighing scale etc. depending on its size, and can be hand-held or kept on a table just like a printer.

SPECTROPHOTOMETER – BASIC PRINCIPLE

Spectrophotometer consists of 2 instruments, a *spectrometer* for producing the light of any selected wavelength (colour), and a *photometer* for measuring the intensity of light. Both these are arranged in such a way that liquid in a cuvette is placed in between the spectrometer beam and the photometer. The amount of light passing through the tube is measured by the photometer, and this delivers a voltage signal to a display device, a galvanometer. The signal changes as the amount of light absorbed by the liquid changes. If the development of colour is linked to the concentration of a substance in solution, then that concentration can be determined by determining the extent of absorption of light at the appropriate wavelength. When a light of a specific wavelength (monochromatic light) passes through a solution there is usually a quantitative relationship (Beer`s law) between the solute concentration and the intensity of the transmitted light.

The device is generally employed to measure the amount of light that a sample absorbs, and operates by passing a beam of light through the sample and measuring the intensity of light reaching a detector. The light beam consists of a stream of photons, and when a photon encounters an analyte molecule (molecule under study) it will absorb the photon. This absorption reduces the number of photons in the light beam and thus reduces the intensity of the beam.

Two laws express the relationship between the absorption of radiant energy and the absorbing medium. They are: Bouguer`s law (also called as Lambert`s law) and Beer`s law.

a. Beer-Lambert Law

The law states that *absorbance is the negative log_{10} of the transmittance* or *the absorbance of a solution is directly proportional to the concentration of the absorbing species in the solution and the path length.* The intensity of light (**I_o** – the number of photons per second) passing through a blank (identical to the sample solution except that it does not contain the solute that absorbs light) is measured. The intensity of light (**I**) passing through the sample is measured (the instrument measures the power – energy per second, which is the product of the intensity [photons per second] and the energy per photon). The data is used to calculate the transmittance (**T**) and the absorbance (**A**) by the equations: $\mathbf{T = I / I_o}$ and $\mathbf{A = - \log_{10} T}$. The law is useful for characterizing many compounds but does not hold as a universal relationship for the concentration and absorption of all substances.

For a given sample, the absorbance depends on 6 factors, namely, the identity of the absorbing substance, its concentration, the path length, and wavelength of light, the identity of the solvent and the temperature. Not all solution species absorb light in the same way. For a given species, a more concentrated solution (with more absorbing substance present in it) results in more light being absorbed. The longer the path length, the more the absorbing substance the light will interact with, resulting in greater absorbance. The solutions with higher absorbance appear darker (more intensely coloured) than the solutions with a lower absorbance. Most of the solvents are colourless including water and do not absorb visible light to any appreciable degree. The effect of the temperature is minimized since most solutions are studied at room temperature. For solutions which are dilute, all the above factors combine together to yield a linear relationship between absorbance, concentration, and path length, known as the Beer-Lambert law equated as: $A = e.\ l.\ c$, where A is absorbance (unitless), l is the pathlength of the sample (cm), c is the concentration of the absorbing species (in molarity) and e is an experimentally determined constant known as the molar absorptivity (if the concentration is in molarity) or the extinction coefficient. The proportionality constant accounts for the diversity of absorbing strengths of different chemical species as well as variable absorbing strength for different wavelengths of light. In other words, the molar absorptivity (e) is unique for a given solution species and a particular wavelength of light.

b. Practical Limitations

The law has implicit assumptions that must be met experimentally for it to apply otherwise there is a possibility of deviations from the law to be observed. For example, the chemical makeup and the physical environment of the sample can alter its extinction coefficient. Hence, the chemical and physical conditions of a test sample must match reference measurements for conclusions to be valid. If a solution is too concentrated, the simple linear Beer-Lambert relationship no longer applies. Less concentrated solutions obey the Beer-Lambert law even more closely. The experimental parameters are to be manipulated so that a species being studied with spectrophotometry are both dilute enough (so that the Beer-Lambert law is valid) and strongly absorbing enough (with A between -0.16 and 0.92). The solutions which are cloudy or inhomogeneous will generally not produce acceptable results. Thus, the techniques of spectrophotometry are limited to typical homogeneous solution samples. The solutions that are not homogeneous can show deviation from this law because of the phenomenon of absorption flattening. The deviations will be most noticeable under conditions of low concentration and high absorbance.

c. Calibration Curve

There is a linear relationship between absorbance and concentration for dilute solutions. Normally, similar sample holders (cuvettes) are used for each measurement so that the path length is constant. So, a plot of absorbance versus concentration gives a straight line at a particular wavelength and temperature. Solutions of various known concentrations are prepared followed by the measurement of their respective ***%Ts*** which are then converted to absorbance. If Beer-Lambert`s law holds good, these experimental data points should fall along a nearly perfect straight line. By using the linear regression, the straight line which best fits these experimental points is obtained, and this line is called the calibration curve or calibration line. The graph`s origin is not included as a data point nor is it included in the regression analysis. The slope of the best fit line is a very good approximation to the quantity (***el***) which is the proportionality constant in Beer-Lambert`s law. Since the path length (***l***) is easily measured (typically 1.0 cm), the value of (***e***), the molar absorptivity, is easily determined.

d. Absorption Spectra

Most solution species which absorb light do not absorb just a single wavelength of light but instead absorb all of the wavelengths to varying degrees across a broad range of the electromagnetic spectrum. The absorption spectrum of an absorbing species is simply a plot of the absorbance versus wavelength. This spectrum is useful for choosing the wavelength at which a particular solution species is to be studied.

USING SPECTRONIC-20 SPECTROPHOTOMETER

This device is widely used in teaching laboratories. The principles remain though the specific instructions will differ with the model. Prior to its use, the instrument must have been warm for at least 15 minutes. The power switch also acts the zeroing control. The wavelength knob is used to set the desired wavelength. Special filters, light sources and/or cuvettes (for holding the sample) are required for extreme wavelengths, in the UV and IR ranges. The zero control is used with the sample cover closed to adjust the meter needle to zero on the percentage transmission scale (with no sample in the instrument, the light path is blocked and the photometer reads no light). The tube containing the reference solution is wiped clean with a tissue and placed in the sample holder. The cover is closed and light control knob is used to set the meter needle to zero on the absorbance scale. The reference tube is removed, the standard tube (first sample) is wiped off, inserted and the cover is closed. The absorbance is read and recorded. The sample tube is removed; the zero is readjusted and recalibrated before checking the next sample.

SPECTROPHOTOMETERS – DIFFERENT TYPES

1. Single Beam Spectrophotometers

In these devices, all the light passes through the sample straight to the detector. An initial reference is required to standardize the instrument before analysis can begin. To measure the intensity of the incident light, the sample must be removed so that all the light can pass thorough. These are usually very simple (less complicated) and economical to purchase (less expensive).

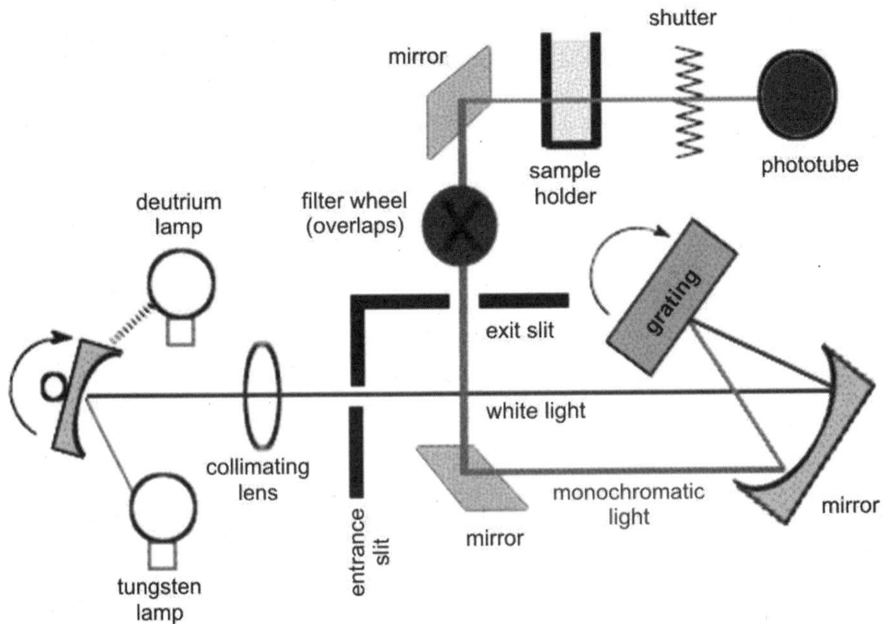

Single beam Spectrophtometer

2. Split Beam Spectrophotmeters

In these devices, the light from the source is split into two paths, with approximately 30% of the energy being diverted from the main path into a feedback detector. 70% is then passed through a monochromator, through the sample compartment to a detector. The feedback detector is used to correct variations in the energy emitted by the lamp. A reference sample is required at the start of each run to correct for cuvette and solvent absorption.

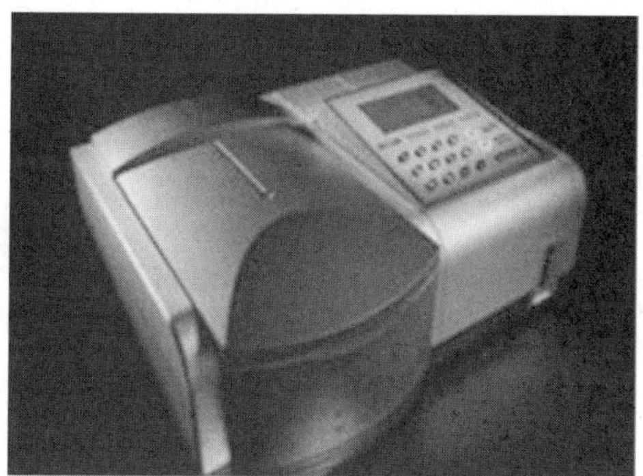

Split-beam Spectrophotometer

3. Double Beam (Dual Beam) Spectrophotometers

In these devices, the light is split into two separate paths, each of which passes through a cell holder on to its own detector. Of these, one cell holder is for the sample and the second is for a reference. The reference should contain the same cuvette type and the solvent as the sample. The absorption of the reference path is substracted from that of the sample path. Every measurement has its own reference, giving more reliable results. This is advantageous because both the reference and sample readings can be read at the same time.

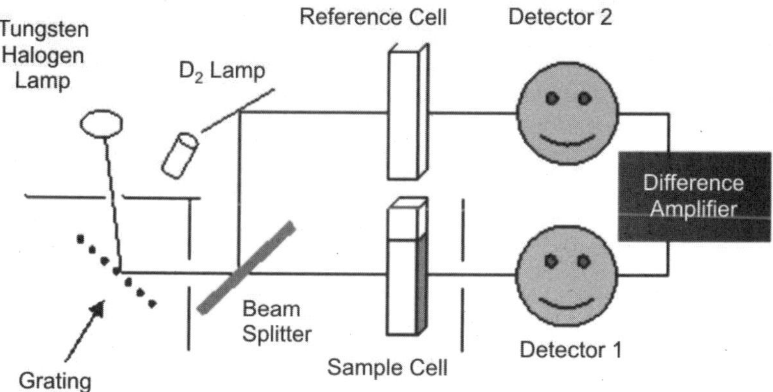

Double-beam Spectrophotometer

4. Visible Light Spectrophotometers

These devices can use incandescent light, halogen lamps, LEDs or a combination of these light sources (400-700 nm). These vary in their accuracy and plastic and glass cuvettes can be used. Spectro-radiometer devices operate like visible region spectrophotometers, and are designed to measure the optical density of illuminants. The important application is the evaluation and categorization of lighting.

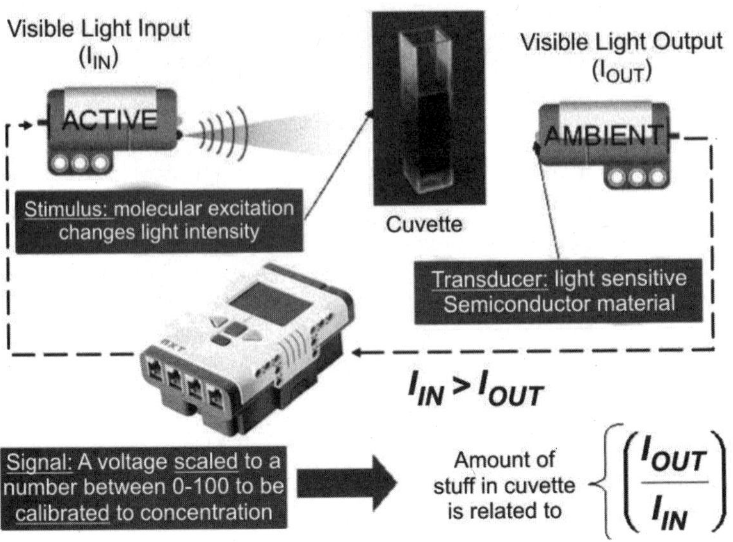

Visible light Spectrophotometer

5. Ultra-violet Light Spectrophotometers

These devices are used for fluids and even solids. Cuvettes only made of quartz are used to place the samples. Ultraviolet–visible Spectrophotometer is used extensively in colorimetry with reference to the manufacturers of ink, printing companies, textile industries etc. The readings are taken for every 5-20 nm along the visible region of the spectrum to produce a spectral reflectance curve which can be used to test a new batch of colorant to match ISO printing standards. The samples are prepared in cuvettes depending on the region and are constructed for glass, plastic (visible region) or quartz (far-UV spectrum).

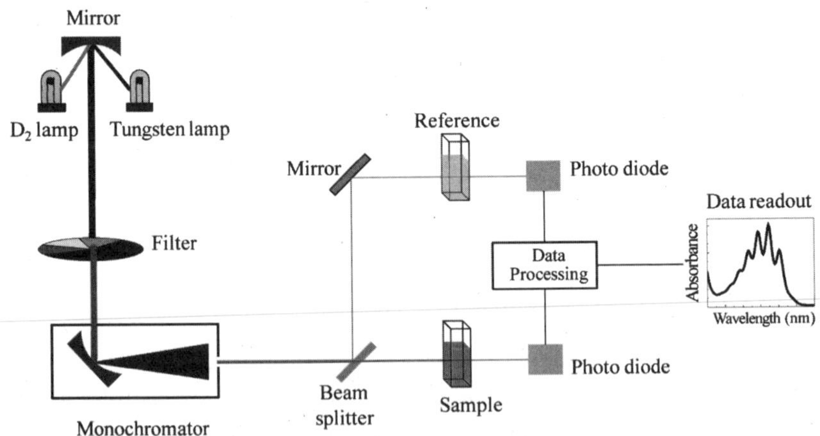

UV-visible light Spectrophotometer

6. Infra-red light Spectrophotometers

Different chemical substances vibrate differently due to the variation of energy associated with each wavelength. These devices help to study different structures of molecules and their vibrations. Mid-range and near infra-red (higher energy waves) tend to cause rotational as well as harmonic vibrations respectively. Infra-red Spectrophotometer devices designed for the infra-red region are different from the normal devices because of technical difficulty. The major factor is the type of photosensors. Another factor is that materials like glass and plastic absorb infrared light and make them incompatible as an optical medium. The samples are smeared between 2 discs of potassium bromide (sometimes ground with potassium bromide and made into a pellet) and insoluble silver chloride is used to construct the cell to measure aqueous solutions.

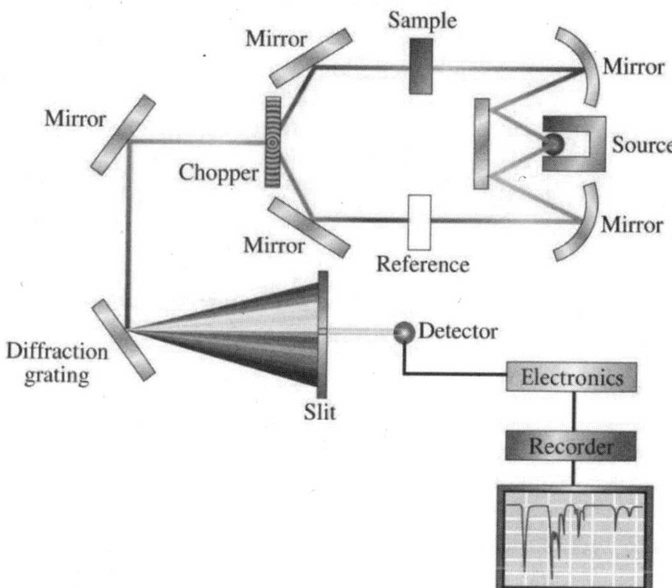

Infra-red Spectrophotometer

SPECTROPHOTOMETER - SOURCE OF LIGHT

The source of light may be:

a. Deuterium Arc Lamps

These lamps provide powerful and stable light in the UV region, 190-370 nm. In meters of higher specification, a specially designed lamp is used to maximize life without output, called `press to read' technology (pulse technology). This technology allows the lamp to go into a lower power mode, where the lamp is kept warm but is not actually on. This increases the lamp life up to three times that of a conventional deuterium lamp. Sometimes, these lamps are paired with tungsten lamps to allow the coverage of the full UV and visible spectrum.

b. Tungsten Lamps

These lamps give stable light in the visible and near infra-red regions of the spectrum, covering 320–1100 nm. These lamps also employ `press to read` technology (pulse technology) to extend the lamp life.

c. Xenon Flash Lamps

These lamps are of high energy light source emitting light across the UV, visible and near infra-red spectrum. These lamps flash up to 80 times per second. These have a very long life span but do not provide the optical stability of the deuterium/tungsten lamps set up in the systems of higher specification.

d. LED (Light Emitting Diode)

These are used for single wave length applications like cell culture measurement. LEDs are stable, of a low cost and offer a long life.

SPECTROPHOTOMETER - METHODS OF MEASUREMENT

The methods are different such as:

1. Single Wavelength

A simple method allows the measurement of transmittance or absorbance of a substance at a single fixed wavelength (260 nm for DNA).

2. Wavelength Scan

The measurements for a range of wavelengths are taken, allowing the system to plot absorbance against wavelength and thus produce a graph. This is particularly useful for analyzing an unknown sample. The wavelengths (start and finish) can be set to target the scan to a region or the available spectrum.

3. Kinetics

A single wavelength measurement is repeated at set intervals for a set number of times. The absorption is then plotted against time and this can be used to measure the reaction progress, or for enzyme activity studies when combined with a temperature-controlled cell holder.

4. Quantitative Analysis

A single wavelength measurement of a number of known standards is taken, from which a standard curve is created by the instrument, to which future measurements of unknown samples can be compared to establish the concentration.

5. Cell Density

The light scattered from cell cultures (bacteria) can be used as an indication of cell concentration, and this is plotted over time to see whether the culture is ready to be harvested or not. The measurements are generally taken at 600 nm.

SPECTROPHOTOMETER – USES

The pharmaceutical companies use bench-top spectrophotometers for the colour measurement in a variety of applications. One important use is to ensure that the colour is consistent in the dosage pills, for example, the pills need to have unique colours and shapes so that they are not mistaken for other medicines. The spectrophotometer is used in the measurement of solids, liquids, powders, pastes and creams. The CM-5 spectrophotometer is used for measuring the samples. Spectrophotometry has its applications in DNA and RNA quantitization, protein concentration determination and cell density measurement. The instrument is also used for measuring the colour consistency and accuracy in paints and coatings, for measuring the freshness and quality in food products and for determining when a step during processing is complete in chemical manufacturing. This is also used in forensic sciences, especially for determining the facts and evidences, and to determine the concentration of particular substances. In medical field, it is used to measure the minutest changes in the chemicals and enzymes. In molecular biology, it is an essential device used to measure the growth of microorganisms. It also finds it use in textile industry, printing industry and ink manufacture etc. Optical emission spectrophotometer is used to analyze many metal bases like ferrous metals of Fe, Co, Ni and Ti base and non-ferrous metals of Cu, Al, Pb, Mg, and Zn base. Infrared spectrophotometry allows scientists to perform experiments in bulk, and this is utilized in many agricultural and pharmaceutical organizations. These instruments apart from being cost effective are extremely reliable, durable and can accommodate several samples simultaneously. Infrared spectrophotometry not only covers chemical analysis, but even helps in the process of drying and coating pills; and it also facilitates the chemists to achieve chemical blend easily and produces accurate results for active pharmaceutical ingredient studies. Thus, spectrophotometer is an essential instrument in many research and development organizations as well as industrial or medical laboratories. The instrument can perform analysis, kinetics, spectrum scanning and multiple components, and protein testing proficiently.

Spectrophotometry is one of the most useful methods of quantitative analysis in various fields like chemistry, physics, biochemistry, molecular biology, material and chemical engineering, industrial applications and clinical applications. This technique allows one to identify substances without ever having to actually touch them. A substance in a sealed container can be identified as long as light is able to shine through it (the substances that are dangerous or highly toxic). This is also useful for the identification of gases, kept in sealed containers. It can be used to discern between the different pigments (chlorophyll-a and chlorophyll-b) found in plant cells. UV Spectrophotometry is useful in the detection of colourless substances in solution and measuring their concentration. IR Spectrophotometry is used in the study of molecular structures of complex organic compounds.

SPECTROPHOTOMETER - APPLICATIONS

Spectrophotometer is directly used to measure light intensity at different wavelengths. It is also used to determine the unknown concentration of a solution. These can be used to determine the equilibrium constant of a reaction involving ions. This can also be used to estimate DNA or RNA concentration and to analyze the purity of the preparation. The proteins absorb light at a specific wave length, allowing direct measurement with a spectrophotometer. Many spectrophotometers come with pre-programmed methods for the analysis of fluorescent dyes for DNA or protein quantification, and this is useful for measuring the yields and brightness of fluorescently labeled *in-situ* hybridization probes. The special cell culture spectrophotometers are used to measure cell density, which can be used to measure the growth curve of a cultured (typically bacteria) population to allow harvesting of cells at the optimum point. These are used in a wide range of industries from waste water analysis, to pharmaceutical quality control and food analysis (such as the grading of olive oil). In medicine, within the last ten years or more, spectrophotometry has been an analytical tool of great versatility and power. The role of this method in research on vitamins and hormones has been spectacular, and the application of these methods to analytical procedures promises to be equally impressive. These methods are advantageous particularly in toxicological investigations. The results are subject to qualitative interpretations having fundamental structural significance. At the same time, they are quantitative and the procedure is direct and simple, and the results are reproducible. Chemically, UV-visible spectrophotometry is useful in the identification of the concentration of the unknown substance (quantitative analysis), identification of an unknown substance (qualitative analysis), and enzyme assay; while structural applications include: control of purification, studies of cis-trans isomerism, determination of molecular weight, and turbidometry.

UVspectrophotmeter can identify the exact levels of compounds within a particular spectrum sample. Single-beam devices are less costly to purchase or operate. They are less complex, so they may introduce fewer operational inconsistencies. Single beam spectrophotometer is designed for increased stability and reliability. Compact equipment can more easily be transported for field use and on-site applications, such as monitoring CO_2 concentrations in greenhouses.

COLORIMETER AND SPECTROPHOTOMETER – DIFFERENCES

Both these are similar devices which play a critical role in the colour management for photographers and other creatives from ensuing true colour from image capture to final print. The only difference lies in the way we see colour and each device measures the colour. A colorimeter quantifies colour by measuring 3 primary colour components, red, green and blue (RGB). On the other hand, a spectrophotometer offers precise specification of any colour – spectral reflectance or transmittance of an object - across the entire spectrum of human visible light wavelengths. Spectrophtometer allows "high resolution" colour capture, ten times greater than that of a colorimeter. As Leonardo Da Vinci first noted, "*the colour of every object is affected by the colour of the light which illuminates the object*". A spectrophotometer allows the separation of the real colour of an object from the colour of the light that illuminates it, while a colorimeter cannot. The exposure to UV light can change the way a colour appears, and only a spectrophotometer is designed to accommodate these variables, while the colorimeter cannot.

■■■

CHAPTER - 11

COLORIMETRY

Introduction

The objective of colorimetry is to incorporate the properties of the human colour vision system into the measurement and numerical specification of visible light. Though the colorimetric representations do not directly predict the colour sensation, they do provide the foundation for the scientific study of colour appearance. This can be applied successfully in practical applications and the foremost among them is colour reproduction.

The colour and colour perception are limited at the first stage of vision by the spectral properties of the layer of light sensitive photoreceptors that cover the rear surface of the eye, upon which an inverted image of the world is projected by the optics of the eye. The photoreceptors transduce arriving photons to produce the patterns of electrical signals that finally lead to perception. Photopic (day time) colour vision depends mainly upon the three classes of cone photoreceptors, each with different spectral sensitivity, which are referred to as long-wavelength, middle-wavelength and short-wavelength sensitive cones (L, M and S cones), according to the part of the visible spectrum to which they are most sensitive. Scotopic (night time) vision, by contrast, depends on a single class of photoreceptor, the rod.

Colorimetry (the science of measuring colours and assigning them numerical values) is the measurement of the wavelength and the intensity of the electromagnetic radiation in the visible region of the spectrum, which is used for the identification and determination of the concentrations of substances that absorb light. This is a method of specifying the optical properties of light sources, transparent and opaque materials, in numerical terms, from the point of view of their capacity to make up our coloured environment. This method also supplies the references for the standardization of colour materials in connection with different techniques used for production and reproduction of colour such as textiles, paints, printing, photography, television, computer graphics etc, as well as colour vision research.

History and Colour/Colorimetry

The correlation between physics and physiology with reference to colour perception was recognized very early on. Isaac Newton (1704), the physicist, made a key contribution to spectrophotometry and colorimetry with his attempts to split the white light into spectral colours; and Thomas Young, the physician, actually defined the effect of colours on the human eye 100 years later. Young was the first to describe how the colour was perceived with 3 receptors in the human eye for the primary colours, red, green and blue. It is now known that every colour impression can be fully described by the 3 values.

The theoretical ground work was laid down by Bougouer (1729). Lambert law was published in 1760, and Beer`s adaptation was reported in 1852. The first colorimeter is believed to have been constructed by Muller (1853) and the double path system of Dehm became known in 1863. Dubosco`s well known colorimeter dates back to 1870.

In 1992, Prof. Arnold Wilkins, Department of Psychology, University of Essex, UK, invented the *Intuitive Colorimeter*, a simple device which was used to investigate the effects of colour on the perception of text, and used the instrument to carry out his scientific research into placebo effect. The device is used to calculate a precise and individual coloured tint to relieve perceptual distortions in visual stress. By using colorimeter, the optometrist can find the optimal precision tint for an individual, this tint can then be used in a pair of spectacles. The instrument is designed to help find the specific colour that most helps an individual have more fluent, efficient, comfortable reading. This was successfully for the first time in the world used by Barnard Levit Optometrists. The colorimeter was invented by Jan Szczepanik (1906), Poland, a colour-control tool, which applies the Beer-Lambert law when determining the concentration.

Detailed experiments carried out in the 1920s showed that the RGB (red, green and blue) primaries could indeed match all visual colors within a certain range called a gamut, but that they could not match all the spectral colors, particularly in the green range. It was found that if a certain amount of red light was added to the color being matched, then all colors could be matched. The quantitative results were expressed in terms of tristimulus values for the RGB primaries, but it was necessary to allow negative values for the red tristimulus values in order to match all colors. In 1931, the Commission International de l'Eclairage (CIE – International Commission on Illumination) defined the Standard Observer for Colorimetry and the 1931 CIE System of Colorimetry. The basics of this system are still valid and many industrial colour evaluation methods are based on this. There were a number of recommendations for the past 75 years and CIE has published the updates of its fundamental colorimetric publication, CIE Publication 15, the latest edition in 2004. Several other CIE publications and CIE standards provide the necessary recommendations to perform the laboratory and industrial colour measurements.

Colorimetry - Definition

Colorimetry is defined as `*a method of determining the concentration of a substance by measuring the relative absorption of light (usually visible) with respect to a known concentration of the substance*`. Colorimetry or Colourimetry also called as visible-spectrophotometry in physical chemistry and analytical chemistry, is a "technique used to determine the concentration of coloured compounds in solution".

It is similar to spectro-photometry but is distinguished by its interest in reducing spectra to the physical correlates of colour perception, most often the CIE colour space tristimulus values and related quantities. The International Optical Committee standardized colorimetry and adopted as the ultimate reference for the standardization of colour materials in connection with various techniques for colour production and reproduction

The CIE has provided recommendations about the precise way in which the basic principles of colour measurement should be applied and CIE 15 (2004) represents the latest edition of these recommendations and contains the information on standard illuminants, standard colorimetric observers, the reference standard for reflectance, illuminating and viewing conditions, the calculation of tristimulus values, chromaticity coordinates, colour spaces and colour differences, and various other colorimetric practices and formulae. The CIE always recommends the using measurement intervals under 5 nm, even for smooth spectra. The CRT (cathode ray tube)-display (in televisions sets) uses colour mixing by superposition of the light emitted by 3 given phosphors, R (orange-red light), G (yellow-green) and B (deep blue) which result in colours of different hues.

COLORIMETER– THE INSTRUMENT

Colorimeter - Definition

The word *colorimeter*, had its origin during 1860 – 1865, from *colour* and *meter* and its first known use was in 1872. Colorimeter is also called as *Colourimeter* or a *Tintometer.* Colorimeter is defined as *`an apparatus for determining the concentration of a solution of a coloured substance by comparing the intensity of its colour with that of a standard solution or standard colour slides`* The device measures the absorbance of particular wavelengths of light by a specific solution. This is most commonly used to determine the concentration of a known solution in a given solution by the application of the Beer-Lambert Law according to which *`the concentration of a solute is proportional to the absorbance`*.

The name *colorimeter* is given to any instrument used for the determination or specification of colours. One type is an instrument used to find the concentration of a substance in solution, based on the intensity of the colour of the solution. If the solution to be tested is colourless, a reagent is added to it that reacts with the substance in the solution producing a colour. This has a range of applications.

Colorimeter - Components

Colorimeter consists of: a light source (an ordinary low-voltage filament lamp or LEDs for red [635 nm], green [565 nm] and blue [470 nm] light), an adjustable aperture, a set of coloured filters, a cuvette (to hold the working solution – reference or control solution), a detector (a photoresistor to measure the transmitted light), and a meter (to display the output from the detector), In addition to these, there is a voltage regulator (to protect the device from the fluctuations in voltage) and a second light path, a cuvette and a detector [to enable the comparison between the working solution and a blank (pure solvent)] to improve its accuracy. Now-a-days, many commercial colorimeters and the open source versions are used in research.

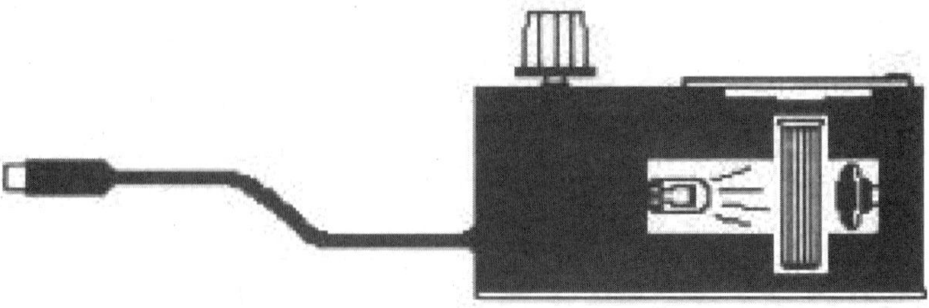

Colorimeter

To maximize the accuracy, changeable optic filters are used at present. These help to select the wavelength of light which the solute absorbs the most. The usual range is from 400 to 700 nm. To operate the device in UV range, some modifications are essential. The lamp and the filters are replaced by light-emitting diodes of different colours in modern colorimeters. The cuvettes are normally inserted and removed by hand in manual colorimeter, while an auto-analyzer (automated colorimeter) is generally fitted with a flow-cell through which the solution flows continuously. The output is displayed by a digital meter and shown as transmittance (a linear scale from 0 to 100%) or as an absorbance (logarithmic scale from zero to infinity). The useful range is from 0 to 2. It is always desirable to keep the range from 0 – 1 because above 1 the results become unreliable due to light scattering. The output can also be sent to a chart recorder, data logger or a computer.

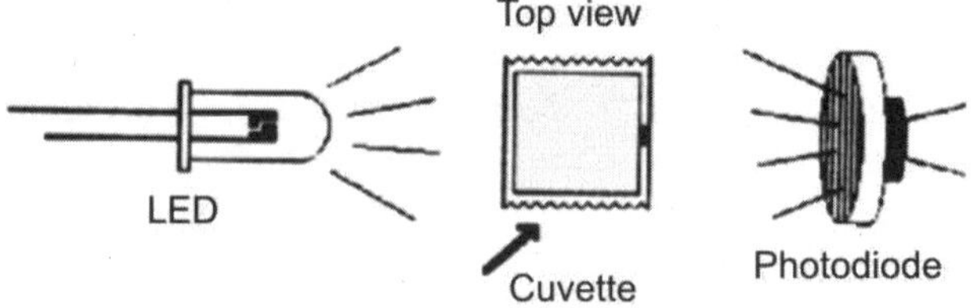

Placing a cuvette in the Colorimeter

The light from a suitable source is passed through a light filter to select the most appropriate wavelength of light, some of which is then absorbed by the solution held in a special glass cuvette (similar to a test tube). The amount of light absorbed is called, and measured as, the absorbance which is a function of the coloured solute concentration. The expensive instruments use a double beam system of two cuvettes, one is a blank of water and the other the actual coloured solution under test, two photocells and sophisticated optics of lenses and mirrors. The cheaper colorimeters allow putting in a cuvette of colourless water, set the instrument to read zero absorbance, replace with a cuvette of the coloured solution and read the absorbance. The zeroing is necessary to eliminate the error because even the apparently colourless blank of glass cuvette and water can absorb a tiny amount of light. The filter is chosen to select the band of wavelengths which are more strongly absorbed by the coloured solution. The colours of the filters (according to complementary colour relationship) to be used are: yellowish-green (if the solution is violet at 400 – 435 nm wavelength), yellow (if the solution is blue at 435 – 480 nm wavelength), orange (if the solution is greenish-blue at 480 – 490 nm wavelength), red (if the solution is bluish-green at 490 -500 nm wavelength), purple (if the solution is green at 500 – 560 nm wavelength), violet (if the solution is yellowish-green at 560 -580 nm wavelength), blue (if the solution is yellow at 580 – 595 nm wavelength), greenish-blue (if the solution is orange at 595 – 610 nm wavelength) and/or bluish-green (if the solution is red at 610 – 750 nm wavelength). But, in practice, it is always better to try several filters on a typical concentration of the solution under test to see which filter gives the highest absorption value, that is, maximum sensitivity and hence, maximum accuracy in the measurements. When the production of a more intense colour is established, a calibration curve is derived. This is done by measuring the absorbance of solutions of known concentration and plotting the calibration curve/graph. In general, at low concentrations, the calibration curve is linear (obeys Beer`s law – a solution of absorbance is directly proportional to the concentration of the solute).

Colorimeter - Types

There are different types of colorimeters such as: *color densitometer* (for measuring the density of primary colours), and *color photometer* (to measure the reflection and transmittance of colour). These may be digital, laboratory and portable types. The digital versions of colorimeters are mostly used in a laboratory for sampling or in the class room for educational purposes, while the portable [battery operated] versions can be carried anywhere regardless of the environment to test water or soil samples etc. Colorimeters in *microscope-like configurations* are used for accurate measurements of small objects such as pills and drug reagents. *Non-contact colorimeters* provide on-line colour measurements of textiles, paper and coated coils on the production lines. Small *palm-top instruments* have been designed for very specific applications like contrast measurement of baked, fried and processed food. The colorimeter generally measures only red, green and blue colours of light.

PRINCIPLE OF COLORIMETRY

Beer-Lambert Law

Beer-Lambert law also referred to as Lambert-Beer Law or Beer-Lambert-Bouguer Law is preferable for most applications, to measure the absorbance rather than transmittance since there is a linear relationship between absorbance and the concentration of the coloured substance being measured, though there are qualifications to be considered.

In colorimetry, two laws are applied: Lambert`s law (proposed by Pierre Bouguer, a French scientist) is related to *the amount of absorbed light and the distance travelled by it through an absorbing medium*, and Beer`s law is related to *the absorption of light and the concentration of the absorbing substance*. There is a logarithmic relationship between the transmission of light through a sample and the product of the absorption coefficient of the absorbing substance times the distance of the light path through the sample. This means that there is a directly proportional straight line relationship between concentration and absorbance. There are certain conditions that must be met for the law to apply. The most important is that the incident light should be as near monochromatic as possible, or atleast should have been equally absorbed throughout its range of wavelengths. Otherwise, the relationship will be non-linear. Most substances will give good straight-line responses in the range of 0 – 0.5 absorbance units.

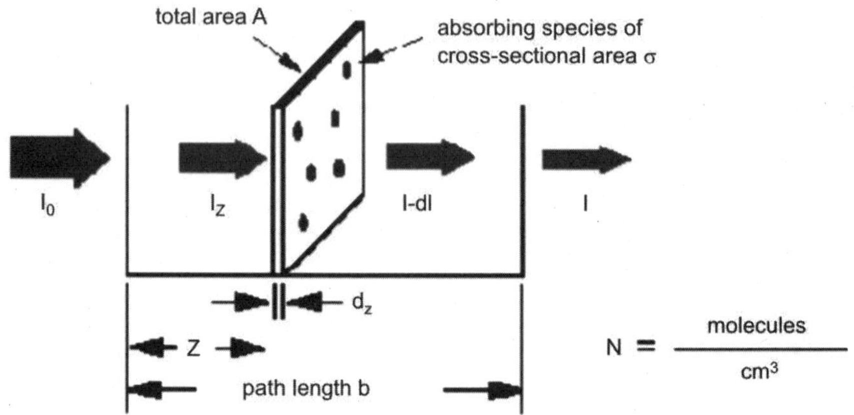

Beer - Lambert`s Law

According to Beer`s law, when the monochromatic light passes through the coloured solution, the amount of light transmitted decreases exponentially with an increase in concentration of the coloured substance. This is equated as $\mathbf{I_t = I_o\ ^{e-KC}}$.

According to Lambert`s law, the amount of light transmitted decreases exponentially with an increase in the thickness of the coloured solution. This is equated as $\mathbf{I_t = I_o\ ^{e-kt}}$.

Together, Beer-Lambert`s law is equated as $\mathbf{I_E/I_o = e^{-KCT}}$ where I_E = the intensity of emerging light, $\mathbf{I_o}$ = the intensity of incident light, **e** = base of neutral logarithm, **K** = a proportionality constant, **C** = concentration and **T** = thickness of the solution.

The ratio of the intensity of transmitted light (**I**) to the intensity of incident light ($\mathbf{I_o}$) is called transmittance (**T**). In mathematical terms, $\mathbf{T = I/I_o}$. The absorbance (**A**) of the solution (at a given wavelength) is defined as equal to the logarithm (base 10) of **1/T**. That is, **A = log (1/T)**. The absorbance of the solution is directly proportional to the concentration (**c**) of the absorbing material in solution. That is, **A = kc**, where **k** is a proportionality constant.

CHEMISTRY OF COLORIMETRY

Majority of the molecules in solution (water) do not have any colour, as these molecules do not absorb light in the visible region of the spectrum. Hence, to measure the absorbance of these colourless molecules, a reaction must be necessary which will produce a colour that can be measured. The most used approaches are: The particles are made to react with a reagent to produce a new compound which has either one or more chromophores. The particles are chelated to form a complex with a different colour than either the particles or the original chelant. A coloured compound is used which is bleached by the particles being analyzed. An intermediate compound is made to be formed which can be either oxidized or reduced afterwards to give a coloured compound. The development of a test should invariably include finding the optimum pH for the reaction and colour development, masking of the possible interferences and determining the optimum time for measurement. e. g: Orion tests.

COLORIMETRY - METHOD

In this method, the entire visible spectrum (white light) is used and the complementary colour of the one absorbed is observed as transmitted light. If monochromatic light (a narrow band of radiation) is used, the instrument is called a spectrophotometer, and this has now-a-days replaced the colorimeter. Majority of the chemical elements and a large number of compounds may be determined colorimetrically at concentrations smaller than one part of the constituent in several hundred million parts of solution.

The most important idea is that colour intensity which is proportional to the concentration which is not always true (at high concentrations and higher colour intensities). When working with colorimetry, many important factors are to be kept in mind such as: which colours are absorbed and how intensely they get absorbed (which depends on the chemical being used as well as the arrangement of the electrons and energy levels within it), the colour which we see (which is complementary to the colour absorbed), the mechanism (the light of a certain wavelength is absorbed by the coloured chemical substance), the report of the instrument (report -A and report %T [absorbance and transmittance respectively], and report A is to be used always), the assumption (the amount of light absorbed is directly proportional to the concentration of the chemical substance through which the light passes), and that the amount of light absorbed is directly proportional to the thickness (the path length) of the solution. All these factors (assumptions) can be summarized as Beer`s law which can be equated as $A = abc$, where A is the absorbance that is measured with an instrument, a is the proportionality factor (molar absorptivity – the amount of light that will be absorbed by 1 cm of a 1 M solution of the chemical whose value depends on what the chemical is and on what colour of light is being absorbed), b is the path length measured in centimeters and c is the molar concentration measured in molarity.

The device is used to test the concentration of a solution by measuring the absorbance of a specific wavelength of light is called the colorimeter. e.g., Duboscq colorimeter, Nessler tubes. The simplest and the most common colorimeters in use are Spectronic-20 and Spectronic-21. To use these, different solutions must be prepared which include a control or reference solution of known concentration. The control solution is usually a mixture of distilled water and another solution, which is first filled into a cuvette and placed inside a colorimeter to calibrate the machine. Only after calibration, the device can be used to find the densities and/or concentrations of other solutions. The filter on a colorimeter must be set to red if the liquid is blue. The size of the filter is extremely important as the wavelength of light that is transmitted has to be the same as that absorbed by the substance. The assays use reagents that undergo a measurable colour change in the presence of the analyte.

COLORIMETER – OPERATIONAL PROCEDURE (EXPERIMENTAL METHOD)

a. Calibration

Calibration of colorimeter is necessary before its use because the aqueous solution absorbs light in addition to the chemical of interest present within it. A blank solution is prepared first to overcome this, and it contains everything that the coloured solution does except the chemical whose absorbance is to be measured. The modern colorimeters have built-in calibration features. Calibration begins when the blank solution is inserted and the meter records this absorbance as zero absorbance.

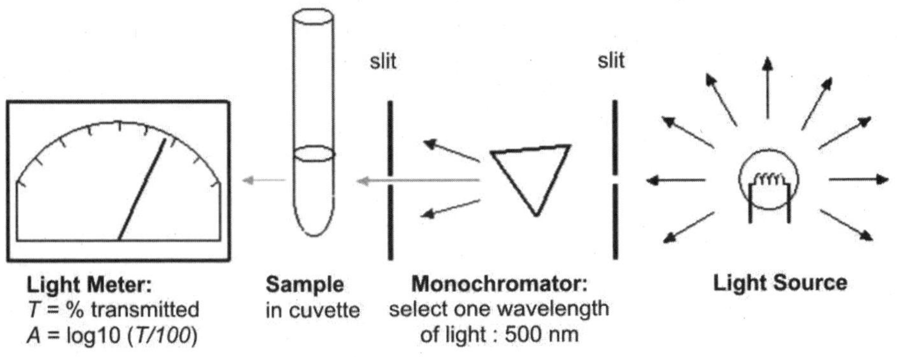

Procedure of Colorimetry

b. Determination of Maximum Absorbance Wavelength

It is always necessary to determine the maximum absorbance wavelength for the chemical in solution before the meter is used. This determination is done either manually or with the help of a graphing utility. For manual determination, a cuvette containing the coloured solution is prepared and it is inserted into the meter at minimum wavelength and the absorbance is recorded. This is repeated changing the wavelength and the wavelength that gives the highest absorbance is recorded. For graphic determination of the highest absorbance, LoggerPro software is used to initiate the calculation of an absorbance spectrum and the highest peak of the spectrum is the maximum absorbance wavelength. If these options are not feasible, a common chemical is used for which the maximum absorbance is available on line.

c. Measuring the Absorbance

Following the calibration and determination of the maximum absorbance wave length, a cuvette filled with the coloured solution is added to the colorimeter. The absorbance data obtained is used to determine an unknown concentration or molar absorptivity constant.

COLORIMETRY – CHEMICAL METHOD

This is a technique used to determine the concentration of coloured compounds in solution, in physical and analytical chemistry. The colorimeter is a device used to test the concentration of a solution by measuring its absorbance of a specific wavelength of light (and this is not to be confused with the tristimulus colorimeter which is used to measure colours in general).

Different solutions are made to use the colorimeter including a control or reference solution of known concentration. With Duboscq colorimeter (a visual colorimeter), the length of the light path through the solution can be varied while the filtered light transmitted through them is compared for a visual match. The concentration times path length is taken to be equal when the colours match, so the concentration of the unknown solution can be determined by simple proportions. Nessler tubes also work on the same principle.

Before these, electronic automated colorimeters were used wherein they had to be calibrated with a cuvette containing the control solution. The concentration of a sample was calculated from the intensity of light before and after it passes through the sample by using the Beer-Lambert`s law. In the 1960s the photo-electric analyzers came into the picture to dominate the scene.

The colour or the wavelength of the filter that has been chosen for the colorimeter is extremely important, as the wave length of light that is transmitted by the instrument has to be the same as that absorbed by the substance being measured (for example, the filter might be set to red if the liquid is blue).

ABSORPTION COLORIMETER

A colorimeter is used to test the concentration of a solution by measuring its absorbance of a specific wavelength of light. During use of this device, different solutions must be made and a control (usually a mixture of distilled water and another solution) is first filled into a cuvette and placed inside the colorimeter to calibrate the machine. Following calibration, the device is used to find out the densities and/or concentrations of the other solutions. This can be done by repeating the calibration, except with cuvettes filled with other solutions. The filter must be set to red if the liquid is blue. The size of the filter is extremely important as the wavelength of light that is transmitted by the colorimeter has to be the same as that absorbed by the substance.

PHOTO-ELECTRIC COLORIMETER - OPERATION

The different steps for operating the photo-electric colorimeter include: the recommended glass filter is chosen and inserted in the filter. The cuvette is filled with the blank solution to about ¾ and placed in the cuvette slot. The instrument is switched on and allowed to warm up for 4-5 minutes. The optical density is adjusted to zero. The test solution is taken in another cuvette and its optical density is read. The standard solution in varying concentration is taken and the optical density is read and recorded as S-1, S-2, S-3, S-4 and S-5 and so on. Turbidity is an important indicator of quality of water for the suspended solids like silt, clay, algae, microbes, organic matter and other minute particles, which can interfere with many chemical and biological tests. This is measured in units of NTU (Nephelometric Turbidity Units) or in units of FTU (Formazine Turbidity Units) depending on the calibration standard used. A graph is plotted taking the concentration of the standard solution versus the optical density. The concentration of the test solution (unknown solution) is calculated from the graph.

COLORIMETREIC ASSAYS

The assays use reagents that undergo a measurable change in their colour in the presence of the analyte. These are widely used in biochemistry to test for the presence of the enzymes, specific compounds, antibodies, hormones and many more anlaytes. For example, para-Nitro-phenylphosphate is converted into a yellow product by alkaline phosphatase. Similarly, Coomassie Blue binding to the proteins elicits a spectrum shift, allowing quantitative dosage. A similar colorimetric assay (Bicinchoninic acid assay) uses a chemical reaction to determine the protein concentration. In the same way, enzyme-linked immunoassays use enzyme-complexed-antibodies to detect the antigens. The binding of the antibody is often inferred from the colour change of the reagents like TMB.

PRECAUTIONS TO BE TAKEN IN COLORIMETRY

Useful precautions that can be taken in practice include:

1. It is to be ensured that all the samples are properly selected and are clean and uniform.
2. All the working standards are to be stored in protective containers when not in use and whenever used the working standards are to be ensured to be clean and in good condition.
3. It is also ensured that the equipment is maintained and operated according to the manufacturer`s instructions.
4. All the samples are mounted and measured at the same orientation.
5. To check precision, repeated measurements are to be made on working standards.
6. The calibrated samples are to be measured to check accuracy.

FACTORS THAT AFFECT/INFLUENCE COLORIMETRIC MEASUREMENTS

The choice of wavelength is the most important factor that affects the measurement by a colorimeter. It is also important to try to get as close as possible to an absorbance peak to get the greatest sensitivity and selectivity. The exact wavelength is not available if LEDs are used as light sources. The sample blanks and zeros (the untreated samples) are also important. It is always a good practice to run a zero to correct for stray light, sample colour or turbidity or absorption of the measured wavelength by some constituent of the sample, and to correct for internal reflection of light by the walls of the container.

ADVANTAGES AND DISADVANTAGES OF A COLORIMETER

The advantages of a colorimeter are as follows:

1. Being a simple instrument, colorimeter requires little maintenance.
2. Many devices are portable and allow on-site testing.
3. Some devices don`t even require a cuvette, but a colour strip is used.
4. The standard is sometimes a mere colour wheel or a document that displays shades of colour and intensity (the best example is the use of litmus paper).
5. The quantity of the material used is very small, and if the material is costly, the dilutions in a solvent like water can be used.
6. The hand-held devices vary in price and are cheaper.
7. The sample testing using the easy to use hand-held devices can take minutes or even seconds to run.
8. The time consumed in running the test is far less and the transport time to the lab can be avoided as well as the cost of the transport.
9. To accomplish the workload, fewer employees are required.
10. Because of their simplicity of application, even an untrained individual can catch up its operation very quickly.
11. The features like temperature setting and zeroing knob are self-explanatory.

The only disadvantage of this method is that it requires a lot of sample for analysis, and the standard solution needs to be repeatedly diluted and in some cases these have to be heated.

USES OF COLORIMETER

The small and portable devices can be used to analyze the colour contrast and the brightness on a television or computer screen, allowing the user to adjust the settings to obtain the best quality picture. In the printing industry, this device is a basic element in a colour management system, in addition to checking the electronic components and quality of pulp paper and measuring the quality of the printing ink. The colorimeter is also used by diamond merchants to measure the optical properties of precious stones. The instrument is also used in cosmetology, to measure the sun protection factor (SPF) of the products that are applied to the skin. These can analyze the skin tones and tooth colour to help diagnose the dermal and dental diseases. In real life, the hospitals use some types of these devices to test the concentration of haemoglobin in blood (is necessary to identify substandard and counterfeit drugs). When the Hb is oxygen-rich, the blood appears red. A high concentration of Hb within the blood is an indicator of good health, while the low levels suggest that the muscles, internal organs and brain are not receiving sufficient supply of oxygen. The colorimeter is also used to test for water quality (for impurities in tap water or elements present in water supply) in real life, by screening for chemicals such as chlorine, fluoride, cyanide, dissolved oxygen, iron, molybdenum, zinc and hydrazine.

The digital colorimeters can register even very small colour changes indicating the presence of these impurities. It is also used to determine the concentration of plant nutrients (like glucose, chlorophyll, phosphorus, nitrate and ammonia) in the leaves which is helpful to farmers to determine if fertilizer or other materials need to be introduced to crops. Colorimeters also help to test for nitrogen, potassium and phosphorus levels in the soil as these are necessary for plants to grow and produce higher yields. Many devices also have pH meters that determine the pH level of the soil to verify it meets the crop`s comfort zone for maximum growth. It is also used in food industry and by the manufacturers of paints and textiles.

COLORIMETRY –APPLICATIONS IN PLANT SCIENCES

In biology lab, colorimeter can be used to monitor the growth of a bacterial/yeast culture. As the culture grows, the medium becomes increasingly cloudy and absorbs more light. Colorimetry is mostly used in testing different food products and the ingredients to determine their safety. Colorimeter can be used to test the concentration of glucose, chlorophyll, ammonium and nitrates in the leaves of a plant (by crushing the leaves and adding a small amount of acetone to form a liquid which is measured) and the information collected is used to determine if any fertilizer or any other material need to be added to the crop for better growth and yield.

COLORIMETRY –APPLICATIONS IN OTHER FIELDS

Colorimetry plays an important role in many industries and is used in industries like manufacture of textiles, colour printing, manufacture of paints and food industry. It is also used in the production of aspirin. It is used in laboratory research, environmental analysis in water quality (testing for impurities in tap water and waste water is the key in testing the harmful components present in water like chlorine, fluoride, cyanide, iron, zinc and others), analysis of soil components (plant nutrients like ammonia, phosphorus and others), monitoring of haemoglobin content in the blood (in addition to the traces of drugs of inferior quality and to determine the health of an individual), and also in the analysis of chemicals used in different industrial settings (measurement of the exact consistency of the materials useful for the production of textiles and paints). It can be used in glass industry and is used in the laboratory to test for phosphorus, ammonia or chlorine.

The first industrial use of colorimetry was in the paint industry with the main aim to try and keep batches of paint as similar as possible and to minimize differences when touching up. Similarly, during the end of 1980s, this method made its way into printing industry. In printing, the absorption behavior of the paper and the ink are not only taken into account, but also other parameters like paper white, surface structure of the paper and the hue of the ink. A colorimeter can also be used to eliminate personal opinion from the assessment of colour in the plumage of birds. It is also used for the quality control of foods and beverages, alpha amylase activity and quality of milk. The method is also used in miscellaneous quality tests (medical and chemical tests) which include alpha-amylase in blood/urine, bilirubin, cholesterol and cholinesterase activity in blood, haemoglobin content in blood, iron in serum, lead in urine/ faeces, pH value of blood/urine, phenol-red (phenolsulpho-phthalein) excretion test, phenolpyruvic acid and phosphatase in blood, inorganic phosphorus in blood, proteins and salicylate in blood, sugar/glucose in blood, sulphobromophthalein retention, sulphonamides in blood/urine, trichloroacetic acid and urea in blood and uric acid in serum etc. In biochemistry, this is used to test for the presence of the enzymes, specific compounds, antibodies, hormones, and many more analytes, e.g., para-Nitrophenylphosphate (changes into a yellow product by alkaline phosphatase enzyme), Coomassie Blue (on binding to the proteins, exhibits a spectral shift), Enzyme-linked immunoassays (use enzymes – complexed antibodies to detect antigens) etc.

■■■

CHAPTER - 12

HYDROGEN-ION CONCENTRATION (pH)

Introduction

Almost all chemical/biochemical processes containing water have a need for pH measurement. Most of the living things depend on a proper pH level to sustain life. All humans and animals rely on their internal mechanisms to maintain the pH level of body fluids like blood. The blood which flows through our veins must have a pH between 7.35 and 7.45 and if it exceeds this range (as little as one-tenth) it could prove fatal. The soil must have an optimal pH to enable the plant growth. The pH of rain is typically around 5.6 and if it increases to harmful levels in its pH due to atmospheric pollution, it results in acid rain. The burning of fossil fuels (coal) releases gases into the upper atmosphere which when combined with rain water change its composition and causes the rain water to become more acidic. The proper pH levels keep the milk from turning sour, make the strawberry jelly gel, and prevent shampoo from stinging the eyes. pH levels play an important role the neutralization of effluent in steel, manufacture of pulp and paper, chemicals and pharmaceuticals, destruction of hexavalent chromium, destruction of cyanide, reverse osmosis, manufacture of petrochemicals and control of the cooling towers etc. On the whole, pH is an integral part of our daily life.

The hydrogen-ion concentration of most solutions is extremely low. In 1909, Sorenson introduced the term pH (pH = -log $[H^+]$ where $[H^+]$ is hydrogen ion concentration in mol/l) as a convenient way of expressing the hydrogen ion concentration which avoids the use of the cumbersome numbers. A change of one pH unit represents a 10-fold change in concentration of hydrogen ion. The pH of the solution equals the exponential form of the [H+] with the minus sign changed to a plus and it is much easier to write "10pH" than it is to communicate "a hydrogen ion concentration of 0.0000000001 mol/L"

EARLIEST pH METER – LITMUS

1. History and Litmus

The earliest pH meters were the strips of litmus paper (simple pH paper). This is the most widely used type of pH paper. The schools generally use it in educational science classes and for science and chemistry classes. The litmus was first used about 1300 AD by Amaldus de Villa Nova, the Spanish alchemist. From the 16th century on, the blue dye was extracted from some lichens, in Netherlands. The term "*litmus*" actually comes from Norse, meaning "*to colour* or *to dye*". This is fitting since the lichens used to make litmus have also been used to dye cloth for hundreds of years. Very little information is available about the beginnings of litmus. Some data suggests that the pH paper was first invented or developed by J. L. Gay-Lussac, a French chemist, in the early 1800s. By noting naturally occurring pH indicators, such as pigments in certain plants, an idea for the paper was formed. Litmus can be found in different species of lichens such as: *Rocella tinctoria* (S. America), *R. fuciformis* (Angola and Madagascar), *R. pygmeaea* (Algeria), *R. phycopsis*, *Lecanora tartarea* (Norway and Sweden), *Variolaria dealbata*, *Ochrolechia parella*, *Prmotrema tinctorum* and *Parmelia*. But, the main source is *Rocella montagnei* (Mozambique) and *Dendrographa leucophoea* (California). The basic materials needed for the composition of pH or litmus paper are wood cellulose, lichens and adjunct compounds. This is still used when as high degree of accuracy is not required and just manual readings are sufficient (if the pH is less than 4.0).

2. Raw Materials for Litmus

The primary raw materials used for making litmus paper are wood cellulose, lichens, and adjunct compounds. Litmus paper, as its name implies, is primarily composed of paper. The paper used to make litmus paper must be free of contaminants that could change the pH of the system it is measuring. Litmus paper is made from wood cellulose like most paper. The wood is treated with solvents prior to paper manufacturing in order to remove resinous material and lignin from the wood. One of the most common solvents in the United States is a sulfate—either sodium sulfate or magnesium sulfate.

The ability of litmus paper to change color when exposed to an acid or base is a result of litmus paper being infused with lichens. Lichens are commonly used as gauge for environmental quality because they are sensitive to various pollutants. Several varieties of lichen are used to produce litmus including *Rocella tinctoria,* native to the Mediterranean, and *Lecanora tartarea,* common lichen in the Netherlands. In fact, the Netherlands is one of the largest producers of litmus paper products.

3. Design and Manufacture of Litmus

Most litmus paper and other types of pH indicators are sold through scientific supply houses. Litmus paper is available in both red and blue varieties. The natural color for litmus paper is blue. When put in an acidic solution the blue paper turns red. Red litmus paper is first mixed with an acid when it is made. This causes the paper to appear red. When put in the presence of a base, the paper returns to its natural blue color. The production of litmus paper has many features in common with paper manufacturing. In this process, the wood pulp is converted to paper, the paper is infused with the lichen solution, and the paper is dried and packaged.

4. Future of Litmus

Litmus paper will most certainly continue to be used extensively in education due to its reasonable cost and ease of use. However, some varieties of lichens are becoming extinct. As a result, it is possible that manufacturers of litmus paper may switch to synthetic materials in the future. This is already being done by manufacturers of other types of pH papers. Additionally, because litmus cannot give quantitative results, it cannot replace other pH papers and pH meters. In fact, the trend is to make pH indicators that are even more accurate and less subjective. One such trend is to utilize fiber optic probes in pH meters in order to make them even more sensitive.

A strip of litmus paper is coated with any one of various dyes that alter the colour within a narrow range of pH values. When dipped into a solution, the change in the colour is immediate and this colour can be compared to the colour-coded chart supplied with the paper. The dyes that are coated on to the litmus are chlorine, iron, copper and peroxide. Litmus strips are inexpensive but they do have disadvantages. If the viscosity is high or if the product has a significant colour, the strip may be difficult to read. Accuracy with the litmus paper is poor or the paper strips are difficult to read. The paper strips or liquid drops used to measure the pH value of a liquid do have a shelf life, do not provide pinpoint accuracy and the colour matching technique is open to interpretation. pH affects most of the chemical reactions that take place in water and so it is important to test the pH of the source and treated water. Pure water has a hydrogen ion concentration of 10 moles per litre at 25^0C, resulting in a pH of 7.0. The main use of litmus is to test whether a solution is acidic or basic. Wet litmus can also be used to test water-soluble gases (ammonia gas which is alkaline turns red litmus blue). Other chemical reactions also cause a colour change for example; chlorine gas turns blue litmus white, due to bleaching (hypochlorite ions). This reaction is irreversible and is not acting as an indicator. Because of a wide recognition the litmus has, it has also become a cultural reference in our society and it is common to use the term litmus test when referring to a test in which a single factor determines the outcome.

5. Universal Indicators

Universal indicators are composed of a variety of materials, each changing different colours at different pH values which allows the observer to determine more precisely where the solution in question falls on the pH scale. Universal indicators can be impregnated onto paper and made into pH paper or they can be used in the liquid form. A reference colour card is provided with each universal indicator that correlates a particular colour with a pH range. Generally speaking, most universal indicators are accurate to within two values on the pH scale. For example, a green result could indicate a pH from 8-9. This means universal indicators can determine the pH of a sample quantitatively within a certain range.

HYDROGEN ION CONCENTRATION (pH)

a. Definition

We know that kilometre is a measure of distance and the hour is the measure of time and so also the pH unit measures the degree of acidity or alkalinity of a solution. pH is the abbreviation for "*potential of hydrogen*" which indicates the acidic or basic properties of a substance. pH is defined as follows: *the lower case letter "p" in pH stands for the negative common (base ten) logarithm, while the upper case letter "H" stands for the element hydrogen. Thus, pH is a logarithmic measurement of the number of moles of hydrogen ions (H+) per litre of solution.* To be exact, pH is *the measurement of the hydrogen ion concentration*, $[H^+]$. pH (hydrogen ion concentration) is strictly defined as: "*the negative logarithm of the hydrogen ion activity*" or "*the negative logarithm of the concentration of hydrogen ions (hydronium ions) in a substance, in moles per litre*". But, in practice, the hydrogen ion concentration is usually taken and this is virtually the same as the activity except in strongly acidic solutions. pH measures *the amount of acidity or alkalinity in a solution using numerical scale between 1 and 14.* A pH value of 1 is most acidic, a pH value of 7 is the centre of the measurement scale, it is neither acidic nor basic, and therefore called as neutral and the values above 7 are referred to as basic or alkaline or caustic. The results of a pH test are applied to an internationally recognized pH scale.

The pH of a solution measures the degree of acidity or alkalinity relative to the ionization of water sample. Pure water dissociates to yield 10^{-7} M of $[H^+]$ and $[OH^-]$ at 25°C, and thus the pH of water is neutral i.e., 7.0. Most of the reading range from 0 to 14. The solutions with a higher $[H^+]$ than water are acidic (pH less than 7.0) and solutions with a lower $[H^+]$ than water are basic or alkaline (pH greater than 7.0).

b. History and pH

The history of measuring the acidity of liquids electrically began with Max Cremer (1906). In his studies of liquid interfaces (interaction between liquids and solids), he discovered that the interface between the liquids could be studied by blowing a thin bubble of glass and placing one liquid inside it and another outside. He found that it created an electric potential that could be measured. This idea was taken by Fritz Haber (1906 - inventor of the synthesis of ammonia and artificial fertiliser) and Zygmunt Klemsiewicz (1906) who discovered that the glass bulb (he named it as glass electrode) could be used to measure hydrogen ion activity and that it followed a logarithmic function.

pH is an abbreviation for "*power of hydrogen*" where `**p**` is short for the German word for power or exponent of, in this 10, *potenz* and **H** is the element symbol of hydrogen. Other symbols such as p_H have been used in the past. The H is capitalized because it is standard to capitalize element symbols. The abbreviation also works in French, with *pouvoir hydrogene* translating as "the power of hydrogen".

Dr. Soren Peter (Peder) Lauritz (Pauritz) Sorenson (1868-1939), the Danish Biochemist and the Director of Carlsberg Laboratory`s Chemical Department in Copenhagen, Denmark, which was supported by the beer company of the same name, invented and developed (1909) the pH scale (the standard of measurement of acidity – until 1924, the scale was known as the Sorensen scale and was later revised and renamed as pH scale) during his pioneering research `*on the effect of ion concentration in the analysis of proteins, amino acids and enzymes*` and subsequently became a leader in the application of thermodynamics to protein chemistry, assisted by his wife, Margrethe Hoyrup Sorensen. Basically meaning '*the power of hydrogen*", the scale provides a simple and universal measurement of the amount of hydrogen ions in a solution, which affects its acidity and how it reacts chemically. The concentration of hydrogen ions is determined by measuring the current generated in an electrochemical cell when the ions migrate to oppositely charged electrodes. Sørensen used a negative logarithm of the hydrogen concentration to create a scale from 0-14, where a pH of less than 7 is an acid, 7 is neutral and higher than 7 is an alkali. The applications of the pH scale have been countless, ranging from foodstuffs and cosmetics to chemicals and pharmaceuticals. Just about every liquid has had its pH measured at sometime to determine how it will react and interact with living organisms. Until Sorensen developed the pH scale, there was no widely accepted way of expressing hydrogen ion concentrations.

In a publication of 1909, Sorenson discussed the inadequacy of measuring acidity by the total amount (normality) of acid additions to a particular solution. The added amount of acid would not necessarily be a true measure of its dissociation, depending on chemical interactions with other chemical species. He proposed that the actual degree of acidity should be rationally measured by hydrogen ion concentration and proposed the pH scale for expressing the hydrogen ion concentration and quoted: " *I will explain here that I use the name 'hydrogen ion concentration" and the designation pH for the numeric value of the exponents of this power*". He was involved in work testing the acidity of beer. The pH variable is obviously a very important variable in aqueous corrosion studies and often monitored as a process variable, together with corrosion signals.

The voltage of the glass electrode could not be measured accurately until the invention of the electron tubes. The measurement was made possible with the invention of FETs (field-effect transistors) and ICs (integrated circuits) with temperature compensation. The voltage produced by one pH unit (from a pH of 7.0 – 8.0) is about 60 mV (milli-volt). At present, the pH meters have microprocessors that can make necessary corrections for temperature and calibration. Even then, the modern pH meters suffer from drift (slow changes) as such they have to be calibrated frequently. The reference electrode which used silver chloride traditionally has been superseded by kalomel (mercurous chloride) electrode which uses mercuric chloride in potassium chloride solution as a gel (like gelatine). But these have no eternal life and need to be replaced when they drift unacceptably or take unusually longer time to settle.

Danish Radiometer in Denmark and Arnold Orville Beckman, Assistant Professor of Chemistry, California Institute of Technology, USA (1936) built the *first commercial pH meters*. Beckman was asked by the California Fruit Growers Exchange (Sunkist) to devise a quick and accurate method for measuring the acidity of lemon juice, and the result is the pH meter. This invention helped him to launch Beckman Instruments Company (Beckman Coulter at present). Beckman pH meter was designated as an ACS (American Chemical Society) *National Historic Chemical Landmark* (2004) in recognition of its significance as the *1st commercially successful electronic pH meter*. The *first portable digital pH meter* was designed and manufactured by Jenco Electronics of Taiwan (1970s). This was sold under Cole-Parmer`s label. Although 46 years passed, almost all the pH meters follow the same general idea – external pH and reference electrodes (often in the same housing), high gain amplifier and an amperometer, all in one portable box.

MEASUREMENT OF pH

The measurement of pH involves comparing the potential of solutions with unknown [H+] to a known reference potential. pH meters convert the voltage ratio between a reference half-cell and a sensing half-cell to pH values. The voltage on the outer membrane surface changes proportionately to the changes in [H+] in acidic or alkaline solutions. The pH meter detects the changes in the potential and determines [H+] of the unknown by the Nernst equation (Hermann Walther Nernst, a German Chemist [1864-1941]):

$$\mathbf{E = E^{o} + (2.3\ RT)\ /n\ F\ \log a_i\ \{unknown\ [H^+]/internal\ [H^+]\}}$$

where: E = the total potential difference (measured in mV) between the two electrodes, E^0 = the reference (standard) potential of the ion, R = the universal gas constant (in Joules/mol-Kelvin), T = the absolute temperature (in Kelvin), n = the number of electrons or the charge of the ion, F = Faraday`s constant (in Coulombs/mol) and [H+] or a_i = Activity of the ion (hydrogen ion concentration). The term, "2.3RT/nF" is called Nernst factor or the slope factor and this provides the amount of change in total potential for every 10-fold change in ion concentration. For hydrogen ion activity where n=1, the Nernst factor is 59.16 mV for every 10-fold change in activity at 25°C, which means that for every pH unit change, the total potential will change 59.16 mV. As T is not constant, the Nernst factor will change when temperature changes. The milli-volt output of the glass electrode will change with temperature in accordance with the Nernst equation. As the temperature increases, so does the milli-volt output and it is the slope that changes. The change in the electrode output versus temperature is linear which can be compensated in the pH meter. The linear function for temperature versus pH change can be expressed as 0.003 pH error/pH unit/°C.

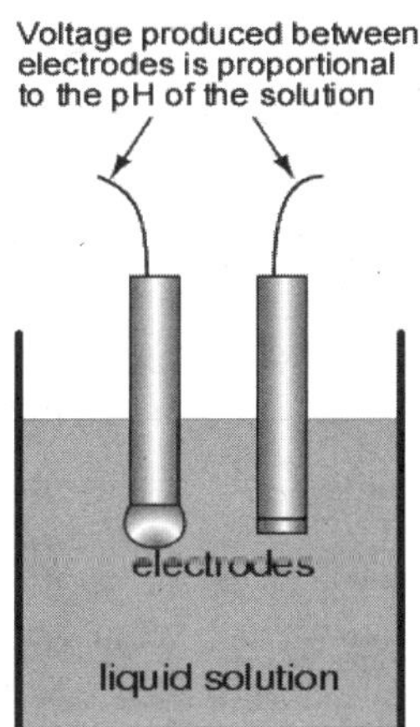

1. pH - Temperature Compensation

The pH of any solution is a function of its temperature. The voltage output (from the electrode) changes linearly in relation to changes in the pH and the temperature of the solution determines the slope of the graph. One unit of pH corresponds to 59.16 mV at 25°C, the standard voltage and temperature to which all calibrations are referenced. The electrode voltage decreases to 54.20 mV/pH units at 0.0°C and increases to 74.04 mV/pH unit at 100.0°C. Since the pH values are temperature dependent, pH applications require some form of temperature compensation to ensure standardized pH values. The meters with automatic temperature compensation (ATC) automatically correct the pH value (as they receive a continuous signal from a temperature element) based on the temperature of the solution. Manual compensation requires the user to enter the temperature of the solution in order to correct pH readings for temperature. ATC is considered to be more practical for most of the pH applications.

2. pH - System

pH reading is dependent upon all its operational components, namely, electrode, pH meter and buffer and any problem with any of these three components will give poor readings. The pH electrode consists of 2 half-cells, an indicating electrode and a reference electrode, both in one body and the measurement problems (over 90%) are related to their improper use, storage or selection of the electrodes. The pH meter is a sophisticated voltmeter which can read small milli-volt changes from the system and the meter itself is the source of problems in the measurement. At present, the pH meter has either manual or automatic temperature compensation to correct the changes in temperature. The micro-processor technology has created convenient features for measurement of pH, auto-buffer recognition, calculated slope and percentage efficiency, log tables for the concentration of ions etc. The buffer solutions of the known pH value allow the user to adjust the system and thus enable accurate measurements of pH. The fresh buffer solutions should be standardized each time. The buffer should frame the range of pH for the samples to be tested. The buffers should be at the same temperature as that of the samples (for example, if the samples are at 50°C, the buffers are to be warmed to 50°C in a water bath). The buffer values are dependent upon temperature.

pH METER

The pH meter is *an electronic device that is used for measuring the acidity or alkalinity* (pH) *of a liquid* (sometimes special probes are used to measure the pH of semi-solid substances). Typically, pH meter consists of a special measuring probe (a glass electrode) connected to an electronic meter that measures and displays the pH reading. This is *a device that is generally used to analyse and control pH values, which identify the solution as acidic, basic or neutral.* The pH of a substance varies with temperature, so for total accuracy, pH is recorded along with temperature. This always includes an internal temperature monitoring system. The meter`s *probe* (essential part of the pH meter which is rod-like made of glass) is generally placed into the fluid sample and at the bottom of the probe is a thin glass *bulb* (the sensitive part of a probe and contains the *sensor*) that contains a *reference electrode* or *standard electrode*, while the glass bulb itself acts as the pH *sensitive glass electrode* or *measuring electrode* or *membrane electrode* or *active electrode.* The bulb should not be touched by hand and to be always cleaned softly with the help of a tissue paper. The probe is fitted with an arm, the probe arm. When placed in the sample, both the electrodes are immersed and an electrical circuit is completed, extending a milli-volt potential across the glass sensor. The voltage is a function of the hydrogen activity in the sample or its pH. The voltmeter in the probe registers the voltage and converts it to a pH value. The meter must be calibrated before a measurement. So, the probe is removed and washed with de-ionized water before it is placed into a buffer of known pH. The meter is calibrated to read the pH of the buffer. The electrode is removed from the buffer, washed and placed back into the process. pH can change due to temperature differences and carbon dioxide absorption. So, it is recommended to use other buffers and also to be cautious when buffers pH 10.0 or greater are used. A wide range of pH meters are available in their size and shape. They vary from the smallest to biggest, and they may be hand-held, pen, pocket, and bench top. There is also stick and water-proof pH meters, Pen, pocket and hand-held meters are meant for field study.

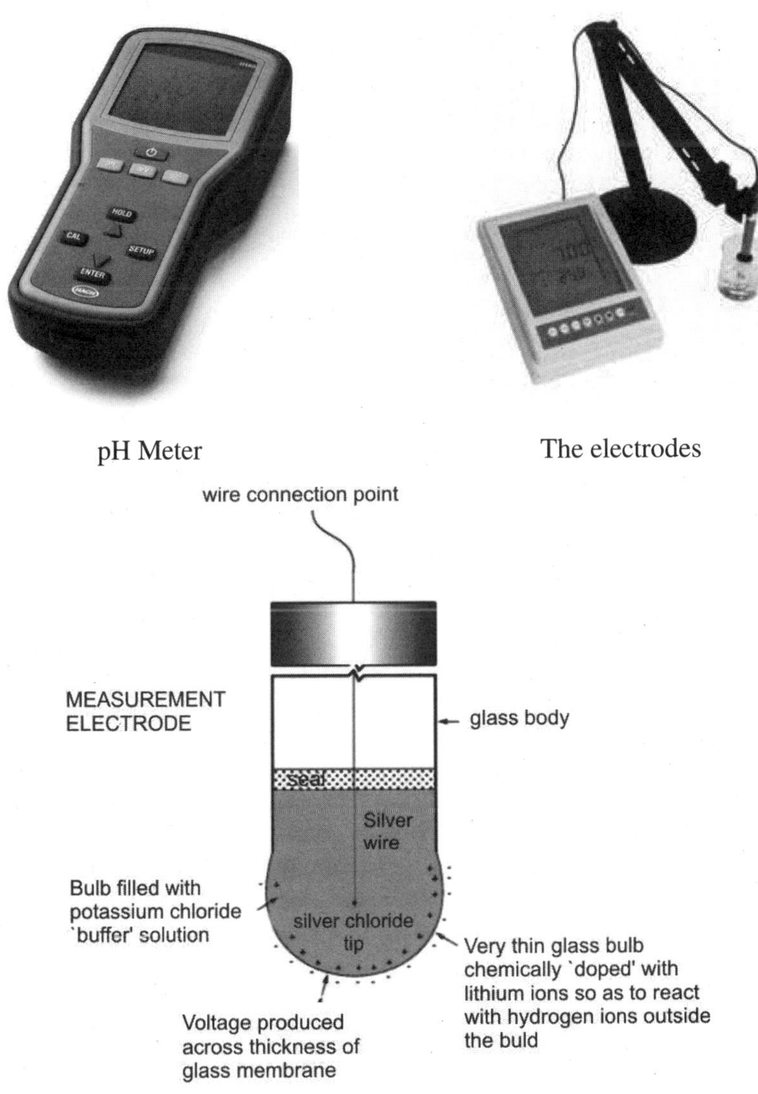

pH Meter

The electrodes

The measurement electrode

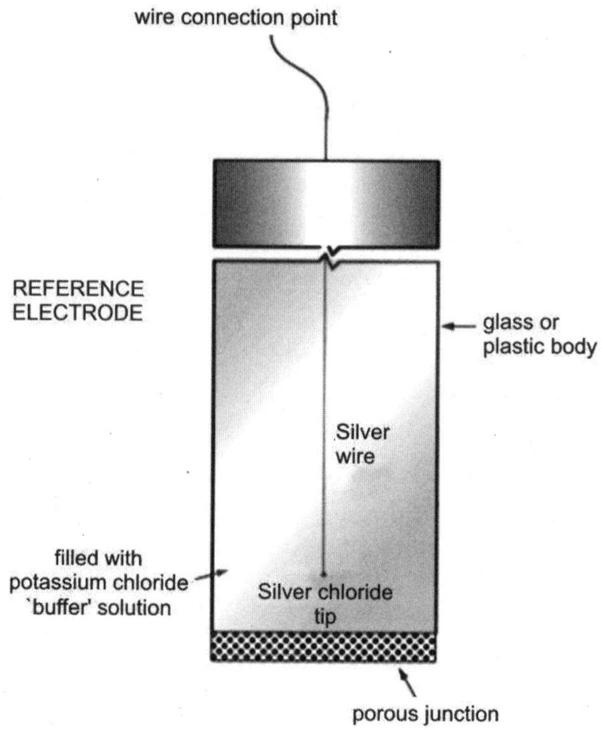

The reference electrode

a. Calibration and Use – Step-by-step

Normally, the pH meter should be calibrated (checked against a known standard) at the beginning of each day because the glass electrode does not give a reproducible e.m.f (electro-motive force) over long periods of time. This should be performed with at least 2 standard buffer solutions that span the range of pH values (at pH 4.0 and pH 10.0) to be measured. The meter has one *calibrate* (first control) to set the reading equal to the value of the first buffer and a *slope* (second control) used to adjust the reading to the value of the second buffer. There is a third control which allows the temperature to be set. Some pH meters allow a 3-point calibration (zero point [7.0], 4.0 or 10.0) for most accurate work. The meters of higher quality will have a provision for temperature coefficient correction while still more advanced meters are provided with built-in temperature probes. After each measurement, the probe is rinsed with distilled water or deionized water to remove the traces of solution if any. It is blotted with a scientific wipe to remove water completely and is quickly immersed in another solution.

The pH meter can be calibrated and used in an easy step by step method as follows:

- Turn on pH meter.
- Wait for 30 minutes for the electrodes to warm up.
- Take the electrode out of the storage solution.

- Rinse the electrode with distilled water or de-ionized water. Do not wipe the electrode membrane but instead dap the electrode with a quality tissue paper.
- Immerse the electrode into the pH 7.0 buffer. The buffers and the samples should be read at room temperature. The readings are noted.
- Press the calibrate button.
- Wait until the pH icon stops flashing and press the calibrate button again.
- Rinse the electrode with distilled water and wipe with a quality tissue paper.
- Immerse the electrode into the pH 4.0 buffer. Do not use pH 10.0 buffer due to its poor stability as a result of buffer reaction with atmospheric CO_2.
- Wait until the pH icon stops flashing and press the measure button.
- Rinse the electrode with distilled water and dab dry with a quality tissue paper.
- Immerse the electrode into the sample. The samples and the buffers should be read at room temperature. The readings are recorded.
- Press the measure button.
- Wait until the pH icon stops flashing and record the pH of the sample.

b. pH Meter - Warnings

Follow the safety regulations when working with hazardous samples and wear hand gloves when testing pH.

c. Storage of Glass Probes

The tip of the glass probe must be kept always wet when not in use, to avoid the dehydration of pH sensing membrane and subsequent dysfunction of the electrode. The glass electrode (without reference electrode) is stored immersed in an acidic solution (pH 3.0). In case of emergency, acidified tap water can be used for storage. The distilled water or deionized water should not be used for long term storage of the probe as it gets degraded. Combined electrodes (both glass electrode and reference electrode) can be stored immersed in the bridge electrolyte (KCl – 3M).

d. Cleaning of the Glass Probes

The probe may be cleaned (once a month) with pH electrode cleaning solution (0.1 M solution of HCl is used with a pH of 1.0). A dilute solution of ammonium fluoride (NH_4F) can be used alternatively. In case of doubt, it is always better to refer to the recommendations of the manufacturer or a standard Text Book of Analytical Chemistry.

pH METERS - TYPES

pH meters range from simple and inexpensive pen-like devices to complex and expensive instruments with a computer interface. These possess inputs for indicator and temperature measurement. The meters and the probes are also available for use in special applications and harsh environments.

In order to properly control pH, one must correctly measure it. There are various ways that pH can be measured including pH paper, digital readout pocket testers, portable and bench top pH meters.

A fast and affordable way to measure pH is by using pH paper. Strips of paper are soaked with a chemical indicator. This indicator, when placed in a sample liquid, turns the paper a certain color. The liquid's pH is found by comparing the color of the paper to a standard chart that is provided. Although these indicators are offered for every pH level, they have limits to their accuracy. It is very difficult to decipher the change of color when working with colored or cloudy samples.

Another way to determine the pH level of a solution is with the pocket tester. They are extremely handy and easy to carry. pH tests can be made very fast and there is no need to take a sample back to the lab. Pocket pH testers are accurate enough for common applications and are reasonably priced.

If your measures need to be more accurate, portable pH meters are a better choice over the pocket tester. The majority of portable meters are bigger than pocket pH testers. They are long-lasting and ideal for accurate pH readings in the field. They use an array of replaceable pH electrodes for individual applications and conditions.

Bench top pH meters are mainly used in testing or research labs. In general, they are more expensive but provide precise accuracy and more testing options. Also, bench top meters use a wide array of replaceable pH electrodes.

ACCURATE MEASUREMENT OF pH

pH is typically measured using a pH meter. This is one of the simplest tests performed in water in the lab. Basically, this consists of calibrating the instrument, placing the electrodes in a well-mixed sample and reading the pH directly from the meter. Measuring the pH of a solution does not tell us which acids and bases are present in the sample, nor does it tell us their concentration. pH is merely an indicator of the overall balance of hydrogen and hydroxide ions in the solution. The precision and accuracy of pH measurement depends on the type of and condition of the instrument and on the care used in its calibration and operation. Under normal conditions, $\pm$ 0.1 pH unit represents the limit of accuracy of the pH meter. The most convenient and reliable method for measuring pH is by the use of the pH meter, which measures the electromotive force (*emf*) of a concentration cell formed from a reference electrode, the test solution, and a glass electrode sensitive to hydrogen ions. Some pH meters have the glass and reference electrodes combined in one unit and not as separate entities.

pH INDICATORS

The indicators are materials that are specifically designed to change the colour when exposed to different pH values. The colour of a wetted sample paper is matched to a colour on a colour chart to infer the pH value. The pH paper is typically used for preliminary and small volume measuring. It cannot be used for continuous monitoring of a process. The different types of indicators are as follows:

1. Liquid pH indicators

pH indicators are substances, usually a weak acid or base, that changes colour depending on the pH of the solution it is mixed with. Indicators are often made from plant extracts. Indicators don't change colour sharply at one particular pH level. They change over a narrow range of pH. And this colour change interval is different for each indicator. For optimal accuracy, the narrower the pH range of the colour change, the better.

2. Common Indicator of pH - Litmus

A commonly used acid-base indicator is litmus. Litmus is a natural indicator obtained from certain lichens. It is red when acidic and blue when basic and the approximate pH range is 5.0 to 8.0. Generally one can see a mixture of red and blue. Even the litmus colour change happens over an unusually wide range, it is useful for detecting acids and bases because it changes colour around pH 7. Phenolphthalein is colourless when acidic and pinkish violet when basic and the approximate pH range is between 8.3 and 10.0. This is generally used to test the pH of highly alkaline solutions, more than 10.0. Methyl orange is red when acidic and yellow when basic and the approximate pH range is 3.1 to 4.4. The colour which one sees is a mixture of two. This is used generally to test the pH of highly acidic solutions, less than 3.0. A universal indicator is actually a mixture of several indicators. The advantage of the universal indicator is that it displays a variety of colors over the whole pH scale. This makes it possible to determine an approximate pH of a solution over a range from 1-14.

3. pH Indicators - Advantages and Disadvantages

These are mainly used for titrations and easy to use. These are cheaper than a pH meter and an electrode. These are well suited for pH measurements in drinking water, waste water, surface water, aquarium water and swimming pool water etc. The solution to be measured should be clear and colourless. The accuracy of the measurement is lower when compared to that with an electrode. The process is manual.

4. Common Organic Indicators

An approximate idea of the pH of a solution can be obtained by using the indicators. The indicators are the organic compounds of natural or synthetic origin whose colour is dependent upon the pH of the solution. The indicators are usually weak acids which dissociate in solution. For example, Bromocresol purple is the best indicator in use. A number of natural or synthetic organic compounds undergo definite colour changes in well-defined pH ranges. These indicators assume different hues within the specified pH range and these are used as either liquid solutions or as pH papers. Some of the indicators with their colour and pH values are as follows:

Indicator	*Acid colour*	*Alkali colour*	*pH range*
Methyl orange	Red	Yellowish-orange	3.1 – 4.6
Methyl red	Red	Yellow	4.4 – 6.2
Litmus	Red	Blue	4.5 – 8.3
Thymol blue	Yellow	Blue	8.0 – 9.6
Phenolphthalein	Colourless	Pink	8.2 – 9.8
Alizarin yellow	Yellow	Lilac	10.1 – 11.1

5. Disadvantages of Common Indicators

As the colour change occurs over a wide pH range, the indicators will give only an approximate indication of pH. The indicators are affected by oxidizing agents, reducing agents, salt concentration and protein. A final precaution is that when using the indicators, only a small quantity is to be added, to the solution under examination. Otherwise, the acid-base equilibrium of the test solution may be displaced and the pH gets changed.

The indicators are probably of more value in determining the end point of a titration and some of those commonly employed are: Thymol blue, Bromophenol blue, Methyl red, Bromocresol purple, Phenol red and Phenolphthalein.

BUFFERS AND BUFFER SOLUTIONS

Buffers

A buffer is defined as *an acid-base conjugate consisting of a proton donor and the corresponding proton acceptor*. They help in maintaining a constant pH in the separating medium. The common acid-base conjugates are represented by: acetic acid (H^+ and CH_3COO^-), phosphoric acid (H^+ and HPO_4^{2-}), ammonium (H^+ and NH_2) and water (H^+ and OH^-). Water is a simple buffer which resists changes in the pH. Buffers attract/release protons according to the environment and maintain a constant pH. Each buffer has a characteristic affinity for a proton relative to the proton affinity of OH^- and this can be best revealed by titration curves. The shape of the titration curves is similar for many acids, but they are displaced vertically along the pH scale. The pH at the midpoint of the curve indicates the equal concentration of proton donor and proton acceptor. Each curve possesses a flat zone extending 1.0 pH unit on either side of the midpoint and in this zone, pH of the system changes very little. This represents the zone where the conjugate acid-base pair acts as a buffer, and outside this zone, the buffer has a less capacity to resist the changes in the pH. The buffering capacity of a buffer is *the maximum of the pH of the exact midpoint of the titration curve at which the concentration of proton acceptor equals that of the proton donor* and the pH = pK.

Buffer Solutions

A buffer solution is the solution that resists the pH change on the addition of an acid or an alkali. Such solutions are used in many biochemical experiments where the pH needs to be accurately controlled. The term, buffer is defined as "*a substance that absorbs an impact, protects against a shock, resists change, or maintains a relative acid/base concentration within a specific range*". With regard to pH, buffers are solutions in which the pH remains unchanged when small amounts of an acid or a base are added. A buffer is *a solution of either a weak acid or a base and one of its salts which must resist a change in its pH value.* A buffered solution contains chemicals that will not readily allow a change in the pH value when being neutralized or changed by diffusion. Tri-sodium phosphate is a poor buffer solution and an example of good buffer (more stable) would be a solution of sodium sesqui-carbonate.

Buffer solutions vary in the extent to which they resist pH changes, and in order to compare different buffer solutions, Van Slyke introduced the term – *Buffer value.* Some of the more common buffer solutions used in biology are: Phosphoric acid, Citric acid, Carbonic acid, Glycyl glycine, Acetic acid, Barbituric acid, TRIS and HEPES.

The actual buffer chosen for a particular experiment needs to be selected with care as the experimental results may sometimes be due to specific ion effects. Examples of the buffers which may exhibit undesirable reactions include: Borate, Citrate, Phosphate and TRIS.

Buffer Solutions of known pH

The buffer solutions of known pH value can be prepared fresh in the laboratory by dissolving their salt (as powder) in 1000 ml of freshly boiled and cooled distilled water.

1. Buffer solution - 0.05 M Potassium hydrogen phthalate

   pH at 25°C – 4.0

   Amount of salt to be dissolved in 1 litre of water – 10.12 g $KHC_8H_4O_4$

2. Buffer solution – 0.025 M Potassium di-hydrogen phosphate + 0.025 M Disodium hydrogen phosphate

   pH at 25°C – 6.86

   Amount of salt to be dissolved in 1 litre of water – 3.387 g KH_2PO_4 and 3.533 g Na_2HPO_4

3. Buffer solution – 0.01 M Sodium borate decahydrate

   pH at 25°C – 9.18

   Amount of salt to be dissolved in 1 litre of water – 3.80 g $Na_2B_4O_7.10H_2O$

DETERMINATION OF pH

The Electrometric method is the most accurate method for the determination of pH of a solution. This method measures electromotive force (*emf*) set up due to the existence of H+ and OH^- in the solution. The instrument which measures pH in this manner is known as the pH Meter.

The other method is the Colorimetric method which is also fairly accurate and can be used in the absence of a pH meter. The basis is that, certain organic dyes (indicators) change colour with the change in pH. Each indicator in its ionised state gives one colour and in an unionised state gives another colour. Each has its useful range of pH with in which it can indicate change in pH by change of colour and beyond which there is no further colour change.

It is possible to prepare an indicator whose pH range covers almost the entire pH scale. Such an indicator is known as the `universal indicator`, which enables us to determine the approximate pH of a solution, and thus helps us to choose an appropriate indicator and buffers for subsequent and more exact determination of pH of the solution.

Permanent standards made from glass discs of appropriate colour and shade for each indicator may also be used in place of a standard buffer and indicator. Hellige`s Comparator with its permanent standard discs made of coloured glass is a very convenient instrument for this purpose.

1. Determination of pH of a Colourless Solution

The pH of a colourless solution can be determined as follows:

1. First determine the approximate pH of the solution with the help of a universal indicator and colour chart.
2. Take a small piece of universal indicator paper and just dip its one end in the given solution. After about 30 seconds, compare the colour of the paper with the standard colour chart provided and know the approximate pH.
3. If universal indicator solution is to be used in place pH paper, put a drop of it in about 2.0 ml of the solution and compare the colour of the solution with that of the colour chart.
4. After having known the approximate pH of the solution, choose from the set of indicators, an indicator which more closely covers the pH range of the solution under examination.
5. Add a drop of this indicator to approximately 2.0 ml of the solution and match the shade of the colour produced against the series of coloured standard buffers.
6. The standard colour solutions are prepared by adding one drop of the same indicator to 2.0 ml of each of the buffer solutions (chosen in the pH range of the unknown).
7. The pH of the unknown solution is that pH of the buffer with which the colour matches.

8. In case, the colour of the unknown is intermediate between the two colours produced by two consecutive buffers, the pH of the unknown lies between two known standards.
9. The accuracy of this method can be increased by employing a set of buffer solutions with closely differing pH values.

2. Determination of pH of a Coloured Solution (Urine)

Place a test tube containing Plain Urine or coloured solution in front of the coloured buffer standards. View the 2 test tubes together and then match this against the urine or coloured solution containing the same indicator in front of which a test tube containing water is to be placed. The use of comparator makes the matching easier. Otherwise, a double rowed test tube rack is also convenient. The reagents required are a set of indicators and standard buffer solutions.

3. Electrometric Method of pH determination

The pH of a solution is determined electrometrically using a pH meter as follows:

1. The switch of the pH meter is put on.
2. After a few minutes when the instrument is warm enough, the pointer is adjusted to 0 mV or pH 7.0 by Zero Control of the set.
3. A standard buffer solution is taken in a clean glass or polythene beaker.
4. The electrodes are lowered so that they are immersed in the solution to a depth of about one inch.
5. The temperature of the buffer solution is measured by a thermometer and the instrument is again adjusted to this value by Temperature Compensate Control.
6. The pointer is again set at pH 7.0 by means of Set Control.
7. The Selector Switch is now turned to proper range i. e., 0.0 to 7.0 or 7.0 to 14.0.
8. The pointer will now move to show the pH of the buffer and by the help of the Set Buffer Control, the pointer is set to the exact pH value of the buffer.
9. Now, the pH meter is standardized and is ready for the determination of pH of any given solution.
10. To measure the pH of a given solution, the Selector is brought back to the zero position and the electrodes are cleaned in distilled water.
11. The given solution is taken in a clean beaker and the electrodes arc immersed in it.
12. The Selector Switch is put on expected pH range and the reading of the pointer on the pH scale is noted. This value is the pH of the given solution.

CARE AND USE OF pH METER

The following points are to be noted each time the pH meter is used:

1. New glass electrodes must be soaked in 0.1 mol./1HCl or distilled water for several hours before use.
2. The solution must be thoroughly stirred before measuring the pH. A Magnetic Stirrer is the most convenient for class work. A stirring rod must not be used since it breaks the fragile electrodes.
3. The beaker should be thermostated.
4. The electrodes should be thoroughly washed in distilled water before and after use and must not be touched by hand.
5. The pH meter is standardized or calibrated before use by a standard solution whose pH is close to that under test.
6. The pH meter should be switched to zero, but the mains switch left on when not in use. After use, the electrodes are stored in distilled water and must never be allowed to dry out.

pH ELECTRODE / pH METER

The most convenient and accurate way of pH determination is by the use of a glass electrode (pH electrode). Though it was developed in 1919, it became popular only during 1930s when the amplifiers became available.

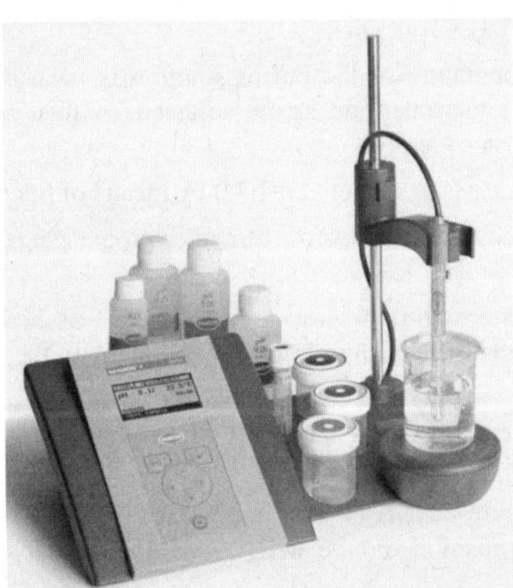

Apparatus for determination of pH

This device depends upon ion exchange in the hydrated layers formed on the surface of the glass electrode. The glass consists of a network of silica wherein the metal ions attached to the oxygen atoms get exchanged with H^+. The electrode acts like a battery the voltage of which depends on H^+ activity of the solution in which it gets immersed. The glass electrodes are useful as they do not interfere with the components of the solution. These are also not readily contaminated by the molecules in the solution. At higher pH, the electrodes do respond to Na and under highly acidic conditions, inaccuracies do occur. The pH electrode is used in conjunction with a pH meter. This helps to record the potential (due to hydrogen ion concentration) by taking a little current from the circuit. The changes in the pH are prevented by high resistance. All the components (pH meter, the glass electrode and the reference calomel electrode) are designed in such a way that pH 7.0 gives a zero potential.

a. Operation of pH electrode/pH meter

The pH meters are available in a variety of shapes and sizes suitable for different applications. Microelectrodes (miniature probes) help to measure intracellular pH. All of these depend on the same principle and are operated almost in the same way.

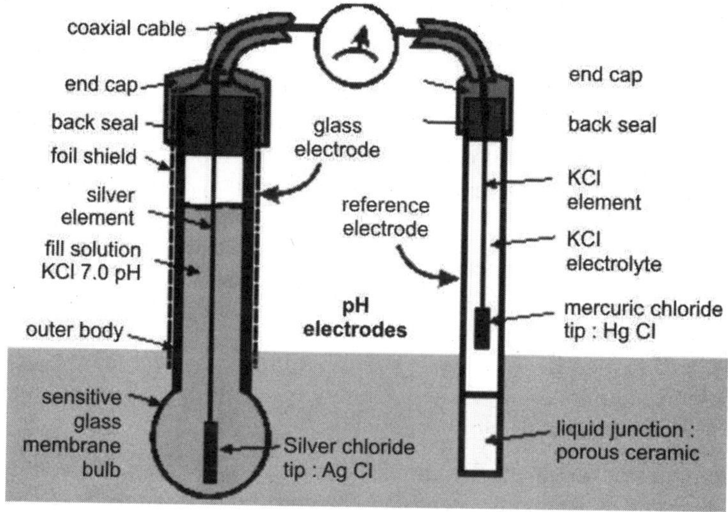

pH Electrodes

The outer layer of glass of the electrode remains hydrated being immersed in solution. The thin glass membrane is highly fragile, and every care is taken not to break/scratch or rub it. To protect it from any damage, the modern devices are encased in plastic. The solutions that contain gelatine or protein should not be allowed to dry on it. The temperature compensation dial must be correctly set before the meter is calibrated. The calibration is first carried out with a pH 7.0 buffer followed by a pH 4.0 (acidic sample) or a pH 9.0 buffer (basic sample). This is followed by immersion of the electrode in the sample solution to be measured for accurate estimate of the pH.

The electrode is the portion of the pH meter which senses the pH and it is very sensitive and fragile which is susceptible to scratches. The electrode is always blotted dry when cleaning it. Light and temperature can also harm it, and hence, it should be stored in a buffer solution at 10-25^{0}C and protected from light. The electrode is free from interference due to colour, turbidity, colloidal matter, oxidants, reductants or high salinity except for sodium at higher pH. This error at a pH above 10 may be reduced by using special `low sodium error` electrodes if water that is used is high in sodium and its pH is also high. The errors can also arise from cracked electrodes, faulty connections or fouling of the electrodes with grease and/or solid residues. The fault can be detected by 2-point calibration (with a second buffer), If faulty, the electrode must be immediately replaced.

b. Reagents required

The reagents that are used to calibrate and store the pH meter are the buffer solutions. These can be procured directly from the market or prepared and stored for at least a week in the lab. It is always best to use a fresh beaker of buffer solution for each calibration. The buffers are prepared in the lab by following the procedure outlined in standard methods.

c. Procedure

The following steps are to be followed in the lab to carry out the measurement of pH by using pH meter.

- The pH meter is turned on and allowed to warm up. Each brand of pH meter operates slightly differently. So, it is better to refer to the instrument`s manual.
- The temperature compensation is essential as pH is slightly temperature dependent. It is always better to bring all (buffer solution and the sample as well as the pH meter) to the same temperature (simply bring them to room temperature). If the meter does not have automatic temperature compensation, measure the temperature of the buffer solutions and adjust the temperature compensation dial manually.
- Calibrate the pH meter by using the buffers. The meter is calibrated before each use to get the most accurate pH reading of the sample.
- The sample solution is prepared. The sample should be contamination-free and is well mixed to get a good reading. The sample is mixed with the help of a magnetic stirrer/stir bar in such a way that the reading represents the entire sample. The sample is not vigorously stirred as it allows the entry of CO_2 into water forming carbonic acid which lowers the pH. The source of the sample is recorded. Prevent contamination. Place the magnetic stirrer into the sample and turn it on.

The pH of the sample is measured as follows: Determine the temperature of the sample and record it. Adjust the compensation dial of the meter accordingly. Remove the electrodes from the storage buffer, rinse with distilled water and blot them dry. Place the electrodes in the sample. Continue stirring of the sample. Record the pH of the sample. Place the meter in standby mode and remove the electrodes. Rinse them with distilled water and blot them dry. Test all the samples, place the electrodes in tap water or a solution with a pH of 7.0 for storage and shut off the pH meter. The data shall include the source of the sample, temperature at which the measurement is carried out and the pH of the test sample.

d. pH-Stat

This is an automatic titrator which is used to maintain a constant pH during a reaction that involves the production or removal of H+ from the solution, by the use of which the rate of a reaction can be determined against time, and also to avoid the danger of overshoot by allowing the addition of lesser reagent at the approach of the end point. A glass electrode (to be very accurate and stable), pH meter, recorder, controller (to break the current to the burette motor at the end point), burette (an ordinary burette with a magnetic valve or a motor-driven burette syringe) and a magnetic stirrer are necessary. The most important limitation of this device is that the solution in the reaction vessel has to be stirred constantly.

e. Ion-selective Electrodes (ISEs)

The glass pH electrode is a type of ion-sensitive electrode sensitive to H^+. Similar electrodes which respond to other ions like Na^+, NH_4^+, Cl^- and NO_4^- have been developed recently. The material of these electrodes may be glass, an insoluble organic salt or an ion-exchange material. Instead of glass, modified alumina-silicate glass responding to mono-valent cations like Na^+, Li^+ and NH_4^+ can also be produced. Similarly, by the use of insoluble inorganic salts like silver sulphide, electrodes which respond to Cu_2^+, Pb_2^+ and Cd_2^+ can be generated. Likewise, lanthanum fluoride is used to produce the electrodes that respond to F^- can be generated. ISEs are easy to use, easily transportable and more economical, use little power and are capable of continuous monitoring without any hazard. These also respond to a wider range of concentrations, rapid and at the same time do not destroy the test sample. These generally find use in the analysis of sea water, serum, beer, soil and skin; analysis of food, drinking water, fertilisers, blood and also air pollution.

pH SCALE

The pH scale was invented to help ordinary people (illiterates and with no science education) to make beer on a more constant level in the earlier years. The scale (chart) has all the colours of a rainbow, from red to blue. The *standard pH scale*, occasionally referred to as the *acidity or alkalinity scale* goes from 0 to 14 although it is possible to exceed these levels sometimes. This was devised by S. P. L. Sorensen, a Danish botanist who defined pH as *the negative logarithm of hydrogen* $[H^+]$ *ion concentration*. The scale is based on the ionic product of water, and designates the actual concentration of $[H^+]$ and $[OH^-]$ ions between 1.0 M H^+ and OH^-. As the scale is logarithmic, if two solutions differ by a pH of 1.0, it means that their actual ionic concentration is 10 times different. The scale is convenient to use, because it converts some odd expressions into a single number. The scale expands the division between zero and one in a linear scale or a compact scale into a larger scale for comparison purpose. The using of a log scale certainly converts infinite small quantities into infinite large quantities. Because the negative log of $[H^+]$ is used in the pH scale, the scale usually have positive values and furthermore, the larger the pH, the smaller the $[H^+]$. It is also interesting to note that higher the pH lower is the hydrogen ion concentration. At neutral pH (7.0), the solution has equal number of hydrogen and hydroxyl ions. The pH of the aqueous solutions is generally measured by indicators and electrodes. The pH scale usually extends from 0 to 14. When an acid fully ionizes in water, it will have pH of 0.0. Acidity describes a substance with a pH under 7.0 at standard conditions. When an alkali or a base fully ionizes in water, it will have a pH of 14.0. Alkalinity describes a substance with a pH above 7.0 at standard conditions. Since pure water has a pH of exactly 7.0, it is neither acidic nor basic, but a neutral solution. The pH values of some common solutions are: Battery acid or 5% Sulphuric acid - 0.0, 0.1 N Hydrochloric acid – 1.1, Vinegar – 2.4 to 3.4, Wine – 3.5 to 2.7, Apple juice,– 3.8, Rain water – 5.6, Fish – 6.7 to 7.0, Chicken – 6.4 to 6.6, Blood – 7.1 to 7.4, Sea water – 8.1, Baking soda – 9.0, Milk of Magnesia – 10.0, Household ammonia – 11.0, Household bleach – 12.0, Oven cleaner – 13.0 and Sodium lye (NaOH – 1mol/litre) – 14.0.

USES OF pH METERS

The pH meters can be combined and used with other meters like ion concentration, ORP, conductivity, dissolved oxygen, viscosity, or bacteria meters, within a single instrument resulting in more efficient measurement. The pH meters are used in the process of water treatment. These are also used in measuring ORP in ultra-pure water in pharmaceutical and micro-electronics applications which require high accuracy when measuring the pH. These are also used in the manufacture of organic chemicals and also in the field of medicine, used to measure the pH of the chemicals produced by the body or that are to be introduced into the body.

pH METERS - ADVANTAGES AND DISADVANTAGES

Advantages

The hand-held pH meters can easily make up to 4000 readings on 4 AA batteries. The pocket pH meters are cheaper than other types of meters. The meters are often programmed with common buffer values to eliminate the need for constant calibration.

Disadvantages

The pH meters are normally slow to register and drift with final values. The deposits gather on the bulbs of the electrodes and disrupt the process of recording. The meters are to be calibrated very often dealing with solutions of variable pH. The calibration can be affected by temperature and carbon dioxide absorption.

pH CONTROLLER

The controllers are those devices which are used to monitor and treat the pH levels in industrial processes. Without proper pH control, the mineral deposits can build up and corrode the machinery in a plant. The controller will monitor the pH level of a fluid and the pH reaches a certain value, it releases a substance that restorers the pH to its desired value. The controller monitors the fluid in a system with a pH meter and measures the amount of free hydrogen ions in the stream. When once the system reaches a certain pH, this will signal an on-line computer or the control valve which controls the addition of an acid or a base either to lower or raise the pH level back to the desired levels. The substance commonly used to lower the pH of an aqueous stream is CO_2, which when added to the stream will form carbonic acid thus lowering the pH of the stream. Sulphuric acid is also commonly used and it can damage the pipes at higher concentrations. Ferric sulphate, aluminium sulphate, chlorine and sulphur dioxide are also helpful to lower the pH value. The simplest and the cheapest method to raise the pH value of a stream is with lime which is a strong base. Many power plants also use ammonia which is a still stronger base but ammonia can damage the pipes at higher concentrations.

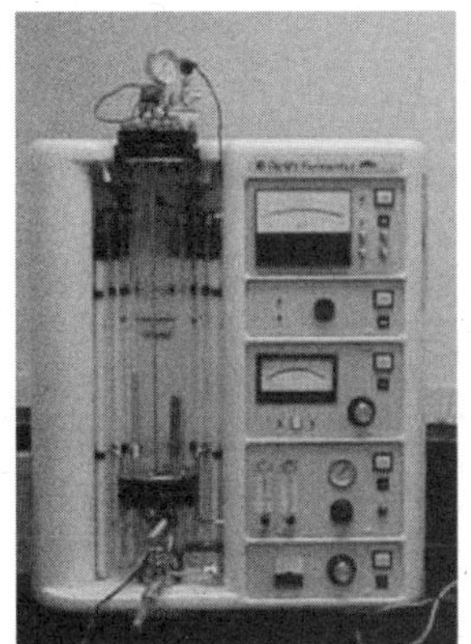

The controller

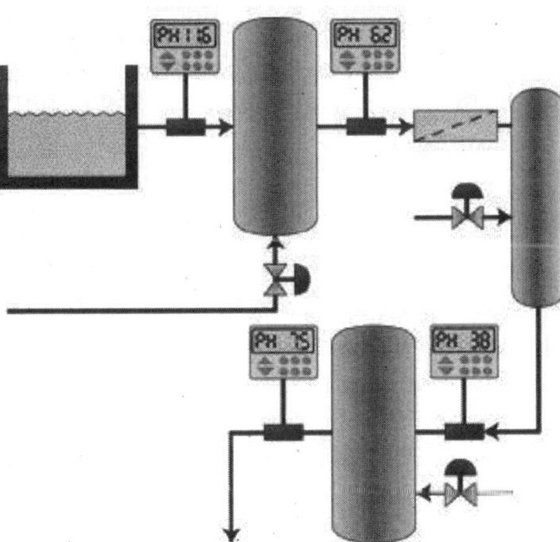

The flow sheet of the design of a controller

1. Use of controllers

The pH control is very important in waste water treatment, pulp and paper industries, chemical processes and biochemical processes. In bioreactors, the controller monitors the pH level of the yeast solution and also provides information to on-line computers to maintain a constant pH.

2. Advantages

The controllers are fast and simple. In case of an emergency and any operating problem, the available alarm sounds. The overshooting of the desired pH value is prevented by the pump that helps to slow down the introduction of an acid or a base.

3. Disadvantages

The inaccurate pH control leads to wear on the equipment and the process is also lost. It is very difficult to add the acid or the base to total neutralization (pH – 7.0) of a substance. The controller can become dysfunctional with a change in the chemical process.

CALIBRATION OF pH METER - WHY?

From time to time, the electrodes must be calibrated to maintain the accuracy of the measurement. The performance of a pH sensor degrades over time and the time period and the related loss of sensor performance varies with each application and the conditions under which measurements are carried out. This is performed to compensate for the changes in the potential within the measuring and reference electrodes, as well as any change of potential between them. The electrodes when put in a zero solution (7.0 pH buffer solution) provide a zero mV output.

The changes or the differences in potential can be caused by one or more of the following factors: the contamination of the reference electrolyte solution, the evaporation/depletion of the electrolyte, the chemical attack of the silver/silver chloride wire, the junction potential and the aging of the measuring electrode. Calibration also involves checking the slope of the measuring electrode. The slope defines *the ability of the measuring electrode to change its output by 59.16 mV per pH unit at 25°C*. The slope adjustment is generally made using a buffer that has a value at least 3 pH units from the zero buffer (7.0 pH). During calibration, the buffers are used to compensate for any inherent offsets of the sensor. While adjusting the zero and slope, one has to carefully follow the instructions given by the equipment manufacturer to avoid inaccurate results. During calibration of an instrument without temperature compensation ability, the temperatures of the buffers are needed to be known. For instruments with automatic temperature compensation, it is better to allow the sensor and buffer temperatures to equalize and to rinse the sensor in distilled water between the buffers. The rinsing removes residual contamination carryover and prolongs the use of buffer solution. The used buffer portion can never be reused or mixed with the bottled buffers. It is better always to discard the buffers after use.

CARE OF pH ELECTRODES

a. Dehydration

The glass membrane of pH electrode will get dehydrated if it is left out of solution, as a result of which the sensor shows a slower response and a higher than normal impedance when it is put back into operation. The repeated dehydration and reuse will greatly reduce the life of the pH sensor, and if dehydration is prolonged, the glass membrane completely fails.

If the reference electrode becomes dehydrated, it will no longer operate properly. The electrolyte leaches out of the electrode cavity through the junctions forming salt crystals on the junction surface. This leaching will weaken the electrolyte potential and may also cause a bridging effect. These two conditions will increase the output impedance making the electrode output unstable. If the dehydration continues the impedance will rise to a level that becomes unusable to the pH meter.

b. Factors Detrimental to Electrode Life

The pH glass is susceptible to attack by the chemicals. The changes in the temperature can alter the rate of this attack, and for every 30°C rise in temperature, the rate of attack increases 10-fold. The strong acids, and to a great extent, strong alkaline solutions attack the glass membrane considerably. Even neutral solutions with high concentration of alkali ions particularly sodium ions attack the glass. HF (hydrofluoric acid) will readily poison the glass membrane when the pH is below 6.0. The greater the fluoride ion concentration, the faster the electrode will fail in its activity. The fluoride strips away the gel layer of the glass membrane rendering it to be inoperative. For measuring the pH in solutions containing HF, a special electrode manufactured from antimony is available.

c. Transportation

Freezing, extreme heat, vibration, mechanical shock etc., are to be avoided when transporting the electrodes and one should always try to use the original box and packing material to transport the electrodes.

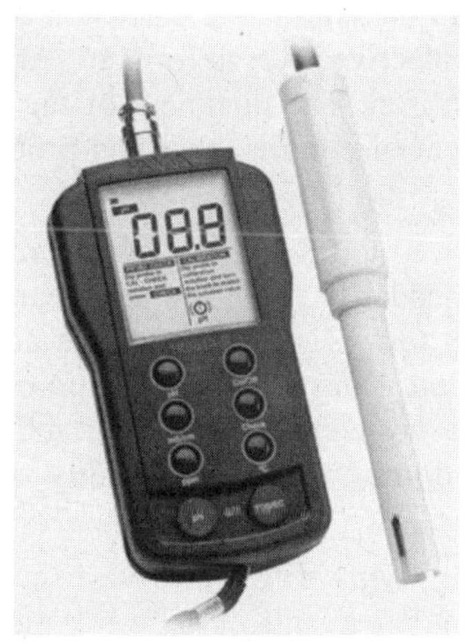

Water-resistant pH meter – Hand-set

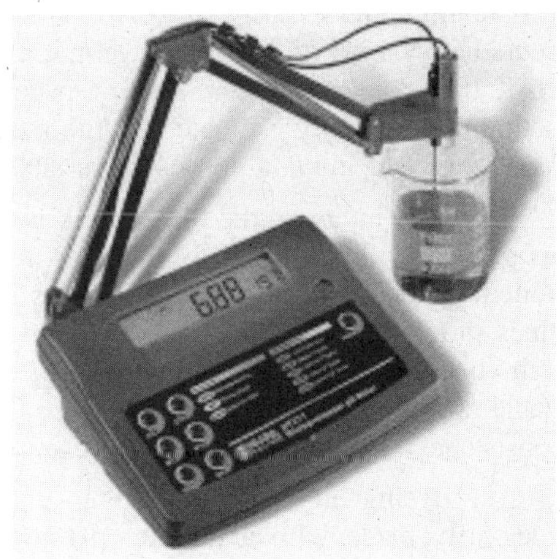

A digital pH meter

d. Storage

The pH sensors should be stored in suitable conditions and temperatures of 10 to 30°C. All caps should be kept intact and installed on to the end of the sensor. 3.0 to

3.5 M Potassium chloride solution is the best solution for storage purpose. This provides a neutral to a lightly acidic environment for the glass electrode. If this solution is not available, pH 4.0 buffer, distilled water or tap water are to be substituted in order of preference, under which condition, both the electrodes can be stored for about 3-5 years. The storage solution is periodically checked not to have evaporated.

pH MEASUREMENT - APPLICATIONS

The applications of pH measurement are useful in different fields like – agriculture, brewing, prevention of corrosion, dairy industry, dyeing, electrical equipment, fermentation reactions, fertilizers, flour milling, manufacture of gelatine and glue, iron and steel, manufacture of jam and jelly, laundries, leather, processing of meat and fish, metal finishing, neutralization, printing, pharmaceuticals, sewage, swimming pools, tropical fish breeding, water etc.

In agriculture, soil pH is important as the plants grow best within a narrow pH range and the optimum pH varies with each plant. pH control is more important in hydroponics as variable pH in the soil (too high or too low) can cause precipitation of some chemicals. The beneficial bacteria also grow best in slightly acidic soil. If the soil pH is too high, the plant nutrients may transform into insoluble compounds and if pH is too low in the soil, the toxic amounts of some metals are made available to plants.

In brewing, pH is important for proper ageing and also for all stages of the process. A decrease in pH decreases the solubility of the bitter parts of the plant material used in brewing. Also to ensure its stability while on shelf in bottles, the beer should be necessarily at a pH of 3.9 to 4.1. In industries, the corrosion of iron occurs below a pH of 4.3 but above 4.3, a semi-protective layer is formed. Above a pH of 10.5, a more resistant coating is formed on iron. The thickness of the oxide coating can be easily estimated by the measurement of e.m.f as such the protective value (complete and non-porous) is variable.

In dairy industry, the pH of milk is not allowed to drop to 4.7 as it curdles at this pH value. A pH of 4.9 is right for making cheddar cheese. The ice cream gets spoiled by the addition of fruits/juices with a too low pH. The pH of the cooling brines should be controlled to prevent pipe corrosion. During the production of soft, fresh cheese and mature, hard cheese, the pH measurements are essential and these enable production and control of cheese fermentation as well as the quality control (consistency, colour and taste) of the cheese.

In dyeing industry, bleaching and dyeing with different dyes must be necessarily made with a definite pH to obtain good results. The wool must be neutralized to a definite pH for effective dyeing. The acid content of the dye also has to be determined before dying.

During the manufacture of chemical products, in order to prevent corrosion of pipes and boilers, the pH of feed water should be controlled as minimum corrosion occurs between a pH of 7.4 and 8.0. The process of softening of soda lime requires a pH of 9.4 for the removal of calcium and a pH of 10.6 for the removal of magnesium.

During the fermentation, for best results, the fermentation processes require a specific pH. During fermentation, pH changes by itself and so it must be adjusted periodically to prevent the formation of unwanted or even harmful by-products.

In fertilizer industry, the pH of acid fertilizers is controlled in order to prevent waste of acid and to ensure a uniform product. In flour milling, by pH measurement, the quality of the flour can be determined and at lower pH (5.9 to 6.5) the better flour can be produced. In the manufacture of gelatine and glue, pH needs to be controlled accurately to ensure the production of a consistent product.

During the production of iron and steel, the measurement of pH determines the effectiveness of pickling baths and the neutralization of waste pickle baths.

In the manufacture of jams and jellies, pH is controlled by the addition of citric acid or tartaric acid. Proper gelling requires a pH of 3.3 (narrow range is best) and it becomes stiff and quite tender at a pH of 3.5.

In laundries, the efficiency of soaps and detergents can be improved by proper control of pH (pH of 11 for un-dyed cottons, 10.0 for wool, 9.6 for coloured clothes and 9.2 for silks). The process of washing will be efficient at a higher pH and this should not exceed the maximum for the material. At a proper pH the sticking of starch during pressing is prevented. In leather industry, during processing of leather, pH control is necessary to obtain maximum efficiency without damage. The texture and colour of the finished product is determined by pH control of tanning and dyeing baths. The de-hairing of leather is done at a pH of 12.3 and is to be completely neutralized to assure good quality. A pH of 4.7 is essential for the minimum swelling of collagen. In processing of meat and fish, the required pH helps keep the quality and freshness of the product.

In metal finishing, the pH of plating baths determines the quality and speed of the process. Some alloys can be plated easily with pH control. The neutralization of acids and bases is best controlled electrometrically with a pH meter where a colour indicator is of no use. In printing, the pH of the paper and ink must be controlled to assure proper penetration and drying of the ink as too high pH causes gumminess and a too low pH slows up drying.

In pharmaceutical industry, an incorrect pH can produce a poison rather than a medicine (from antibiotic molds). In sewage industry, to assure an efficient coagulation of the sludge, the pH must be controlled (pH of 3.4 or 7.4 is used for good results).

To prevent contamination, the pH of the effluent water must be controlled (a pH of 6.8 to 7.6 is the best digestion range). The pH of filtration also depends upon the chemicals used (3.4 for ferric chloride and 4.4 for alum).

In swimming polls, the pH levels should be maintained near the neutral range or slightly alkaline to prevent irritation of skin of swimmers. Higher pH is known to accelerate the solid salt deposition in heater lines and filters while lower pH tends to cause corrosion of iron pipes in the pools.

In tropical fish breeding, the thriving of expensive tropical fish depends upon definite ranges of pH as each species has its own best pH environment which is critical during breeding season. For example, the Neon Tetra fish prefers a pH of 7.0,

Angel fish requires a pH of 6.8, fresh water fish needs a pH of 6.0 to 8.0 and salt water aquaria should be maintained at a pH of 8.3 etc. The institutes of oceanography, fish and wildlife service and water authorities often measure the pH of water sources like rivers, lakes and oceans to study natural conditions of wildlife. Such measurements also assist to determine the extent of pollution in domestic and industrial supplies.

pH IN NATURE

The pH measurements are important in medicine, biology, chemistry, agriculture, forestry, food science, environmental science, oceanography, civil engineering, chemical engineering, nutrition, water treatment and water pollution and many other applications.

The pH dependent plant pigments can be used as pH indicators and occur in plants like *Hibiscus*, red cabbage (anthocyanin) and red wine. The juice of red cabbage will function over a wider pH range, from as low as 1.0 up to 12.0 and the juice of red beet will change from red to yellow somewhere between the pH of 11.0 and 12.0. The citrus fruit is acidic because of citric acid. Carboxylic acids do occur in many living systems. Lactic acid is produced by muscular activity. ATP is pH dependent. Haemoglobin is affected by pH in a process called the *root effect*. Distinct pH scales do exist in chemical oceanography (pH of sea water). The pH of different cellular compartments, body fluids and organs is regulated in acid-base homeostasis. The most common disorder is acidosis (acid overload in the body falling below 7.35) and alkalosis (high pH in the blood). Enzymes and other proteins have an optimum pH range. The pH in foods and drinks is often called as acid-ash or alkaline-ash, and the ash refers to the pH value of a food after it is metabolized by the body. The foods like oranges, lemons, grapes and apple will test acid before metabolizing but turn alkaline afterwards. The turmeric, a common spice used in Indian cooking can be used to test its pH in foods or drinks. The active ingredient of turmeric, curcumin is a pH indicator and when turmeric is added to foods with a pH value under 7.4, the turmeric will turn yellow, and in food with a pH above 8.6, it turns red. The alkaline food is essential for the body and should be consumed in abundance to keep the pH level balanced. The food like vegetables, fresh fruits, raw nuts and seeds, and fresh juices create alkaline residues which strengthen health, vitality and on-demand energy essential for the body.

In nature, the pH of natural waters generally depends on the relative concentrations of carbonate ions, hydrogen carbonate ions and dissolved carbon dioxide. For example, the rain water is usually slightly acidic (pH – 5.7) due to the reaction of water and dissolved carbon dioxide from the atmosphere that the rain has fallen through, while the ground water is usually slightly alkaline (pH – 8.0) due to the reaction of water with dissolved carbonate ions from rocks that water has passed over. The water in the rivers and lakes can be acidic or alkaline depending on whether it is precipitation dominated, rock dominated or evaporation dominated. Under acidic conditions (pH less than 6.3), the dominant species is carbon dioxide, whereas under strongly alkaline conditions (pH more than 10.3), the carbonate ions dominate, and under the intermediate conditions (pH - 6.3 to 10.3), the hydrogen-carbonate ion dominates. Although the pH of pure water is 7.0, drinking water and natural water exhibits a pH range because it contains dissolved minerals and gases. The surface waters typically range from pH 6.5 to 8.5, while the ground water ranges from pH 6.0 to 8.5.

pH AND LIFE

Most cells can only function within very narrow limits of pH and require buffer systems to resist the changes in pH that would otherwise occur in metabolism. The main buffer systems in living material are protein, bicarbonate and phosphate. The relative importance of each depends on the type of the cell and organism.

The most important buffering system in mammalian plasma is bicarbonate. The kidneys also play an important role the maintenance of acid-base balance by adjusting the excretion of acid or base in the urine, so that the pH of the urine can normally vary from 4.8 to 7.5 in man. The extracellular fluid is usually slightly alkaline at pH 7.4, while the average cytoplasmic pH of many animal cells is about pH 6.8 which at 37°C is neutral. The average cytoplasmic pH of plants is similar to that of animals. But, the cell sap of many and majority of land plants is acidic and lies within the range of pH 5.2 to 6.5. The internal pH of bacteria appears to be round about neutrality although many grow well at pH 6.0 or 9.0, some even at a pH of 11.0. It is very clear that, the simpler forms of life appear to be able to survive a wider pH range of the external medium.

For example, the pH of common substances is as follows: Alkaline sodium hydroxide and liquid drain cleaner is 14.0, caustic soda is 13.9, household lye is 13.5, an oven cleaner is 13.0, household bleach is 12.5, Soap water is 12, household ammonia is 11.5, and, hand soap is 9.0 to 10.0, milk of magnesia is 10, baking soda is 9.0, sea water is 7.7 to 8.3, and human blood is 7.7 to 7.45. The pure water is neutral with a pH of 7.0, distilled water is 7.0, while milk acid is 6.5, cow`s milk is 6.0, urine is 6.0, tea is 5.5, coffee is 5.0, acid rain is less than 5.0, that of soft drinking water, black coffee and pure rain is 5.0, while beer is 4.5, and of tomato juice, acid rain, orange juice and apple juice is 3.5, wine is 3.0, vinegar is 2.9, cola is 2.5, lemon juice is 2.4, gastric acid is 1.5 to 2.0, hydrochloric acid (gastric acid) secreted by the lining of stomach is 1.0, and battery acid is 0.5, and strong hydrofluoric acid is 0.0.

The pH values of some of the common food stuffs we use daily are as follows: The pH of *Aloe vera* is 6.1, apples is 3.3 to 3.9, apricots is 3.3 to 4.8, artichokes is 5.5 to 6.0, asparagus is 6.0 to 6.7, avocados is 6.3 to 6.6, bananas is 4.5 to 5.2, beans is 5.6 to 6.5, beets is 5.3 to 6.6, blackberries is 3.9 to 4.5, blueberries is 3.1 to 3.4, white bread is 5.0 to 6.2, broccoli (cooked) is 5.3, butter is 6.1 to 6.4, buttermilk is 4.4 to 4.8, cabbage is 5.2 to 5.4, cactus is 4.7, carrots is 5.9 to 6.3, celery is 5.7 to 6.0, cheese is 4.8 to 6.4, cherries is 3.2 to 4.5, chilli sauce is 2.8 to 3.7, coconut is 5.5 to 7.8, coconut milk is 6.1 to 7.0, corn is 5.9 to 7.3, curry sauce is 6.0, dates is 6.5 to 8.5, eggs (fresh) is 7.6 to 8.0, wheat flour is 5.5 to 6.5, gooseberries is 2.8 to 3.1, grape fruit is 3.0 to 3.7, and grapes is 3.5 to 4.5.

The pH of horse-radish is 5.4, fruit jams is 3.5 to 4.0, fruit jellies is 2.8 to 3.4, ketchup is 3.9, lemons is 2.2 to 2.4, lemon juice is 2.0 to 2.6, limes is 1.8 to 2.0, lime juice is 2.0 to 2.4, mangoes is 5.8 to 6.0, maple syrup is 4.6 to 5.5, melons is 6.0 to

6.7, milk is 6.4 to 6.8, molasses is 4.9 to 5.4, mustard is 3.5 to 6.0, olives (green and fermented) is 3.6, black olives is 6.0 to 7.0, oranges is 3.0 to 4.0, peaches is 3.4 to 4.1, peanut butter is 6.3, pears is 3.6 to 4.0, peas is 5.8 to 6.4, sour pickles is 3.0 to 3.4, plums is 2.8 to 3.0, potatoes is 5.6 to 6.0, pumpkin is 4.8 to 5.2, raspberries is 3.2 to 3.6, rhubarb is 3.1 to 3.2, soft drinks is 2.0 to 4.0, soya bean milk is 7.0, soya sauce is 4.4 to 5.4, spinach is 5.5 to 6.8, squash is 5.0 to 5.4, strawberries is 3.0 to 3.9, strawberry jam is 3.0 to 3.4, sweet potatoes is 5.3 to 5.6, tea is 7.2, tomatoes is 4.3 to 4.9, tomato juice is 4.1 to 4.6, tomato puree is 4.3 to 4.5, turnips is 5.2 to 5.6, vegetable juice is 3.9 to 4.3, vinegar is 2.4 to 3.4, drinking water is 6.5 to 8.0, water melon is 5.2 to 5.6, wine is 2.8 to 3.8, and yams (cooked) is 5.5 to 6.8.

Similarly, water which occurs in different states of matter in the atmosphere, and which occurs in different water bodies in nature, in the bodies of plants, animals and humans and which is available in different forms and is essential to the existence of life on the planet varies in its pH. The pH of sea water plays an important role in the ocean`s carbon cycle, and there is evidence of ongoing ocean acidification caused by carbon dioxide emissions. Similarly, the pH of different cellular compartments, body fluids, and organs is usually tightly regulated in a process called acid-base homeostasis. The most common disorder in acid-base homeostasis is acidosis, which means an acid overload in the body, generally defined by pH falling below 7.35, and alkalosis is the opposite condition, with blood pH being excessively high. The pH of blood is usually slightly basic with a value of pH 7.365 and this value is often referred to as physiological pH in biology and medicine. Plaque can create a local acidic environment that can result in tooth decay by demineralization. The enzymes and other proteins have an optimum pH range and can become inactivated or denatured outside this range.

■■■

CHAPTER - 13

MICROTOMY AND MICROTOME

Introduction

Microtomy is a method for the preparation of thin sections for materials (bones, minerals and teeth, an alternative to electro-polishing and ion milling), and the sections can be made thin enough to section a human hair across its breadth, with section thickness between 50 nm and 100 μm.

The word microtome comes from two Greek words, *micros* meaning `small` and *temnein* meaning 'to cut'. It is a mechanical instrument used to cut biological specimens into very thin (extremely thin) segments (sections) for microscopic examination. Sectioning paraffin wax embedded tissue blocks is the commonest way of achieving this. Microtomes are important in science and are used in microscopy allowing for the preparation of the samples for observation under transmitted light or electron radiation. These use steel, glass or diamond blådes depending upon the specimen being sliced and the desired thickness of the sections being cut. The steel blades are used to prepare the sections of animal or plant tissues for light microscopy histology, while glass knives are used to slice the sections for light microcopy and to slice very thin sections for electron microscopy. Industrial grade diamond knives are used to slice hard materials (bone, teeth and plant matter) for both light and electron microscopy while gem quality diamond knives are used for slicing thin sections for electron microscopy.

History and Microtomy

The general opinion seems to be that the microtome is of fairly recent origin. According to Thome, the greater number of microtomes goes back to two fundamental types, the Ranvier and the Rivet microtome. Minot is of the opinion that the first microtome which resembles the modem microtome is the instrument of His (invented in 1839), the Valentine double, (Valentine's double-bladed knife – has been of limited service in animal histology) being regarded as a forerunner of the modem microtome. Another instrument, called a microtome, is The Straus-Dnickheim microtome.

In the beginnings of the development of light microscopy, sections from plants and animals were manually prepared with the use of razor blades (sections by hand), and it was found that to observe the structure of the specimen under observation, it was important to make clean reproducible cuts on the order of 100 μm, through which the light can be transmitted, and this allowed the sample observation using light microscopes in transmission mode. The earliest form of the microtome enabled free hand sectioning of fresh or fixed material using a sharp razor. A diagram of a microtome drawn by Alexander Cumming (also referred to as Cummings) in 1770, and one of the first devices for making the cuts of tissues, the microtome itself, was first invented by George Adams Jr., the younger, an English optician and a writer and a mathematical instrument maker to George III (1770). The device was hand operated, and the sample was held in a cylinder and sections were created from the top of the sample using a hand crank. This was further developed by Alexander Cumming, a Scottish watch-maker. Cumming's manufacturing company designed manual sectioning instruments (*Cumming-pattern section cutter* or the *Cutting Engine* – two microtomes appear to be on public display at England`s Museum of Science History and The Science Museum), made from wood which were exclusively used in botany for cutting plants, which were subsequently improved upon by Robert Custance whose plant sections further stimulated the excitement over the microscope. Later, a table based model was developed by Andrew Prichard (1835) which allowed for the vibration to be isolated by attaching the device to the table, separating the operator from the knife. It was Chevalier (1839) who first introduced the term, `*microtome*` and *microtomy* in scientific terminology and introduced it in print, for these devices. In 1843, Oschatz invented an instrument for cutting sections which he called a microtome (this was made in two forms, either a simple hand microtome or a table form that had a three-legged base. This term was generally accepted by German histologists but the English microscopists used the term *cutting machine* or *cutting engine*, reserving the name microtome for the Straus-Druckheim instrument. Chevaliar called these sectioning instruments of Cumming, and Adams, "*le couteau micrometrique ou mieux V instrument microtomtquef*" as such he was the coiner of the word microtome and used it to describe any instrument for cutting sections. Welcker`s microtome was very similar that of Oschatz, and other microtomes which were made were those of Smith (1859), Schmidt (1859) and Luys (1868). Other microtomes were also made judging from Harting's statement (1864) that there was a very old microtome in the Utrecht Cabinet, while Linger mentions using a cutting machine made by Plossl of Vienna. The microtome of Ranvier presents no further advancement in the construction of the instrument and in fact this was one of the simplest ever devised. Gudden improved the Ranvier microtome. Rivet`s microtome is the fundamental type on which all improvement of the sliding microtome have been made. Brandt constructed the sliding microtome of metal instead of wood, often called Rivet-Leyser microtome (as this was made by Leyser). The Caldwell and the rocking microtome (both automatic) were introduced by the Cambridge Scientific Instrument Company (1885). The first published description is that of the so-called `Johns Hopkins` microtome (1886) and Minot`s microtome was made in 1887 by Baltzer of Leipzig, whose description appeared in 1888.`

A Heidelberg pathologist, Rudolf Thoma in concert with a precision engineer, Rudolf Jung, designed the first microtome in series production called `*Thoma microtome*` enabling the beginning of a new era in histology. Sometimes, the invention of the microtome is attributed to Wilhelm His Sr. (1865). Some other sources further attribute the development of a microtome to Jan Evangelista Purkyne, a Czech physiologist, and several sources also describe the Purkyne model as the first in practical use. Such obscurities in the origins of a microtome are due to the fact that the first microtomes were simply cutting apparatuses and the developmental phase of these early devices is not documented widely. The development of very thin and consistently thin samples by microtome, together with the selective staining of the important cell components or molecules allowed for the visualization of microscopic detail, only at the end of 1800s. With the hand-held knives barely adequate for cutting small specimens, rotary microtomes were introduced by scientific investigators in 1883 and 1886 for greater precision and uniform section thickness. The Spencer Lens Company, between 1901 and 1910, manufactured the first clinical microtome and a larger, precision laboratory microtome. As on today, a majority of microtomes are a knife-block design with a changeable knife, a specimen holder and an advancement mechanism. In majority of the devices, the cutting of the samples begins by moving the sample over the knife, where the mechanism moves automatically forward such that the next cut of chosen thickness can be made. The thickness of the section is controlled by an adjustment mechanism allowing for precise control. The modern microtomes are precision instruments designed to cut uniformly thin sections of a variety of materials for detailed microscopic examination. With practice, the sections which are quite thin and translucent could be produced. For light microscopy, where the magnifications can reach up to 188x, the thickness of a section can vary between 1 and 10 microns (thin sections); and for electron microscopy, where the magnifications of about 10,000,000x is possible, the thickness of a section is usually of the order of 10 nanomicrons (ultra-thin sections). Since many types of microtomes are commercially available in the market, choosing the right microtome is essential for producing the best results as required.

PARTS OF A MICROTOME

Microtome consists of three main parts, namely, *Base* (body of the microtome), *Knife attachment* and *knife* and *material* or *tissue holder*. With majority of microtomes, a section is cut by advancing the material holder towards the knife whilst the knife is held rigidly in place. The cutting action (which can be either in a vertical or horizontal plane) is coupled with the advance mechanism, so that the material holder is moved after each cut. The distance moved is pre-selected using a scale setting on the body of the microtome and usually extends between 0.5 and 50 microns on microtomes cutting thin sections, and from less than 60 nm to over 500 nm on machines cutting ultra-thin sections.

PROPOSED CLASSIFICATION (TYPES) OF MICROTOMES

Different types of microtomes are available depending on the manufacturer`s specifications and based on the application required for sectioning of the samples. The previous literature does not cite any classification of microtomes, and hence, an attempt is made to classify the commercially available microtomes according to their mode of operation. Microtomes can be classified as: *manual micrtotomes, semi-automatic microtomes* and *automatic microtomes*. The manual microtomes are represented by – *rocking microtome, rotary microtome, sledge microtome, freezing microtome, vibrating microtome, ultra microtome, cryostat, sliding microtome, saw microtome* and *hand microtome*; while the automatic microtomes are represented by – *laser microtome, computer microtome* and *ultra-thin computer microtome*.

The different microtomes mentioned hereunder are used for different purposes.

a. Rocking Microtome

This device is one of the oldest in design, relatively cheap and is exclusively designed for sectioning paraffin blocks in which the material is embedded. The name of the microtome is derived from the rocking action of the cross arm of device. This comprises of three moving parts and is simple to use, extremely reliable and requires minimum maintenance. The knife of the device is fixed and the tissue block moves through an arc to strike the knife (which is slightly biconcave Heiffor knife) causing the sections to be cut in a curved plane in the required thickness by means of a ratchet operated micrometer thread. The steady forward and backward movement of the handle of the device gives ribbons of very good sections. The disadvantages include: the size of the block that can be cut is limited; the sections are cut in a curved plane; and instrument being light in weight, it is to be fit into a tray which can be screwed to the bench or placed on a damp cloth to avoid movement during the process of cutting. In recent times, this has largely been replaced by the more precise rotary microtome.

b. Rotary microtome

Rotary microtome is the first machine designed by Prof. Minot, and hence, this is often referred to as the *Minot Rotary*. This device derives its name as such from the rotary action of the hand-wheel (fly wheel) which actuates the cutting movements. The devices of this sort are general purpose microtomes intended for cutting semi-thin to thin sections for light microscopy. The operation of the instrument is based upon the rotary action of a hand wheel which activates the advancement of a block holder (sample holder) towards the knife which is held rigidly in a horizontal position. The material block moves up and down in a vertical plane in relation to the knife and cuts flat sections. The main advantage is that being heavier, it is more stable than the rocking type, and is ideal for cutting serial sections (in a larger number from a single block) . As the cutting angle of the knife (tilt of the knife) is adjustable, larger blocks of tissue may be cut on this device. There is less vibration when cutting because of the use of heavier and a larger knife with this instrument. Other advantage is its easy adaptation to all types of tissues, whether hard, fragile or fatty, and the device is heavier and more stable. The typical thickness of the section ranges between 0.5 μm to 60μm on most instruments. The embedded tissues (in paraffin wax) are normally cut within the range of 3 to 5μm while those embedded in resin (epoxy) are cut within the range of 0.5 μm to 1 μm.

Rocking Microtome

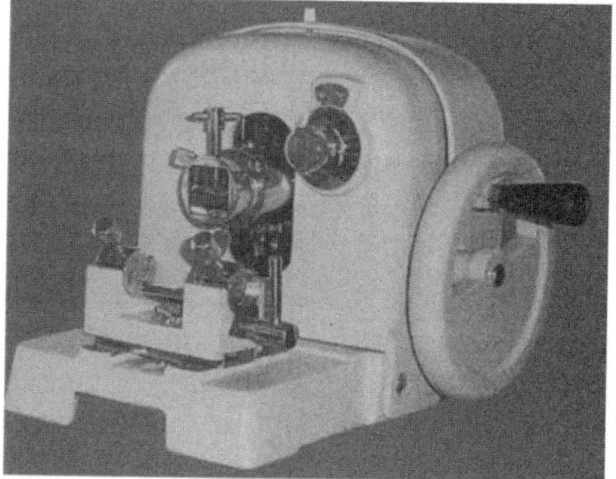

Minot Rotary

c. Sledge microtome

This device was originally designed for cutting the sections of very large blocks of tissue (whole brain) and was used primarily for large blocks, hard tissues, whole mounts etc. The device is especially useful in neuropathology and ophthalmic pathology. The instrument has become a popular machine for routine use since the World War II. The sample is placed into a fixed holder (called *shuttle*), which moves backwards and forwards across a large wedge shaped knife (about 24 cm in length). The block holder is mounted on a steel carriage which slides backwards and forwards on guides against a fixed horizontal knife. The pressure applied to the sample during the cut can be reduced by adjusting the angles between the sample and the microtome knife. Typical application for this microtome is the preparation of large samples (embedded in paraffin) for biological preparations. The thickness achievable is between 1 and 60 µm. The advantages include: the device is heavy, very stable and is not subject to any vibration; the large knife (24 cm in length) produces less vibration; and the knife holding clamps are adjustable and allow the tilt and the angle (slant) of the knife to the block to be adjusted with ease. The disadvantage is that the device is slower in its use than rocking or rotary devices, and the sections from the routine blocks can be cut as quickly as on any other type of microtome.

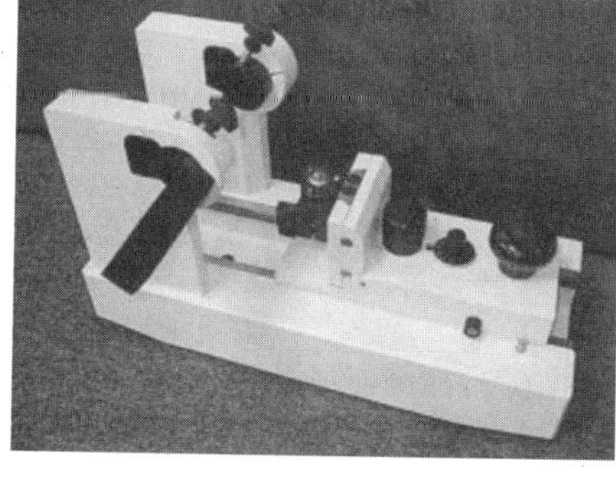

Sledge Microtome

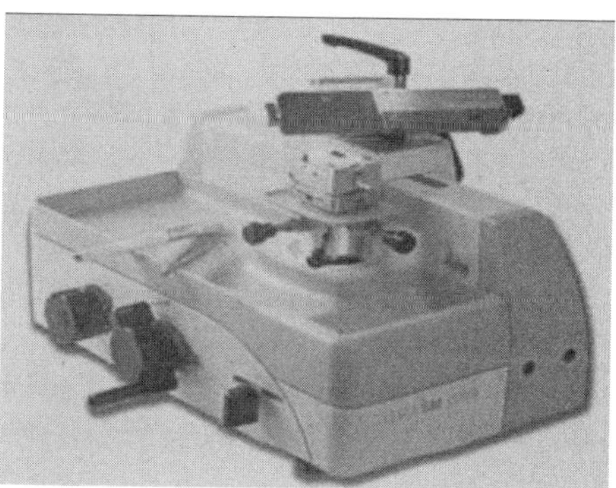

Sliding Microtome

d. Sliding Microtome

This device is designed for cutting the celloidin embedded tissue blocks. The knife is stationary and moves horizontally against a fixed block which moves against it in an inclined plane. This can also be used for paraffin wax embedded sections, although it is meant for other sections.

e. Freezing Microtome

This instrument was known to be efficient in producing the best results and was used universally for cutting the frozen sections. The machine is clamped to the edge of a bench and is connected to a cylinder of carbon dioxide by means of specially strengthened flexible metal tube. The cutting action differs from the other microtomes described above because the knife moves while the tissue blocks remain static (stationary). The block moves by a pre-set adjustment in microns at the end of each cut. Though the device is consistent and of a high quality, thin sections are very difficult to obtain with this device. The instrument enables the tissue to be frozen without the necessity of solid carbon dioxide or liquid nitrogen and as such this method has acquired popularity before the cryostat became widely used for frozen sections. The thickness of the sections ranges from 5 μm instead of 1 μm.

f. Vibrating Microtome

This device was conceived as a microtome which could produce high quality sections of fresh and unfixed material from animal and/or botanical sources and to replace the hand microtome. This has been designed to cut the tissue which has not been fixed, processed or frozen. It has the greatest application in enzyme histochemistry and ultrastructural histochemistry. The name was derived from the high speed vibration produced by a safety razor blade (the thickness of which generally greater than that used for paraffin wax) used for cutting. The amplitude of vibration is adjusted by altering the electrical voltage applied to the knife. The soft material is cut whilst immersed in a fluid (water, saline or a fixative to prevent tearing) which aids in dissipating the heat produced at the vibrating edge of the razor during cutting. The tissues are normally cut at a very slow speed to avoid disintegration. The instrument is used usually for difficult biological samples, and the thickness of the cut is around 30 to 500 μm for live tissues and 10 to 500 μm for fixed tissues.

Freezing Microtome

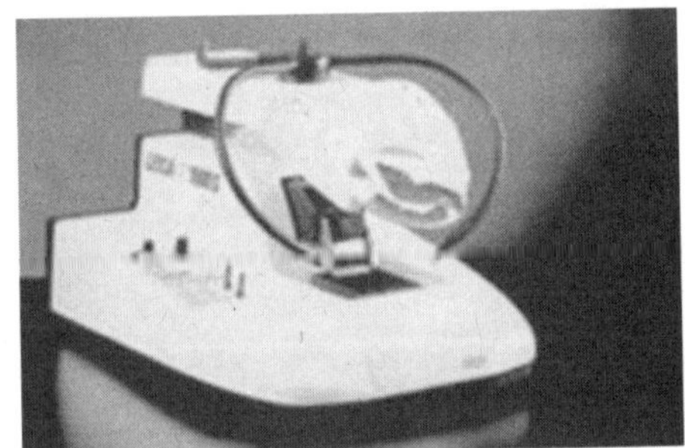

Vibrating Microtome

g. Ultra-microtome

This device is used to prepare ultrathin sections for light microscopy and electron microscopy, and functions in the same manner as a rotary microtome. This has very light tolerance on the mechanical construction as a result of which the linear thermal expansion of the mounting is used to provide very fine control of the thickness of the cuts. Before cutting with the device, very small samples of the tissue or industrial products are usually embedded in hard resin. The extremely thin cuts are important for use with TEM and SBFSEM as well as light microscopy, and the thickness of these cuts lies between 40 to 100 nanometers for TEM, while for SDBFSEM the thickness lies between 30 to 50 nanometers. The sections that are thicker up to 500 nanometers are also taken for some specialized TEM applications or for light microscopy. The thermal mechanism of the device relies upon the heat induced expansion in a bifurcated metal strip, while in the mechanical form, a microprocessor coupled to a precise stepping motor controls the advance mechanism. The cutting stroke is motor-driven to provide a regular smooth motion for cuts of even thickness and constant reproducibility. The knives are usually made from glass, diamond or sapphire, and the block is brought to the edge of the knife under microscopical control and as each section is cut, it is floated on to a water bath or a liquid adjacent to the edge of the knife, and is carefully picked up on to the grid suitable for TEM specimen viewing. The thickness of the section can be estimated by the thin film interference colours of reflected light that are seen as a result of the extremely low sample thickness.

h. Saw microtome

These devices find their use to cut sections from very hard material such as undecalcified bone, teeth, glass or ceramics. The samples are embedded in resins and are moved extremely slowly against a diamond-coated saw rotating at approximately 600 rpm. The recessed rotating saw slices through the sample to produce sections of about 20 μm or greater (around 30 μm for comparatively larger samples), provided the saw blade is in perfect condition. The blade is not capable of producing very thin sections.

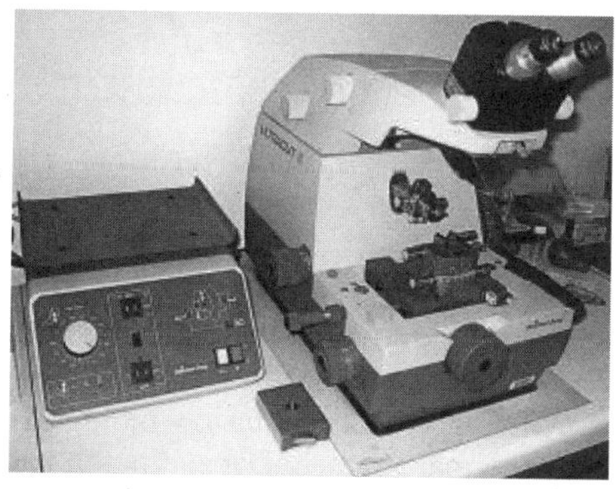

Ultramicrotome

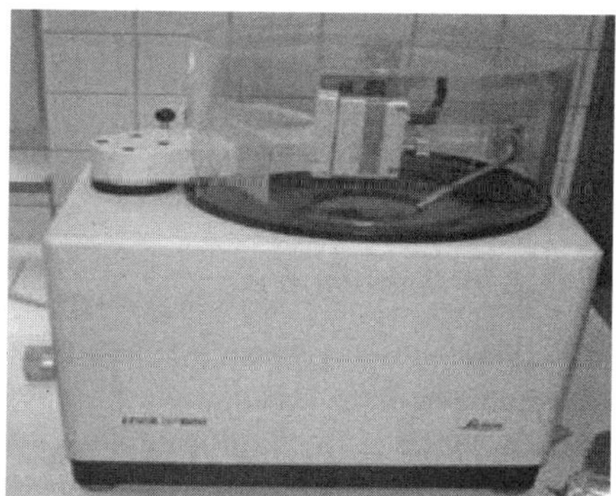

Saw Microtome

Hand Microtome

This device is limited to sectioning intrinsically rigid botanical material and it is difficult to obtain thin sections from animal tissues.

j. Cryostat

The introduction of fluorescent antibody staining technique by Coons, Creech and Jones (1941) led to a need for thin sections (about 3 to 5 μm) of fresh frozen tissue free of ice crystal defects, and to satisfy these criteria the tissue must be snap frozen at a very low temperature. Accordingly, Linderstrom-Lang and Mogensen (1938) designed the first cryostat, followed by redesigning it by Coons and his colleagues (1951). This device consists of any type but preferably rust-proof, which is enclosed and operated within a deep freeze cabinet. The cabinet of the device is fitted with a double glass window, and a door through which the material may be passed in and out. The cabinet is equipped with a fluorescent light and a fan to ensure the circulation of cool air. The temperature may be regulated between -10°C to -40°C. This can be used as an alternative to a freezing microtome for rapid sectioning. Some devices have been designed by incorporating an internal Freon quick freeze stage that holds four block holders which after the tissue is frozen on them, are transferred to the microtome. As the sections adhere firmly to the warm slides, the staining and mounting may be carried out more rapidly. This can be even adjusted to cut the sections from 2 to 16 μm.

Hand Microtome

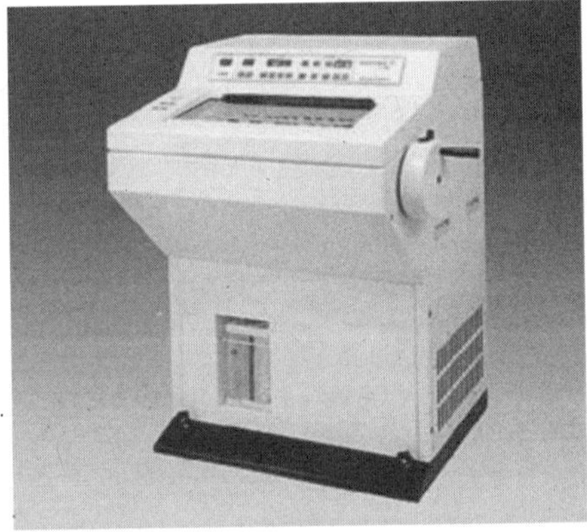

Cryostat

k. Laser Microtome

This device is used for precise, non-contact sectioning, and was designed to slice the samples with high precision. The instrument is equipped with state-of-the-art femtosecond laser technology. This enables non-contact cutting inside biological tissues and various materials without causing any thermal damage. Depending on the material being processed, the slice thicknesses of about 5 to 100 μm are feasible. This is an instrument for contact free slicing and prior preparation of the sample (embedding, freezing or chemical fixation) is not required and this minimizes the artifacts from preparation methods. This can be used for very hard materials (bones and teeth as well as some ceramics). The thickness that is achieved is between 10 and 100 μm depending upon the properties of the material of the sample. The instrument operates using a cutting action of an infra-red laser. The laser radiation is directed on to a fast scanning mirror-based optical system that allows for 3-dimensional positioning of the beam crossover, while allowing for bean traversal to the desired region of interest. The combination of high power with a high faster rate allows the scanner to cut large areas of sample in a shorter time. In this device, the laser micro-dissection of the internal areas in tissues, cellular structures and other types of smaller features are also possible. The main and added advantages of this instrument are: non-contact processing, sub-micrometer precision, cutting of the tissue in its native state, no thermal damage, fewer artifacts and less time consumption in tissue preparation.

Laser Microtome

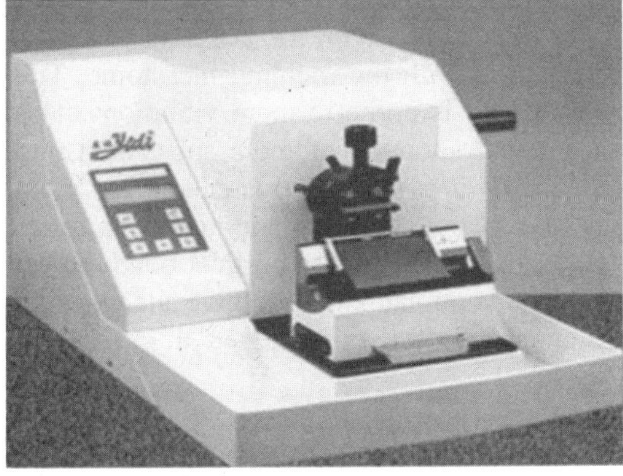

Computerized Microtome

l. Computerized Microtome

This device is equipped with the advanced rapid thermostatic switch, semi-conductor freezing, cryoscalpel and cryoplate. This can carry out the rapid freezing section or routine paraffin section (dual purpose). The thickness of the slice attained is in the range of 1 to 25 μm with least slice adjusting graduation of 1 μm and a maximal slice section of 32x32 mm. The temperature of the cryoscalpel and cryoplate range between 0°C to -18°C and 10°C to -40°C respectively.

MICROTOME – KNIFE DESIGN AND CUT TYPES

The important part of the microtome in the production of good sections is the microtome knife. The very first attempt to produce an adequate cutting surface was a sharp razor blade, and this becomes blunt very quickly. It was Latta and Hartmann (1950) who discovered the edges fine enough by using freshly cut glass. Microtomy virtually begins and ends with a sharp, blemish-free cutting edge. The introduction of the disposable blades has made easier the production of good quality, thin sections, but they are often unsatisfactory for sectioning harder tissues (especially bone). The microtome knives can be classified according to the material used for making the knife or based on the shape of the knife edge. On the basis of the material used to make the knife, they may be: *steel knives, non-corrosive knives for the cryostat, disposable blades, tungsten carbide knives, glass knives, diamond knives* or *sapphire knives*. On the basis of the shape of the knife edge (profile), they may be: of *profile A* (strongly plano-concave or biconcave), of *profile B* (plano-concave), of *profile C* (wedge shaped) or of *profile D* (plane shaped).

The knives have been developed to fit specific types of microtomes and to cope with different degrees of hardness of the tissues and embedding media. The knives made of steel are used to cut paraffin wax embedded tissues while glass knives are made use of to cut the resin embedded tissues. The *wedge shaped knives* (profile C) were originally designed for cutting the frozen sections, as they give a great rigidity to the knife and can be used for cutting all types of sections on any instrument. The *plano-concave knives* (profile B) are used primarily for cutting nitrocellulose embedded tissues and these are available with varying degrees of concavity. The *biconcave knife* (profile A) is the shape of a classical knife first introduced by Heiffor, and is normally used with the rocking microtome. This knife is relatively easy to sharpen, less rigid and is prone to more vibrations. This design has lost its popularity due to the gradual adoption of more substantial microtomes during the development. The *tool edge knife* (profile D) is called as *chisel edge* as its shape is similar to the chisel of a wood worker, and this is primarily used to section exceptionally hard tissue (decalcified dense cortical bone, un-decalcified bone etc.). The blade is stouter than conventional knives to give added rigidity and the edge may be coated with tungsten-carbide for an increased life. The types of blades are as follows:

a. Steel Knives

These are manufactured from high quality carbon or tool grade steel which is heat treated to harden the edge (fully hardened knives are the best). The steel should be resistant to rust and free from any impurities and should contain anti-corrosives. The knives that are only surface hardened lost the cutting edge very quickly once the hardened area is removed through repeated re-sharpening.

b. Non-corrosive Knives for Cryostats

These knives are manufactured from hardened, heat treated stainless steel free from all impurities. These contain 12 to 15% chromium. Teflon coated blades are suitable for use in cryostats as these blades offer reduced cutting resistance and minimal friction.

c. Disposable Blades

These blades are essentially refined, thickened razor blades and when these are held in a specially adapted knife holder, the blades consistently produce high quality sections. In many instances, these blades have replaced the conventional microtome knives. The blades are manufactured from high quality stainless steel, and the edge of these blades can be coated either with platinum or chromium to enhance their strength and prolong their cutting life. The smaller and thinner disposable blade also reached the cryostat chamber temperature more rapidly than a conventional knife, thus minimizing the delay during blade exchanges or temperature adjustments. These blades need to be held rigid in a special holder to prevent vibration during the cutting stroke.

d. Tungsten Carbide

The knives that are manufactured from high quality tungsten are non-magnetic and 100 times harder than hardened tool steel. These have an excellent resistance to wear but are brittle because of their extreme hardness and thus should be handled very carefully. It has been reported that up to 30,000 serial sections of un-decalcified bone embedded in methyl-crylate can be obtained per sharpening of this blade.

e. Glass Knives

The cutting edge of the glass knives (also called Ralph knives with edges of 25 or 38 mm) used for conventional sectioning is parallel to one surface of the glass. But in the knives that are used for ultramicrotomy, this is positioned against/across the thickness of the glass. Different profiles of this knife can be produced very quickly for cutting the sections from different embedding media. The knife holders are available so that the knives can be used also with a rotary microtome. The knives are hard and brittle and require care when being handled. These deteriorate with storage due to the changes in the flow or strain of the glass after fracture and from oxidation impurities remaining in the hardened glass after manufacture. Thus, the knives should be prepared for immediate use. These can be classified according to the profiles of their cutting edges into: *high profile* (with 1.0 to 1.5 cm cutting edge for cutting sections from soft embedding medium like waxes), *medium profile* (with 0.5 to 1.0 cm high cutting edge for sections from soft and hard plastics), *low profile* (with less than 0.5 cm high cutting edge for cutting sections from hard plastics) and *front profile* (with the sloping edge artifact that makes the knife unsuitable for use, caused by inadequately securing the glass strip at each end).

f. Diamond Knives

These are manufactured from gem quality diamonds without flaws. Though these are very expensive, these are extremely durable due to their hardness. These are used primarily for cutting very thin sections.

g. Sapphire Knives

These knives are manufactured from one piece of solid sapphire artificially produced from an alumina monocrystal under computer-controlled thermal conditions. The sapphire is harder than tungsten carbide or glass and ensures high durability of the cutting edge for all types of material. The only restriction during use is the size of the block as the edge of the knife is limited to 11 mm. A special holder for the knife is required for this.

In general, the microtome knives are characterized by the profile of the knife blade, which may be planar concave, wedge shaped or chisel shaped designs. The *planar concave knives* are extremely sharp, but also very delicate. Hence, they are used with very soft samples. The *wedge shaped knives* are more stable and are used in moderately hard materials (epoxy or cryogenic sample cutting). The *chisel shaped knives* have blunt edges and are more stable and require more force to achieve the cut compared to the other two. The *glass knives* and *diamond knives* required for the ultramicrotome, the cut breadth of the blade is on the order of a few millimeters, and smaller than the knives of the classical microtomes. The glass knives are manufactured by the fracture of glass bars using special knife-maker fracturing devices. The glass knives have small troughs made with plastic tape and are filled with water to allow the sample to float for collection later. The diamond blades may also be built into such existing trough allowing for the same type of collection method.

The *disposable blades* are used for routine microtomy as well as cryotomy, as they provide a sharp cutting edge and produce flawless 2 – 4 mm thick sections. The blade holder is incorporated into the microtome or an adapter. The blade is commonly coated with PTFE (polytetrafluoroethylene) as it allows the ribbons to be sectioned with ease. The blade in the clamping device should not be over tightened as it may cause cutting artifact (thin and thick sections).

Many types of microtomes and microtome knives are commercially available in the market. The selection of which device to be used will depend on the type of tissue sample and the requirement that is needed. Understanding the microtomes will be useful for the selection which can minimize the damage or distortion to the tissue sample.

Sharpening of the Knives

The microtome knives can be sharpened either manually or automatically by *honing* (abrasive grinding of the facets) or *stropping* (polishing). *Honing* is generally done with natural slabs of stone possessing variable abrasive properties (like Belgian black vein, arkansas, aloxite and carborundum composites. The blades are often lubricated with soap water or light oil during use. The readily available and cheaper glass plates are used for hand sharpening. Copper and bronze plates which are expensive with superior properties are used in automatic sharpening machines used for honing. The common abrasives in use are alumina (aluminum oxide) jeweler`s rouge (iron oxide), silicon carbide and diamond. *Stropping* refers to the process of polishing an already fairly sharp edge. The best strops are made from the hide from the rump of horse (marked shell horse), and these are of two types, namely, flexible (hanging) and rigid.

Two basic designs are available in *automatic knife sharpeners*, in one type, the knife is held vertically with revolving sharpening wheels grinding the cutting edge of the knife, or the knife is held horizontally against the surface of a slowly rotating flat plate. The plates are made of glass, copper or bronze charged with abrasive. The glass plates need to be roughened before their use to allow the abrasive particle to be held more easily in place, while the copper and bronze plates are used in conjunction with diamond paste (6 micrometer particle size being more appropriate for rough sharpening, and 1 micrometer for fine polishing). The knife is laid on the near end of the strap with the cutting edge towards the operator (in opposite direction to that used in honing), and the knife is held with forefinger and thumb to facilitate easy rotation at the end of each stroke. The action is an exact opposite to that used in honing, using full length of the strap and stropping evenly the whole of the blade.

Method of Sectioning by a Microtome

The biological materials are usually placed in a more rigid fixative (embedding) prior to cutting by a microtome, and this is achieved by the inflow of a liquid substance around the sample (paraffin [wax] or epoxy) which is placed in a mold and later hardened to produce a block which is ready to be cut. If the blade of the knife is at right angles (declination is equal to 90° – the angle of contact between the sample vertical and knife blade), the cut is made directly using a pressure based mode, and the forces are proportionally larger. If the microtome knife is tilted, the relative motion of the knife is increasingly parallel to sample motion, allowing for a slicing action. This type of behavior is very important for larger or harder samples. For an optimal result, the inclination (the angle between the knife face and the sample) and the sample must be chosen appropriately. This depends upon the knife geometry, the cut speed and many other parameters. If the angle is adjusted to zero, the knife cut can become erratic, and a new location of knife must be used to smooth this out, and the angle is too large, the sample can crumple and the knife can induce periodic thickness variations in the cut, and finally by further increase in the angle so that it is too large, one can damage the knife blade.

MAINTENANCE OF A MICROTOME

The maintenance of a microtome includes:

1. Cleaning the Microtome

This can be done by wiping off the sectioning debris etc, with a paper towel. Alcohols (such as industrial methylated spirits, ethanol or methanol) may be used for cleaning the microtome. Xylene can also be used sparingly to remove the built-up wax, but excess xylene is wiped off immediately. Harsh solvents like acetone are to be avoided for cleaning the microtome.

2. Lubrication of the Microtome

After cleaning, both the sliding surfaces of the sledge base are to be oiled with microtome sledge oil. During lubrication, the sledge is moved several times over its full length of travel to distribute the oil all along its length.

3. Decontamination of the Microtome

It is the responsibility of the user to always ensure that a decontamination procedure is employed which is appropriate to the nature of the work carried out while using a microtome.

APPLICATIONS OF A MICROTOME

The most common applications of a microtome are:

1. Traditional Histology Technique

In this method, the tissues are hardened by replacing water with paraffin, and the tissue is then cut at thicknesses varying from 2 to 50 μm. The tissue is mounted on the microscopic slide, stained with appropriate aqueous dyes after prior removal of paraffin and examined using a light microscope.

2. Frozen Section Procedure

In this method, the water-rich sections are hardened by freezing and cut in this state with a freezing microtome (microtome cryostat), and the sections are stained and examined with a light microscope. This is much faster than traditional histology method (5 minutes x 16 hours) and is used in conjunction with medical procedures to achieve a quick diagnosis. The cryo-sections can also be used in histochemistry as freezing the tissues stops degradation of the tissue faster than using a fixative and does not alter or mask its chemical composition much.

3. Electron Microscopy Technique

In this method, after embedding the tissues in epoxy resin, a microtome (*ultramicrotome*) equipped with a glass or a gem-grade diamond knife is used to cut very thin sections (typically 60 to 100 nm), and these sections are stained with an aqueous solution of an appropriate heavy metal salt and examined with a transmission electron microscope. This is also used with its glass knife or an industrial grade diamond knife (generally gem quality diamond knife) to cut survey sections (generally 0.5 to 1 μm thick and are mounted on a glass slide and stained to locate areas of interest under a light microscope prior to thin sectioning for the TEM) prior to thin sectioning. The ultra-microtomes are found mounted inside a SEM chamber so that the surface of the block face can be imaged and then removed with the microtome to uncover the next surface for imaging. This method is called as SBFSEM (*Serial Block-Face Scanning Electron Microscopy*).

4. Botanical Microtomy

In this method, the hard materials such as wood, bone and leather require a sledge microtome and these have heavier blades and cannot cut as thin as a regular microtome.

5. Spectroscopy (FTIR or Infra-red Spectroscopy)

In this method, thin polymer sections are needed in order that the infra-red beam will penetrate the sample under examination and it is normal to cut the samples to between 20 and 100μm in thickness. *FTIR microscopy* can be used for sample inspection and for more detailed analysis of much smaller areas in a thin section.

6. Fluorescence Microscopy

In this method, the samples can be made into thin slices to be viewed under a fluorescent microscope.

A more recent development is the *laser microtome*, which cuts the target specimen with a *femtosecond laser* instead of a mechanical knife, and this method is contact-free and does not require sample preparation techniques. This also has the ability to slice almost every tissue in its native state, and depending on the material being processed, the slice thicknesses of 10 to 100 μm are feasible.

Conclusion

This chapter has been written to cover the use of the microtome and to give the reader the benefit of the experience of the research laboratory. The process of section cutting with the microtome depends on no single item. The formula for obtaining excellent sections is a precision microtome, a sharp knife, properly prepared specimens at the right temperature, and a skillful operator. Microtomy is an art which is learnt mainly from the experience, and the beginner who understands and profits from the mistakes is bound to become an expert technician with the device.

■■■

CHAPTER - 14

BIOSTATISTICS

Introduction

Statistics is both an art as well as science, and is a way of thinking about the data that is variable, and implies both data and statistical methods. This can neither prove nor disprove anything, and it is just a tool which without scientific application has no roots. Statistics may be defined as *the discipline concerned with the treatment of numerical data derived from a group of individuals, and these may be human beings, animals, plants or other organisms*. Biostatistics may be defined as a *branch of statistics applied to biological or medical sciences*. This branch covers applications and contributions not only from health, medicines and, nutrition but also from fields like genetics, biology, epidemiology and others. This mainly consists of various steps like generation of hypothesis, collection of data, and application of statistical analysis.

Biostatistics also called as Biometry is the application of statistics to a wide range of topics in biology, and it encompasses the design of the biological experiments (in medicine, pharmacy, agriculture and fisheries), the collection, the summarization, and analysis of data from the experiments, and the interpretation of and inference from the results. This is an applied branch of statistics and refers to the application of statistical principles to questions and problems in biology, medicine or public health. It also provides stochastic models and methods, algorithms and graphical tools for the analysis of data from genetics, bioinformatics and the medical, biological, agricultural and environmental health sciences. This includes methodology and models for data at the sub-cellular level (genetics, genomics, proteomics and bioinformatics), cellular level (cancer models, neuron models, cell distribution, cell kinetics and survival), tissue and/or organ level (pharmacokinetic and toxicological modeling, imaging techniques), individual level (clinical studies, life history, growth, aging and survival), and the population level (bio-mathematical modeling in ecology, epidemiological, demographic and public health studies) etc. The core problems and methodologies include the survival analysis, clinical trials, longitudinal studies, generalized linear models, dose-response and estimating equations, mixed and random effects modeling, and the analysis of molecular sequence data as well as the statistical methodology of bioinformatics.

History and Biostatistics

The major applications of biostatistics started in the middle of the 17th century in the analysis of vital statistics. After the early developments in vital statistics, the field of genetics was the next area that benefited most from the new statistical ideas emerging in the works of Charles Darwin (1809-1882), Francis Galton (1822-1910), Karl Pearson (1857-1936), and Ronald A. Fisher (1890-1962). Now, the fields of application and areas of concern of biostatistics include, among others, bioassay, demography, epidemiology, clinical trials, surveys of human populations, community diagnosis, bio-mathematical modeling, etc.

Biostatistical reasoning and modeling were of critical importance to the foundation theories of modern biology like the rediscovery of Mendel`s laws of inheritance (1900s) and Darwin`s theory of evolution which led to vigorous debate among the biometricians (Walter Weldon and Karl Pearson and Charles Davenport, William Bateson and Wilhelm, Johannsen respectively). The models built on statistical reasoning had helped to produce the neo-Darwinian theory of modern evolutionary synthesis. The leading figures in biology who relied entirely on statistics and developed its use were: Ronald Fisher (who developed many basic statistical methods in his field experiments at Rothamsted Research and in his book, *The Genetical Theory of Natural Selection* [1930]); Sewall G. Wright (who developed *F-Statistics* and computing methods) and J.B.S. Haldane (who reestablished natural selection as the primary mechanism of evolution and explained it in terms of the mathematical consequences of Mendelian genetics, in his book, *The Causes of Evolution*). The work of these scientists and the work of other statisticians, mathematical biologists and statistical geneticists helped to bring out evolutionary biology and genetics into a consistent, coherent whole. Associated with this development, the pioneering work of D`Arcy Thompson, (and his publication, *On Growth and Form*), helped to add quantitative discipline to biology. Despite the frequent necessity of statistical reasoning, there have been tendencies among biologists to distrust results which are not qualitatively apparent. The best example comes from Thomas Hunt Morgan banning the *Friden Calculator* from his department at Caltech.

Sir Francis Galton (1822 – 1911), a cousin of Charles Darwin, a 19th century English Victorian statistician, polymath, sociologist, psychologist, anthropologist, eugenicist, tropical explorer, geographer, inventor, meteorologist, proto-geneticist and psychometrician, considered as the *Father of Biostatistics*, was the founder of biometry and the origins of many statistical methods can be seen in his pioneering efforts, such as branching processes, the normal probability distribution, and regression. He was the first to apply statistical methods to the study of human differences and the inheritance of intelligence, and introduced the use of questionnaires and surveys for the first time for the collection of data on human communities, which he needed for his genealogical and biographical works and also for his anthropometric studies, and also created the statistical concept of correlation and widely promoted regression toward the mean. In 1866, the Royal Society conferred on him a Royal medal, in recognition of his statistical inquiries into biological phenomena. His first contribution to the Journal of Statistical Society was *The relative supplies from town and country families to the population of future generations* (1873), followed by *A common error in statistics* (1885), and *Application of the method of percentiles to Mr. Yule`s data on the distribution of pauperism* (1896).

Galton was able to place his researches about heredity on a scientific basis by applying novel statistical concepts, and can be claimed as one of the very first social scientists. This paved the way for the development of statistics as a discipline, through Galton's follower Karl Pearson. It also laid the foundation for the study of differential psychology: through Charles Spearman, who was heavily influenced by Galton and made important contributions of his own to the development of factor analysis; and through Cyril Burt and his pupils, who moulded the study of intelligence and hereditary ability into a formidable discipline. There are few aspects of modern social science that do not (or at least, should not) rely on the statistical innovations that Galton introduced. Galton first discovered *regression*, which he initially called *reversion*, by experimenting with sweet peas. His sweet peas produced seeds with a (normal) variation of sizes that regressed from the distribution of their parents. Some years later, he formulated the statistical correlation coefficient by another indirect route, painstakingly graphing and re-graphing his data about bivariate normal distributions until he realized that the formulae for elliptical curves (a topic popular in 19th century mathematics but almost entirely moribund today) could provide him with a method for summarizing with a number the graphical relationships he saw. This number could then be used to reason about the relationship, and form a basis for comparisons. Galton has left an excellent description of his *discovery of correlation and regression*, in an article he published for the North American Review in 1890, *Kinship and Correlation*. His innovations in statistics include: historiometry, the Lexical hypothesis, the questionnaire, variance and standard deviation, experimental derivation of the normal distribution, bivariate normal distribution, correlation and regression, regression to mediocrity etc.

Halbert L. Dunn (1896 – 1975) known as the *Father of the wellness movement, established* a national vital statistics system in the USA. In 1929, he was the first biostatistician hired by the Mayo Clinic and established its computer coding system for deriving medical statistics. In 1929, he published a huge paper on the development of statistics in medical research in a physiology journal. Sir Austin Bradford Hill (1897 – 1991), an English epidemiologist and a statistician, pioneered randomized trial and is widely known for his *Bradford Hill criteria* also known as *Hill`s criteria for causation*, for determining a causal association. In 1937 he published about 17 articles on statistical methods (*The Lancet*, 1937) as chapters of the book, *The Principles of Medical Statistics*, which became renowned worldwide amongst physicians, epidemiologists and medical statisticians. He launched randomized control trials (RCT) in preventive medicine to test a pertusis (whooping cough) vaccine, and the other in clinical medicine, to assess the effect of streptomycin in the treatment of pulmonary tuberculosis (TB) and these trials were followed by a stream of articles published in 1948, and the application of statistics in medicine grew eight-fold from 1952 to 1982.

Scope of Biostatistics

The biostatisticians are trained in the skilled application of statistical methods to the solution of problems that are encountered in medicine and public health. As such they collaborate with the scientists in every area related to health and have made major contributions to the understanding of AIDS, cancer, genetics, bioinformatics and immunology etc, and they spent a considerable amount of their time in development and evaluation of the statistical methodology in the specific projects they are associated with.

PRINCIPLES OF EXPERIMENTAL DESIGN AND STATISTICAL INFERENCE

The design of an experiment can be compared to the plans of the architect for a structure, whether it is a giant skyscraper or a modest home. The basic requirements for the structure are given to the architect by the owner of the structure. It is the task of the architect to fill the basic requirements, and he has ample scope for exercising his ingenuity, and several plans may be drawn up to meet all the basic requirements; some plans may be very costly than others, given two plans having the same cost, one may offer potential advantages that the second does not.

In the design of an experiment, the designer has the role of the architect, and the experimenter has the role of the owner, and these two roles are not necessarily mutually exclusive, as the experimenter may do considerable portion of the design work. The basic requirements and the primary objectives of the experiment are formulated by the experimenter, and he may or may not be aware of the possible alternative approaches that can be followed in the conduct of the experiment. It is the function of the designer to make the experimenter aware of these alternatives and to indicate the potential advantages as well as the disadvantages of each of the alternative approaches. It is the task of the experimenter to reach the final decision about the conduct of the experiment.

The individual qualified to design an experiment is the one who is – most familiar with the nature of the experimental material–most familiar with the possible alternative methods for designing the experiment and – most capable of evaluating the potential advantages and disadvantages of the alternatives. The roles of the experimenter and designer are one where the individual possesses all these qualifications. In many experimental fields, on some research problems the experimenter is capable of making all the necessary decisions without seeking extensive assistance. The experimenter may turn to his experienced colleagues on more complex research problems who are equally or more familiar with the subject matter for assisting in formulating the basic requirements and primary objectives of his/her experiment.

Planning an experiment to obtain appropriate data and drawing inference out of the data with respect to any problem under investigation is called the *design and analysis of experiment*. This may range from the formulations of the objectives of the experiment in clear terms to the final stage of the drafting reports incorporating the important findings of the enquiry. The structuring of the dependent and independent variables, the choice of the their levels in the experiment, the type of experimental material to be used, the method of manipulation of the variables on the experimental material, the method of recording and tabulation of the data, the mode of analysis of the material, the method of drawing sound and valid reference etc, are all intermediary details that go with the design and analysis of an experiment.

Most empirical studies fall into two groups, namely, *experimental studies* and *observational studies*. In an experiment, a researcher manipulates a variable (imposes some treatment) to study its effect. The three basic principles of experimental design are *randomization* (to ensure groups being compared are similar apart from the treatment), *replication* (to minimize the variation that results from taking the samples) and *local control* (to control and compare, to study the difference with and without treatment), and these make a valid test of significance possible. In some cases, a matched pair/blocked design controls the likely sources of variation in the data. These three principles are complementary to each other in trying to increase the accuracy of the experiment and to provide a valid test of significance, retaining at the same time the distinctive features of their roles in any experiment.

Before the conduct of the experiment, an *experimental unit* is to be defined. For example, a leaf, a tree or a collection of adjacent trees may be an experimental unit. This is also sometimes referred to as a *plot*. A collection of plots is termed a *block*. The observations made on experimental units vary considerably and these variations are partly produced by the manipulation of certain variables of interest called *treatments*, built in and manipulated deliberately in the experiment to study their influences. For example, clones in clonal trials, levels and kinds of fertilizers in fertilizer trials etc, can be called as treatments. Besides the variations produced in the observations due to these known sources, the variations are also produced by a large number of unknown sources like uncontrolled variation in extraneous factors related to the environment, genetic variations in the experimental material because of their undesirable influences are called *experimental error*, meaning not an arithmetical error but variation produced by a set of unknown factors beyond the control of the experimenter. Such errors may be *systematic or random* in their mode of incidence. The errors arising due to an equipment (like a spring balance that goes out of calibration due to continued use or due to the observer`s fatigue) are examples of *systematic error*. The unpredictable variation in the amount of leaves collected in litter traps under a particular treatment in a related experiment is a *random error*. It is clear that any number of repeated measurements would not overcome systematic error while the random errors would cancel out with repeated measurements. The three basic principles namely, randomization, replication and local control are the devices to avoid the systematic error and to control the random error.

1. Randomization

This is the first principle of an experimental design, which is a random process of assigning treatment to the experimental units. This is defined as the random allocation of experimental units to experimental treatments, and to allocate at random, each unit must have had the same chance of receiving any of the possible treatments. The process implies that every possible allotment of treatments has the same probability. The experimental unit is the smallest unit (division) of the experimental material and a treatment means an experimental condition whose effect is to be measured and compared. The main purpose of this process is to remove bias and other sources of extraneous variation, which are not controllable. Another advantage of this process is that it forms the basis of any valid statistical test. As such the treatments must be assigned at random to the experimental units. This is usually done by drawing numbered cards from a well shuffled pack of cards, or by drawing numbered balls from a well shaken container or by using tables of random numbers. For example, to randomize 20 units between 2 experimental groups: assign each experimental unit a unique number from 1 to 20; use a random numbers table or computer software, for example, the function RAND () in MS EXCEL to also assign a random number to each unit; and sort the data in order of the random number; and the first 10 units in the ordered list are allocated to group 1 and the remaining 10 units to group 2.

2. Replication

This refers to the repetition of the basic experiment, or it is a complete run for all the treatments to be tested in the experiment. In all the experiments, some variation is introduced because of the fact that the experimental units (individuals or plots of land in agricultural experiments) cannot be physically identical. This variation can be removed by the use of a number of experimental units. Hence, the experiment is to be performed more than once (basic experiment is repeated). An individual repetition is called a replicate, and the number, the shape and the size of replicates depend upon the nature of the experimental material. A replication is normally used: to secure more accurate estimate of the experimental error, a term which represents the differences that would be observed if the same treatments were applied several times to the same experimental units; to decrease the experimental error and thereby to increase precision, which is a measure of the variability of the experimental error; and to obtain more precious estimate of the mean effect of a treatment (since $\sigma2/y=\sigma2/n$ where n denotes the number of replications).

3. Local Control

It was a general observation that all the extraneous sources of variation are not removed by randomization and replication, and this necessitates a refinement in the experimental technique. It is necessary that one needs to choose a design in such a manner that all extraneous sources of variation are brought under control. For this purpose, it is always better to make use of local control, a term which refers to the amount of balancing, blocking and grouping of the experimental units. Balancing means that the treatments should be assigned to the experimental units in such a way that the result is a balanced arrangement of the treatments. Blocking means that the experimental units should be collected together to form a relatively homogeneous group. A block is also a replicate. The main purpose of the principle of local control is to increase the efficiency of an experimental design by decreasing the experimental error. The main point to remember is that the term local control should not be confused with the word control, the word in experimental design is used for a treatment, which does not receive any treatment but one should need to find out the effectiveness of other treatments through comparison.

For example, a chemist wants to obtain the calibration curve of the ability of an instrument to detect the concentration of a chemical in a sample. He will test the response of the instrument for a series of samples with known concentrations of a chemical in a random order. A well-designed experiment can often be analyzed using simple statistical methods (a t-test) and if randomization and control has been sufficient then there is no need to worry about any additional variables affecting the results. Therefore, one can attribute any change in the response to be due to the experimental treatment. If we have sufficient replication, we should be able to detect an effect if it exists.

Statistical Inference and Analysis

In an observational study, we are only investigating associations, which we cannot assume are causal relationships because of possible confounding variables, and we need to investigate the effects of possible confounding variables by analysis of data, for example, multiple regression, to infer the factors likely to be causing the patterns in the data.

Internal versus External Validity

The well designed experiments are less biased than the observational studies because they eliminate the effects of additional variables in the data and allow the estimation of the true effect in the context being studied. Because of this, the experiments can be considered to have a high internal validity, and by contrast, the external validity of a study is determined by whether its results can be generalized to other contexts of interest. For example, if we are interested in the effects of increasing air temperature due to climate change on plant growth in forests, we might perform an observational study and collect data on air temperature and forest growth from a sample of plants mapped to unique locations in a forest.

TABULAR DESIGN

Tabular presentation represents the culmination of a long series of survey operations and the presentation can reflect only all the results of the previous stages, and this makes the sound presentation vitally important. Good presentation transmits the quality of all the previous operations and also assists in making the clear meaning of the material. Poor presentation not only fails to clarify the data and it tends to destroy or impair the effect of the work that has been done. In the design and the preparation of the statistical tables the basic objective is to arrange and present the data in such a way that their meaning and significance can readily be grasped by the user. The statistical table is an efficient and ideal way of presenting the data and its usefulness increases with multiplicity of classification. The table design should be kept as simple as possible taking into account the subject matter and the available space. The dividing line between the good and the bad presentation may be best determined by the answer to the question: "*which method is simple, focuses the attention of the reader on the data* (rather than on the form), *and makes clear the meaning and the significance of the information*?". The table which clarifies the meaning of statistics and concentrates on the attention upon them is *good presentation*; and that which obscures the meaning of the data or distracts the attention from them is *poor presentation.*

Tables are commonly used in collecting and organizing raw data during an experiment, and also for representing final data to be included in a research paper or report. Most of the raw data is recorded in tabular form in a spreadsheet, or a laboratory note book, or a laboratory manual; but once recorded, the data need to be reorganized, summarized and reshaped into a final table or a graph. In some cases, the laboratory experiment will require sketches from observation and may or may not require a table to go along with them. The representation of data in a table is referred to as tabular representation, and this allows the data to be organized for further analysis, and allows large amounts of raw data to be sorted and reorganized in a net format, and allows the inclusion of the most important data which is relevant. It also facilitates a dialogue between the text and the exact numbers in your results, so that you don`t have to describe all the specific numerical values in your report. One should never put data in a table if it can be described efficiently in one or two sentences. The tabular representation lets the researcher to place his/her results in an organized display of rows and columns that enable him/her to group his/her data by different classifications so that he/she can make comparisons and better understand the data. When using a table as a final representation of data to communicate the results in the report, the specific data values should be listed, and comparisons between variables should be drawn by listing the subtotals, totals, averages, percentages, frequencies, statistical results etc. The tables are not the choice when one wants to show a trend or relationship between the variables. These can be best represented by the graphs. Good tables should be easy to read across rows and down columns, easy to understand, and easy to refer to in the text. They should always include only the relevant data from the results.

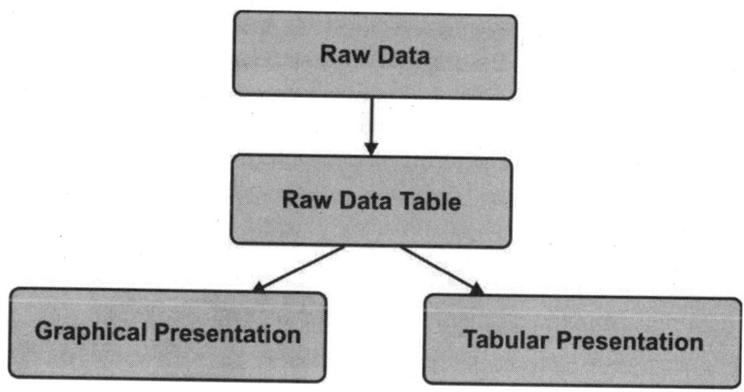

Parts of a Table

1. Title of the Table

The title provides a brief description of the contents of the table. It should be concise and include the key elements that are shown in the table (groups, classifications, variables etc.). It should never be more than two lines. Most titles should be underlined or italicized, and thc first letter of each word should be capitalized following the rules for any title or the entire title can be in capitals. The periods are left out at the end of the title. If the title is long (more than two lines), the lines can be either single-spaced or double-spaced according to the style one uses. The title (often referred to as title legend) should always go above the table.

2. Table Number

The tables should necessarily be numbered in the order that they are referred to in the report (Table 1, Table 2, Table 3 and so on). The number has a period at the end and a space to separate it from the title, which normally follows the number.

3. Headings and Subheadings

The headings and subheadings allow one to establish an order to the data by identifying the columns while the data itself forms the body of a table. They should be written singular in form unless they refer to groups (men, women etc.) and the first letter of the first word should be capitalized. The headings should be key words that best describe the columns beneath them. They should not be longer than the longest entry in their columns.

a. Column Headings

Each column has a heading in order to identify what data are listed below in a vertical arrangement. When this heading is above the leftmost column, it is often referred to as the *stub head*, and the column is the *stub column*. The column usually lists the independent variable. The data that follow the stub column are known as the *stub*. All other column headings are simply referred to as *column heads*. The units must be specified in column headings when applicable.

b. Column Spanner

This refers to a heading that sits above two or more columns to indicate a certain classification or grouping of the data in those columns. A *column spanner* may also specify units, when appropriate.

4. Table Body

The actual data in a table that occupies the columns such as percentages, frequencies, statistical test results, means, `N` [number of samples) etc, constitutes the table body.

5. Table Spanner

This is located in the body of the table in order to divide that data without changing the columns. Spanners go to the entire length of the table and are often used to combine two tables into one in order to avoid repetition. This may be written in plural form.

6. Dividers

These are the lines that frame the top and bottom of the table and/or mark the different parts of a table. They are often used for division or emphasis within the body.

7. Table Notes

The table notes is used to explain anything in the table that is not self-explanatory. The basic symbols and abbreviations such as SD (standard deviation), N (sample size) and % (percentage) are commonly used. For explanation of the other technical terms or other issues, an *asterisk* (*) is placed for the first note after the specific data value. The asterisk would be placed below the table followed by the note or explanation required for that value. Other data values requiring notation would get two, three asterisks, or a *stacked cross* in that order. Notes following these additional items would follow the first note using the same format, and the notes that apply to the table in general should be listed after the word `*Note:*` under the table.

Tabular presentation of the Raw Data

When collecting the data during a laboratory experiment, it is very important that one must record it in a tabular format in a spreadsheet or a laboratory manual or laboratory note book or word processing software. The raw data table should include all the data collected during an experiment as well as all necessary calculations. The final version of the data presentation is a table or a figure, the initial tabulation will make it easy for one to read and interpret the data. The first thing the table needs is a title. Make sure that the title is descriptive of the data that one is going to collect. There should be a place for the data and the name of the recorder(s). In the laboratories dealing with the multiple variables, the table may have both the headings and subheadings. The columns should be titled, with the name of the variables followed by the units of measure in parentheses, and extra columns should be made to allow room for the observations, calculations, and notes. Each extra column must be labeled appropriately and the independent variable(s) is recorded in the first column, and the dependent variable(s) is recorded in the subsequent column(s). The number of observations taken during the course of the work determine the number of rows.

Tabular Presentation of Data

When the recorder is ready to include all the results in the report, he/she may decide that the best representation for the data is a tabular presentation. One may be working with the raw data that he/she collected himself/herself during an experiment, and he/she may want to revise it so that it includes only the relevant data and so that it is organized in a proper fashion. One may also be working with longer lists of raw data collected by other people or given to him/her by his/her laboratory instructor. In such a case, it is up to him/her to sort through all the data and find a suitable way to best represent it all in a tabular form, and to accomplish this task, he/she must be familiar with the basic rules for tabular presentation.

The basic rules for the tabular presentation are as follows: always limit the table to data that is relevant to the hypotheses in the experiment; be certain that the table can stand alone without any explanation; make sure that the table is supplementary to the text and does not replicate it; refer to all the tables by numbers in the text (for examples, Table 1, 2, 3, and so on); discuss only the highlights of the table in the text; always give the units of measurement in table headings; align the decimal places if any; round the numbers as far as possible, and try to round them off to two decimal places unless more decimals are required; unless using a specific format style that requires to place the tables separately at the end of the report, then place the tables near the text that refers to them; decide on a reasonable amount of data to be represented, but too little so that the reader does not understand the results, but not too much so that the reader gets overwhelmed and confused; it is always better to include only the necessary number of tables in the report not to confuse the reader; do not use the tables if there are two or fewer columns and rows, and replace it by means of a textual description; organize the tables neatly so that the meaning of the table is obvious at first glance, and if the reader spends too much time deciphering the table, then it is too complicated and not efficient; remember that too many rows or columns could make it difficult for the reader to understand the data, and one may need to reduce the amount of the data or separate the data into more number of additional tables; if there are identical columns or rows of data in two or more tables, it is always better to combine the tables; it is better to provide column/row totals or other numerical summaries that can make it easier to understand the data; and one should be consistent with his/her tabular presentation and he/she must always use consistent table, title and heading formats.

Types of Tables

The tables are of three types, namely, word tables, statistical tables and numerical tables.

A. Textual (Word) Tables

Often one may need tables that have textual data in the body, and this is the case when dealing with qualitative data. These serve the same function as any table to make comparisons of items easy. These are also used when one wants to present examples, which may be grouped in a certain way to show categories of different items.

B. Statistical Tables

These tables present descriptive or inferential statistics or both. The former are tabulations such as mean, standard deviation, mode, range, or frequency, while the latter refer to statistical tests, and in such tables, the test values are presented.

C. Numerical Tables

These refer to the most common types of data which typically represent quantitative data, but sometimes may represent a combination of quantitative and qualitative data, and the body of the table mostly consists of specific number values.

Best Practice of Presenting the Table

In tabular presentation, the best practice refers to designing the tables that can be read easily and quickly. This can be accomplished by manipulating *contrast, alignment, spacing* and *ordering*. All these elements help to achieve clarity so that the reader can pick out specific data and also understand the discussion of the results achieved.

a. Contrast

One can group or distinguish the data from each other by making the key elements of the table stand out from one another. The title, dividers and/or the headings can be represented in bold font. Different sizes, styles or letter cases of the font may also be used for the different elements of the table. The colour can also be used to emphasize the background or the text. If one chooses to create the contrast, it should be remembered that less is more to create an effective table.

b. Alignment

This is important for keeping the table neat and clear. All the numbers in the columns should be lined up with each other and also with their headings. The table is to be structured so that all the elements seem to be properly aligned with each other – titles, headings, data, dividers, and notes.

c. Ordering

The similar items are to be necessarily grouped to give a sense of structure and meaning to the table. This also helps to break up the data to make easier to the eye. Another way is to indent subordinate data when it falls below specific column data.

d. Spacing

The white areas around the table are to be manipulated which can help clarify and organize the table. It is always better to have enough space around and between the texts so that it stands out prominently, and the space can also be used to separate the groups or to emphasize them.

GRAPHICAL DESIGN

Statistics is a subject that deals with usually large numerical data, and the data can be represented graphically, and such representation is an essential step during statistical analysis.

Statistical surveys and experiments provide valuable information about numerical scores and for better understanding and to make conclusions as well as interpretations; the data should be managed and organized in a systematic form. A graph is the representation of data by using graphical symbols such as lines, bars, pie slices, dots etc. The various types of graphical representations are as follows:

1. Dot Plots

This is one of the simplest ways of graphical representation of the statistical data, and as the name itself suggests, a dot plot uses the dots. It is a graphic display which usually compares the frequency within different categories, and the dots are plotted on a graph paper. In the dot lot, every dot denotes a specific number of observations belonging to a data set and one dot usually represents one observation. These dots are to be marked in the form of a column for each category, and in this way, the height of each column shows the corresponding frequency of some category. The dot plots are quite useful when small amounts of data is given within the small number of categories.

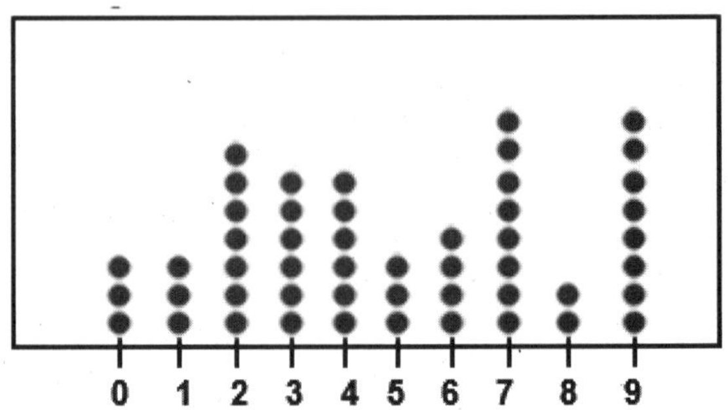

Dot Plot of Random Values

2. Bar Graph

A bar graph is very frequently used graph in statistics. This is a type of graph which contains rectangles or rectangular bars. The lengths of these bars should be proportional to the numerical values represented by them. In the bar graph, the bars may be plotted either horizontally or vertically, but a vertical bar graph (also called as column bar graph) is used more than a horizontal bar graph. The rectangular bars are separated by some distance in order to distinguish them from one another. The bar graph shows comparison among the given categories. in most of the cases, the horizontal axis of the graph represents specific categories and vertical axis shows the discrete numerical values

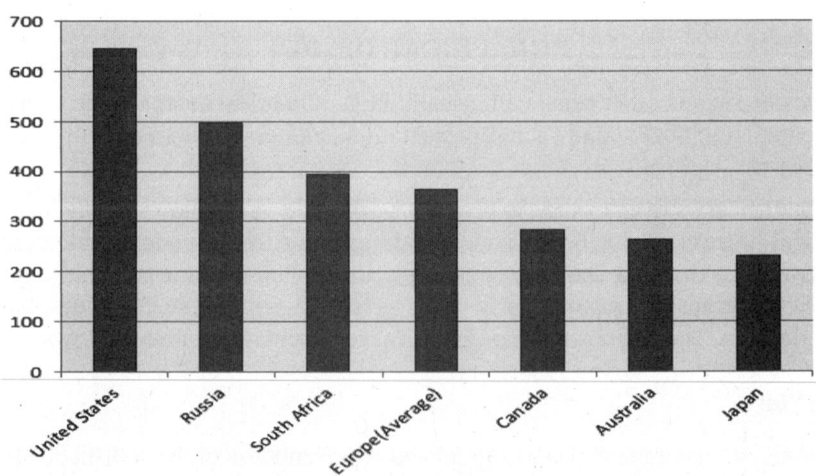

A vertical bar graph representing the number of students who went to different states in different countries for their study.

3. Line Graph

This is a kind of graph which represents data in a way that a series of points are to be connected by the segments of straight lines. In a line graph, the data points are plotted on a graph and they are joined together with straight line. The line graphs are generally used in science, statistics and media, and they are easy to create. These graphs are quite popular in comparison with other graphs since they visualize characteristics revealing data trends very clearly, and also a line graph gives a clear visual comparison between two variables which are represented on X-axis and Y-axis.

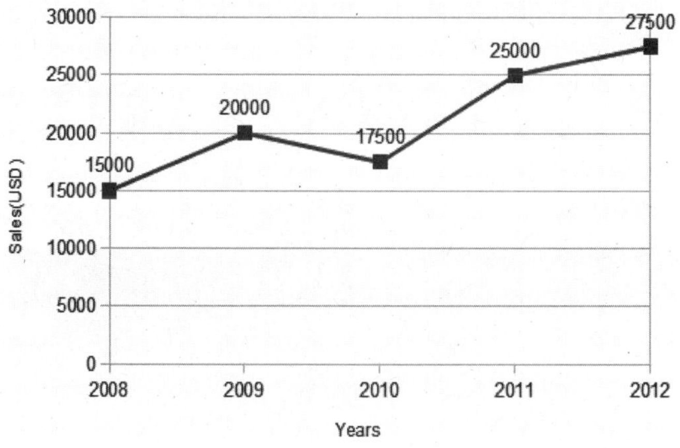

A sample line graph

4. Circle Graph

This graph is also known as a *pie graph* or *pie chart*, and it is so called since it is similar to the slice of a pie. This is defined as a graph which contains a circle which is divided into many sectors, which illustrate the numerical proportion of the data. The arc lengths of the sectors in a pie chart are proportional to the numerical value they represent. The circle graphs are quite common in mass media as well as the business world.

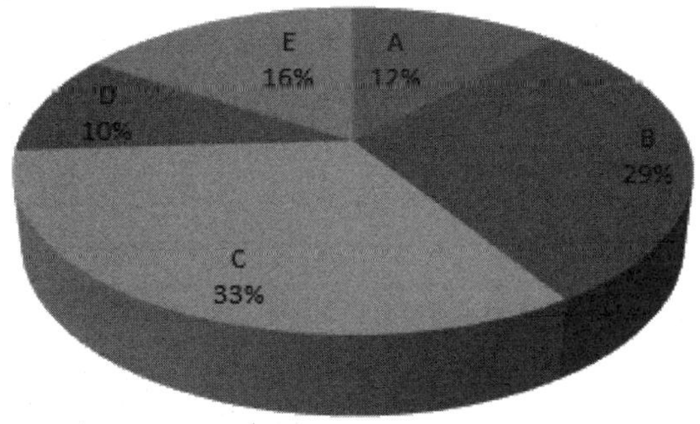

A sample of a Pie Chart

5. Histogram and Frequency Polygon

Both of these representations are very common graphs in statistics. A *histogram* is defined as a graphical representation of the mutually exclusive events, and this is similar to the bar graph. Both are made up of rectangular bars, but the difference is that there is no gap between any two bars in the histogram. This is normally used to represent the continuous data. The *frequency polygon* is a type of graphical representation which gives us better understanding of the shape of the given distribution, and these serve almost the similar purpose as histograms do. This is quite helpful for the purpose of comparing two or more sets of data. These are said to be the extensions of the histograms. When the midpoints of tops of the rectangular bars are joined together, the frequency polygon is generated.

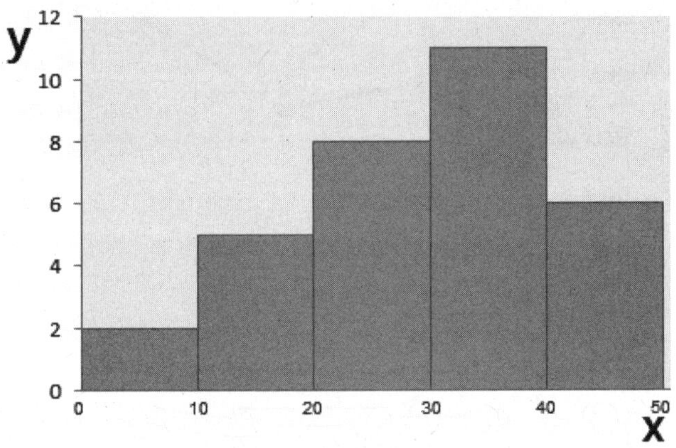

The appearance of a Histogram

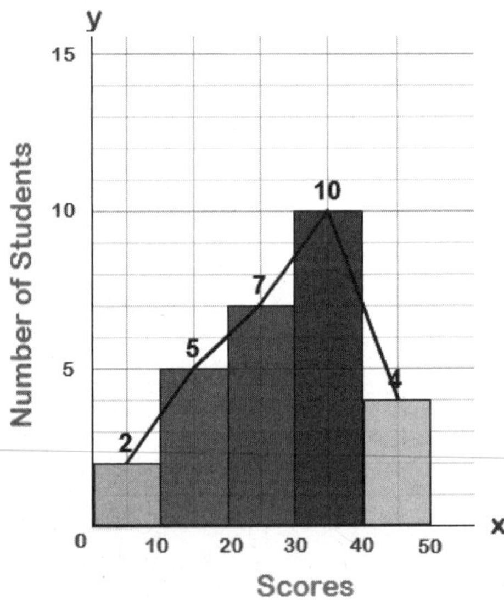

The appearance of a Frequency Polygon

DIAGRAMMATIC REPRESENTATION OF DATA

Although tabulation is very good technique to present the data in statistics, the diagrams are an advanced technique to represent the data. As a layman cannot understand the tabulated data easily, with only a single glance at the diagram, one gets a complete picture of the data presented, because the `*diagrams register a meaningful impression almost before we think*` (M.J. Moroney).

Utility or Importance of Diagrams

The utility of diagrams is manifold such as:

1. They give a very clear picture of the data and even a layman can understand it very easily and within a short time;
2. The comparison should be made between different samples very easily and there is no need to use any statistical technique to compare;
3. The technique can be used universally at any place and at any time, almost in all the subjects and different fields of study;
4. The diagrams have much impressive value and even a common man is impressed easily by good diagrams;
5. The technique can be used for numerical type of statistical analysis (to locate mean, median or other statistical values);
6. This type of representation not only saves time and energy but also is economical, and not much money is needed to prepare good diagrams;
7. The diagrams give us more information compared to tabulation (the technique of tabulation has its own limits);
8. The diagrammatic data is easy to remember as the diagrams leave their lasting impression much more than other data techniques; and
9. The data can be condensed with the diagrams and even a simple diagram can present what cannot be presented by even 10,000 words.

Diagrammatic Representation – General Guidelines

The general guidelines to be followed in presenting the diagrams are as follows:

1. The diagram should be drawn properly, and the pith and the substance of the subject matter must be made clear under a broad heading which properly conveys the main purpose of the diagram;
2. The size and the scale of the diagram should neither be too big (long) nor too small (short), the former will look ugly while the latter may not convey the full meaning, and in each diagram, the size of the paper must be taken note of as it will help to determine the size of the diagram;
3. Some notes should be added at the foot of the diagram for the clarification of certain ambiguities and this shall provide the visual insight of the diagram;

4. The diagrams should be always neat and clean, and there should be no vagueness or overwriting on the diagram;
5. Simplicity is preferred, this means that the diagram should convey the meaning clearly and easily;
6. Scale must be presented always with the diagram;
7. The diagram must be self-explanatory, it must indicate nature, place and the source of data presented therein;
8. To make the diagrams more easily understandable, different shades and colours can be used;
9. Always vertical diagrams should be preferred to horizontal diagrams; and
10. The diagrams must be accurate, and accuracy must not be done away with to make it more attractive or more impressive.

Diagrammatic Representation – Limitations

The limitations to be followed in the representation of the diagrams are as follows:

1. The diagrams do not present the small differences in a proper manner;
2. The diagrams can be easily misused;
3. Multi-dimensional diagrams can only be drawn by an artist;
4. Diagrams are of no use in statistical analysis;
5. The diagrams are just a supplement to tabulation;
6. Only a limited set of data can be presented in the form of a diagram;
7. The diagrammatic presentation of the data takes more time than tabulation;
8. Only preliminary conclusions can be presented through diagrams; and
9. The diagrammatic presentation of the data shows only an estimate of the actual behavior of the variables.

Diagrams – Types

a. Line Diagrams

In these, only a line is drawn to represent one variable, and these may be vertical or horizontal. The lines are drawn in such a way that their length is the proportion to the value of the terms or items so that comparison may be done very easily.

b. Simple Bar Diagrams

These figures are used where only single dimension (length can present the data). The procedure is almost the same, only one thickness of lines is measured, and these can also be drawn either vertically or horizontally. The breadth of the lines or bars should be equal, and similarly, the distance between these bars should be also equal.

The breadth and the distance between them should be taken according to the space available on the paper.

c. Multiple Bar Diagrams

These are used when one wants to make a comparison between more than two variables. The number of variables may be 2, 3, 4 or more, and in case of 2 variables, a pair of bars is drawn, in case of 3 variables, triple bars are drawn etc. The bars are drawn on the same proportionate basis as in the case of simple bars, and the same shade is given to the same item, but the distance between pairs is always kept constant.

d. Component Bar Diagrams / Sub-diagrams

These charts are very useful in comparing the sizes of different component parts (the elements or parts which a thing is made up of) and also for throwing light on the relationship among these integral parts. For example, the sale proceeds from different products, expenditure pattern in a typical Indian family (the components being food, rent, medicine, education, power etc.), budget outlay for receipts and expenditures, components of labour force, population etc. These diagrams are usually shaded or colured suitably.

e. Sub-divided Bar Diagrams

The data which is presented by the multiple bar diagram can be presented by the sub-divided bar diagram. The different variables are added for a period and drawn on a single bar. This diagram is more efficient if the number of components is less (3 to 5).

f. Percentage Bar Diagrams

In this case also, the data of one particular period or variable is put on a single bar, but in terms of percentages. The components are always kept in the same order in each bar for an easy comparison.

g. Duo-directional Bar Diagrams

The diagram is on the sides of the base line (to left and right or above and below sides) in this case.

h. Broken Bar Diagrams

This is used when the value of some variable is very high or low as compared to others. The bars with bigger terms or items may be shown broken in this case.

The diagrammatic presentation of the data provides the quickest understanding of the actual situation to be explained by data in comparison to tabular or textual presentation. The presentation of the data by this method translates quite effectively the highly abstract ideas contained in numbers into more concrete and easily comprehensible form. Though the diagrams are less accurate, they are much more than the tables in presenting the data. There are various types of diagrams in common use and amongst them the more important ones are as follows:

A. Geometrical Diagram

The bar diagram and pie diagram come into this category for the presentation of data. The bar diagrams are of three types, namely, simple, multiple and component bar diagrams.

B. Frequency Diagram

The data in the form of grouped frequency distributions are generally represented by frequency diagrams like histogram, frequency polygon, frequency curve (cumulative frequency curve which may be less than or more than curves).

C. Arithmetic Line Graph

This is also called time series graph and is a method of diagrammatic representation of data. In this, time (hour, day/date, week, month, year etc.) is plotted along x-axis and the value of the variable (time series data) along y-axis. A line graph obtained by joining these plotted points, is called arithmetic line graph (time series graph), and it helps in understanding the trend, periodicity etc, in a long term time series data.

Now, one can make an appropriate choice of the form of data presentation as well as the type of diagram to be used for a given set of data, and can make the presentation of data meaningful, comprehensive and purposeful. When classification is done according to qualitative characteristics like social status, physical status, nationality etc, it is called qualitative classification.

Statistical Method - Types

This has two major branches, namely, *descriptive* and *inferential.*

1. Descriptive Statistics

This explains the distribution of population measurements by providing the types of data, estimates of central tendency (like mean, mode and median), and measures of variability (like standard deviation, correlation coefficient).

2. Inferential Statistics

This is used to express the level of certainty about the estimates and includes hypothesis testing standard error mean, and confidence interval.

Data (Information) – Types

The data constitutes the information or observations recorded during the research. The methods of analysis mainly depend on the type of data collected. In general, the data shows a picture of the variability and central tendency; as such the understanding of the types of data is very important. The data is of three types, namely, *nominal data*, *ordinal data* and *interval data.*

a. Nominal Data or Categorical Data

This is synonymous with the categorical data where the data is simply assigned *names* (categories) based on the presence or absence of certain *attributes* (characteristics without any ranking between the categories). For example, the patients can be categorized by their *gender* such as males or females; by their *religion* such as Hindu, Muslim or Christian. The nominal data also includes *binomial data* which refers to two possible outcomes. For example, the outcome of cancer may either be death or survival, the drug therapy with a drug (X) will show improvement or no improvement at all.

b. Ordinal Data or Ordered Data or Graded Data

This type of data is generally expressed as *scores* (*ranks*). The different categories show a natural order amongst them, and they can be ranked (arranged) in an order. For example, the pain may be classified as mild, moderate and severe (as there is an order between these three grades of pain, this is called as *ordinal*), and to indicate the intensity of pain, it may also be expressed as scores (mild = 1, moderate = 2 and severe = 3), and the data can be arranged in an *order* (*rank*).

c. Interval Data

This is characterized by an equal and definite interval between the two measurements. For example, the weight is expressed as 10, 11, 12, 13 or 14 kg, where the interval between 10 and 11 is same as that between 13 and 14. This can be either *continuous* or *discrete*. A *continuous variable* can take any value within a given range. For example, the levels of haemoglobin may be taken as 10.3, 11.5 or 12.4 gm% while a *discrete variable* is usually assigned integer values (which do not have fractional values). For example, the values of blood pressure are in general discrete variables or the number of cigarettes smoked per day by a person.

Sometimes, certain amount of data may be converted from one form to another to reduce skewness (sleekness) and make it to follow the normal distribution. For example, the doses of drug are converted to their log values and plotted in dose response curve to obtain a straight line so that the analysis becomes easy. The data can also be transformed by taking the logarithm, square root or reciprocal; and logarithmic conversion is the most common data transformation used in medical research.

Central Tendencies – Measures

The three measures of central tendencies are *mean*, *median* and *mode*. *Mean* is the common measure most widely used in the calculations of averages, and is least affected by sampling fluctuations. The mean of a number of individual values (X) is always nearer to the true value of the individual value itself, and mean shows less variation than that of individual values, as such they give confidence in using them. This is calculated by summing up the individual values (Σx) and dividing the sum by the number of items (n). Suppose the height of 6 children is 60, 70, 80, 80, 90 and 100 cm, and the addition of height of 6 children is 490 cm, the mean (X) is $\Sigma x/n$ = 490/6 = 80.

Median is an average which can be obtained by getting the middle values of a set of data arranged (ordered) from the lowest to the highest (or *vice versa*). In this process, 50% of the population has the value smaller than and 50% of the samples will have the larger value than the median, and it is used for the ranks (scores). Median is a better indicator of the central value when one or more of the lowest or the highest observations are wide apart or are not evenly distributed. Median in case of an even number (4, 6, 8 or 10) of observations is taken arbitrarily as an average of two middle values, and in case of odd number 5, 7, 9 or 11) of observations, the central value forms the median.

Mode is the most frequent value, or is the point of maximum concentration. The most fashionable number which occurs repeatedly contributes mode in a distribution of quantitative data (80 in the above example). The mode is used when the values are widely varying and is rarely used in medical studies. The mode and the median are useful for skewed distribution samples when there is wide variation.

Even after calculating the mean, it is necessary to have some index of variability among the data. The range or the lowest and the highest values can be given, but this is not very useful if one of these extremes is far off from the rest. At the same time, it does not tell how the observations are scattered around the mean. Hence, following the indices of variability (*standard deviation* and *correlation coefficient*) play a key role in biostatistics.

Averages

In statistics, an *average* (the mean – the average value) is defined as the number that measures the central tendency of a given set of numbers. There are a number of different averages including but not limited to *mean, median, mode* and *range*. The *mean* is the average you are used to, where you add up all the numbers and then divide by the number of numbers. The *median* (the midpoint between the lowest and the highest value of the set) is the middle value in the list of numbers, and to find the median, your numbers have to be listed in numerical order, so you may have to rewrite your list first. The *mode* (most frequently occurring value) is the value that occurs most often (repeated most often), and if no number is repeated, and then there is no mode for the list. The *range* (the span of values over which your data set occurs) is just the difference between the largest and the smallest values.

When performing the statistical analysis on a set of data, the mean, median, mode and SD are all helpful values to calculate. The mean, median, and mode are all estimates of where the middle of the set of data is, and these values are useful when creating groups or bins to organize larger sets of data. SD is the average distance between the actual data and the mean.

The mean (also called as average) is obtained by dividing the sum of observed values by the number of observations (n). Although data points fall above, below or on the mean, it can be considered as a good estimate for predicting subsequent data points. The formula for the mean is given below as equation (1); the excel syntax for the mean is AVERAGE (starting cell: ending cell).

$$\overline{X} = \frac{\sum \frac{w_i x_i}{w_i} X_i}{n} \tag{1}$$

However, this equation (1) can only be used when the error associated with each measurement is the same or unknown, otherwise, the weighted average, which incorporates the SD, should be calculated using the equation (2) given below.

$$X_{wav} = \frac{\sum w_i x_i}{\sum w_i} \tag{2}$$

where $w_i = \dfrac{1}{\sigma_i^2}$ and *xi* is the data value.

The median is the middle value of a set of data containing the odd number of values, or the average of the two middle values of a set of data with an even number of values. The median is especially helpful when separating data into two equal sized bins. The excel syntax to find the median is MEDIAN (starting cell: ending cell).

If all the data points are relatively close together, the average gives you a good idea as to what the points are closest to. If on the other hand, almost all the points fall close to one or a group of close values, but occasionally a value that differs greatly can be seen, then the mode might be more accurate for describing the system, whereas the mean would incorporate the occasional outlying data. The median is useful if you are interested in the range of values your system could be operating in. Half the values should be above and half the values should be below, so you have an idea of where the middle operating point is.

There are different statistical means or the measures of central tendency for the data points, and each one has its own utility. The *arithmetic mean, geometric mean* and *harmonic mean* are some of the most commonly used measures of statistical mean. They make sense in different situations, and should be used according to the distribution and nature of the data. The arithmetic mean is frequently used in scientific experimentation, while the geometric mean is used in finance to calculate compounding quantities.

The most common type of average is the arithmetic mean. If n numbers are given, each number denoted by a_i (where i = 1, 2, 3, ... n), the arithmetic mean is the sum of the *a* as divided by *n* or

$$AM = \frac{1}{n} \sum_{i=1}^{n} a_i = \frac{1}{n} \left(a_1 + a_2 + \ldots + a_n \right)$$

The *arithmetic mean*, simply called the *mean*, of two numbers, such as 2 and 8, is obtained by finding a value *A* such that 2 = 8 = *A* + *A*. One may find that *A* = (2 + 8/2 = 5. Switching the order of 2 and 8 to read 8 and 2 does not change the resulting value obtained for *A*. The mean 5 is not less than the minimum 2 or greater than the maximum 8. If we increase the number of terms in the list to 2, 8, 11, the arithmetic mean is found by solving for the value of A in the equation 2 + 8 + 11 = *A* + *A* + *A*. One finds that *A* = (2 + 8 + 11)/3 = 7.

The geometric mean of the n -non-negative numbers is obtained by multiplying them all together and then taking the nth root. In algebraic terms, the geometric mean of a_1, a_2, a_3 ..... a_n is defined as

$$GM = n\sqrt{\prod_{i=1}^{n} a_i} = n\sqrt{a_1 a_2 \ldots a_n}$$

The *geometric mean* can be thought of as the antilog of the arithmetic mean of the logs of the numbers. For example, GM of 2 and 8 is GM = √2.8 = 4.

The harmonic mean for a non-empty collection of numbers a_1, a_2, a_3 ..... a_n, all different from *0* is defined as the reciprocal of the arithmetic mean of the reciprocals of a_i`s:

$$HM = \frac{1}{\frac{1}{n}\sum_{i=1}^{n}\frac{1}{a_1}} = \frac{n}{\frac{1}{a_1}+\frac{1}{a_2}+\ldots+\frac{1}{a_n}}$$

For example, harmonic mean is useful when examining the speed for a number of fixed-distance trips. If the speed for going from point A to B was 60 km/h, and the speed for returning from B to A was 40 km/h, then the harmonic motion speed is given by

$$\frac{2}{\frac{1}{60}+\frac{1}{40}} = 48$$

A well-known inequality concerning the AM, GM and HM means for any set of positive numbers is:

$$AM \geq GM \geq HM$$

It is easy to remember noting that the alphabetical order of the letters A,G, and H is preserved in the inequality. Thus, for the above harmonic mean, AM = 50, GM = 49 and HM = 48 km/h.

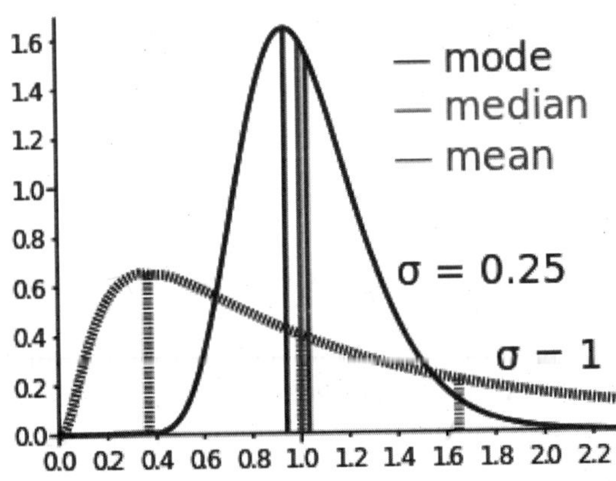

Comparison of AM, median and Mode of two long normal distributions with different skewness.

Standard Deviation (SD)

The degree of variability of responses in addition to the mean has to be indicated since *the same* mean may be obtained from different sets of values. SD describes the variability if the observation about the mean. Most useful measurer of variability to describe the scatter of the population is SD. The summary measures of variability of individuals (i.e., mean, median and mode) are further needed to be tested for reliability based on the samples from population variability of the individuals.

To calculate the standard deviation (SD), its square called variance is to be needed. Variance is the average square deviation around the mean and is calculated by

$$Variance = \Sigma(x\text{-}x\text{-})\ 2/n \text{ or } \Sigma(x\text{-}x\text{-})\ 2n\text{-}1, \text{ now } SD = \sqrt{Variance}.$$

SD helps us to predict how far the given value is away from the mean, and therefore, one can predict the coverage of values. This is more appropriate only if the data is normally distributed. If the individual observations are clustered around a sample mean (M) and are scattered evenly around it, the SD helps to calculate a range that will include a given percentage of observation. Example: If $N \geq 30$, the range $M \pm 2(SD)$ will include 95% of observation and the range $M \pm 3(SD)$ will include 99% of the observation. If the observations are widely dispersed, the central values are fewer representatives of data, hence the variance is taken. While reporting mean and SD, the better way of representation is `mean (SD)` rather than `mean ± SD` to minimize the confusion with confidence interval.

Correlation Coefficient

The *relationship between two variables* is called *correlation*, and this is used to *measure the degree of linear relationship between two continuous variables*, and is represented by `*r*`. In chi-square test, one cannot get the degree of association, but can know whether they are dependent or independent of each other. This may be due to some direct relationship between two variables, and may also be due to some inherent factors common to both the variables. The correlation may be expressed in terms of coefficient, and the values are always between -1 and +1. The correlation coefficient is zero (0) of the variables which are not correlated. The maximum value of 1 is obtained if there is a straight line in scatter plot and is considered as *perfect positive correlation*. The association is positive if the values of x-axis and y-axis tend to be high or low together. On the contrary, the association is negative (-1), if the high y-axis values tend to go with low values of x-axis and considered as *perfect negative correlation*. If the correlation coefficient is larger, stronger is the association. A *weak correlation* may be statistically significant if the numbers of observation are large. The correlation between the two variables does not necessarily suggest the cause and effect relationship. It normally indicates the strength of the association for any data in comparable terms. Example: The correlation between height and weight, age and weight, weight loss and poverty, parity and birth weight, socio-economic status and haemoglobin. It always requires x and y variables to be normally distributed while performing these tests. This is generally used to form a hypothesis and also to suggest areas of future research.

Standard Error of Mean (SEM) and its Applications

Since one studies some points/events (samples) to draw conclusions about all the patients or population, and use the sample mean (M) as an estimate of the population mean (M1), one must need to know how far M can vary from M^1, if repeated samples of size (N) are taken. A measure of such variability is provided by Standard Error of Mean (SEM). SEM can be calculated as (SEM = SD/√n). SEM is always less than SD, and what SD is to the sample; the SEM is to the population mean.

The applications of Standard Error of Mean include: 1) to determine whether a sample is drawn from same population or not when it`s mean is known and 2) to work out the limits of desired confidence within which the population mean should lie.

Standard Deviation (SD) and Standard Error (SE)

The terms standard deviation and standard error are often confused. The contrast between the two terms reflects the important distinction between data description and inference. SD is a valid measure of variability regardless of the distribution while the SE of the sample mean depends on both the standard deviation and the sample size (SE = SD), and SE falls as the sample size increases, as the extent of chance variation is reduced. The SE is a statistical term that measures the accuracy with which a sample represents a population. In statistics, a sample mean deviates from the actual mean of a population, and this deviation is the SE. The smaller the SE, the more representative the sample will be of the overall population. The SE is also inversely proportional to the sample size, the larger the size of the sample, the smaller the SE because the statistic will approach the actual value. SD is a measure of dispersion within the data set whereas the SE is considered as the level of error (dispersion) of the data from a population mean. SE is important to calculate confidence interval for the population mean. SE is used to compute confidence intervals and margins of error in statistics.

Confidence Interval (CI) or Fiducial Limits

These refer to two extremes of measurement within which 95% observations would lie, and describe the limits within which 95% of mean values if determined in similar experiments are likely to fall. The value of `t` corresponding to a probability of 0.05 for the appropriate degree of freedom is read from the table of distribution. By multiplying this value with the standard error, the 95% of confidence limits for the mean are obtained as per the formula:

Lower Confidence Limit = mean – ($t_{0.05}$ x SEM)

Upper Confidence Limit = mean + ($t_{0.05}$ x SEM)

If n > 30, the interval M ± 2 (SEM) will include M with a probability of 95% and the interval M ± 2(SEM) will include M with a probability of 99%. These intervals are, therefore, called the 95% and 99% confidence intervals respectively. The difference between the `p` value and confidence interval is the clinical significance, while the `p `value indicates the statistical significance. Therefore, in many clinical studies, the confidence interval is preferred instead of `p` value. In medicine, mean and SEM are used to describe variability within the sample. The SEM is a measure of precision for estimated population mean, while SD is a measure of data variability around the mean of a sample population. Hence, SEM is not a descriptive statistics and should not be used as such. The correct use of SEM would be only to indicate the precision of estimate mean of population.

Distribution – Types

Even though the universe is full of certainty and variability, a large set of experimental and/or biological observations always tend towards a normal distribution, and this unique behavior of data is the key to entire inferential statistics. The distribution is of two types, namely *Gaussian* (normal) and *Non-Gaussian* (non-normal).

1. Gaussian Distribution or Normal Distribution

The distribution of data is called normal or Gaussian (named after the noted statistician Prof. Gauss who developed this), if the data is symmetrically distributed on both sides of mean and form a bell-shaped curve in frequency distribution plot. The normal curve describes the ideal distribution of continuous values, like heart rate, blood sugar level and Hb percent level. Whether the data is normally distributed or not can be checked by putting the raw data of study directly into computer software and applying distribution test. Statistical treatment of the data can generate a number of useful measurements, the most important of which are mean and standard deviation of mean. In an ideal Gaussian distribution, the values lying between points 1 SD below and 1 SD above the mean value (± 1 SD) will include 68.27% of all values. The range, mean ± 2 SD includes approximately 95% of values distributed about this mean, excluding 2.5% below the range. In an ideal distribution of the values, mean, mode and median are equal within population under study. Even if the distribution in original population is far from normal, the distribution of sample averages tend to become normal as size of the sample increases. This is the single most important reason for the curve of normal distribution. Various methods of analysis are available to make the assumptions about normality, including `t` test and analysis of variance (*ANOVA*). In normal distribution, skew is zero, and if the difference (mean-median) is positive, the curve is positively skewed, and if it is (mean-median) negative, the curve is negatively skewed, and therefore, the measure of central tendency differs.

2. Non-Gaussian Distribution or Non-normal Distribution

The distribution is non-normal or non-Gaussian, if the data is skewed on one side, it may be binomial distribution (Poisson distribution). In such a case, the event can have only one of two possible outcomes such as *yes/no, positive/negative, survival/ death and smokers/non-smokers.* When the distribution of data is non-Gaussian, different tests like *Wilcoxon, Mann-Whitney, Krusksal-Wallis* and *Friedman test* can be applied depending on the nature of data.

Null Hypothesisμ

The important object of statistical analysis is to find out whether the effect produced by a compound under study is genuine or not, and is not due to chance. So, the analysis usually attaches a test of statistical significance. In such a test, the first step is to state the *null hypothesis* (statistical hypothesis). In this an assumption is made that no differences exist between the two groups. An *alternative hypothesis* (research hypothesis) states that there is a difference between two groups. For example, a new drug 'A' is claimed to possess analgesic activity and one wants to test it with a placebo. Here, the null hypothesis would be '*drug A is not better than the placebo*', an alternative hypothesis would be '*there is a difference between new drug A and placebo*'. When the null hypothesis is accepted, the difference between the two groups is not significant. It means that, both the samples were drawn from a single population, and the difference between two groups is statistically different. A difference between drug 'A' and placebo group, which would have arisen by chance is less than 5% of the cases, that is less than 1 in 20 times is considered as statistically significant ($P < 0.05$).

Errors – Types

In any experiment, there is a possibility of occurrence of two errors, *false positive* (type-I error) and *false negative* (type-II error).

a. Type I Error or False Positive Error (α-error)

This is the probability of finding a difference; when no such difference actually exists, which results in the acceptance of an inactive compound as an active compound. This error, which is not unusual, may be tolerated because in subsequent trials, the compound will reveal itself as inactive and thus finally rejected. For example, it was proved in our trial that new drug 'A' has an analgesic action and accepted as an analgesic. If type I error is committed in this experiment, then subsequent trial on this compound will automatically reject our claim that drug 'A' is having analgesic action and later on drug 'A' will be thrown out of market. Type I error is actually fixed in advance by choice of the level of significance employed in test. The error can be made small by changing the level of significance and by increasing the size of sample.

b. Type II Error or False Negative Error (β-error)

This is the probability of inability to detect the difference when it actually exists, thus resulting in the rejection of an active compound as an inactive. This error is more serious than type I error because once the compound is labeled as inactive, there is possibility that nobody will try it again. Thus, an active compound will be lost. This error can be minimized by taking larger sample and by employing sufficient dose of the compound under trial. For example, one claims that drug 'A' is not having analgesic activity after suitable trial. Therefore, drug 'A' will not be tried by any other researcher for its analgesic activity and thus drug 'A', in spite of having analgesic activity, will be lost just because of type II error. Hence, researcher should be very careful while reporting type II error.

Level of Significance

If the probability (***P***) of an event (outcome) is high, one refers to it that it is not rare or not uncommon. But, if the ***P*** is low, it is referred to as rare or uncommon. In biostatistics, a rare event (outcome) is called significant, whereas a non-rare event is called non-significant. The '***P***' value at which an event (outcome) is enough to be regarded as significant, it is called the significance level. In medical research, most commonly ***P*** value less than 0.05 or 5% is considered as significant. However, on justifiable grounds, one may adopt a different standard like $P < 0.01$ or 1%. Whenever possible, it is always better to give actual ***P*** values instead of $P < 0.05$. Even if one has found the true value (population value) from sample, one cannot be confident as he/she is dealing with a part of population only; howsoever big the sample may be. One would be wrong in 5% cases only if the population value is placed within 95% confidence limits. Significant or insignificant indicates whether a value is likely or unlikely to occur by chance. '***P***' indicates probability of relative frequency of occurrence of the difference by chance.

Outliers

Sometimes, during the analysis of the data, one value is very extreme from the others and this is referred to as *outliers*. This could be due to two reasons: firstly, the value obtained may be due to chance; and in that case, the value should be kept in final analysis as it is from the same distribution and secondly, it may be due to a mistake. The causes may be listed as typographical errors or measurement errors, and these values should be deleted to avoid invalid results.

One - tailed and Two - tailed Test

When comparing two groups of continuous data, the null hypothesis is, that there is no real difference between the groups (A and B), while the alternative hypothesis is that there is a real difference between the groups (A and B). This difference could be in either direction, either A > B or A < B. When there is some sure way to know in advance that the difference could only be in one direction, A > B and when a good ground considers only one possibility, the test is called *one-tailed test*. Whenever both the possibilities are considered, the test of significance is known as a *two-tailed test*. For example, when we consider that English boys are taller than Indian boys, the result will lie at one end, which is one tail distribution; as such one-tail test is used. When one is not absolutely sure of the direction of difference, which is usual, it is always better to use two-tailed test. For example, a new drug 'X' is supposed to have an antihypertensive activity and one wants to compare it with atenolol, the exact direction of effect of drug 'X' is not known, and in such a case, one should prefer two-tailed test. When one wants to know the action of particular drug is different from that of another, but the direction is not specific, always two-tailed test is to be used. At present, most use two-sided ***P*** values as a standard norm in biomedical research.

Determination of Sample Size - Importance

The sample is a fraction of the universe as studying the universe is the best parameter. But, when it is possible to achieve the same result by taking fraction of the universe, a sample is taken. Such an application can save time, manpower, cost, and at the same time, increasing efficiency. Hence, an adequate sample size is of prime importance in biomedical studies. If the sample size is too small, as it will not give valid results, validity in such a case is questionable, and therefore, the whole study will be a waste. A good small sample is much better than a bad large sample. Hence, appropriate sample size will be ethical to produce precise results.

Factors Influencing Sample Size

a) *Prevalence of particular event or characteristics* - If the prevalence is high, small sample can be taken and *vice versa.* If prevalence is not known, then it can be obtained by a pilot study.

b) *Probability level considered for accuracy of estimate* - If more safeguard about conclusions on data is required, we need a larger sample. Hence, the size of sample would be larger when the safeguard is 99% than when it is only 95%. One needs a larger sample if only a small difference is expected and if we need to detect even that small difference.

c) *Availability of money, material, and manpower.*

d) *Time bound study* curtails the sample size as routinely observed with research work.

Sample Size Determination and Variance Estimate

To calculate the sample size, the formula requires the knowledge of standard deviation or variance, but the population variance is unknown. Therefore, standard deviation has to be estimated. Frequently used sources for estimation of standard deviation are:

a) A pilot or preliminary sample may be drawn from the population, and the variance computed from the sample may be used as an estimate of standard deviation. Observations used in pilot sample may be counted as a part of the final sample.

b) Estimates of standard deviation may be accessible from the previous or similar studies, but sometimes, they may not be correct.

Sample Size - Calculation

Calculation of sample size plays a key role while doing any research. Before calculation of sample size, following points are to be considered. First of all, the minimum expected difference between the groups is to be assessed. Then, one has to find out standard deviation of variables. Then, the level of significance (alpha level, generally set at $P < 0.05$) is to be set as well as the power of study (1-beta = 80%). After deciding all these parameters, one has to select the formula from computer programs to obtain the sample size (software is available free of cost). Finally, appropriate allowances for non-compliance and dropouts are given, and the final sample size for each group in study is determined.

This can be explained with two examples to understand sample size calculation.

a) In the first example, the mean (SD) diastolic blood pressure of a hypertensive patient after enalapril therapy is found to be 88(8). It is claimed that telmisartan is better than enalapril, and a trial is to be conducted to find out the truth. For convenience, let us consider the minimum expected difference between the two groups is 6 at significance level of 0.05 with 80% power, and the results will be analyzed by unpaired 't' test. Here, minimum expected difference is 6, SD is 8 from previous study, alpha level is 0.05, and power of study is 80%. After putting all these values in the computer program, sample size comes out to be 29. If we take allowance to non-compliance and dropout to be 4, then final sample size for each group would be 33.

b) In the second example, the mean haemoglobin (SD) of a newborn is observed to be 10.5 (1.4) in a pregnant mother of low socio-economic group, and it was decided to carry out a study to decide whether iron and folic acid supplementation would increase haemoglobin level of newborn. If so, there will be two groups, one with supplementation and other without supplementation. The minimum difference expected between the two groups is taken as 1.0 with 0.05 level of significance and power as 90%. In this example, SD is 1.4 with minimum difference 1.0. After putting these values in computer-based formula, the sample size comes out to be 42 and with allowance of 10%, final sample size would be 46 in each group.

Power of Study

This is a probability that study will reveal a difference between the groups if the difference actually exists. A more powerful study is required to pick up the higher chances of existing differences. Power is calculated by subtracting the beta error from 1. Hence, power is (1-Beta). Power of study is very important while calculating the sample size. Power of study can be calculated after the completion of study, and it is called as *posterior power calculation*. This is very important to know whether study had enough power to pick up the difference if it existed. Any study to be scientifically sound should have at least 80% power. If power of study is less than 80% and the difference between groups is not significant, then one can say that difference between groups could not be detected, rather than no difference existed between the groups. In this case, power of study is too low to pick up the exiting difference, and it means probability of missing the difference is high and hence the study could have missed to detect the difference. If the power of study is increased, then sample size also increases. So, it is always better to decide power of study at initial level of research.

Choosing an Appropriate Statistical Test

There are number of tests in biostatistics, but the choice of the appropriate test mainly depends on the characteristics and type of analysis of data. Sometimes, one has to find out the difference between means (medians) or association between the variables. The number of groups used in a study may vary; therefore, study design also varies. Hence, in such situation, one has to make the decision which is more precise while selecting the appropriate test as inappropriate test will lead to invalid conclusions. Statistical tests can be divided into *parametric* and *non-parametric tests.* If the variables follow normal distribution, the data can be subjected to *parametric test,* and for non-Gaussian distribution, one should apply *non-parametric test.* Statistical test should be decided at the start of the study.

The following are the different parametric tests used in the analysis of various types of data.

1) Student's 't' Test

It was W. S. Gosset, a civil service statistician, who introduced 't' distribution of small samples and also published his work under the pseudonym '*Student.*' This test is one of the most widely used tests in pharmacological investigations that involve the use of small samples. The 't' test is always applied for analysis when the number of samples is 30 or less and it is usually applicable for graded data like blood sugar level, body weight, height etc. If the sample size is more than 30, then 'Z' test is applied. The `t` test is of two types namely, *paired* and *unpaired.*

Application of paired and unpaired `t` test

a) When a comparison has to be made between two measurements in the same subjects after two consecutive treatments, paired 't' test is used. For example, when we compare the effect of a drug A (decrease blood sugar) both before treatment (baseline) as well as after 1 month of treatment.

b) When a comparison has to be made between two measurements in two different groups, unpaired 't' test is used. For example, when we compare the effects of drug A and B (i.e. mean change in blood sugar) after one month from baseline in both groups, unpaired 't' test' is applicable.

2) ANOVA

When we want to compare two sets of unpaired or paired data, the student's 't' test is applied. However, when there are 3 or more sets of data to be analyzed, we need the help of well-designed and multi-talented method called as *analysis of variance* (ANOVA). This test compares multiple groups at one time. In ANOVA, we draw assumption that each sample is randomly drawn from the normal population, and also they have same variance as that of population.

ANOVA is mainly of two types namely, *one-way ANOVA* and *two-way ANOVA.*

A) One-way ANOVA

This compares three or more unmatched groups when the data are categorized in one way. For example, we may compare a control group with three different doses of aspirin in rats. Here, there are four unmatched groups of rats. Therefore, we should apply one way ANOVA. We should choose repeated measures ANOVA test when the trial uses matched subjects. For example, effect of supplementation of vitamin-C in each subject before, during, and after the treatment. The matching should not be based on the variable you are comparing. For example, if you are comparing blood pressures in two groups, it is better to match based on age or other variables, but it should not be to match based on blood pressure. The term repeated measures applies strictly when you give treatments repeatedly to one subject. ANOVA works well even if the distribution is only approximately Gaussian. Therefore, these tests are used routinely in many fields of science. The *P* value is calculated from the ANOVA table.

B) Two-way ANOVA or Two –factors ANOVA

This determines how a response is affected by two factors. For example, you can measure a response to three different drugs in both men and women. This is a complicated test.

Post-hoc Test - Importance

Post-test is the modification of 't' test. This accounts for multiple comparisons as well as for the fact that the comparisons are interrelated. ANOVA only directs whether there is a significant difference between the various groups or not. If the results are significant, ANOVA does not tell us at what point the difference between various groups subsist. But, post-test is capable of pinpointing the exact difference between the different groups of comparison. Therefore, post-tests are very useful as far as statistics is concerned.

The post-hoc test is of five types, namely; Dunnett's, Turkey, Newman-Keuls, Bonferroni, and test for linear trend between mean and column number.

Selection of a Post-hoc test

a) *Dunnett's post-hoc test* is selected if one column represents control group and we wish to compare all other columns to that control column but not to each other.

b) The *test for linear trend* is selected if the columns are arranged in a natural order (i.e. dose or time) and we want to test whether there is a trend so that values increase (or decrease) as you move from left to right across the columns.

c) *Bonferroni*, *Turkey's*, or *Newman's test* is selected if we want to compare all pairs of columns.

Following are the non-parametric tests used for analysis of different types of data.

1. Chi-square Test

The Chi-square test is a non-parametric test of proportions and this test is not based on any assumption or distribution of any variable. Though different, this test follows a specific distribution known as *Chi-square distribution*, which is very useful in research. It is most commonly used when data are in frequencies such as number of responses in two or more categories. This test involves the calculations of a quantity called *Chi-square* (x^2) from Greek letter '*Chi*'(x) and pronounced as '*Kye*' and was developed by Karl Pearson.

Applications of Chi-square Test

a) *Test of Proportion*: The test of proportion is used to find the significance of difference in two or more than two proportions.

b) *Test of Association*: The test of association between two events in binomial or multinomial samples is the most important application of this test in statistical methods. It measures the probabilities of association between two discrete attributes. Two events can often be studied for their association such as smoking and cancer, treatment and outcome of disease, level of cholesterol and coronary heart disease etc. In these cases, there are two possibilities (*binomial sample*), either they influence or affect each other or they do not. In other words, one can say that they are dependent or independent of each other. Thus, the test measures the probability (*P*) or relative frequency of association due to chance and also if two events are associated or dependent on each other. Varieties used are generally dichotomous (improved/not improved). If data are not in that format, the investigator can transform data into dichotomous data by specifying above and below limit. *Multinomial sample* is also useful to find out association between two discrete attributes. For example, to test the association between numbers of cigarettes equal to 10, 11-20, 21-30, and more than 30 smoked per day and the incidence of lung cancer. Since, the table presents joint occurrence of two sets of events, the treatment and outcome of disease, it is called *contingency table* (*Con* meaning `together', and *tangle* meaning `to touch').

2 × 2 Contingency Table or Four cell Table – How to prepare?

When there are only two samples, each divided into two classes, it is called as *four cell* or *2 × 2 contingency table*. In contingency table, one needs to enter the actual number of subjects in each category, and fractions or percentage or mean cannot be entered. Most contingency tables have two rows (two groups) and two columns (two possible outcomes). The top row usually represents the exposure to a risk factor or treatment, while the bottom row is mainly for control. The outcome is entered as a column on the right side with the positive outcome as the first column and the negative outcome as the second column. A particular subject or patient can only be in one column but not in both.

Even if sample size is small (< 30), this test is used by using *Yates correction*, but frequency in each cell should not be less than 5. Though, Chi-square test tells an association between two events or characters, it does not measure the strength of association. This is the limitation of this test and it only indicates the probability (*P*) of occurrence of association by chance. Yate's correction is not applicable to tables larger than 2 X 2. When the total number of items in 2 X 2 table is less than 40 or number in any cell is less than 5, *Fischer's test* is more reliable than the Chi-square test.

2) Wilcoxon – Matched - Pairs Signed - Ranks Test

This is a non-parametric test and is used when data are not normally distributed in a paired design. It is also called *Wilcoxon-Matched Pair test.* It analyses only the difference between the paired measurements for each subject. If *P* value is small, the idea that the difference is coincidence can be rejected and that the populations have different medians can be concluded.

3) Mann - Whitney Test

It is a student's 't' test performed on ranks and for large numbers, it is almost as sensitive as student's 't' test. For small numbers with unknown distribution, this test is more sensitive than Student's 't' test. This test is generally used when two unpaired groups are to be compared and the scale is ordinal (ranks and scores), which are not normally distributed.

4) Friedman Test

This is a non-parametric test which compares three or more paired groups. In this, one has to rank the values in each row from low to high. The goal of using a matched test is to control the experimental variability between subjects, thus increasing the power of the test.

5) Kruskal - Wallis Test

It is a non-parametric test which compares three or more unpaired groups. Non-parametric tests are less powerful than parametric tests. Generally, *P* values tend to be higher, making it harder to detect the real differences. Therefore, first of all, one has to try to transform the data. Sometimes, simple transformation will convert non-Gaussian data to a Gaussian distribution. Non-parametric test is considered only if outcome variable is in rank or scale with only a few categories. In this case, population is far from Gaussian or one or few values are off scale, too high, or too low to measure.

The following is the summary of statistical tests applied for different types of data.

Common Problems Faced by the Researcher

Whenever any researcher thinks of any experimental or clinical trial, the number of queries arises before him/her. To explain some common difficulties, one example can be taken and try to solve it. Suppose, one wants to perform a clinical trial on the effect of supplementation of vitamin-C on blood glucose level in patients of type II *Diabetes mellitus* on metformin; two groups of patients will be involved, where one group will receive vitamin-C and other placebo.

a) Determination of the Size of the Sample

In such a trial, the first problem is to find out the sample size. The sample size can be calculated if S.D. minimum expected difference, alpha level, and power of study are available. S.D. can be taken from the previous study and if the previous study report is not reliable, one can do a pilot study on few patients to get S.D. Minimum expected difference can be decided by the investigator, so that the difference would be clinically important. In this case, Vitamin-C being an antioxidant, the difference between the two groups in blood sugar level can be taken as 15. Minimum level of significance may be taken as 0.05 or with reliable ground it can be increased, and lastly, power of study is taken as 80% or one can increase power of study up to 95%, but in both the situations, sample size will be increased accordingly. After putting all the values in computer software program, the sample size for each group can be arrived at.

b) The Correct Test to be applied

After calculating the sample size, next comes application of a suitable statistical test. One can apply either parametric or a non-parametric test. If data are normally distributed, parametric test should be used, otherwise non-parametric test is applied. In this trial, the blood sugar level in both groups is being measured after 0, 6, 12 weeks, and if data are normally distributed, then repeated measure ANOVA can be applied in both the groups followed by *Turkey's post-hoc test* if we want to compare all pairs of column with each other, and *Dunnet's post-hoc test* for comparing 0 with 6 or 12 weeks observations only. If one wants to see whether the supplementation of vitamin-C has any effect on blood glucose level as compared to placebo, then one has to consider the change from baseline i.e. from 0 to 12 weeks in both the groups and apply unpaired ‘t’ with two-tailed test as directions of results is non-specific. If one is comparing effects only after 12 weeks, then paired ‘t’ test can be applied for intra-group comparison and unpaired ‘t’ test for inter-group comparison. If one wants to find out any difference between basic demographic data regarding gender ratio in each group, then the Chi-square test is to be applied.

c) Correlation between the Variable

To see whether there is any correlation between age and blood sugar level, or gender and blood sugar level, Spearman or Pearson correlation coefficient test is applied, depending on Gaussian or non-Gaussian distribution of data. If all these questions can be answered satisfactorily before start of the trial, then it becomes painless to conduct research in an efficient manner.

SOFTWARE FOR BIOSTATISTICS

Now-a-days, the statistical computations are made very feasible owing to the availability of computers and suitable software programs. At present, the computers are mostly used for performing various statistical tests as it is very tedious to perform it manually. Commonly used software's include: *MS Office Excel, Graph Pad Prism, SPSS, NCSS, Instant, Data plot, Sigma stat, Graph Pad In stat, Sys stat, Gen stat, MINITAB, SAS, STATA,* and *Sigma Graph Pad.* The free website for statistical software is *www.statistics.com* and *http://biostat.mc.vanderbilt.edu/wiki/Main/PowerSampleSize*. The statistical methods are necessary to draw valid conclusions from the data. The post-graduate students should be aware of different types of data, measures of central tendencies, and different tests commonly used in biostatistics, so that they would be able to apply these tests and analyze the data themselves. The above discussion provides background information, and an attempt is made to highlight the basic principles of statistical techniques and methods for the use of post-graduate students.

STATISTICAL APPLICATIONS IN BIOLOGICAL RESEARCH

With the increasing application of the statistical methods to new fields of research work, the problem of the handling of small samples has become more and more important. It is true that the larger the sample, the more trustworthy are the inferences which can be drawn from it, but there are certain problems, whether biological or industrial, in which the time and the cost involved in obtaining even a moderately large sample would be quite prohibitive. Such a need for the development of small sample theory has emphasized the importance of placing the methods of inference on a clearly defined and logical basis. The main aim of the statistician must be to bring the simple assumptions of theoretical analysis into correspondence with the varied and complex forms of situations of practical work in biology and medicine. In view of its applied aspects, statistics serves as a separate mathematical science and in that respect; biostatistics is the application of statistical concepts and methods in biology, public health and medicine.

The important areas in biological research include: flexible Bayesian non-parametric models for the analysis of array CGH data (there are essentially three technological platforms for detecting copy number variation – *array-based technology* [including array comparative genomic hybridization, oligonucelotide array or bacterial artificial chromosome array]. *SNP genotyping technology* and *next generation sequencing technology*), *screening for active compounds in drug discovery* (mass automated screening which can be supplemented by other forms of data, involving similarity matrices of compounds as calculated in a variety of ways), *statistical genetics for complex disease studies* (developing novel models and methods for analyzing human genetic variation in relation to complex diseases and different drug response), *integrative analysis of `omics` data* (*omics* span a wide range of fields including *genomics, transcriptomics, proteomics, metabolomics, pharmacogenomics, nutrigenomics, phylogenomics, interactomics,* etc, which are responsible for the generation of larger quantities of data – and how best statisticians can model these data in order to understand the underlying biological systems), *shape analysis in biology* (to identify whether certain parts of DNA molecule can be matched by parts of another one, and other interesting applications of shape analysis in bioinformatics and on the molecule matching algorithms), *stochastic models in Yeast Prion research* (to understand better the ability of the prions to self replicate and complexity of these processes studying them by simulation) etc. The examples of biological statistical applications include clustering and significance of gene expression data from microarrays, mapping of quantitative trait loci, clinical trials, health care expenditures, and design of agricultural field trials. The specific problems in computational biology include the detection of genomic regions underlying complex traits, the detection of positive selection, the evolutionary genomics of plant and animal domestication, the discovery of new human genes, and the identification and characterization of functional non-coding elements in mammals. In genetics and molecular biology, the applications also include linkage mapping, association studies, gene finding and sequence alignment, protein structure prediction, design and analysis of microarray data, molecular evolution and phylogenetic trees, DNA topology and data base search strategies etc.

Starting from the ancient age to the modern times not a single area can be found where statistics is not playing a vital role. This has now been recognized and universally accepted as an essential component of research in every branch of science. This is being extensively used in biological sciences. Biostatistics is a branch of applied statistics that concerns with the application of statistical methods to medical, genetics, and biological problems. In this sequel, one important step is the appropriate and careful analysis of the statistical data to get precise results, and it is pertinent to mention that the majority of statistical tests and techniques are applied under certain mathematical assumptions. Hence, it is necessary to realize the importance of relevant assumptions. In this connection, the assumption of normality of population(s) and variance homogeneity etc, are the most important, and if these assumptions are not satisfied, the results may be potentially misleading. Therefore, it is suggested to check the relevant assumption(s) about the data before applying statistical test(s) to get valid results.

Statistical data analysis and statistical inference (estimation and hypothesis testing) are essential for interpreting the rapidly increasing databases of numerical measurements collected from biological processes using modern biotechnology (microarray technology). In current life science research, there are many open statistical problems, including issues such as experimental design of gene chips, signal filtering, data analysis, and inference of cellular networks, and inference of genome-wide profiles etc. Genome-wide studies describe the genomic landscape of disease, as well as molecular mechanisms of genome evolution and function. The biologists can identify important biological problems, while the statisticians can develop statistical modeling and analytic methods to analyze the data produced by biological observations and experiments, and together they will be able to validate the results biologically and statistically. Statistics is unavoidable in biology and medical research and this is due to the complexity of these areas involving large population, and many factors related to outcome variables, measurement aspects, clustering, and other things that invite statistics to play its role in biology and medicine. Statistics for the life sciences is almost synonymous with biostatistics, as it incorporates quantitative modeling and methods of data analysis for clinical and epidemiological research. It also includes the methods used in genetic research and genomics, which have a classical foundation, but are rapidly developing in answer to the present-day opportunities provided by data from new experimental platforms, like microarrays or whole genome scans. It is not an exaggeration to say that all empirical research in the social and behavioural sciences relies predominantly on statistical analysis. Similar sub-domains that have emerged more recently are the *scientometrics* and *bibliometrics*, (quantitative study of the development of science and technology), *stylometrics* (quantitative study of stylistic forms and patterns in the use of the language), *sensometrics* (quantitative study of the taste and smell), *cliometrics* (quantitative study of history), and *jurimetrics* (empirical approach to the law) etc.

BIOSTATISTICS – APPLICATIONS

The applications include: the design and analysis of clinical trials in medicine; the assessment of severity state of a patient with progress of the outcome of a disease; use of population genetics and statistical genetics to link variation in genotype with a phenotypic variation, in agriculture to improve the crops and farm animals (animal breeding), and in biomedical research, this can assist in finding the candidates for gene alleles that can cause or influence predisposition to disease in human genetics; the analysis of the genomic data (experiments related to microarray or proteomics) often associated with the diseases or different stage of a disease; ecological forecasting; biological sequence analysis; and systems biology for the analysis of pathways or for gene network inference etc. Biostatistics is well recognized as an essential tool in medical research, clinical decision making, and health management.

Deficient basic biostatistical knowledge adversely affects research quality. Biostatistical methods make research scientific if they are used from the stage of planning of the research itself. Unbiased, consistent, and efficient parameter estimates are provided by correct use of statistics. This is possible by applying statistics from the planning stage until the end of the study. Improvements in teaching statistics to medical students should improve their understanding of statistical concepts and reduce the incidence of misconceptions among clinicians and medical researchers. Statisticians help medical researchers to design studies, decide what data to collect, analyze data from medical experiments, help interpret the results of the analyses, and collaborate in writing articles to describe the results of medical research

RECENT DEVELOPMENT IN MODERN BIOSTATISTICS

The advent of modern computer technology and relatively cheap computing resources has enabled the computer-intensive bio-statistical methods (boot-strapping and re-sampling methods). In addition, bio-medical technologies (microarrays, next generation sequences for genomics – and mass spectrometry for proteomics) generate enormous amounts of data that can only be analyzed with bio-statistical methods. The classical statistical techniques (*linear or logistic regression* and *linear discriminant analysis* do not work well for high dimensional data (that is, when the number of observations [n] is smaller than the number of features or predictors [p] – $n<p$). In such cases of high dimensionality, one should always consider an independent *validation test set* and the corresponding *residual sum of square* (RSS)

and R^2 of the validation test set, not those of the training set. In recent times, the *random forests* (the technique was invented by Leo Breiman, a statistician – used for clinical decision support systems) have gained popularity and generates a lot of decision trees randomly and uses them for the classification (the response is on a nominal or ordinal scale, as opposed to regression where the response is on a ration scale). Even with a very basic understanding of mathematics and statistics, one can draw and interpret the decision trees. GSEA (Gene Set Enrichment Analysis) is a new method for analyzing the biological high throughput experiments, and with this method, one does not consider the perturbation of single genes but of whole gene sets which are functionally related. The main advantage of this approach is that it is more robust, and it is more likely that a single gene is found to be falsely perturbed than it is that a whole pathway is falsely perturbed. The researcher can easily integrate the accumulated knowledge about the biochemical pathways using this approach.

■■■

CHAPTER - 15

COMPUTERS IN LIFE SCIENCES

Introduction

The computers and the computer technology have become an indispensable part of our modern life and their widespread usage is changing the way we live, but computing is still relatively new. The first true computers were electro-mechanical giants developed by the governments and institutions driven by the desperate circumstances of World War II. For decades after the war, these remained in the hands of the universities, governments and business institutions, and as the technology improved, they became smaller, more affordable and more accessible until they came into our homes and finally our pockets.

COMPUTER – BASIC PRINCIPLES

Definition

A computer is a device that can be programmed to carry out a finite set of arithmetic or logical operations. A computer is also defined as a programmable machine which responds to a specific set of instructions in a well-defined manner and can execute a pre-recorded list of instructions (or a progamme). As the sequence of operation can readily be changed, the computer can solve more than one kind of problem. The computer refers to an object that can accept some input and produce some output. The human brain itself is a sophisticated computer and our most common use of the word, computer, describes an electronic device containing a microprocessor (a small electronic device that can carry out complex calculations within so short a time such as the blink of an eye).

The computers come in many forms, from the traditional desktop PC to portable devices such as laptops, tablets and even smart-phones. However, a number of basic elements are generally found within most computing devices used by us. There are many additional types of computer used only within certain specialized contexts such as super-computers and servers. The modern computers are electronic and digital and consist of hardware (the actual machinery, the wires, transistors and circuits) and software (the instructions and the data). The basic principles of how a computer operates are generally the same whatever its purpose is.

1. Operating System

The operating system on a computer bridges the gap between the hardware and the user. The common operating systems include Windows, Mac OS and Linux, all of which are available in varying forms. The operating system in a computer communicates with and controls hardware including the hard disk, memory, processor, input and output devices. At the opposite end, the operating system presents an interface which allows the users to control the computer.

2. Input Devices

These represent the conduit through which data and the instructions enter a computer keyboard and a mouse. The input devices on a computer vary depending on the type of computing device in question. For desktop and laptop PCs, the input is typically provided through a keyboard and a mouse. Most keyboards used in the West adopt a QWERTY layout for text input, as well as number pads and other function keys. A mouse may be provided as part of a laptop, in the form of a touch-pad just below the keyboard area. Alternatively, a mouse may be a separate component which plugs into the computer and controls applications through movement and button clicks. Some smart-phones and tablet PCs use touch-screen interfaces, in which the screen is used for both display and user interaction.

3. Output Devices

These let one see what the computer has accomplished (a display screen or a monitor, printer or any other device). The output devices in computers typically include visual and auditory elements. The main output component in a computer tends to be the monitor which presents a graphical display to users. The monitor may be a separate element connected to the computer through cables, or may be part of the same hardware item as in a laptop or tablet PC. The audio output may be provided through built-in speakers which form part of the computer or monitor element. Alternatively, audio output can be captured through headphones plugged into an audio output port.

4. Networking

The computers are regularly networked together, allowing users to share data and applications. Within workplaces and organizations, it is common for computers to be linked via a network, providing access to shared programs and databases. The Internet is also a way in which computers are networked, connecting vast numbers of people around the world from domestic, public and work environments. The internet connections allow computers to access websites and use communication techniques such as emailing, instant messaging and social networking.

5. Applications of a Computer

Most computers have a number of applications installed on their operating systems. These can facilitate a very wide range of functions, including those for work, personal life and entertainment. For office and organizational use, programs such as Microsoft Office and Open Office include applications for word processing, spreadsheets and calendars. Graphic design and image editing software is also common, as are email clients and web browsers for internet access.

COMPUTER - PARTS

The computer is a collection of many devices that function as a unit. A personal computer can take on many guises. The most common type available in majority of schools is the IBM compatible model. Some schools may opt for the Macintosh platform which can be all-in-one unit or a modular unit. The modular units usually consist of the following components. The basic parts of a desktop computer are the computer case (with a central processing unit or CPU), a monitor, a keyboard, a mouse, and power cord. Each part plays an important role whenever you use a computer.

i) Actual Computer

The computer CPU (central processing unit) is a metal and plastic rectangular box which contains the main components of the computer and sits on the desktop (desktop case) or next to computer under the desk (tower case). This is actually a small electronic device inside the case and the term is often used to refer the whole collection of electronics inside the box. The case houses the motherboard, central processing unit (CPU – the heart of the computer which actually executes the instructions), the power supply etc. The computer cases come in different shapes and sizes. A desktop case lies flat on a desk, and the monitor usually sits on top of it. A tower case is tall and sits next to the monitor or on the floor. The front of the case usually has an on/off switch and one or more optical drives. Most of the personal computers of today include tower cases, rather than desktop cases; however, some computers are being made with all of the internal components built into the monitor, which completely eliminates the tower.

Computer case

Monitor

ii) Monitor

The monitor is the user`s window into the working of the computer and works with a video cord located inside the computer case, to display the images as well as the text on the screen. The conventional television can be used as a monitor if a translation device is used to connect them. The quality of the picture leaves something to be desired. Latest monitors usually have liquid crystal display (LCD) or light-emitting diode (LED) display, and these can be made very thin and are often called flat panel displays. The older monitors generally use cathode ray tube (CRT) displays and such monitors are much larger and heavier and also take up more desk space. Most of the monitors have control buttons which allow to change the monitor`s display settings and some also have built-in speakers. LED displays often allow for greater contrast than a traditional LCD display.

iii) Power cord

This is the link between the power outlet and the power supply unit in the computer case. This is always plugged in and to protect the computer from voltage spikes, this is plugged into a surge protector. Uninterruptable power supply (UPS) can also be used to act as a surge protector which also provides temporary power if there is a blackout.

iv) Key board

This is the primary input device used to communicate with the computer and enter data. This closely resembles a conventional typewriter key board provided with additional keys used specifically for computing functions. These are of many types such as wired, wireless, ergonomic, multimedia etc. Almost all the key boards are similar and allow accomplishing basically the same tasks though there may be differences in the location of some keys and features.

Key board

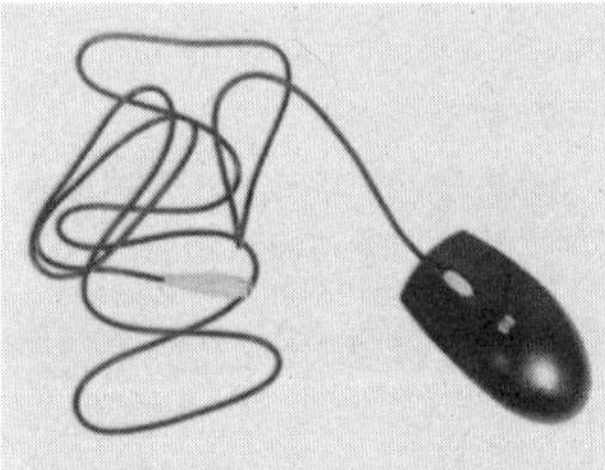

Mouse

v) Mouse

The mouse is a peripheral that is called a pointing device. It derives its name as such because of the resemblance of the wire coming out of it (mouse tail) and this was introduced in the early 1980s when Macintosh created its graphical. It lets one to point to objects on the computer screen, to click on them and also to move them. The mice are of two main types, namely, optical and mechanical. The former uses an electronic eye to detect the movement and is easier to clean while the latter uses a rolling ball to detect the movement. A mechanical mouse is cheaper and to keep it working proper, it requires regular cleaning. A mouse connects to the computer using a USB or PS/2 connection. A wireless mouse can reduce clutter on the desktop.

vi) Mouse Alternatives

The alternatives are other devices which can do the same thing a mouse can do but with a difference in their look and feel. These are easier to use and also require much less desk space than a mouse. The most common mouse alternatives are trackball and touch pad (track pad). The former has a ball on its top that can rotate freely. The ball can be rolled simply with the fingers or thumb (as in mobile devices) to move the pointer. The latter is a touch-sensitive pad that lets one to control the pointer by making a drawing motion with the finger. The touch pads are very common on laptop computers.

Touch pad on a Laptop

Floppy Diskette Drive

vii) Floppy Diskette Drive

The disk drives and tape drives allow a computer to permanently retain large amount of data (mass storage devices). Earlier, the floppy diskettes were the most advanced storage devices that are normally used as temporary storage containers or transportation media for data. A standard diskette can hold 1.44 MB of computer data which amounts to a rather larger number of pages if translated to the paper standard for textual information. The diskettes are not as reliable or fast as the internal storage drives on the computer and are also the primary vector of virus infection in the computer world as such they have been now replaced by CD-ROM disks.

vii) CD-ROM Drive

This is the modern miracle which gained prominence in the late 1960s which has become the primary distribution medium for software consumers. The Compact Disk – Read Only Memory (CD-ROM) disk itself is a collection of concentric circles containing millions of pits and plateaus which correspond to on/off bits of data. This can be read with an optical laser similar to the one used to scan the groceries at the supermarket. Most disks are Read-Only. The memory enables the computer to store data and the programmes at least temporarily.

CD-ROM Drive

viii) Computer Peripherals

Computer peripherals are the electronic devices that can be hooked up to a computer other than the standard input-output devices like monitor, keyboard and mouse. These also include the speakers, microphones, printers, scanners, digital cameras, plotters and modems. These often require special software packages called drivers.

Computers – Types according to Size and Power

The computers are classified according to their size and the power they utilize into:

1. **Personal Computer –** This is a small and single user computer based on a micro-processor in addition to a key board (for entering the data), a monitor (for displaying the information) and a storage device (for saving the data).
2. **Workstation –** This is a powerful, single user computer which has a more powerful micro-processor and a monitor of high quality and can be used as a personal computer.
3. **Minicomputer –** This is a multi-user computer which is capable of supporting from 10s to 100s of users simultaneously.
4. **Mainframe –** This is a powerful multi-used computer capable of supporting many 100s of 1000s of users simultaneously.
5. **Supercomputer –** This is an extremely fast computer that can perform 100s of millions of instructions per second.

Brief History of Computing

The first use of the word, computer was recorded in 1613 in a book, "The young man's Gleanings" by Richard Braithwait, an English writer, who wrote: "I have read the truest computer of Times, and the best Arithmetician that ever breathed, and he reduceth thy days into a short number". This referred to a person who carried out calculations (or computations), and the word continued with the same meaning until the mid-20th century. From the end of the 19th century, the word began with to take on its more familiar meaning, "a machine that carries out computations".

The history of the modern computer began with 2 separate technologies, namely *automated calculation* and *programmability*. No single device can be identified as the earliest computer because of the inconsistent application of the term. A few are worth mentioning which were very successful and survived for centuries until the advent of the electronic calculator. These include: *Abacus* (the oldest and the ancient calculator in existence for mathematical calculation and to transact business since 2700 BC), the *slide rule/slide ruler* (also known as *slipstick* in USA, a mechanical analog computer invented around 1620 – 1630 after the publication of the concept of logarithm by John Napier, and were carried on Project Apollo space missions and also flew with Buzz Aldrin to the moon on Apollo 11), the *astrolabe* (an elaborate inclinometer used by astronomers, navigators and astrologers for locating and predicting the positions of the sun, moon, planets and stars and also for determining the local time, surveying and triangulation), and the *Antikythera mechanism* (an ancient astronomical computer built by the Greeks around 88 BC and generally referred to as the first known analog computer designed to predict the astronomical positions and eclipses for calendrical and astrological purposes as well as the cycles of Olympic games).

The Renaissance saw the invention of the mechanical calculator (1642), a device that could perform all 4 arithmetic operations (addition, deletion, multiplication and division) without relying on human intelligence. This was at the root of the development of computers in 2 ways, namely, that the first computer was theorized by Charles Babbage and then developed, and the development of a low-cost electronic calculator (successor to the mechanical calculator) resulted in the development of the first commercially available microprocessor integrated circuit developed by the Intel.

The first electronic digital computers were developed between 1940 and 1945 in the UK and USA. In the beginning, the computers were of the size of a large room, and consumed the power of several hundred modern personal computers. The mechanical analog computers were mainly used for the military applications. The modern computers (based on the integral circuits) are millions to billions of times more capable than the early machines, and these occupy only a fraction of the space compared to the original ones. The simple computers are small enough to fit into mobile devices, and the mobile computers can be powered by small batteries. The personal computers (PCs) are the icons of the information age and these are what most people think of as computers. The embedded computers found in many devices from MP3 players to fighter aircraft and from toys to the industrial robots are the most numerous.

Joseph Marie Jacquard (1801) made an improvement to the textile loom by introducing a series of punched paper cards as a template which allowed his loom to weave intricate patterns automatically. Charles Babbage (1837) was the first to conceptualize and design a fully programmable mechanical computer, his *analytical engine*. His son, Henry Babbage (1888) completed a simplified version of the analytical engine's computing unit, the *mill*, and he gave a demonstration (1906) of its use in computing tables, which was given to the Science Museum in South Kensington in 1910. Ada Lovelace, an analyst of Babbage's Analytical Engine, between 1842 and 1843, translated an article by Luigi Menabrea, an Italian Military

Engineer, on the engine supplemented with an elaborate set of notes of her own, called *Notes*, which contained the first computer programme – an *algorithm* encoded for processing by a machine. She also developed a vision on the capability of a computer to go beyond mere calculating or number-crunching. Alan Turing (1936), the *Father of Modern Computer Science*, provided formalization of the concept of algorithm and computation with his *Turing Machine* and provided a blueprint for the electronic digital computer.

Herman Hollerith (late 1880s) invented the recording of data on a machine-readable medium. The initial trials were with the paper tape but he settled on punched cards. He invented the tabulator and the key punch machine to process the punched cards. These (3 inventions) were the foundation of the modern information processing industry. By the end of the 19th century, a number of technologies like Boolean algebra, the vacuum tube (the thermionic valve), punched cards and tape and the tele-printer had begun to appear. Many scientific computing needs were met by sophisticated analog computers, but these lacked the accuracy of modern digital computers. *ABC* (Atanasoff-Berry Computer) was the world`s first electronic digital computer (which is not programmable) and is considered to be *one of the fathers of the computer*. This was conceived by Prof. John Atanasoff (1937), Physics professor, Iowa State College, and built with the assistance of his graduate student, Clifford Berry. It was designed only to solve systems of linear equations but not programmable and did parallel computation. *ABC* used vacuum tube based computation, binary numbers, and regenerative capacitor memory

During 1930s and 1940s, a succession of steadily more powerful and flexible computing devices was constructed. The very first programme-controlled computer – Z3, was invented by a German inventor, Konrad Zuse (1941). This was the first working, programmable electro-mechanical computing machine, featuring binary arithmetic including floating point arithmetic and a measure of programmability and was the world`s first operational computer. Hence, Zuse was regarded as the *inventor of the computer*. *Colossus* was the first programmable electronic computer built by Tommy Flowers (1943). The *Father of the Modern Digital Computer*, George Stibiltz (1937) while working at Bell Labs invented and built a relay-based calculator, *Model-K* (for Kitchen Table on which it was assembled) which was the first to use binary circuits to perform an arithmetic operation. The secret British Computer, *Colossus* (1943) used for breaking the German wartime codes) was constructed followed by the development of a large scale electro-mechanical computer by Harvard Mark (1944). The first general purpose electronic computer, *ENIAC* (the *Electronic Numerical Integrator and Computer*) was developed by US Army`s Ballistic Laboratory (1946). This was followed by designing of *SSEM* or *Baby* (the Manchester *Small Scale Experimental Machine* (1948) on the basis of a design described by John von Neumann in his paper: "*First Draft of a Report on the EDVAC*". A year later (1949), *EDSAC* (the *Electronic Delay Storage Automatic Calculator*) was completed at the Cambridge University and EDVAC (the *Electronic Discrete Variable Automatic Computer*) was completed shortly thereafter. Sergei Sobolev and Nikolay Brusentsov (1950s), the Soviet scientists at Moscow State University, designed *Setun*, a functional ternary computer. During 1960s, the computers using vacuum tubes (as electronic elements) were replaced by transistor-base machines which were smaller, faster, cheaper to produce, required lesser power and more reliable. The first transistorized computer was demonstrated at the University of Manchester (1953). Integrated circuit technology and the creation of microprocessors like Intel-4004 (1970s) decreased the size and cost and increased the speed and reliability of computers. The home computers and the personal computers were produced during 1980s. The smart phones of the modern day (the present day) are fully programmable computers and the most common form of computers in existence.

COMPUTERS – APPLICATIONS IN LIFE SCIENCES

In life sciences, particularly genetics and community ecology, the digital computers are becoming increasingly valuable. This is mainly because of the increasing complexity of the data that must be processed, and is the value of the computers in simulating the changes that occur in populations over long periods of time. Increase in the parameters that must be taken into account to describe the relationships between the populations and their environment. The computers can solve complex problems in a short period of time that the biologists can solve within a longer time from their existing facilities.

The plants serve as the conduit of energy into the biosphere, provide food materials used by human beings and they also shape our environment. The future of plant science is the key that addresses the major challenges that face the humanity in the 21st century and the development of new technology is the key to transforming plant biology in order to meet the needs of the human being. The three major challenges that face humanity in our times are food, energy and environmental degradation which are plant related. All the food is produced by the plants, either directly or indirectly via the animals that eat them (producers and consumers). Plants are also a source of energy production and they are also intimately involved in climate change and a variety of environmental concerns including the expansion of agriculture and its impact on the destruction of the habitat and pollution of water bodies. The changing climate is responsible for the additional stress on the supply of food and also other habitats. Hence, research in plant sciences is instrumental in satisfactorily addressing all these problems which is critically associated with the application of advanced technologies both to test the existing hypotheses and also to gain new information and generate fresh hypotheses.

The technology can be divided into three categories namely, existing technology that is not being applied for all of its potential uses, new and readily envisioned technology and the technology we would like to have but don`t know how to create. The technological overview includes the expanding existing technologies such as DNA sequencing, RNA cataloguing, mass spectroscopy, fluorescence-based microscopy and electron microscopy as well as advanced imaging technologies among many others.

Dramatic improvements in DNA sequencing technology have led to a near exponential increase in the volume of sequenced data being generated worldwide. The current generation of sequencing platforms is capable of generating millions of sequenced genomic fragments in a single overnight run. The size of data generated in a typical sequencing experiment is in the order of several gigabytes to a few terabytes. Besides the need for developing novel methods/algorithms that can efficiently analyze, interpret and extract biologically meaningful information from the generated data, it is imperative to simultaneously develop methods that enable efficient storage / archival and dissemination / transmission of such huge volumes of data.

Working with the Center for Computational Biology at the University of California at Berkeley, the scientists are developing methods for the interpretation of the genome variation and they have developed a pipeline for the automated execution for the exome and whole genome sequencing starting from the mapping of raw reads to the calling of variants (Human genome analysis).

The project in computational structural biology uses knowledge-based and high performance computational methods to understand the role of the structure, interaction and dynamics of biological molecules to their function to identify druggable targets for pathogenic diseases (Computational structural biology). The Drug Development R&D Labs activities are focused on pharmaco-kinetic and pharmaco-dynamic (PK-PD) modeling of safety and efficacy data, obtained from pre-clinical and clinical experiments.

The researchers in life sciences routinely deal with large data to be analyzed by applications that need considerable computing power, including advanced parallel and distributed infrastructures, such as grids/clouds, clusters, and many-cores, multi-cores and FPGA architectures. Managing these data, typically terabytes or petabytes in size (usually referred to as Big Data), and optimizing the applications for these architectures is one of the challenges that must be tackled. At the same time, Big Data issues are three-dimensional, covering their increasing volume, variety and velocity. Therefore, there is a clear need for design and implementation of new scalable approaches for dealing with the associated information overload and cognitive complexity issues. High-throughput techniques for DNA sequencing, gene expression analysis, healthcare data analysis, etc., have led to an exponential increase in the volume of biological data, which in turn are geographically spread in several laboratories and data repositories.

Grid for the Life Sciences is an environment that allows sharing of resources, in which heterogeneous and dispersed health data as well as applications can be accessed by all users as a tailored information providing system according to their authorization.

Cloud computing is emerging as a model for enabling convenient, on demand network access to a shared pool of configurable resources that can be rapidly provisioned and released with minimal management effort or service provider interaction. Institutions and medical professionals who frequently do not have enough storage and computing resources can manage their biomedical information through applications built on top of these types of services, accessing advanced computing infrastructures that they could not afford otherwise. While many companies, like Google, IBM, Amazon, and Microsoft, were early adopters of cloud computing, its application to biomedicine has been only recently proposed, mainly for bioinformatics applications. Several reviews and research papers have reported on these, while in contrast, applications in medical informatics have been infrequent. Now, medical applications are increasingly proposed and implemented, though reports are still scarce from a research and academic perspective.

The main goal of the track is to exchange ideas and results related to ongoing big data, grid and cloud computing research in Biomedicine, Life Sciences and correlated disciplines, such as Climate Change, that impact on human health, focusing on different aspects of middleware, technologies and applications. Bio-computers use systems of biologically derived molecules, such as DNA and proteins, to perform computational calculations involving storing, retrieving, and processing data. The development of bio-computers has been made possible by the expanding new science of nano-biotechnology. Bioinformatics is an interdisciplinary scientific field that develops methods for storing, retrieving, organizing and analyzing biological data. A major activity in bioinformatics is to develop software tools to generate useful biological knowledge.

Computational neuroscience is the study of brain function in terms of the information processing properties of the structures that make up the nervous system. It is an interdisciplinary science that links the diverse fields of neuroscience, cognitive science, and psychology with electrical engineering, computer science, mathematics, and physics. Computational neuroscience is distinct from psychological connectionism and from learning theories of disciplines such as machine learning, neural networks, and computational learning theory in that it emphasizes descriptions of functional and biologically realistic neurons (and neural systems) and their physiology and dynamics. These models capture the essential features of the biological system at multiple spatial-temporal scales, from membrane currents, proteins, and chemical coupling to network oscillations, columnar and topographic architecture, and learning and memory. These computational models are used to frame hypotheses that can be directly tested by biological and/or psychological experiments.

Medical imaging is the technique and process used to create images of the human body (or parts and function thereof) for clinical purposes (medical procedures seeking to reveal, diagnose, or examine disease) or medical science (including the study of normal anatomy and physiology). Although imaging of removed organs and tissues can be performed for medical reasons, such procedures are not usually referred to as medical imaging, but rather are a part of pathology

Computers in pharmaceutical research and development help in understanding diseases: mining complex systems for knowledge, scientific information handling and enhancing productivity, computers in drug discovery, computers in preclinical development, computers in development decision making, economics, and market analysis, computers in clinical development and future applications and future development. The use of computers in all phases of drug discovery, development, and marketing explains how applications are used at various stages, including bioinformatics, data mining, predicting human response to drugs, and high-throughput screening.

COMPUTERS IN AGRICULTURAL RESEARCH

Agricultural research has benefited from technological advances and the industrial age has brought mechanization and synthetic fertilizers. The information age has also brought the potential for integrating the technological industrial advances into sustainable agriculture production. The application of computers in research helped to exploit the conversion of statistical formulae or complex models in digital form for easy and accurate calculation. The computers have been successfully used for mechanization, automation and also to develop decision support system for taking decisions on the agricultural production and protection research. Very recently, remote sensing and geographic information system has a place in major and crucial role in agriculture research especially in prediction of the yield, suitability of soil for a particular crop and site specific resource allocation of agriculture inputs etc. Other questions, such as climate change impact on the life sciences, and the disciplines to address them, are adding to the deluge of data: many prevalent human diseases are linked to climate fluctuations, from heat stress, cardiovascular mortality and respiratory illnesses due to heat waves, to altered transmission of infectious diseases and malnutrition connected to crop failures.

COMPUTERS AND PLANT SCIENCES

Botany is the key to address the major challenges that face the humanity in the 21st century. David Ernhardt and Wolf Frommer of Carnegie University argue that the development of new technology is the key to transforming the plant biology in order to meet the human needs. The plants serve as the conduit of energy into the biosphere and to provide the food and materials used by humans and they shape our environment. The three major challenges that face the humanity in our time are the food, energy and environmental degradation, and all the three are plant-related. All of our food is produced by the plants, either directly or indirectly via the animals that eat them. The plants are a source of energy production and they are intimately involved in climate change and a major factor in a variety of environmental concerns, including agricultural expansion and its impact on habitat destruction and waterway pollution. None of these issues are independent of each other. For example, the climate change places additional stress on the supply of food and on also different habitats. Hence, plant research is instrumental in addressing all these problems and moving into the future, and to move significantly forward, technological development is critical both to test the existing hypotheses and also to gain new information to generate fresh hypotheses.

Such a technology is divided into three categories, namely, existing technology that is not being applied for all of its potential uses; new and readily envisioned technology; and technology we would like to have but do not know how to create. Such a technological overview includes expanding technologies like DNA sequencing, RNA cataloguing, mass spectroscopy, fluorescence-based microscopy, and electron microscopy etc, and the main focus is on the advances possible through advanced imaging technologies.

The researchers of Michigan State University (MSU) through a collaboration between plant and computer scientists figured out how certain genes in a plant (*Arabidopsis thaliana* which contains more than 25,000 genes of which about 3,000 to 10,000 are used to cope with extremes under just one particular stress) turn on and off depending on the stress the plant is experiencing, be it cold or heat, drought or too much water, or bacterial infection.

When a plant goes into defense mode to protect itself against harsh weather or disease, that is good for the plant, but bad for the farmer growing the plant, bad because when a plant defends itself, it turns off its growth mechanism. The researchers of MSU have successfully figured out how the plants can make the decision between growth and defense, a finding that could help the plants strike a balance – keep safe from harm while continuing to grow. Shen Yang He (Professor, Plant biology, MSU) and his team found that the two hormones that control growth (*gibberellins*) and defense (*jasmonates*) come together in a crisis and figure out what to do.

Randy Beaudry and Ryan Warner, AgBioResearch horticulture scientists of MSU, in collaboration with Jones` laboratory, developed a procedure to perform molecular profiling of 1000s of *Stevia* plants using a mass spectrometer. The technique helped to identify the individual plants that consistently and reliably produced sweet compounds and guided plant breeding efforts to improve the quality and yields of *Stevia* sweeteners.

Robert Last and Cornelius Barry, AgBioResearch scientists of MSU and Prof. Eran Pichersky of University of Michigan along with Jones worked on a project that focused on the specialized cells on the surfaces of the tomato plant that produced the scent that lingers after brushing against it, and the chemicals responsible for this smell make up an arsenal of chemical defenses, many of which accumulate in the trichomes that line the leaves and stems. The knowledge of the genes responsible for these chemicals may provide the clues to what regulates the production of these chemicals in other plants. By understanding how the trichomes produce these chemicals, the scientists may be able to breed (engineer) plants that make more of them and some of these genes can be reintroduced from wild relatives of tomato into conventional varieties of tomato to make them more insect-resistant. It is also possible that the discovered genes can be finally used to launch another biotech revolution; generating new industries that make valued components from plants.

The initiative of the Medicinal Plant Consortium funded by the National Institute of Health, USA, was particular to understand what plant genes are responsible for making the medicinal compounds in plants like common periwinkle flower which produces tiny quantities of two compounds (vincristine and vinblastine) used to treat certain cancers. The understanding and the development of the genes that are involved in the production of these compounds offers a promise for improving the medicines and decreasing their cost of production.

The plants are a source of food or medicines or ornamentals that add beauty and colour to our surroundings, and plants are also a crucial component of nature systems that protect both the environment and human health. The plants play a very active role and help to mitigate various pollutants (Dawn Reinhold, Assistant Professor, Department of Biosystems and Agricultural Engineering, MSU). Her research has demonstrated that plants have a direct effect on improving the outcome of environmental fate by decreasing the concentrations of the pollutants (triclosan and triclocarbon) in water and soil which can potentially affect human health. Similarly, the medicinal plant blueprints (genes) can also be decoded to advance drug discovery and development (for example, in plants like foxglove, periwinkle, ginseng and other medicinal plants found in the household gardens represent cornucopias of compounds with diverse medical applications). The discovery of the new gene

(CLMP1) from clumped chloroplasts (a new class of proteins which help chloroplast to divide and multiply) helps to envision more efficient energy production in plants.

IMPORTANCE OF COMPUTER LITERACY AND USEFUL SOFTWARE FOR STUDENTS AND RESEARCHERS IN LIFE SCIENCES

The computer literacy plays a critical role in today's research in life sciences, because without the ability to use the computers to efficiently manipulate and analyze large amounts of data that results for biological experiments and simulations, many of the pressing questions could not be clearly answered. Despite the ubiquity of computers in their lives, today's undergraduates seem to be largely unfamiliar with how the computers are being used to pursue and answer such questions efficiently. For every college student of today, the computer is virtually a must to own or have an easy access to a computer for him/her to succeed in course work. Almost every assignment in the classroom requires some computer work, be it researching literature, performing the calculations, or writing a report. In addition, the students and young people take advantage of a variety of online communication resources like *e-mail*, *instant messengers* or *social websites* like *MySpace*, *Facebook* and *Twitter,* and thus are familiar with many types of computer software. Though many students are far more adaptable to computers and their applications than older generation, they seem to be daunted by the more sophisticated uses of the computer, like data manipulation and analysis for scientific research. This is partially due to the insufficient preparation in terms of basic mathematical skills, especially logic and statistics which are indispensable in life sciences research. The purpose of computer literacy for life sciences is for students to develop a hands-on (and brains-on) working experience in the use of standard computer software tools as well as computer techniques and methodologies used in research. The researchers should be familiar with the operating systems of the computer and its main and most useful applications. Also they must know how to effectively manipulate the documents and data files including those downloaded from the libraries and data repositories available on World Wide Web (WWW). Similarly, the scientists must be able to efficiently manage and query the data that results from wet lab experiments as well as those derived from the simulations of the computer. These data should then be subjected to statistical analysis, or other means of data mining, to answer the underlying scientific questions to justify their research work. Finally, the researchers should be able to present their discoveries as reports or scientific articles as well as audio-visual presentations. The nine (9) important software tools that help boost the productivity of research students are: *Wunderlist* (for task management), *Mendeley* (for literature management), *Evernote* (for handling information deluge), *Dropbox* (for file back-up, sharing and sync), *Virtual Box* (for using software from other operating systems), *Microsoft Word* (for writing), *Microsoft Excel* (for handling data and plots), *Microsoft PowerPoint* (for the preparation of slides, presentations etc.), *Google Chrome* or *Mozilla Firefox* or *Microsoft Internet Explorer* (for browsing the web) etc.

The useful software for studies in molecular biology, biotechnology and bioinformatics include: *AMIADA* (an integrated computer programme for organizing, exploring, visualizing and analyzing microarray data), *Cluster* (to perform a variety of cluster analysis and other types of processing on large scale microarray datasets), *ScanAlyze* (to process fluorescent images of microarrays, semi-automatic definition of grids and complex pixel and spot analyses), *OligoWiz* (to provide intelligent design of oligonucleotides for DNA microarrays), *SeWeR* (**SE**quence analysis using **WE**b **R**esources – to common web-based services in bioinformatics), *JEMBOSS* (**J**ava version of the **E**uropean **M**olecular **B**iology **O**pen **S**ource **S**oftware suite), *StarORF* (the open reading frame finder), *blaSTOR* (to store and analyze the nucleotide and protein sequences), *prefectBLAST* (to allow the researchers to do nucleotide or amino acid sequence similarity searches against public databases locally stored), *BugView* (a genome browser for comparing the arrangement of genes on a pair of related genomes and also to view individual genomes), *CGView* (to generate high quality, zoomable maps of circular genomes complete with labels, a title, legends and footnotes), *ACT* (**A**rtemis **C**omparison **T**ool – to allow an interactive visualization of comparisons between complete genome sequences and associated annotations), *Gepard* (**GE**nome **PA**ir – **R**apid **D**otter – to allow the calculation of dot plots for large sequences like chromosomes or bacterial genomes), *GenoViz* (an open source framework designed for the rapid assembly of visualization software applications for genomics), *DNAPlotter* (an interactive application for generating the circular and linear representation of the genomes), *Picky* (an oligo-microarray design programme to identify the probes that are very unique and specific to input sequences), *GelScape* (to allow the analysis of standard 1-D and 2-D protein gels), *FLICKER* (an open source computer programme for visual comparing 2-D gel images), *MPEx* (**Mem**brane **P**rotein **Ex**plorer – a tool for exploring the topology and other features of membrane proteins by means of hydropathy plots using thermodynamic principles), *PFAAT* (**P**rotein **F**amily **A**lignment **A**nnotation **T**ool – is a protein sequence alignment application designed to facilitate the analysis, curation and annotation of large protein sequence families), and *ISHAN* (a flexible platform for performing fast homology analysis and molecular phylogenetic studies on proteins and DNA sequences) etc.

Similarly, the software that is useful in phylogenetic studies, graphic packages, analysis of DNA, RNA and genomes, protein analysis, etc, include: *TREECON* (for the construction and drawing of phylogenetic trees on the basis of evolutionary distances inferred from nucleic acid and amino acid sequences), *TREEMAP* (for visually comparing the host and parasite phylogenies and interactively create the reconstructions of the history of the host-parasite association, perform randomization test of the tree similarity and compare the branch lengths in two trees), *Tree finder* (which computes phylogenetic trees from the nucleotide sequences), *GeneTree* (an experimental programme for comparing the gene and species trees), *TinyQuant* (a graphical display programme designed for analysis and limited manipulation of images obtained by scanning of gels and autoradiographs), *UTHSCSA ImageTool* (to acquire, display, edit, analyze, process, compare, save and print gray scale and colour images), *GIMP* (a software for photo retouching, image composition and image authoring), *ACD/ChemSketch 11.0* (for drawing chemical structures and graphical images) etc.

The software used in plant studies include: BG-MAP (**B**otanical **G**arden **Map**ping System – gives all the information on computerized mapping of plant collections housed at facilities like botanical gardens, arboreta, zoos and large private sites), WINLABEL (Herbarium Lable Making Software – is a database programme designed to maintain the records of plants collected for placement in a herbarium, and its primary purpose is to maintain detailed records of collected plant specimens in a format that permits to generate one or more herbarium labels for each specimen in a clean, consistent and professional manner, supports database entry for any state, country or province with a header and footer text and also options for printing), PANKEY (a package of programmes which provides specialized tools for problems of diagnosis or identification in biology or medicine; the programmes process the descriptive data that is presented in a standard format [DELTA – **DE**scription **L**anguage for **TA**xonomy] and generates the output suitable for use during the preparation of monographs, handbooks, floras or faunas), NATUREFILE (for wildlife recording and collection management), LUCID (an interactive programme to view and construct the identification keys), GRAFICA NATURA (a real-time programme for the visualization of natural phenomena such as plant growth with specific parameters fed into the computer), BIOPAK (a software for computing plant biomass) and DECODA (**D**atabase for **E**cological **CO**mmunity **Da**ta – a programme for the storage, manipulation, retrieval, and analysis of the data collected by ecologists for the study of the composition or the structure of biological communities) etc.

■■■

CHAPTER - 16

APPLICATIONS OF RADIOISOTOPES IN BIOLOGY/PLANT SCIENCES

Introduction

The use of radioisotopes in life sciences has provided us with the new techniques by which we can unravel many of the scientific complexities of the various basic life processes. In the plant sciences, we have been able to intensify our efforts to understand the selective uptake by roots and leaves of a variety of radioisotopes, the movement of these in the soil, and the factors that modify these things. These efforts have produced much fundamental knowledge of soils, of plant metabolism, plant genetics, and of the ecological relationships of living things in the artificial and natural habitats. Radioisotopes mainly come from 3 sources namely: from nature such as radon in the air or radium in the soil; from machine-produced nuclear interactions in devices like linear accelerators and cyclotrons; or from nuclear reactors. The periodic table currently comprises over 3000 isotopes of 114 elements while another 4000 undiscovered isotopes are theoretically believed to exist. The periodic table will now see four more isotopes being added to its fag end. H. M. Devaraja from MCNS (Manipal Centre of Natural Sciences), Manipal University, Karnataka, in collaboration with an international team at GSI – Giessen, Germany, have discovered, [published in the journal *Physics Letters B*], four new atomic nuclei to be added to the chart of the nuclides in the periodic table – one isotope each of the heavy elements like berkelium (Bk – atomic number 97) and neptunium (Np – atomic number 93) and two isotopes of the element americium (Am – atomic number 95) [Courtesy: The Hindu, Science and Technology, Page 02, 21 September 2015, Monday].

RADIOACTIVE ISOTOPES

More than 100 years ago, the scientists discovered that many elements commonly found on earth occur in different atomic configurations, and these called isotopes, were found to have identical electronically charged particles and identical chemical properties, but different atomic weights and physical properties. Isotopes are the two or more species of atoms of the same chemical elements which contain the same number of protons in the nucleus (same atomic number) and same position in the periodic table and nearly identical chemical behaviour but different number of neutrons (with different atomic masses and physical properties - for example, carbon-12 has 6 neutrons while carbon-14 has 8 neutrons). Radio-active isotopes are the isotopes which are unstable (that is, their nuclei does not hold together well and therefore break down making the compound radio-active) and undergoes radio-active decay while emitting heat and radiation. The practical applications of these isotopes (*aka radioisotopes*) are very many and are used to study the living organisms, to diagnose and treat the diseases, to sterilize the medical instruments, to sterilize the food, to produce the energy for the generation of heat and power, and also to monitor different steps in all types of industrial processes. Some of the common radioisotopes used in medicine are: Molybdenum-99, Technetium-99, Chromium-51, Cobalt-60, Copper-64, Iodine-123 and 125, Iron-59, Iridium-192, Phosphorus-32, Potassium-42, Selenium-75 and 92, Sodium-24, Strontium-89, Xenon-133, Carbon-11, Nitrogen-13, Oxygen-15 and Fluorine-18 etc.

HALF-LIFE ($t_{1/2}$)

Half-life is the amount of time required for a quantity to fall to half its value as measured at the beginning of the time period. While the term – *half-life* – can be used to describe any quantity which follows an exponential decay, it is most often used within the context of nuclear physics and nuclear chemistry – that is the time required, probably, for half of the unstable, radioactive atoms in a sample to undergo radioactive decay. The original term, was *half-life period* (dating to Rutherford`s discovery in 1907 – studies of age determination of rocks by measuring the decay period of radium to lead-206), which was shortened to half-life in the early 1950s. This is constant over the life time of the decaying quantity and is a characteristic unit of the exponential decay equation. Generically, the term may be used to refer to any period of time in which a quantity falls by half, even if the decay is not exponential.

HALF-LIFE IN BIOLOGY AND PHARMACOLOGY

A *biological half-life* or *elimination half-life* is the time it takes for a substance (for example, a drug, radioactive nuclide, or any other) to lose one-half of its pharmacological, physiological, or radiological activity. In medical context, the half-life may also describe the time that it takes for the concentration in blood plasma of a substance to reach one-half of its steady-state value (the *plasma half-life*). The relationship between the biological and plasma half-lives of a substance can be complex, due to factors including accumulation in tissues, active metabolites and receptor interactions. While a radioisotope decays almost perfectly according to the `first order kinetics` where the rate constant is a fixed number, the elimination of a substance from a living organism usually follows more complex chemical kinetics (for example, the biological half-life of water in a human being is about 9 to 10 days, though this can be altered by the behavior and various other conditions.

RADIOACTIVE TRACERS

Because the chemical properties and reactions of a radioisotope are exactly the same as those of a nonradioactive isotope of the same element, a radioisotope can be substituted for a stable isotope of the same element in a molecule or compound without changing the chemical properties of the compound. Such a compound is said to be `tagged` or `labeled`. Because the radiation emitted by the radioisotopes can be detected some distance from the radiating atom, the progress of those atoms through the body can be followed. Hence, such labeled compounds are called `tracers`.

The most common application of radio-isotopes is as tracers which are radio-active isotopes of an element whose pathway through which a chemical reaction can be followed. These are commonly used in the medical field and also in the study of plants and animals. The tracers can be used within a plant or animal, to follow the movement of certain chemicals, while in medicine, they can be used to identify abnormal bodily processes (imaging), testing of new drugs and for conducting research into cures for diseases.

TRACER STUDIES/TRACER TECHNIQUES

The tracer technique involves the substitution of a relatively few radioisotopes of an element for the stable isotope of the element. By means of radiation detection apparatus like Geiger-Muller counter the radioisotope can then be followed through the various biological processes involved. The radioisotope is chemically the same as the stable element and reacts in the same biological manner. The physical difference relates to its radiation properties which make it useful.

The simplest radioactive tracer studies consist of the tagging of a biological entity with a radioisotope and the entity is then tracked by following the radiation from the isotope. The operation becomes more complex when a large number of biological particles are labeled (RBCs or bacteria). When the labeled substance is injected into an animal, it is impossible to follow individual labeled particles, but their average movement can be tracked by observations of the radiation. Finally, a radioisotope of a particular element can be used to tag that element. For example, phosphorus-32 can be introduced into the soil where a plant is growing, and the amount of phosphorus absorbed and its distribution throughout the plant can be studied. In most biological tracer experiments, the radioisotope is introduced into the system and its radiation subsequently measured with Geiger-Muller counter or scintillation detector. The low intensity (extremely soft) radiations can be detected by the use of photographic film.

In medical applications, a radioactive atom can be attached to a molecule or more complex substance, which can then be used to examine a chemical reaction in a test tube, or it can be administered to a patient by ingestion or injection and subsequently incorporated into a biochemical process. The radioactive emissions from the radioactive atom can be used to track the behavior of the labeled molecule or substance in biological processes by means of medical imaging utilizing the techniques such as PET (Positron Emission Tomography) or SPECT (Single-Photo-Emission Computed Tomography). The branch of medicine that uses the radioactive tracers is called *nuclear medicine*. The radiotracers of practically every element can be produced in nuclear reactors or cyclotrons. The radioactive tracers are used as a part of the diagnostic process and 3 radionuclides – carbon-14, hydrogen-3 and phosphorus-32 remain the backbone of modern biomedical sciences.

The radioactive tracers are the substances that contain a radioactive atom to allow easier detection and measurement (Radioactivity is the property possessed by some elements of spontaneously emitting energy in the form of particles or waves by disintegration of their atomic nuclei). For example, it is possible to make a molecule of water in which one of the two hydrogen atoms is a radioactive tritium (hydrogen-3) atom which behaves in almost the same way as a normal molecule of water. The main difference between the tracer molecule containing tritium and the normal molecule is that the tracer molecule continually gives off radiation that can be detected with a Geiger-Muller counter or some other type of radiation detection instrument. One important application for the tracer molecule (described earlier) would be to monitor plant growth by watering the plants with it. The plants would take up the water and use it in leaves, roots, stems, flowers, fruits and seeds in the same way it does with normal water. It would be possible to find out how fast the water moves into any one part of the plant. One would simply pass a Geiger-Muller counter over the plant at regular intervals and see where the water has gone.

Another application of tracers might be in scientific research on plant nutrition. If a scientist wants to find out how plants use some nutrient such as phosphorus, he could feed a group of plants with a fertilizer that contains radioactive phosphorus. As the plant grows, the location of phosphorus can be detected by the use of a Geiger counter. Another way to trace the movement of phosphorus would be to place a piece of photographic film against the plant. The radiation from the phosphorus tracer would expose the film, in effect taking its own picture of its role in plant growth.

INSTRUMENTS TO MEASURE IONIZING RADIATION

The two common instruments that are used to measure ionizing radiation in use are Geiger-Muller counter and Scintillation counter.

a. GEIGER-MULLER COUNTER (GM COUNTER)

This is also called a Geiger counter/Geiger-Muller tube and this is a device normally used for the detection and measurement of all types of ionizing radiation, i.e., alpha (a single sheet of paper will stop the alpha particle because it has very little penetrating power though it is heavy), beta (a thin sheet of metal will stop a beta particle as it is more penetrating than alpha particle though it is lighter) and gamma (have no charge or weight and can be extremely energetic and highly penetrating) radiation. It detects radiation using the ionization produced in a Geiger-Muller tube, and this gives the name of the instrument as such, and is one of the world's best known radiation instruments. The original detection principle was discovered in the year 1908, but, it was not until the development of the Geiger-Muller tube (1928) that it became a popular instrument for use in radiation dosimetry, radiological protection, experimental physics and the nuclear industry.

1. History and G-M Counter

Hans Geiger (1908) under the supervision of Ernest Rutherford at the then Victoria University of Manchester (at present the University of Manchester) developed an experimental technique for detecting the alpha particles that would later be used in the G-M tube. This (*early alpha particle counter*) was only capable of detecting the alpha particles and was part of a larger experimental apparatus. The fundamental ionization mechanism used was discovered by John Sealy Townsend (between 1897 and 1901), and is known as the *Townsend discharge* which is the ionization of molecules by ion impact. It was in 1928 that Geiger and Walther Muller (his student) developed the sealed G-M tube (relatively small and rugged and could detect more types of ionizing radiation) which developed the basic ionization principles previously used experimentally. A G-M counter was born thus which could be produced relatively cheaply. The modern versions of G-M counter use the halogen tube (invented by Sidney H Liebson in 1947) and this supersedes the earlier Geiger tube because of its much longer life and lower operating voltage (between 400 and 600 volts).

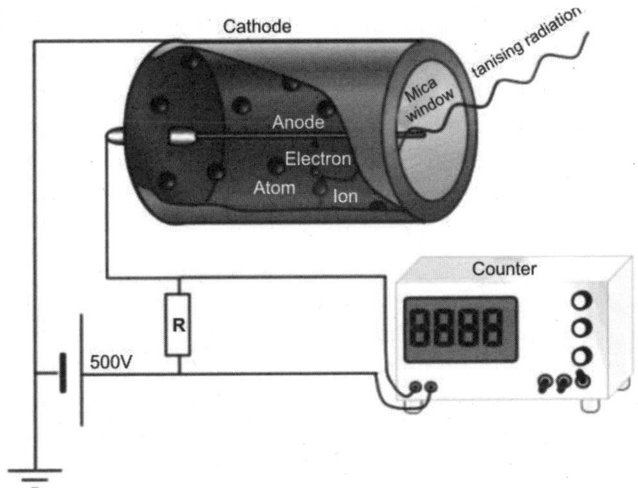

G-M Counter

2. Principle of Operation

A Geiger-Muller tube consists of a sealed metallic tube filled with argon or another noble (inert) gas (helium, neon or argon) mixed with a small amount of alcohol vapour or bromine gas. The argon gas is called the *detecting gas* whereas the bromine gas or alcohol vapours are referred to as the *quenching gas*. The gas mixture inside the tube is at a pressure below atmospheric pressure. A thin metal wire runs through the centre of the tube. An electric potential of up to 1 kilovolt is maintained between the metal wire (the anode) and the cylinder (the cathode). In the absence of any radiation no current flows between the wire and the cylinder.

When a radioactive particle enters the tube it ionizes an argon atom. The resulting electron is accelerated towards the metal wire or anode. As the electron approaches the metal wire it experiences an increasing electric field strength which in turn applies a greater accelerating force on the electron. The accelerating force becomes so strong that on collision with other argon atoms the electron can ionize them. The electrons from these ionizations can go on to generate a cascade of further electrons, an effect called the *avalanche effect*. The ionization by one particle can result in millions of electrons striking the metal wire. This migration of electrons inside the tube results in an electric discharge. This gives a measurable voltage pulse in the external circuit of the Geiger-Muller tube. The counter registers the number of pulses and converts them into sound signals or displays them as a measure on the screen.

This device basically consists of a pair of electrodes (that have a high voltage across them) surrounded by a gas. The Geiger-Muller tube detects the radiation and processes electrons which display the result. The tube is filled with an inert gas like helium, neon or argon at a low pressure to which a high voltage is applied. For a brief period the tube conducts electrical discharge when a particle (photon) of the incident radiation makes the gas conductive by ionization. The ions (and the electrons) are attracted to the electrodes and an electric current is produced. A *scaler* counts the current pulses and one can obtain a count whenever the radiation ionizes the gas within the tube. The ionization is considerably amplified within the tube by the Townsend avalanche effect to produce an easily measured detection pulse, which is fed to the processing and display electronics. Such a large pulse from the tube makes the G-M counter relatively cheap to manufacture, as the subsequent electronics is greatly simplified. The electronics also generates the high voltage (typically 400-600 volts) that has to be applied to the G-M tube to enable its operation.

3. Readout

The radiation read-out is of two types, namely, counts or radiation dose. The count display is the simplest and is the number of ionizing events displayed either as the count rate (counts per second) or as a total over a set times period (an integrated total). This is normally used when the alpha or beta particles are being detected. The radiation dose is more complex, and this is a display of radiation dose rate, which is displayed in a unit such as the *sievert* which is normally used for measuring gamma or X-ray dose rates. A G-M tube can detect the presence of radiation but not is energy which influences the radiation's ionizing effect. The instruments that measure the dose rate require the use of an energy compensated G-M tube so that the dose displayed relates to the counts detected. The electronics will apply known factors to make this conversion, which is specific to each instrument and is determined by the design and calibration. The readout can be either analog or digital, and increasingly, the modern instruments offer serial communications without a host computer or a network. The instrument usually has an option to produce the audible clicks that represent the number of ionization events detected, and this is the distinctive sound normally associated with hand-held or portable G-M counters, and this allows the user to concentrate on the manipulation of the instruments whilst retaining auditory feedback on the radiation rate.

4. Limitations

The G-M counter has mainly two limitations. The tube cannot differentiate between the radiation types as the output pulse from a G-M tube is always of the same magnitude regardless of the energy of the incident radiation. The second limitation is the inability to measure high radiation due to the *dead time* (an insensitive period after each ionization of the gas during which any further incident radiation will not result in a count, and the indicated rate is lower than the actual) of the tube. The dead time will reduce indicated count rates above about 10^4 to 10^5 counts per second depending on the characteristic of the tube being used.

The application and the use of a G-M counter is dictated entirely by the design of the tube, generally categorize as *end-window* or *windowless thin-walled* or *thick-walled* and sometimes hybrids of these types.

b. SCINTILLATION COUNTER

A scintillation counter is a device for detecting and measuring ionizing radiation by the use of excitation effect of the incident radiation on a scintillator material and detecting the resultant light pulses. The device consists of a scintillator which generates photons of light in response to incident radiation, a sensitive photomultiplier tube which converts the light to an electrical signal and electronics to process this signal.

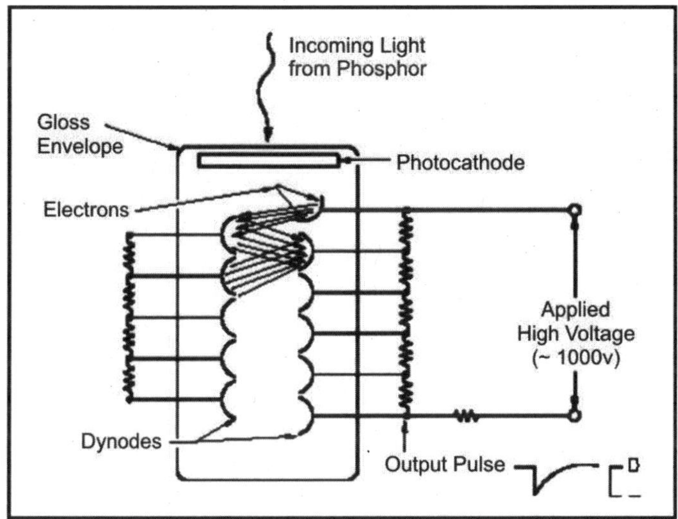

Scintillation Counter

1. History and Scintillation Counter

The modern electronic scintillation counter was invented by Sir Samuel Curran (1944), University of California, Berkeley, to measure the radiation from small quantities of uranium.

2. Operation

The scintillator consists of a transparent crystal, usually a phosphor, plastic (containing anthracene) or an organic liquid that fluoresces when struck by ionizing radiation. Cesium iodide (CsI) in crystalline form is used as the scintillator for the detection of protons and alpha particles. Sodium iodide (NaI) containing a small amount of thallium is used as a scintillator for the detection of gamma radiation, and Zinc sulphide (ZnS – used by Rutherford for his scintillating experiment) is widely used as a detector of alpha particles. Lithium iodide (LiI) is used as a neutron detector. Liquid scintillation counters are an efficient and practical means of quantifying beta radiation.

When a charged particle strikes the scintillator, its atoms are excited and photons are emitted. The photons are directed at the photocathode of the photo-multiplier tube, which emits the electrons by the photo-electric effect. The electrons are electro-statically accelerated and focused by an electric potential so that they strike the first dynode of the tube. Such an impact of a single electron releases a number of secondary electrons which are in turn accelerated to strike the second dynode. Each subsequent impact releases further electrons, and so there is a current amplifying effect at each dynode stage. Each stage is at a higher potential than the previous to provide the accelerating field. The resultant output signal at the anode is in the form of a measurable pulse for each photon detected at the photocathode, and is passed on to the processing electronics. The pulse carries the information about the energy of the original incident radiation on the scintillator. Thus, both the intensity and energy of the radiation can be measured.

3. Applications

These devices are widely used in radiation protection, assay of radioactive materials and physics research because they can be made inexpensively with good quantum efficiency and can measure both the intensity and the energy of incident radiation. These are used to measure radiation in a variety of applications which include hand-held radiation survey meters, personal and environmental monitoring for radio-active contamination, medical imaging, radiometric assay, nuclear security and nuclear plant safety.

RADIOACTIVITY – APPLICATIONS IN LIFE SCIENCES

In life sciences, radioactivity can be used as a radiolabel to visualize components or target molecules in a biological system. The detection of radioactivity may be quantitative, qualitative or by microscopy.

1. *Quantitative Detection* can be achieved by liquid scintillation assay (LSA) or liquid scintillation counting (LSC) during which process a small aliquot, filter or swab is added to scintillation fluid and the plate or vial counter in a scintillation counter. A Geiger-Muller counter is a quick and rough approximation of activity, but the lower energy emitters like tritium cannot be detected.

2. *Qualitative Detection* is possible by autoradiography and phosphor storage screen. During autoradiography, a membrane such as a Northern blot or a hybridized slot blot is put against a film that is then developed. In the second method, the membrane is placed against a phosphor storage screen which is then scanned in phosphor-imager. This is 10 times faster and more precise than film and the result is already in digital form.

3. *Detection by Microscopy* involves the use of electron microscopy and micro-autoradiography imager. In electron microscopy, the sample is not exposed to a beam of electrons but the detectors pick up the expelled electrons from the radionuclei. In the second method, a slide is put against scintillation paper and in a PMT (photo-

multiplier tube). When two different radiolabels are used, a computer can be used to discriminate the two.

RADIOISOTOPES – APPLICATIONS IN PLANT SCIENCES

The radioisotope ^{14}C was the first of many to be used not only in the hospitals but also in the universities and other institutions conducting research in different aspects of biology, chemistry, agronomy and related sciences. Radioisotopes are used extensively in the study of basic physiological processes that occur during plant metabolism. The investigations of the uptake of mineral elements by plants and their subsequent incorporation in plant tissues have been advanced greatly by the application of radioactive tracers. The translocation of organic compounds elaborated during photosynthesis and respiration also has been elucidated by the use of radioisotopes. An example of the many important contributions that have been made in the field of plant physiology is the series of studies by Melvin Calvin and co-workers using ^{14}C to tag and identify the first products of photosynthesis for which Calvin was awarded the Nobel prize in 1961. There are many problems in plant physiology which need to be solved and which will be investigated undoubtedly with the aid of radioactive isotopes. Even though some of these studies will require elaborate facilities and extensive preparation, some of them can be studied with the apparatus normally available in the high school laboratory.

a. Phosphorus Uptake by Plants

The plants take up compounds containing phosphorus from the soil through their roots. By the addition of a small amount of radioactive phosphorus-32 to a fertilizer (as a tracer), and then measuring the rate at which the radioactivity appears in its leaves, it is possible to calculate the rate of uptake of phosphorus from the soil. The information that is gathered helps plant biologists to identify the types of plants which can absorb phosphorus very quickly. These plants give better yields resulting in more food or fibre at less expense.

b. The Levels of Pesticide in Plants

To measure the levels of a pesticide, a pesticide can be tagged with a radioactive isotope (like chlorine-39) and is applied to a field of test plants, and the radioactive measurements were made over a period of time. The estimates can then be made about how much of the pesticide accumulates in the soil, how much is taken up by the plant and how much is carried off in run off surface water.

RADIOISOTOPES AND AGRICULTURE

At present, the radioisotopes are being widely used in agriculture than in any other field of science and their application is leading us to the solution of many agricultural problems in a shorter time and more precisely. The radioisotopes have become very important aid to scientists dealing with the solution of agricultural problems. Radioactive tracers and radiations have become indispensable to all the intricate agricultural research problems. These give us the chance to clear the events in the nutrition and growth of plants and evolution of new crop varieties. They also help us to clear the causal factors which produce ill-effects to plants in different ways.

The radioisotopes are used in the nutritional studies of major and minor elements, milk production, mechanism of photosynthesis studies, plant protection, plant pathology, action of insecticides, uptake of fertilizers, and mobility of ions in the soil and plants, and food preservation.

The advantages of the application of radioisotopes in agricultural research are: The presence of a single atom and the molecule and their movement can be easily located as such they give us an opportunity to follow up step by step different processes that are related to nutrition of the plant right from germination to maturity. Very small quantities of the labeled nutrients can be accurately measured in presence of large quantities of other nutrients. In addition, the location of the materials can be identified by radio-autography. The tracer technique enables to trace those elements taken by the plants accurately and precisely. It also helps to accurately study the effect of one element upon the absorption of another and their interaction by plants (the interaction among the mineral nutrients).

The artificial production of radioisotopes has increased the use of radioisotopes in the field of agriculture. One has to determine the fertility status of the soil which is unproductive to obtain an increased yield by applying fertilizers. Radioactive phosphorus is used in most of these investigations in order to determine the P-status of the soils. Such studies help to solve many problems such as comparison of various fertilizers, influence of particle size, placement, doses and time of application, absorption by plants and the reaction of the applied fertilizer in the soil. By using radioactive phosphorus research workers have succeeded to distinguish between soil phosphorus and the fertilizer phosphorus, taken by the plants. Radioisotopes like Fe, Mn, K, Ca, N, Rb, C, Cs, Si and Sr etc. and other macro- and micro-elements have been used by researchers in order to find out their movement in different soils as well as their position in different fractions of the clay soils. The method proved to be very reliable and helpful in determining fertility level of soil. Radioisotopes have also helped the investigation of the effect of such factors as cultivation and irrigation methods of plants.

Studies with radioisotopes with different crops have shown that the supply of plant nutrients through the leaves is more quick and effective than the application in soil. For example, it has been found that a nutrient, which is hardly absorbed up to 10% by root, can be absorbed up to 90% when applied on leaves. It has also been found that the absorption of nutrients by leaves occurs not only during the day but also at night. Through isotope technique, it is possible to determine the amount of nutrient that is taken up by the plants, in addition to an opportunity to know about their movement and also places of accumulation. Radioisotopes and radiation is used in mutation induction. Mutation is a sudden heritable change of the hereditary factors on the chromosomes of the organisms. The use of radiation to induce hereditary variants is a useful tool of potential value in agriculture (useful in plant improvement). Radioisotopes and ionizing radiations are of an immense value for obtaining an insight into ecological habits of insects. With the aid of radioisotopes we can find out population density, the maturity rate during different stages of the life cycle, modes of dispersal, movement and migration, flight range, hibernating places, egg-laying sites in relation to predators, parasites, feeding and mating habits and disease transmission etc.

RADIOISOTOPES – APPLICATIONS IN AGRICULTURE

The nuclear science and nuclear technology offers many techniques that are being used to improve the productivity while conserving valuable resources needed for everyday life. The radioisotopes are used as a research tool to develop new strains of agricultural crops that are drought and disease resistant, crops of higher quality, crops with a shorter growing time and crops that produce a higher yield. The radioactive elements emit a variety of radiations and energy particles during their decay which are also used in health care, agriculture and physical sciences for basic research and in a wider range of applications too. The radioactive exposure improves the quality and productivity of agricultural products along with insect, pest and disease management. They are also helpful in the study of optimum utilization of fertilizers, insecticides and pesticides in cultivated crops without harmful effects to plants and other living organisms. The radioisotopes have played an important role in improving the productivity in agriculture in a sustainable manner. The ionizing radiation is very useful for the preservation of agricultural and food products. Many products used in our daily life have in some way benefited from radiation during their production. Radioisotopes and controlled radiation are now a days used in a variety of studies like crop improvement, food preservation, to determine the ground water resources, sterilize the medical supplies, analyze the hormones, analyze X-ray pipe lines, control industrial processes and also to study environmental pollution.

IAEA (International Atomic Energy Agency) promotes the wider use of radioisotopes and radiation in research, industry, agriculture and medicine. FAO (Food and Agriculture Organization) and IAEA work to facilitate the development and adoption of nuclear technologies at national and international levels for improving agricultural productivity. The mission of the Joint FAO/IAEA Division is to develop improved technologies for sustainable food security involving nuclear methods. FAO and IAEA jointly support and coordinate research projects throughout the world on the use of isotopes and radiation in the fields of irrigation and crop production, soil fertility, insect and pest control, livestock production, health and food preservation. The improvement of agriculture is one of the most significant contributions that atomic energy can make to meet the challenge of food security for the present and future generations, to conserve natural resources and to protect the environment.

1. Plant Nutrition Studies

The efficient use of fertilizers (which are very expensive) is of great importance to reduce the production cost of agricultural crops, and it is also essential that a maximum amount of fertilizer used during cultivation finds its way into the plant and a minimum is lost. The radioisotopes are very useful in the estimation of phosphorus and nitrogen available in the soil which helps in determining the amount of phosphate and nitrogen fertilizers that should be applied to the soil. Labelled fertilizers with isotopes like phosphorus-32 and nitrogen-15 have been used to study the uptake, retention and utilization of fertilizers and the isotopes provide a means to determine the amount of fertilizer taken and lost to the environment by the plant as excessive use of fertilizer affects biodiversity and damage the environment. Nitrogen-15 also helps in the assessment of nitrogen fixed by the plant from the atmosphere under field conditions. The radioactive isotopes can also be used as `tags` to monitor the uptake and use of the essential nutrients by plants from the soil. The tracer technique also allows the scientists to measure the exact nutrient and water requirements of crop under the given conditions. The nuclear moisture density gauges can be used to monitor and determine the moisture content of the soil to indicate the exact irrigation needs of a particular area. Ionizing radiation is also used to sterilize the soil to eradicate the soil of microbes which cause the diseases and are harmful for plants.

2. Insect and Pest Management

Insect pests throughout the world are responsible for a significant reduction in the crop production. These pests are a serious threat to agricultural productivity and they not only reduce the crop yield but also transmit diseases to cultivated crops. Labeled pesticides were used to monitor the persistence of their residues in items of food, in soil, in ground water and also in the environment. The studies with the radioactive isotopes have helped to trace and also minimize the side effects of both the pesticides and insecticides. The nuclear science is being used by IAEA to help develop eco-friendly and environmental friendly alternatives for control of the pests. FAO and IAEA which sponsor and conduct research on the control of insects using ionizing radiation have placed considerable emphasis on SIT (Sterile Insect Technique) proposed by Knipling (1955), which relies on the application of ionizing radiation as a means to effectively sterilize male insects without affecting their ability to function in the field and successfully mate with wild female insects. This method involves the release of large numbers of sterile male insects of the target species in the field crop. The sterile male insects compete with the regular male population during sexual reproduction and the eggs produced from their mating are infertile so that they produce no offspring. This is a highly specific form of `birth control which reduces and eliminates the insect population after 2 or 3 generations`. This technique has been effectively utilized in the elimination of Mediterranean fruit fly from USA, Mexico and Chile; as well as Screw worm infestation in the USA and Mexico. It has also been successfully used to eradicate several insect pests of agricultural significance throughout the world.

3. Crop Improvement

Breeding requires genetic variation (of useful traits) for improvement of the crop varieties. The radiation of different types can be used to induce mutations to develop desired mutants that are resistant to disease, which are of high quality, that allow earlier ripening, and also produce a higher yield. L. J. Stadler, an American scientist (1930) made an initial attempt to induce the mutations in plants by the use of X-rays. This was followed by the use of gamma and neutron radiation. This technique of utilization of radiation energy for inducing the mutations has been widely used to obtain the desired (improved) characters in many plant varieties and it offered the possibility of inducing the desired characters that either cannot be found in nature or have been lost during the course of evolution. A proper selection of mutant varieties from the group can finally lead to improved quality and productivity. The mutations induced by radiation have increasingly contributed to the improvement of crop varieties during the last 2 decades, and it has become an established part of methods of plant breeding. BARC (Bhabha Atomic Research Centre) has developed a number of high yielding varieties of red gram, green gram, black gram, groundnut, jute and rice using radiation energy for inducing mutations, and these have been released and approved for commercial production.

4. Food Processing and Preservation

In the developed as well as developing countries the demand for instant food (wholesome) and the food which has a longer shelf life is growing as 25 – 30% of the food produce is lost due to spoilage by microbes and pests, and this can be best avoided by the use of efficient food preservation techniques. As such radiation can be used to destroy microbes in food and the insect and parasite infestation can be controlled in the harvested food to prevent wastage and spoilage. Similarly, the extension of shelf life (longer shelf life) by irradiation also saves the food. The gentle exposure of certain seeds and canned food to radiation helps to store them for longer periods. Irradiation of food is energy-conserving, can reduce the post-harvest losses, can reduce the use of food additives and fumigants (hazardous), does not heat up the food to keep it fresh, eliminate the microbes, insects etc., form packaged foods and material, kills insects and parasites, inactivates bacterial spores and moulds, prevents reproduction of microbes and insects, inhibits the sprouting of root crops, delays ripening of fruits, and improves the technological properties of food.

5. Agricultural Chemical Residues

Radioisotopes have proved to be a valuable tool in obtaining information of the residues of agricultural chemicals such as pesticides. Many labeled pesticides have been synthesized and used to obtain residue data needed for the establishment of safe limits or tolerances for the pesticides. The isotopes have also been proved useful in studying the mode of action of labeled pesticides in plants and animals.

6. Soil Physics

This comprises the investigation of the mechanical properties of soils, their retention and transmission of water, and the characteristics of soil gases. Applications of research are made in various fields such as erosion control, land drainage, water conservation, watershed management, and soil tillage. Soil moisture content can be measured accurately and rapidly in the field using either a neutron scattering method or a gamma radiation source like cobalt-60 or caesium-137. Precise soil density measurements can also be made using the gamma ray sources. Labeled water (with tritium) is being used as a tracer for studies of the efficiency of irrigation systems. The isotopes have also made possible rapid studies of root distribution in the soil.

7. Soil Chemistry and Plant Nutrition

The use of radioisotopic techniques in research on soils and plants has made substantial contributions to our understanding of the behavior and movement of plant nutrients in soil, their uptake by roots, their essential functions in plants and the relationship of soil water to these processes. These techniques also offer the advantages in that one can distinguish between the uptake of the nutrient from the fertilizer and that from the soil. The isotopes have also proved valuable in assessing the relative economic advantage of the foliar feeding (leaf application of liquid fertilizers) as compared to soil or irrigation water application of fertilizers.

8. Production of New Plant Varieties

A number of new varieties of important crop plants have been released since 1956 as a direct consequence of radiation induced mutations which have imparted new desirable characters to the crops which make them more satisfactory than their predecessors.

STABLE ISOTOPES – USE IN AGRICULTURE

The radioisotopes are of two types namely, *stable isotopes* (show no disintegration and no radioactivity – a stable isotope is any of the two or more forms of an element whose nuclei contains the same number of protons and electrons, but with a different number of neutrons, and remain unchanged indefinitely) and unstable isotopes (show radioactive disintegration or undergo spontaneous disintegration, and emit alpha, beta and gamma radiation. The unstable nuclei have excessive energy and this is released by their disintegration into stable forms of lower energy, the excess being released in the form of radiations, mainly alpha (α), beta (β) and gamma (¥), as such they are called radioactive).

The stable isotopes are the tools used by researchers worldwide in the diagnosis of the disease, to understand metabolic pathways in humans, and to answer fundamental questions in nature. They help better to understand a process, trace a compound from a particular source, measure the concentration of a chemical in a sample, or measure the rate of a related process. These play an important role in research today and will become even more important in the future. The stable isotopes are used in the same way as radioactive isotopes in soil/plant studies. The particles emitted by the radioactive isotopes are captured in photomultiplier tubes and counted stable isotopes are separated from each other by passing a gas containing them through a strong magnetic field, which deflects them differently according to their mass. The most common stable isotope used is ^{13}N (with a half-life of 9.97 minutes) but a large number of other stable isotopes are produced which are increasingly being used in agricultural studies. The uses include the custom synthesis of new and complex labeled compounds to use in agriculture, biology, chemistry, drug testing, geology, health, nutrition, physics as well as diagnosis. Some examples are: Oxygen-17 and Oxygen-18 water, Heavy water, Metal isotopes, Carbon-13 salts, Nitrogen-15 salts, Gases-13C, 2H and 15N, Deuterium gas, Carbon-13CO and $13CO_2$, and Nitrogen gas. There are about 82 stable elements and about 275 stable isotopes of these elements.

The radioisotopes are the isotopes having unstable nuclei. The radioactive atoms are unstable because certain combinations of neutrons and protons produce nuclei of latent instability. The specific average life time will be characteristic of each unstable combination. The disintegration of a given radioactive atom is a matter of chance conditioned only by a set of many requirements being completely fulfilled simultaneously. They disintegrate spontaneously at a characteristic decay rate. Overall there are about 3800 radioisotopes, out of which there are about 200 radioisotopes used on a regular basis and most of them must be produced artificially. These can be produced in different ways, namely, by neutron activation in a nuclear reactor (most common method) which involves the capture of a neutron by the nucleus of an atom resulting in an excess of neutrons (neutron rich) or in a cyclotron in which protons are introduced into the nucleus resulting in a deficiency of neutrons (proton rich). Radioisotopes encountered in everyday life are: Carbon-14 (with a fairly long half-life of 5700 years), Potassium-40 (with a half-life of 1.28×10^9 years), Thorium-232 (with a half-life of 1.41×10^{10} years) and Radon-222 (with a half-life of 3.8 days).

RADIOISOTOPES – USES IN MOLECULAR AND CELL BIOLOGY

Radioactive isotopes such as ^{3}H, ^{14}C, ^{32}P, ^{35}S, ^{86}Rb and ^{125}I have played and still continue to play an important role in the understanding of metabolic aspects in the cells or bacteria, yeasts, plants and animals including humans and in the elucidation of the fundamental properties of the genetic material. The labeled metabolites trace the corresponding stable molecules and autoradiographic or counting measurements provide the information of interest. For example, ^{32}P-Datp is used in the phosphorylation of a protein in order to evaluate its kinase activity. Similarly, ^{35}S or ^{125}I label the proteins in order to evaluate some specific expressions. ^{35}S-, ^{33}P- or ^{32}P-dideoxynucleotides are used for the synthesis of families of DNA labeled molecules (template DNA) in DNA sequence analysis, and as probes for the detection of specific genes. The value of the uses of tracers in genetics will continue to improve, particularly considering that gene therapy is now beginning to find its way into ordinary clinical application.

Conclusion

Gradually we have learned to harness the radioisotopes for our use in this modern, hi-tech world, and as such the radioisotopes have numerous applications in medicine, agriculture, industry and pure research. They are to be used in a proper manner (ensuring safer practices) to be a productive part to today`s world. It is felt that in addition to the marathon efforts of the scientists and engineers who are engaged in developing nuclear science and technology, the sincere efforts of the media in propagating the masses and laymen the beneficial uses of radio-isotopes for the development of the nation will play a major role in realizing the full potential of the atom.

■■■

CHAPTER - 17

BIOINFORMATICS

Introduction

Bioinformatics has become an important part of many areas of biology. The techniques such as image and signal processing in experimental molecular biology allow the extraction of useful results from large amounts of raw data. In genetics and genomics, it helps in sequencing and annotating genomes and their observed mutations. It has also a role in the text mining of biological literature and the development of biological and gene ontologies to organize and query biological data. It also plays a role in the analysis of gene and protein expression and regulation. The tools of bioinformatics aid in the comparison of genetic and genomic data and more generally in the understanding of the evolutionary aspects of molecular biology. It also helps to analyze and catalogue the biological pathways and networks that are an important part of systems biology. It also aids in the simulation and modeling of DNA, RNA and protein structures as well as molecular interactions in structural biology.

This is a field that has been developing over the last 30 years, and it is a discipline that represents a combination between biotechnology and computer technologies and has evolved through the convergence of advances in each of these two fields. As on today, this is a field that encompasses all aspects of the application of computer technologies to biological data, and the computers are used to organize, link, analyze and visualize complex sets of biological data. With the advent of Next Generation Sequencing and Proteomics, this has become very essential to the biological sciences in general. Bioinformatics is essential in examining how data from biological experiments can be used to generate critical information about gene sequence and function, protein structure, molecular evolution, drug targets and disease mechanisms in today's science– and technology–driven world. This emerging field requires individuals with a multi–disciplinary background in biology and computer science.

Bioinformatics is the science of data management system in genomics and proteomics of life forms, and it is a comparatively young discipline in information technology which has progressed very fast in the last few years. This is being practiced all over the world by biotechnologists to access different databases for research and exchange information for comparison, confirmation, storage and analysis. To date, there are a number of databases on specific genes and proteins pertaining to humans, animals, plants, bacteria and other life forms, and these are being updated through continuous research in modern biology with the practice of bioinformatics, and these databases help in new inventions that are useful to mankind. Thus, this branch of science is enabling life sciences to invent discovery of novel drugs as well as drug delivery systems for a greater progress in the field of biotechnology. Such useful inventions attain greater importance in the present scenario of patents and WTO regime. Bioinformatics will have to play a vital role for the future development of biotechnology with the involvement of internet tools and the web (WWW). It is also true that future rDNA research would be guided largely by the databases that are available in generic and specific forms. Hence, bioinformatics and biotechnology have to move hand in hand for their progress. As such bioinformatics can now be branded as a bonafide discipline within information technology.

Definition

Bioinformatics is `*an interdisciplinary field that develops methods and software tools for understanding biological data*`. It combines computer science, statistics, mathematics and engineering to study and process biological data. This is an *umbrella term* for `*the body of biological studies that use computer programming as a part of their methodology, as well as a reference to specific analysis `pipelines` that are repeatedly used, particularly in genetics and genomics*`. The computers are used to gather, store, analyze and integrate biological and genetic information which can then be applied to gene-based drug discovery and development. The need for its capabilities has been precipitated by the explosion of publicly available genomic information from the Human Genome Project, the goal of which is the determination of the sequence of the entire human genome (approximately 3 billion base pairs). Bioinformatics, *the melding of molecular biology with computer science*, is essential to the use of genomic information in understanding human diseases and in the identification of new molecular targets for the drug discovery. Bioinformatics employs a wide range of computational techniques including sequence and structural alignment, database design and data mining, macromolecular geometry, phylogenetic tree construction, prediction of protein structure and function, gene finding, and expression data clustering. The emphasis is on approaches integrating a variety of computational methods and heterogeneous data sources. Finally, bioinformatics is a practical discipline. This also helps to collect, compile, analyze, process and represent the information in order to understand the processes of life in healthy as well as diseased states and also find new processes or better drugs and delivery systems for genes, drugs and other genetic components for the improvement of life. It is an inter-disciplinary scientific tool without any barrier among various disciplines of sciences like biology, mathematics, computer sciences and information technology.

History and Bioinformatics

The roots of bioinformatics can be traced back to a Hungarian Engineer, Karl Ereky (1919) who coined the very term "biotechnology" and at that time, it referred to process of developing products from raw materials with the help of living organisms. The use of advance methods made it possible to establish enormous genetic databases that presently contain genome sequences of a large number of organisms and protein structural data. The size of this information would be impossible to handle without the use of powerful software programmes and computers. This has given birth to the science of bioinformatics which describes the mining of biological sequence information to predict gene function, protein and RNA structure, gene regulation, genome organization and phylogenetic history of genes and gene families. The evolution of biotechnology has helped us understand the evolution of life and with the unraveling of this mystery, the sequencing of human, plant genome and other organisms is continually offering a host of benefits. It is helping us to deal with the problem of diseases, pests, environmental challenges and above all population growth. The development of this science has certainly proved to be one of greatest achievements of the 20th century and promises to be equally and perhaps even more exciting in the new millennium.

The term – *bioinformatics* – was coined by Paulien Hogeweg and Ben Hesper (1970) to refer to the study of information processes in biotic systems, and this definition placed bioinformatics as a field parallel to biophysics or biochemistry. The first known use of bioinformatics dates back to 1988. "The term "bioinformatics" is a relatively recent invention, not appearing in the literature until 1991.

The classical bioinformatics according to Tekaia at the Pasteur Institute, Paris, France, is defined as: the mathematical, statistical and computing methods that aim to solve biological problems using DNA and amino acid sequences and related information. NCBI (National Centre for Biotechnology Information – 2001) defines bioinformatics as: the field of science in which biology, computer science, and information technology merge into a single discipline. Within this field of science, there are three important sub-disciplines, namely, the development of new algorithms and statistics with which to assess relationships among the members of large data sets; the analysis and interpretation of different types of data including nucleotide and amino acid sequences, protein domains and protein structures; and the development and implementation of tools that enable efficient access and management of different types of information. The general definition is: a scientific discipline encompassing all aspects of biologic information acquisition, processing, storage, distribution, analysis, and interpretation that combines the tools and techniques of mathematics, computer science, and biology with the aim of understanding the biologic significance of a variety of data.

The history of computing in biology goes back to the 1920s when scientists were already thinking of establishing biological laws solely from data analysis by induction (e.g. A.J. Lotka, Elements of Physical Biology, 1925). However, only the development of powerful computers and the availability of experimental data that can be readily treated by computation (for example, DNA or amino acid sequences and three–dimensional structures of proteins) launched bioinformatics as an independent field. Today, practical applications of bioinformatics are readily available through the World Wide Web, and are widely used in biological and medical research. As the field is rapidly evolving, the very definition of bioinformatics is still the matter of some debate.

The relationship between computer science and biology is a natural one for several reasons. First, the phenomenal rate of biological data being produced provides challenges: massive amounts of data have to be stored, analysed, and made accessible. Second, the nature of the data is often such that a statistical method, and hence computation, is necessary. This applies in particular to the information on the building plans of proteins and of the temporal and spatial organisation of their expression in the cell encoded by the DNA. Third, there is a strong analogy between the DNA sequence and a computer program

Analyses in bioinformatics focus on three types of datasets: genome sequences, macromolecular structures, and functional genomics experiments (e.g. expression data, yeast two–hybrid screens). But bioinformatic analysis is also applied to various other data, e.g. taxonomy trees, relationship data from metabolic pathways, the text of scientific papers, and patient statistics. A large range of techniques are used, including primary sequence alignment, protein 3D structure alignment, phylogenetic tree construction, prediction and classification of protein structure, prediction of RNA structure, prediction of protein function, and expression data clustering. Algorithmic development is an important part of bioinformatics, and techniques and algorithms were specifically developed for the analysis of biological data.

Bioinformatics has a large impact on biological research. Giant research projects such as the human genome project would be meaningless without the bioinformatics component. The goal of sequencing projects, for example, is not to corroborate or refute a hypothesis, but to provide raw data for later analysis. Once the raw data is available, hypotheses may be formulated and tested *in silico*. In this manner, computer experiments may answer biological questions which cannot be tackled by traditional approaches. This has led to the founding of dedicated bioinformatics research groups as well as to a different work practice in the average bioscience laboratory where the computer has become an essential research tool. Three key areas are the organisation of knowledge in databases, sequence analysis, and structural bioinformatics.

BIOINFORMATICS – CONCEPT AND GOAL

Bioinformatics (Life Science Informatics) is a young and fast developing branch of science that is applied discipline which utilizes computational tools to conceptualize biology, and has emerged as a cutting edge technology and a knowledge revolution. The present age of bioinformatics was ushered by the extraordinary growth of information technology and an unprecedented advance in molecular biology and rDNA techniques. The human kind is on the brink of another revolution and this was made possible by the advances such as proteomics, genomics, combinatorial chemistry, statistics, nanotechnology, spectroscopy and structural and computational biology.

The major goal of bioinformatics is to obtain the complete sequences of as many different genomes as possible (human genome completed in June 2000 is one of the greatest scientific achievements in the history of biology). With this sequence information, the research organizations and manufacturing companies can start `sequence based biology' and also use the information for designing the experiments as well as the analysis, as such a fundamental shift occurs in how the scientists actually approach to solve their questions in molecular biology.

BIOINFORMATICS – RELEVANCE TO BIOTECHNOLOGY

Bioinformatics has evolved mainly to deal with four distinct problems, namely:

- handling and management of the biological data, organization, control, linkages, analysis etc.
- communication among the people, projects and institutions engaged in biological research and applications. The communication includes e-mail, file transfer, remote login, video conferencing, e-bulletin boards and the establishment of web-based information resources.
- organization access, search and retrieval of biological information, documents, and literature, and
- analysis and interpretation of the data through the computational approaches including visualization, modelling and simulation, and development of algorithms for highly parallel processing of complex biological structures.

BIOINFORMATICS - IMPORTANT THRUST AREAS

The important thrust areas of biotechnology which are fully dependant on bioinformatics are six in number such as:

1. Comparative Genomics

The functional correlation is generally implied by the similarity of two sequences, DNA, RNA or protein. For this kind of analysis, several software tools have been developed (such as BLAST) that efficiently perform the sequence-alignments against large databases of known sequences. The comparison of a new sequence against several databases has become a routine task.

2. Functional Genomics

The expression analysis via microarray and DNA-chips is made possible by investigating the genes in their cellular context. The comparison of expression patterns of well defined metabolic states allows the identification of pathological phenotypes of a molecular level. On the basis of such identifications, a reverse engineering of the pathways and localization of pharmaceutical targets is made possible.

3. Proteomics or Proteome Technology

The proteome (a set of proteins coded by a genome) refers to the identification and the analysis of all the proteins (physical separation) of a cell which involves the determination of the structures as well as the identification of protein interactions and their function in biological pathways. This can be elucidated clearly as one gene can give rise to a number of products through mechanisms like splicing or post-translational modifications with the help of the complete genome sequences. The proteins are quantitatively isolated by HPLC and other types of columns and protein identification by N-terminal sequencing by mass spectroscopy, and the measurement of protein activity can be studied by functional assay.

4. Structural Genomics

This covers the calculation of 3-dimensional structures on the basis of the sequence of a macromolecule. In silico-biology, the theoretical basis of the relationship between the sequences and the structures is the fundamental problem, and from a computational point the folding problem is the main objection. But, the knowledge of the structure of a protein can easily provide a deeper understanding of its function.

5. Pharmacogenomics

The genetic variations among all humans are only 0.1% of the total DNA and these differences are mostly point mutations with a phenotypic impact. Such Single Nucleotide Polymorphisms (SNPs) always become good candidates for the development of the drugs and also diagnosis. With the investigation of these SNPs using micro-arrays, the investigation of the entire population becomes feasible; as such it would be very convenient to personalize the drugs for each patient individually.

6. Cellomics (Structural Biology)

More complex investigations can be easily investigated if sufficient data is made available and all the relevant components for life are identified. The simulation of life *in silico* is a future directive for bioinformatics.

BIOINFORMATICS - USES

Bioinformatics derives knowledge from computer analysis of biological data. These can consist of the information stored in the genetic code, but also experimental results from various sources, patient statistics, and scientific literature. Research in bioinformatics includes methods of development for storage, retrieval, and analysis of the data. Bioinformatics is a rapidly developing branch of biology and is highly interdisciplinary, using techniques and concepts from informatics, statistics, mathematics, chemistry, biochemistry, physics, and linguistics. It has many practical applications in different areas of biology and medicine.

The common uses include the identification of candidate genes and nucleotides, and such identification is made with the aim of better understanding the genetic basis of the disease, unique adaptations, desirable properties (in agricultural species especially) or differences between populations. It also tries to understand the organizational principles within nucleic acid and protein sequences.

1. Sequences

The computers became very essential in molecular biology when the protein sequences became available after Sanger determined the sequence of insulin in the early 1950s. The comparison of multiple sequences manually proved to be impractical. Margaret Oakley Dayhoff, a pioneer in the field and referred to as the *`mother and father of bioinformatics`* compiled one of the first protein sequence databases, initially published as books and also pioneered the methods of sequence alignment and molecular evolution. Elvin A. Kabat was another early contributor to bioinformatics who pioneered biological sequence analysis (1970) with his comprehensive volumes of antibody sequences between 1980 and 1991.

2. Genomes

It became clearly evident that the computer-assisted analysis would be insightful as the complete genome sequences became available with the work of Sanger. The very first analysis of this type was applied to the nucleotide sequences of the bacteriophages – MS2 and PhiX174. This work clearly showed that standard methods of cryptology could reveal intrinsic features of the genetic code (like codon length and the reading frame). The term – bioinformatics – was re-discovered and used to refer to the creation of databases like GenBank (1982).

GOALS OF BIOINFORMATICS

The primary goal of bioinformatics is to increase the understanding of biological processes. It focuses on developing and applying computationally intensive techniques to achieve its goal. Some important examples include: *pattern recognition, data mining, machine learning algorithms,* and *visualization*. Major research efforts in this field of biology include: *sequence alignment, gene finding, genome assembly, drug design, drug discovery, protein structure alignment, protein structure prediction, prediction of gene expression, protein-protein interactions, genome-wide association studies* and *the modeling of evolution*. This field also entails the creation and advancements of *databases, algorithms, computational* and *statistical techniques*, and theory to solve formal and practical problems arising from the management and analysis of biological data. The rapid developments in genomic and other molecular research technologies and developments in information technologies over the past few decades have combined to produce a tremendous amount of information related to molecular biology. The *actual process of analyzing and interpreting data* is referred to as *computational biology*, and the important sub-disciplines include: the development and implementation of computer programmes that enable efficient access to the use and management of different types of information, as well as the development of new algorithms (mathematical formulae) and statistical measures that assess the relationships among the members of larger data sets. These *mathematical and computing approaches used to glean understanding of biological processes* have been given the name *bioinformatics*.

APPROACHES OF BIOINFORMATICS

The most common activity includes the mapping and analyzing DNA and protein sequences, aligning DNA and protein sequences to compare them and to create and view 3-D models of protein structures. The modeling of a biological system (such as a living cell) can be done in two fundamental ways – Static and Dynamic approaches. The *static approach* includes *sequences* (of proteins, nucleic acids and peptides) as well as the *interaction data among the entities* (including microarray data and networks of proteins and metabolites). The *dynamic approach* includes *structures* (of proteins, nucleic acids, ligands [metabolites and drugs] and peptides), *systems biology* (which includes reaction fluxes and variable concentrations of metabolites) and *multi-agent based modeling approaches* (which capture the cellular events like signaling, transcription and reaction dynamics). Such a broad sub-category is referred to as *structural bioinformatics*.

BIOINFORMATICS IN RELATION TO OTHER FIELDS

Bioinformatics is similar to but distinct from biological computation and computational biology. *Biological computation* uses bioengineering and biology to build biological computers, while bioinformatics uses computation to understand biology better. Bioinformatics and computational biology both have similar aims and approaches, but they differ in their scale – bioinformatics organizes and analyzes basic biological data, while *computational biology* builds theoretical models of biological systems. Analysis of biological data to produce meaningful information involves writing and running software programmes that use algorithms from *graph theory, artificial intelligence, soft computing, data mining, image processing* and *computer simulation*. The algorithms in turn depend on theoretical foundations (like *discrete mathematics, control theory, system theory, information theory* and *statistics*).

a. Sequence Analysis

Since 1977 (following the sequencing of Phage Φ-X174) the DNA sequences of 1000s of organisms have been decoded and stored in special databases, and this information is analyzed from time to time to determine the genes that encode proteins, RNA genes, regulatory sequences, structural motifs and repetitive sequences. A comparison of genes within a species or between different species shows similarities between their protein functions or relation between the species to help construct the phylogenetic trees (in *molecular systematics*). As on today the computer programmes (like BLAST) are used almost daily to search the sequences from more than 260,000 organisms containing over 190 billion nucleotides as it has become practical (with the growing amount of data) to analyze DNA sequences manually. These compensate for mutations (caused by exchange, deletion or insertion of bases) in the sequence of DNA, to identify not identical but related sequences. Shot-gun sequencing (used to sequence the first bacterial genome of *Haemophilus influenzae* by the *Institute of Genomic Research*) is the method of choice for all the genomes sequenced today. Another aspect of sequencing analysis is annotation which involves computational gene finding to search for protein-coding genes, RNA genes and other functional sequences. Bioinformatics also helps to bridge the gap between the genome and proteome projects.

b. Genome Annotation

Annotation is the process of marking the genes and other biological features in DNA sequence (in genomics) which needs to be automated because most genomes are too large to annotate by hand. This is made possible as the genes have recognizable start and stop regions (initiating and terminating codons) although the exact sequence can vary between the genes. It was Owen White (1995) who designed the first annotation software system to find the genes, and the t-RNAs.

c. Computational Evolutionary Biology

Evolutionary biology is the study of the origin and descent of species as well as their change over time. Bioinformatics has assisted the evolutionary biologists by enabling the researchers to trace the evolution of a large number of organisms by measuring changes in their DNA rather than physical taxonomy or physiological observations. It also helps compare the entire genomes and the prediction of factors important in bacterial speciation. It also helps to build complex computational models of the populations to predict the outcome of the system and also to track and share the information on an increasingly large number of species as well as organisms.

d. Comparative Genomics

The comparative genomic analysis is the establishment of the correspondence between genes in different organisms. The inter-genomic maps make it possible to trace the processes of evolution responsible for the divergence of two genomes. The genome evolution is shaped by a multitude of evolutionary events that act at different organizational levels. The point mutations affect the individual nucleotides at the lowest level; while large chromosomal segments undergo duplication, inversion, transposition, deletion, insertion and lateral transfer at a higher level. Finally, the whole genomes are involved in hybridization, polyploidy and endo-symbiosis which leads to rapid speciation.

e. Genetics of Disease

In the present day, one can obtain enough sequence data needed to map the genes of some complex diseases (like infertility, breast cancer and Alzheimer`s disease) with the advent of next generation sequencing and the genome-wide association studies are very essential to pinpoint the mutations for such diseases.

f. Analysis of Mutations in Cancer

In oncology, the genomes of the affected cells are rearranged in complex or even unpredictable ways. Massive sequencing is used to identify previously unknown point mutations in a variety of genes in cancer. The scientists continue to produce specialized automated systems to manage the volume of sequence data produced, and to create new algorithms and software to compare the results of sequencing. New detection technologies are employed to identify the chromosomal gains and losses and to detect known point mutations. Such methods usually measure several 100,000 sites throughout the genome, and generate terabytes of data per experiment. Such massive amounts and new type of data generate new opportunities.

g. Gene and Protein Expression

1. Analysis of Gene Expression

The expression of genes can be determined by measuring mRNA levels with multiple techniques. These include: microarrays, EST sequencing (expressed cDNA sequence tag), SAGE tag sequencing (**S**erial **A**nalysis of **G**ene **E**xpression), MPSS (**M**assively

Parallel **S**ignature **S**equencing), RNA-Seq. (also known as WTSS – **W**hole **T**ranscriptome **S**hotgun **S**equencing), or various applications of multiplexed *in-situ* hybridization.

2. Analysis of Protein Expression

A snapshot of proteins present in a biological sample can be provided by protein micro-arrays and high throughput (HT) mass spectrometry (MS). Bioinformatics is involved in making sense of protein array and HT MS data. The former approach faces problems as with microarrays targeted at mRNA while the latter involves the problem of matching large amounts of mass data against predicted masses from protein sequence databases, and the complicated statistical analysis of the samples where multiple, but incomplete peptides from each protein are detected.

3. Analysis of Regulation

Regulation refers to the complex orchestration of events starting with an extra-cellular signal (like a hormone) and leading to an increase or decrease in the activity of one or more proteins. The techniques of bioinformatics have been applied to explore the various steps in this process. The expression data can be used to infer gene regulation.

h. Structural Bioinformatics

An important application of bioinformatics is the prediction of protein structure. The amino acid sequence of a protein (primary structure) can be easily determined from the sequence on the gene that codes for it. In the majority of cases, the primary structure determines a structure in its native environment. The structural information is generally classified as secondary, tertiary and quaternary and the knowledge of this structure is vital in understanding the function of protein. The main idea in bioinformatics is the notion of homology and this is used to predict the gene function. This is used to determine which parts of a protein are important in structure formation and interaction with other proteins in the structural branch. This information is used in homology modeling to predict the protein structure when the structure of a homologous protein is known. The best example is the similar protein homology between haemoglobin in humans and the leg-haemoglobin in legumes and both serve the same purpose of oxygen transport in the organism, though both of these proteins have completely different amino acid sequences, their protein structures are identical which reflects their near identical purpose.

i. Network and Systems Biology

The *network analysis* seeks to understand the relationships within biological networks. Though the biological networks can be constructed from a single type of genes, the network biology attempts to integrate different data types (like proteins, small molecules, gene expression data) which are all connected physically, functionally or both. *Systems biology* involves the use of computer simulations or cellular sub-systems to analyze and visualize the complex connections of cellular processes.

j. Molecular Interaction Networks

Many 1000s of 3-dimensional protein structures have been determined by X-ray crystallography and protein NMR (nuclear magnetic resonance) spectroscopy. The question is whether it is practical to predict possible protein-protein interactions based on these 3-D shapes, without performing experiments? Many methods have been developed to tackle the protein-protein docking and it seems there is still much work to be done.

k. Others

1. Literature Analysis

The continuous growth in the number of published literature makes it virtually impossible to read every paper and the literature analysis aims to employ computational and statistical linguistics to mine the growing library of textual resources.

2. Hi-through-put Image Analysis

The computational technologies are used to accelerate or fully automate the processing, quantification and analysis of large amounts of high-information-content *biomedical imagery*, and the modern image analysis systems augment an observer's ability to make the measurements from a large or complex set of images, by improving the accuracy, objectivity or speed. A fully developed analysis system may completely replace the observer. Although these systems are not unique to biomedical imagery, the biomedical imaging is becoming more important for both the *diagnostics* and *research*. Some important examples are: *high-content screening*, *cyto-histopathology*, *bio-image informatics* (for high-throughput and high-fidelity quantification and sub-cellular localization), *morphometrics*, *clinical image analysis* and *visualization*, *determination of the real-time air-flow patterns* (in breathing lungs of living animals), *quantification of occlusion size in real-time imagery* (from the development of and recovery during arterial injury), making of *behavioural observations* (from extended video recordings of laboratory animals), *infra-red measurements* from metabolic activity determination and inferring clone overlaps in DNA mapping (*Sulston score*) etc.

3. Hi-throughput Single Cell Data Analysis

The computational techniques are used to analyze high-throughput, low measurement single cell data (like that obtained from *flow cytometry*). These methods involve finding populations of cells that are relevant to a particular disease state or experimental condition.

4. Biodiversity Informatics

This deals with the collection and analysis of biodiversity data (like taxonomic databases or microbiome data), and such analyses include phylogenetics, niche modeling, species richness mapping or species identification tools.

5. Databases or Biological Databases

The databases are very essential for research and applications in bioinformatics. The huge number of available databases cover almost everything from DNA and protein sequences, molecular structures, to phenotypes and biodiversity. These generally fall into one of the 3 types. Some databases contain data resulting directly from empirical methods like gene knockouts, while others contain predicted data while most contain data from both the sources. The meta-databases incorporate data compiled from multiple databases. Some databases are specialized, specific to an organism, and these vary in their format, way of their accession and whether they are public or not. For a most comprehensive list of databases, the reader is advised to browse the net. The most commonly used databases are: *GenomeNet MOTIF search* (used in Motif finding), *DAVID, FuncAssociate* and *GATHER* (used in Gene Ontology), *Hidden Markov Model* (used in Gene Finding), *PFAM* (used in finding Protein Structure and Family), *FASTQ Format* (the data format used for Next Generation Sequencing), *GEO* (used in Gene Expression Analysis) and *BioGRID, MINT, HPRD, Curated Human Signaling Network* (used in Network Analysis, Interaction Analysis Databases), and *STRING, KEGG* (functional Networks) etc.

6. Software and Tools

These range from simple command-line tools, to more complex graphical programmes and stand-alone web-services from different companies and public institutions.

7. Open-source Bioinformatics Software

Much free and open-source software exist and continue to grow in the present day. The combination of a continued need for new algorithms for the analysis of emerging types of biological readouts, the potential for innovative *in silico* experiments, freely available open code bases have helped to create opportunities for all the research groups to contribute to both bioinformatics and the range of open-source software available, regardless of their funding arrangements. These tools often act as incubators of ideas, or community-supported plug-ins in commercial applications. They also provide *de facto* standards and shared object models for assisting with challenges of bio-information integration.

The range of open-source software packages include: *AMPHORA, Anduril* (developed by the University of Helsinki), *AutoDock, Bioclipse* (developed by the Bioclipse Project), *Bioconductor* (developed by Fred Hutchinson Cancer Research Centre), *BioJava, BioMOBY, BioPerl, BioPHP, Biopython,* and *BioRuby* (all developed by Open Bioinformatic Foundation), *Galaxy* (developed by Pennsylvania State University and Emory University), *GenePattern* and *GenomeSpace* (developed by Broad Institute), *GeWorkbench* (developed by Colombia University), *GENtle* (developed by Magnus Manske), *InterMine* (developed by the University of Cambridge), *LeoTask* (developed by the University College, London), *PathVisio* (developed by Maastricht University), *Staden Package* (developed by Wellcome Trust, Sanger Institute/Medical Research Council), and *UGENE* (developed by Unipro). Other details can be collected from browsing the net. To maintain and create further opportunities the non-profit Open Bioinformatics Foundation supports the annual Bioinformatics Open-Source Conference.

8. Web Services in Bioinformatics

A wide variety of bioinformatics applications allowing an application running on one computer in one part of the world to use algorithms, data and computing resources on servers in other parts of the world have been developed by *SOAP*- and *REST-based interfaces*. The main advantage is that the end users do not have to deal with software and database maintenance overheads. Basic bioinformatics services are classified into 3 categories by the EBI (*European Bioinformatics Institute*), namely, *Sequence Search Services* (SSS), *Multiple Sequence Alignment* (MSA), and *Biological Sequence Analysis* (BSA). The availability of these resources demonstrate the applicability of web-based bioinformatics solutions, and range from a collection of standalone tools with a common data format under a single, standalone or web-based interface, to integrative, distributed and extensible bioinformatics workflow management systems.

9. Networking and Current Status

An information superhighway, the internet has practically compressed the world into a cyber colony through LAN (Local Area Network), WAN (Wide Area Network), MAN (Metropolitan Area Network) and other intranets. As on date, the development of internet and the emergence of WWW as a common vehicle for the communication and instantaneous access to databases are an exciting aspect in bioinformatics. But, searching WWW requires very many tools, software packages and strategies. The tools include servers and computation boxes, cross-platform software, web-based interfaces to tools and databases as well as proprietary and public data sets. To date, there are many bioinformatics software and data access tools available on the internet free of cost, for different computational environments.

The efforts are on to create standardized bioinformatics tools so that it becomes easy for the user to implement them with a standard operating system. But, the development has been limited due to the short supply of biologists and biotechnologists who are skilled in computer science and information technology. Since bioinformatics can be practiced with an internet connection and access anywhere in the world, it helps the users with theoretical research where data complement experimental biology. The bioinformatics has become an essential tool and the backbone for research in biotechnology, involving researches in population biology, environment and ecology as well as many simulations and analytical approaches through various databases on biotechnology.

10. Bioinformatics Workflow Management Systems

This is a specialized form of workflow management system designed specifically to compose and execute a series of computational or data manipulation steps, (or a workflow) in a bioinformatics system. These are designed to provide an easy-to-use environment for individual application scientists themselves to create their own workflows; to provide interactive tools for the scientists enabling them to execute their workflows and view the results in real-time; to simplify the process of sharing and reusing workflows between the scientists; and to enable scientists to track the provenance of the workflow execution results and the workflow creation steps. For example, some platforms like *Galaxy, Kepler, Taverna, UGENE* and *Anduril* provide this service.

11. Education Platforms

There are many software platforms to teach the concepts and methods in bioinformatics. These include *Rosalind* and online courses offered through the *Swiss Institute of Bioinformatics Training Portal*, and the *Canadian Bioinformatics Workshops* which provide videos and slides from training workshops on their website under a *Creative Commons* license.

12. Conferences

Some of the most notable examples of several large conferences that are concerned with bioinformatics are: ISMB (*Intelligent Systems for Molecular Biology*), ECCB (*European Conference on Computational Biology*), RECOMB (*Research in Computational Molecular Biology*) and PSB (*Pacific Symposium on Bio-computing*).

Bioinformatics and Future Challenges

It is impossible to predict the future of bioinformatics keeping in view the present pace of investment and development. The searching of biological databases is becoming increasingly difficult via the WWW, as differences in the structure of databases and nomenclature hinder the efforts of research where the standardizations have met with much resistance. New software is being produced for different biotechnological applications and such developments are important for the future of bioinformatics and the development of biotechnology. The other challenge for the future of bioinformatics is the audit and control of databases which are becoming increasingly larger and larger. The fast development in the internet and use of available information in the databases may also prove to be `library killers` as the subscriptions to scientific journals will reduce. The availability and the demand of outnumbered biotechnologists with computer knowledge can be made possible through many training programmes like EMBLnet and ICCBnet (initiated by UNESCO) which are playing a vital role. Other challenges are the development of various strategies to surf WWW, and for reaching a consensus on coining uniform definitions and to adopt uniform platforms and technologies. The information technology talks in terms of terabytes (trillion bytes), petabytes (1000 terabytes) and exabytes (1000 petabytes), and now a new term zettabyte has been coined for huge databases involving 1000 exabytes and 1000 zettabytes as a yottabyte[6] or 10^{24} bytes. The data which being regularly flooded on to the internet poses a question as to how best the biologists and biotechnologists would be able to use them. Hence, it is envisaged that internet-II (a collaborative effort of more than 120 universities of USA trying to establish GIGAPOPS [gigabyte per second point of presence] will be launched to meet the future challenges and also save energy as well as the resources for the development of researchers in bioinformatics who are capable of using the advanced applications. The NGI (next generation internet) initiative, a multi-disciplinary federal research and development programme that is being developed for the advancing networking technologies and is going to revolutionize the applications that require advance networking, which would perhaps be the new challenge and course for future developments in bioinformatics.

Bioinformatics and the Indian Scenario

The impact of bioinformatics on Indian biological science and biotechnology can be seen by the research and development activities growing in both quality and quantity, and the number of research papers published from our country from time to time. The Department of Biotechnology (DBT), Ministry of Science and Technology, is playing a key role in the advent of bioinformatics by establishing the bioinformatics centers, and at present the network comprises 52 bioinformatics centers with its headquarters in New Delhi. The BTIC (Biotechnology Information Centre) is responsible for coordinating; organizing and providing information services at a national level that cover a wide range of subjects. It has also the mandate to assess continuously the information requirements and also to organize the creation of computer and communication infrastructure to the community of users nationally. It also coordinates the activities between the DICs (Distributed Informatics Centres – about 10 in number) and other sub-centres to provide a nation-wide communication network. The DICs have been located at IISc (Indian Institute of Science), Bangalore; JNU (Jawaharlal Nehru University), New Delhi; MKU (Madurai Kamaraj University), Madurai; Bose Institute, Calcutta; University of Poona, Pune; IARI (Indian Agricultural Research Institute), New Delhi; CCMB (Centre for Cellular and Molecular Biology), Hyderabad; National Institute of Immunology, New Delhi; Institute of Microbial Technology, Chandigarh; and National Brain Research Centre, New Delhi. Thirty eight (38) Research and Developments institutions and universities form chains of Distributed Sub-centres in the country. The DICs act as repository of information in their specialized disciplines, the sub-centres provide an access for the information at the universities, R&D and manufacturing institutions. The main centre coordinates the linkages and cooperation with external resources including documentation and information centres abroad. The network has endeavored in the establishment of national databases in collaboration with many national and international agencies and encourages the bilateral and international collaboration in bioinformatics (for example, with EMBLnet and ICCBnet etc.). The scientists at the University of Pune have developed several databanks on animal viruses, agricultural pests, biological and medical research etc, to provide an updated and accurate information in the area of biotechnology. The URLs for this site are: http://bioinformo.ernet.in and http://bioinfo.ernet.in/~sunita/main.html. The other important Indian site on the web is `Gateway` maintained by the Molecular Biology group of TIFR (Tata Institute of Fundamental Research), Mumbai, in addition to the databases maintained by MKU, Madurai and IISc, Bangalore. Many other important URLs can be found at http://neehow.ym.edu.tw/wonderful/biosites/bioinforg97.html, which is updated periodically, for easy access to researchers.

Conclusion

Bioinformatics has become synonymous with biological research. Many new academic programmes are providing the students in bioinformatics with the necessary background in molecular biology and computer science including database design and analytical approach and these will have a significant implication in medical sciences and betterment of human lives. Soon, this discipline itself may be coined Bio-IT (bio-information technology) at par with IT (information technology).

This is a rapidly growing branch of science that helps the scientists to share the information and provide powerful solutions. The tools of bioinformatics are efficient for the prediction and implication in life sciences and betterment of human lives. This is also helpful for the analysis and designing of new biological parts which can be used in the recreation or reprogramming of cell with predictive tasks, and it can also be helpful in accelerating the research in biotechnology for humankind. This has also provided the opportunities to people interested in inter-disciplinary research, and is benefitted by collective vision, efforts and intellectual pursuits of the researchers from different spheres of scientific disciplines, thus making it a holistic science in itself. The future of bioinformatics is bright and full of possibilities and so are opportunities in this upcoming field.

■■■

CHAPTER - 18

POLYMERASE CHAIN REACTION (PCR)

Introduction

PCR method is an invaluable tool for molecular biology research. This provides a simple and ingenious method exponential amplification of specific DNA sequences by *in vitro* DNA synthesis. It is used daily in labs around the world in a wide array of applications like *cloning, gene expression analysis, genotyping, sequencing*, and *mutagenesis*. PCR (*Polymerase Chain Reaction*) nicknamed now as *People`s Choice Reaction*, invented by Dr. Kary Banks Mullis (1983), an American Biochemist (La Jolla, California, USA), an author and a Lecturer, Cetus Corporation, a California Biotechnology Firm, Emeryville, California, and a Nobel Prize winner (1993) who shared the prize in Chemistry, with Prof. Michael Smith, a British-born Canadian biochemist (University of British Colombia, Vancouver, Canada), for his fundamental contributions to the establishment of oligonucleotide-based, site-directed mutagenesis and its development for protein studies. Mullis conceptualised and got the basic idea of this reaction during a 3-hour moonlight drive in the Californian mountains in the spring of 1983, while he was driving his vehicle, on Highway 128, on a late Friday Spring night in April 1983, with his girl friend, Jennifer Barnett, [both are DNA chemists at Cetus], to a cabin he had been building in Northern California, when she was asleep up a long and winding road between Cloverdale and Booneville, into Northern Californias Redwood county, USA. That night, he realised that a pair of primers could be used to copy a specific target DNA using DNA polymerase and that by raising the temperature of the sample, the extended oligonulceotides could then be separated from the DNA targets, thus becoming the templates for the next cycle in a chain reaction. The procedure could be repeated many times, doubling the amount of DNA target at each step. Through such iterative cycles, millions of copies of the desired DNA segment could be exponentially generated within a few hours. This made an intellectual leap and he reasoned that by using a pair of opposed primers, one complementary to the upper strand and the other to the lower, and performing multiple cycles of denaturation, annealing and polymerization, he could exponentially amplify the piece of DNA between the primers, to bracket the desired DNA sequence and to copy it using DNA polymerase; a technique which would allow rapid amplification of a small strand of DNA which later became a standard procedure in molecular biology laboratories. He concentrated on PCR full-time and succeeded in demonstrating the same on 16th December 1983. In 1985, he came up with the idea of using polymerase isolated from the *extremophile bacterium*, found in hot-springs, *Thermophilus acquaticus*, known as *Taq polymerase* (the prestigious journal, **Science** [USA] considered PCR as the major scientific development of the year 1989, and had chosen *Taq DNA polymerase* , the enzyme used in PCR as the *molecule of the year - 1989*) which has an optimal activity at 72^{0}C and can withstand the 94^{0}C required for the denaturation of DNA, and that many reaction cycles can be performed without being replenishing the enzyme. This breakthrough made PCR cost-effective and convenient. This made the technique a highly original and significant, which replaced the gene cloning with vectors completely, in due course of time. Mullis described how he got the basic idea of his reaction, while driving through the Redwood mountains of California, USA, in 1985, in his article published in **Scientific American**, April 1990 issue. The chemical methods that Mullis and Smith have each developed for studying the DNA molecules (the PCR method and site-directed mutagenesis) have hastened the rapid development of genetic engineering. These methods have greatly simulated the basic biochemical research and also opened a way for new applications in medicine and biotechnology. As the researchers made improvement to and variations on the process, there are now 100s of PCR based-applications in use, in a variety of fields. Gene cloning, sequencing of complete genomes, DNA finger printing and DNA-based diagnostics have become either inefficient, crude or plain techniques before PCR. The distinctive characteristic of PCR is its extraordinary versatility. The biological and biotechnological arena has been transformed by massive scale genomic sequencing. Within 20 years of its discovery, this became the basis for several molecular biology protocols and also formed the foundation of the HGP (Human Genome Project). PCR has become a quick and an easy way to create unlimited copies of DNA from just one original strand. Millions of copies of a section of DNA are made in just a few hours. The copied DNA can be then used reliably in a wide variety of tests to diagnose or monitor diseases, using a minimal amount of sample, such as blood or tissue, or for basic molecular biology research.

PCR – What it is?

Sometimes called as "*molecular photocopying*", PCR is an efficient and cost-effective biochemical technique or method to copy or amplify small segments of DNA or RNA into 1000s to millions of copies of a particular DNA sequence or segment in molecular biology. Though PCR occurs *in vitro* or outside of the body in a lab, it is based on the natural process of DNA replication. In its simplest form, the reaction occurs when a DNA sample and a DNA polymerase, nucleotides, primers and other reagents (man-made chemical compounds) are added to a sample tube. The reagents facilitate the reaction needed to copy the DNA code. In addition to detecting the diseases in a sample, PCR enables the monitoring of the amount of a virus present (viral load) in a person`s body.

PCR – Principle

The PCR involves the primer-mediated enzymatic amplification of DNA. PCR is based on using the ability of DNA polymerase to synthesize new strand of DNA complementary to the offered template strand. Primer is needed because DNA polymerase can add a nucleotide only onto a pre-existing 3′-OH group to add the first nucleotide. DNA polymerase then elongates its 3 end by adding more nucleotides to generate an extended region of double stranded DNA.

PCR – GENERAL PROTOCOL

The protocol describes how to amplify a segment of double-stranded DNA in a chain reaction catalyzed by a thermo-stable DNA polymerase. This is the foundation for all subsequent variations of the PCR.

Materials Required

The materials required for PCR are as follows:

***Buffers and Solutions*:** 10x amplification buffer, Chloroform, dNTP solution (20 mM) containing all the four dNTPs (pH - 8.0)

***Enzymes and Buffers*:** Thermo-stable DNA polymerase, Nucleic acids and Oligonucleotides, Forward primer (20 μM) in water, Reverse primer (20 μM) in water, and Template DNA. The template DNA is dissolved in 10 mM Tris-Cl (pH – 7.6) containing a low concentration of EDTA (0.1 mM) at the following concentrations – mammalian genomic DNA, 100 μg/ml; yeast genomic DNA, 1 μg/ml; bacterial genomic DNA, 0.1 μg/ml; and plasmid DNA, 1–5 ng/ml.

Method

1. In a sterile 0.5 ml microfuge tube (Fig. 2), mix in the following order: Deionized water – 37.5 μl; *Taq* Assay Buffer (10×) – 5 μl; Template DNA – 1 μl; dNTPs mix – 2 μl; Forward Primer – 2 μl; Reverse Primer – 2 μl; *Taq* DNA Polymerase – 5 μl. The following table provides standard reaction conditions for PCR are Mg^{2+} (1.5 mM); KCl (50 mM); dNTPs (200 μM); Primers (1 μM); DNA Polymerase (1-5 units); Template DNA (1 pg to 1 μg). The amount of template DNA required varies according to the complexity of its sequence. In the case of mammalian DNA, up to 1.0 μg is used per reaction. Typical amounts of yeast, bacterial and plasmid DNAs used per reaction are: 10 ng, 1 ng, and 10 pg.
2. If the thermal cycler (Fig. 1) is not fitted with a heated lid, overlay the reaction mixtures with 1 drop (approximately 50 μl) of light mineral oil. Alternatively, place a bead of wax into the tube if using a hot-start protocol. Place the tubes or the micro-titer plate in the thermal cycler.
3. Amplify the nucleic acids using the denaturation, annealing, and polymerization times and temperatures listed below.
4. Withdraw a sample (5–10 μl) from the test reaction and the four control reactions, analyze them by electrophoresis through an agarose gel, and stain the gel with ethidium bromide or SYBR Gold to visualize the DNA. A successful amplification reaction should yield a readily visible DNA fragment of the expected size. The identity of the band can be confirmed by DNA sequencing, Southern hybridization, and/or restriction mapping. If all is well, lanes of the gel containing the samples of the two positive controls (tubes 1 and 2) and the template DNA under test should contain a prominent band of DNA of the appropriate molecular weight. This band should be absent from the lanes containing samples of the negative controls (tubes 3 and 4).

5. If mineral oil was used to overlay the reaction (step 2), remove the oil from the sample by extraction with 150 µl of chloroform. The aqueous phase, which contains the amplified DNA, will form a micelle near the meniscus. The micelle can be transferred to a fresh tube with an automatic micro-pipette. Do not attempt chloroform extractions in micro-titer plates. The plastic used in these plates is not resistant to organic solvent.

Recipes

The main recipes necessary for PCR are as follows:

***Amplification Buffer*:** The chemicals required for this buffer are: 500 mM KCl; 100 mM Tris-HCl (pH 8.3 at room temperature); and 15 mM $MgCl_2$. Autoclave the 10x buffer for 10 minutes at 15 psi (1.05 kg/cm^2) on liquid cycle. Divide the sterile buffer into aliquots and store them at -20°C.

***KCl*:** Dissolve an appropriate amount of solid KCl in water, autoclave for 20 minutes on liquid cycle and store at room temperature. Ideally, this 4 M solution should be divided into small (approximately 100 µl) aliquots in sterile tubes and each aliquot thereafter used one time.

***Tris-HCl (Tris [hydroxymethyl] aminomethane [THAM] hydrochloride)*:** Dissolve 121.1 g of Tris base in 800 ml of water. Adjust the pH to the desired value by adding concentrated HCl (pH – HCl: 7.4 – 70 ml: 7.6 – 60 ml; and 8.0 – 42 ml) - (1 M). Allow the solution to cool to the room temperature before making final adjustments to the pH. Adjust the volume of the solution to 1 litre with water. Dispense into aliquots and sterilize by autoclaving. If the 1 M solution has a yellow colour, discard it and obtain Tris of better quality. The pH of Tris solutions is temperature-dependent and decreases approximately 0.03 pH units of each 1°C increase in temperature. For example, a 0.05 m solution has pH values of 9.5, 8.9, and 8.6 at 5°C, 25°C and 37°C, respectively.

***dNTP Solution*:** Dissolve each dNTP (deoxyribonucleotide triphosphates) in water at an approximate concentration of 100 mM. Use 0.05 M Tris base and a micropipette to adjust the pH of each of the solution to 7.0 (use pH paper to check the pH). Dilute an aliquot of the neutralized dNTP appropriately, and read the optical density at the wavelengths given below. Calculate the actual concentration of each dNTP. Dilute the solutions with water to a final concentration of 50 mM dNTP. Store each separately at 70°C in small aliquots. For PCRs, adjust the dNTP solution to pH 8.0 with 2N NaOH. Commercially available solutions of PCR-grade dNTPs require no adjustment.

The wavelength (*nm*) and the extinction coefficient (*E*) ($M^{-1}cm^{-1}$) of all the four bases of DNA are as follows: Adenine (A) has a wavelength of 259 and an extinction coefficient of 1.54×10^4; Guanine (G) has a wavelength of 253 and an extinction coefficient of 1.37×10^4; Cytosine (C) has a wavelength of 271 and an extinction coefficient of 9.10×10^3; and Thymine (T) has a wavelength of 267 and an extinction coefficient of 267×10^3, respectively. For a cuvette with a path length of 1 cm, absorbance = EM. 100 mM stock solutions of each dNTP are commercially available.

PCR - Precautions to be taken

Chloroform ($CHCl_3$) is irritating to the skin, eyes, mucous membranes, and respiratory tract. It is a carcinogen and may damage the liver and kidneys. It is also volatile. Hence, avoid breathing the vapours. Wear appropriate gloves and safety glasses. Always wear a chemical fume hood.

PCR - Three Main Steps

PCR is a relatively simple and inexpensive tool that can be used to focus in a segment of DNA and copy it billions of times over. It is used in everyday life to diagnose disease, identify bacteria and viruses, match criminals to crime scenes etc.

Each PCR cycle has three clear and simple steps and each cycle doubles the amount of target DNA. The reaction is exponential and more than one billion copies (*amplicons*) of the original or target DNA are generated in 30 to 40 PCR cycles.

a. Preparation of the Sample

Before initiation of PCR, DNA must be isolated from a sample. The extraction of DNA is a multi-step process that may be done manually or with an instrument such as COBAS AmpliPrep instrument, and this is the first instrument that prepared the samples automatically without human intervention.

Fig. 1: A thermal cycler (PCR machine)

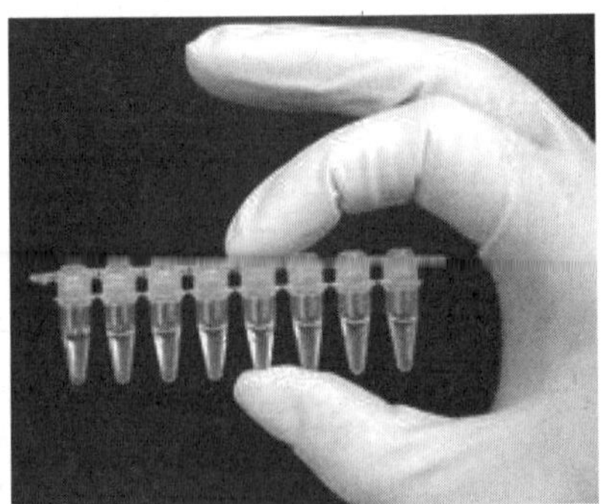

Fig. 2: A strip of 8 PCR tubes (each contain 100µl reaction mixture)

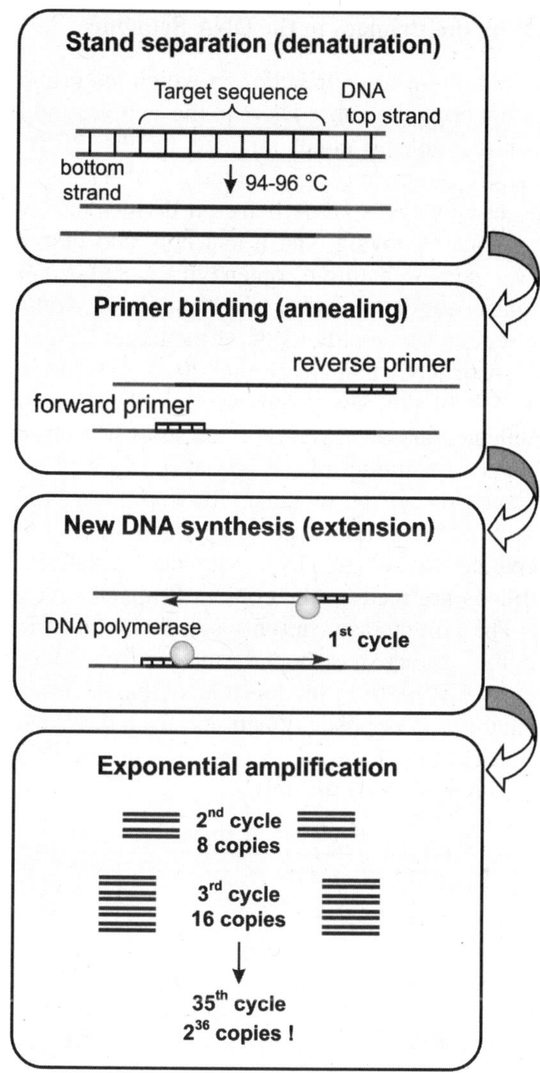

Fig. 3: Components of PCR

PCR – the Process

Following the preparation of the sample, the 3-step process gets initiated.

1. *Denaturation* - Separating the target DNA

Denaturation is the first step of PCR in which the tube containing the sample DNA is heated to more than 90°C (194°F) which separates the double-stranded DNA into two separate strands. The high temperature breaks the relatively weak bonds (hydrogen bonds) between the nucleotides that form the DNA code, and two single strands of DNA act as templates for the production of the new strands of DNA (Fig 3). This usually takes 15–30 seconds.

2. *Annealing* – Binding the Primers to the DNA Sequence

Annealing is the second phase of the reaction in which the primers bind to the DNA sequence. PCR does not copy the entire DNA in the sample, and it copies only a very specific sequence of the genetic code, targeted by the PCR primers. These are necessary because DNA polymerase, is a naturally occurring complex of proteins whose function is to copy a cell`s DNA before it divides into two, cannot attach at just any old place and start copying, and it can only add onto an existing piece of DNA. The primers are made in nature by an enzyme called *primase*. To do this, PCR uses primers, man-made oligonucleotides (short pieces of synthetic DNA) that bind (anneal) only to the sequences on either side of the target DNA region. Two primers (single strands of DNA or RNA that are around 20 to 30 bases in length) are used in this step – one for each of the newly separated single DNA strands. The primers (designed to be complementary in sequence to the short sections of DNA on each end to be copied) bind to the beginning of the sequence (by way of hydrogen bonding) that will be copied, marking off the sequence for step three. During the second step, the tube is cooled and primer binding occurs between 40° to 60°C (104° to 140°F). This yields two separate strands of DNA with the sequences marked off by the primers. The two strands are ready to be copied. Primers serve as the starting point for DNA synthesis. The polymerase can only add DNA bases to a double strand of DNA. When once the primer has bound, the enzyme starts making the new complementary strand of DNA from the loose DNA bases. The two strands of DNA are complementary and run in opposite directions (from the 5' end to the 3' end); as a result there are two primers – a forward primer and a reverse primer (Fig. 4, 5, 6, 7). The entire step takes about 10 to 30 seconds.

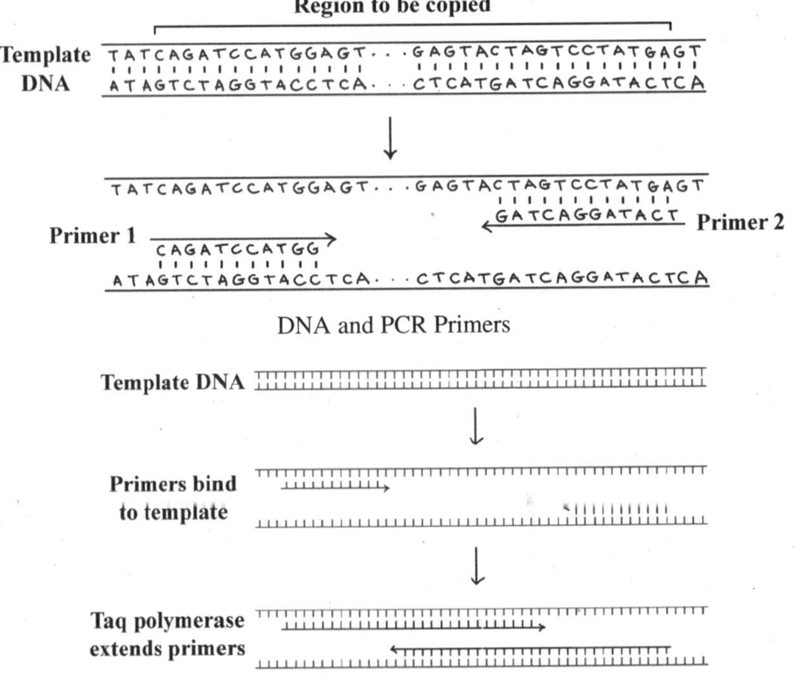

Fig. 4: Primers – Directionality

3. *Extension* or *Elongation* – Making a Copy

Extension or elongation is the third phase of the reaction in which the temperature is increased to approximately 72°C (161.5°F). Beginning at the regions marked by the primers, single nucleotides (dNTPs or deoxynucleotide triphosphates – essentially the building RT-PCR blocks for new strands) in the solution are added to the annealed primers by the DNA polymerase (Taq DNA polymerase, from *Thermis aquaticus*, a heat-loving bacterium, which lives in hot springs of Yellowstone National Park, USA, and can tolerate the temperatures above 80°C [survive near boiling temperatures]. Pfu DNA polymerase (from *Pyrococcus furiosus*, is used widely because of its higher fidelity when copying DNA. The enzyme from most other organisms would not be able to withstand high temperatures, e.g. human polymerase works ideally at body temperature – 37°C) to create a new strand of DNA complementary to each of the single template strands. After completion of the extension, two identical copies of the original DNA have been made (Fig. 4, 5, 6, 7). The duration of this step depends on the length of DNA sequence being amplified but usually takes around one minute to copy 1000 DNA bases (1Kb).

After making the two copies of the DNA through PCR, the cycle begins again, using the new duplicated DNA. Each duplicate creates two new copies and after approximately 30 or 40 PCR cycles (Fig. 3), more than one billion copies of the original segment have been made. As the entire process has been automated, this can be completed in just a few hours.

Results after running a PCR

The most common way is by running a gel electrophoresis. The PCR product is pipetted out into a special agar that will separate the DNA fragments according to their weight by using electricity. The sugar backbone in DNA is negatively charged in a neutral to basic solution, the buffer solution (1x TBE). After running the gel, the fragments can be seen with a UV light (due to the addition of Ethidium bromide dye to the gel). The dye will adhere to the DNA in the gel and when the gel is finished, it can be viewed under a UV light and can also be photographed. The gel will be run under the fume hood and gloves must be worn because the dye is a known mutagen and a carcinogen.

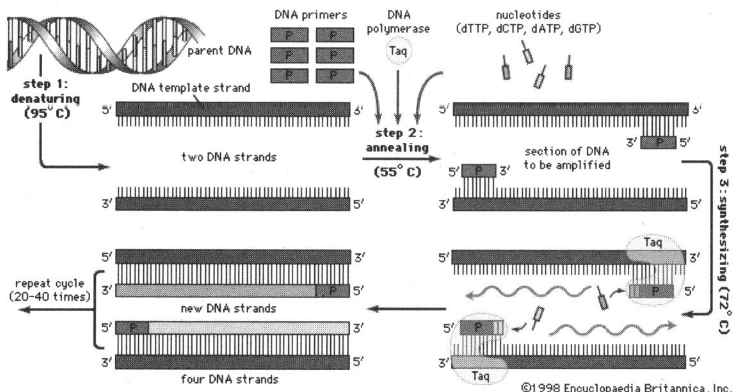

Fig. 5: PCR – Protocol – Flow-sheet

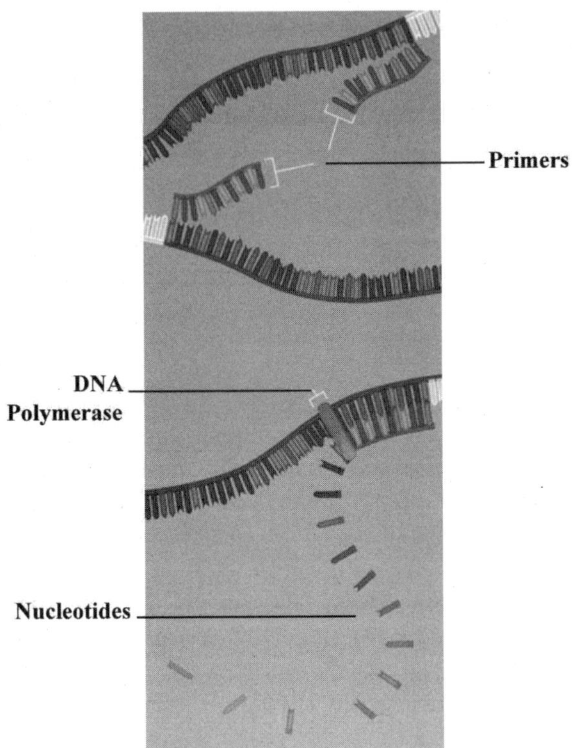

Fig. 6: How copies are made by PCR?

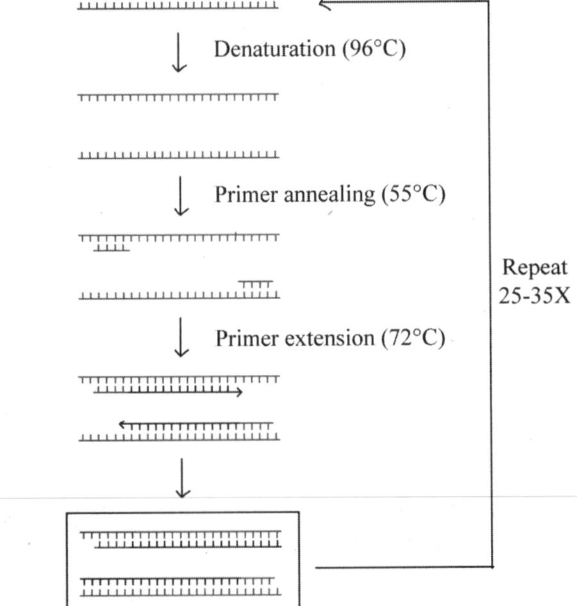

Result after 1 cycle: # of DNA molecules dobuled

Fig. 7: Basic steps of PCR – Diagrammatic

PCR - Important Points to Remember

- All the reagents of PCR should always be kept on ice.
- See that the centrifuge tubes and other glassware are not contaminated.
- Centrifuge tubes are not to be flamed during making cell suspension because they might melt.
- Make all the tubes closed to prevent cross-contamination with pipettes (Fig. 2).
- Make sure that you use the correct pipette and tip for the known amount you pick up.
- Labelling everything is always important or else you have to start over.

Limitations of PCR

This reaction starts to generate copies of the target sequence exponentially. Since the inhibitors of the polymerase reaction found in the sample (reagent limitation, accumulation of pyrophosphate molecules, and self-annealing of the accumulating product) the PCR ceases eventually to amplify target sequence at an exponential rate and a plateau effect occurs, making the end point quantification of PCR products unreliable.

PCR – DIFFERENT SCHEMES AND TYPES

These are represented as top 11 strategies on the basis of modifications of PCR as per the necessity.

1. Hot-Start PCR

This is commonly used to enhance the specificity in PCR amplification. These methods employ an enzyme modifier like an antibody, affibody, aptamer or chemical modification to inhibit DNA polymerase activity at room temperature. Such a modification prevents non-specific amplification due to primers binding to template sequences with low homology (mis-priming) and primers binding to each other (primer-dimers) during the setup. As the activity of DNA polymerase is blocked at room temperature, the hot-start strategy enables the convenience of setting up multiple reactions at ambient temperature (as in high-throughput experiments) without significantly compromising the specificity and amplification.

After the reactions are set up, the DNA polymerase is activated in an initial heating step or hot-start in which the enzyme modifier is released at a high temperature (usually above 90°C) (Fig 8). The activation time and temperature vary depending on the DNA polymerase and the nature of the hot-start modifier. For some polymerases, activation and initial denaturation steps may be combined into a single step.

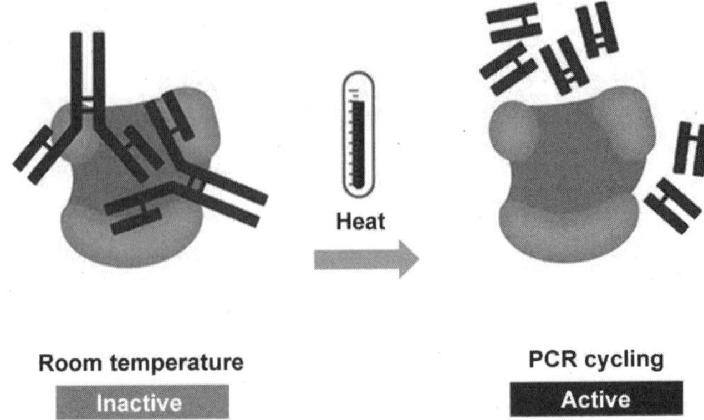

Fig. 8: Antibody-based hot-start technology with DNA polymerase

2. Touchdown PCR

A different approach to promote specificity is to modify the PCR cycling parameters. In this method, the annealing temperature of the first few cycles is set to be a few degrees higher than the highest melting temperature (T_m) of the primers. The higher temperatures help to destabilize the formation of primer-dimers and non-specific primer-template complexes thus minimizing the undesirable amplification. As such, the higher annealing temperatures reduce the non-specific PCR products and promote specific amplification at the start of PCR.

While preventing primer-dimers and non-specific primer binding, the higher annealing temperatures may result in lower PCR yield due to an increased dissociation of primers from their intended target. To overcome this challenge, the annealing temperature is often decreased 1°C at every cycle of the initial few cycles to produce a sufficient yield of the desired amplicon. Once the annealing temperature reaches, or touches down, at optimal temperature (usually 3–5°C lower than the lowest primer T_m), it is maintained throughout the remaining cycles for primer annealing. In this manner, the desired PCR products are selectively increased with little or no amplification of non-specific targets over the course of PCR.

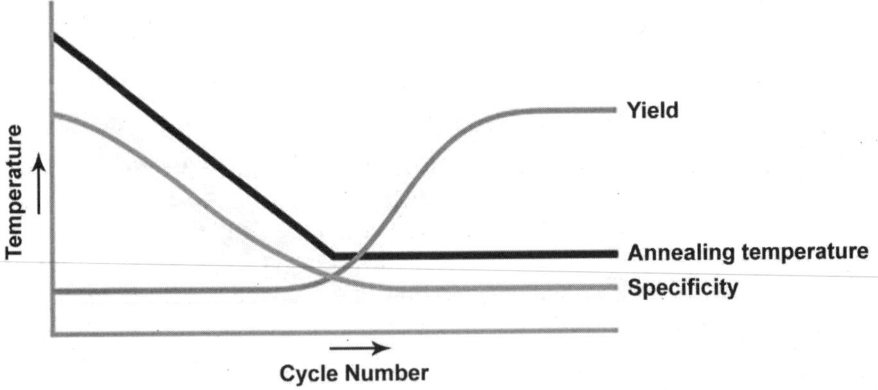

Fig. 9: Touch-down PCR – Graphical representation

3. Nested PCR

This method is a variation of standard PCR that enhances the specificity and yield of the desired amplicons. In this method, two pairs of PCR primers are designed. One set (outer primers) flanks a region of DNA containing the amplicon of interest, while the second set (nested primers) corresponds to the precise region of DNA to be amplified. The outer primers are used in a first round of PCR to amplify the target with extended flanking regions. The product of this first round (Fig. 10) serves as a template in a second round of PCR with the nested primers.

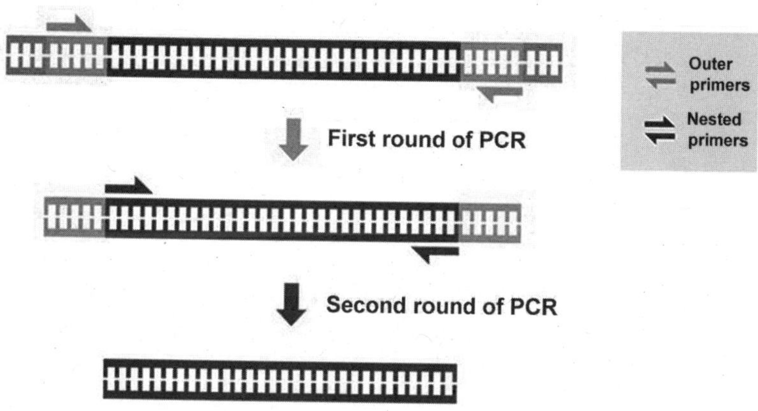

Fig. 10: Nested PCR

4. Fast PCR or Two-step PCR Protocol or Fast-Cycling PCR

In this method, the durations of PCR steps are shortened to complete the amplification faster without impacting the yield or efficiency. Fast cycling conditions are especially suitable with highly processive DNA polymerases as these enzymes are capable of incorporating a greater number of nucleotides during each binding event. These enzymes can maintain high amplification efficiency with PCR extension times that are ½ to 1/3 the duration of the extension times needed for Taq polymerase which has a low processivity. PCR times can be further shortened combining the primer annealing and extension into a single step, if their temperatures are within a few degrees of each other.

When a DNA polymerase with low processivity like Taq polymerase is used, fast cycling conditions may be possible for short targets of <5000bp (Fig. 11). The amplicons of this size generally do not require extended polymerization times, and so it may be possible to shorten the extension step in the PCR protocol. To find the shortest elongation time that docs not compromise product yield. PCR may be performed with a series of extension time decrements (in seconds). Each target and primer set will likely produce variable results and thus this will need optimization for fast PCR under specified conditions. Another modification of fast PCR is to shorten the denaturation time, while increasing the temperature to 98°C to compensate. The caveat to this strategy is that enzymes that are not highly thermo-stable will easily become denatured at such high temperatures. The use of a thermal recycler with fast cycling times and thin-walled PCR plastics will also greatly accelerate PCR via the fast temperature ramping and efficient heat transfer respectively.

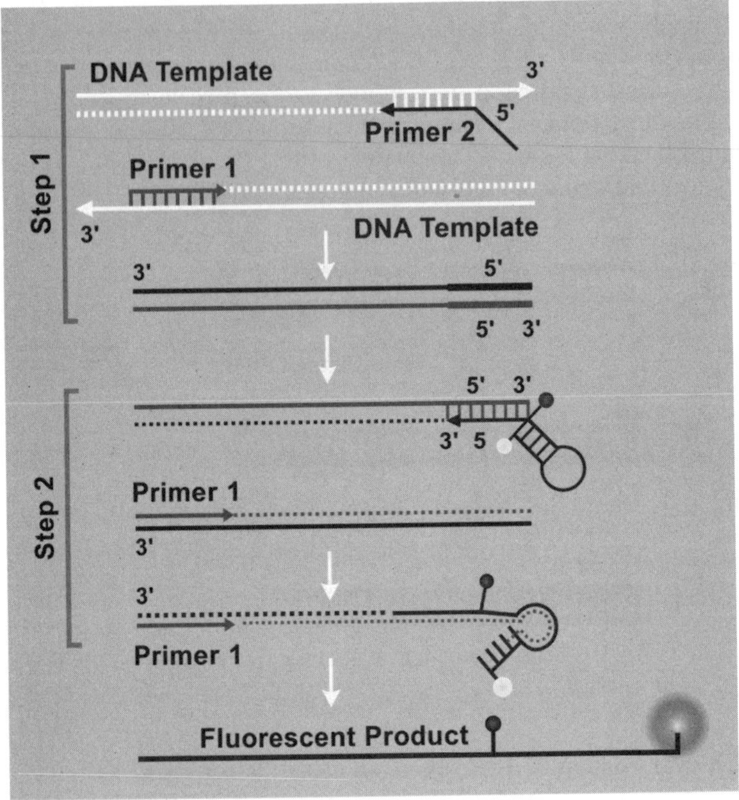

Fig. 11: Fast PCR - Steps

5. Direct PCR

This method refers to amplification of the target DNA directly from the samples without nucleic acid isolation (Fig. 12). In this method, the samples like cells or tissues are lysed in specifically formulated buffers to release DNA during high temperature denaturation steps. Hence, this method simplifies work-flows, saves hands-on-time and also prevents DNA loss caused by the steps of purification.

The enzymes, DNA polymerases with high processivity are often recommended. Cellular debris proteins, lipids, and polysaccharides are released into the lysates along with the DNA and they can inhibit PCR. The polymerases with high processivity tolerate such inhibitors and make direct PCR possible. The enzymes often display higher sensitivity and can be used to successfully amplify low amounts of DNA from unpurified samples.

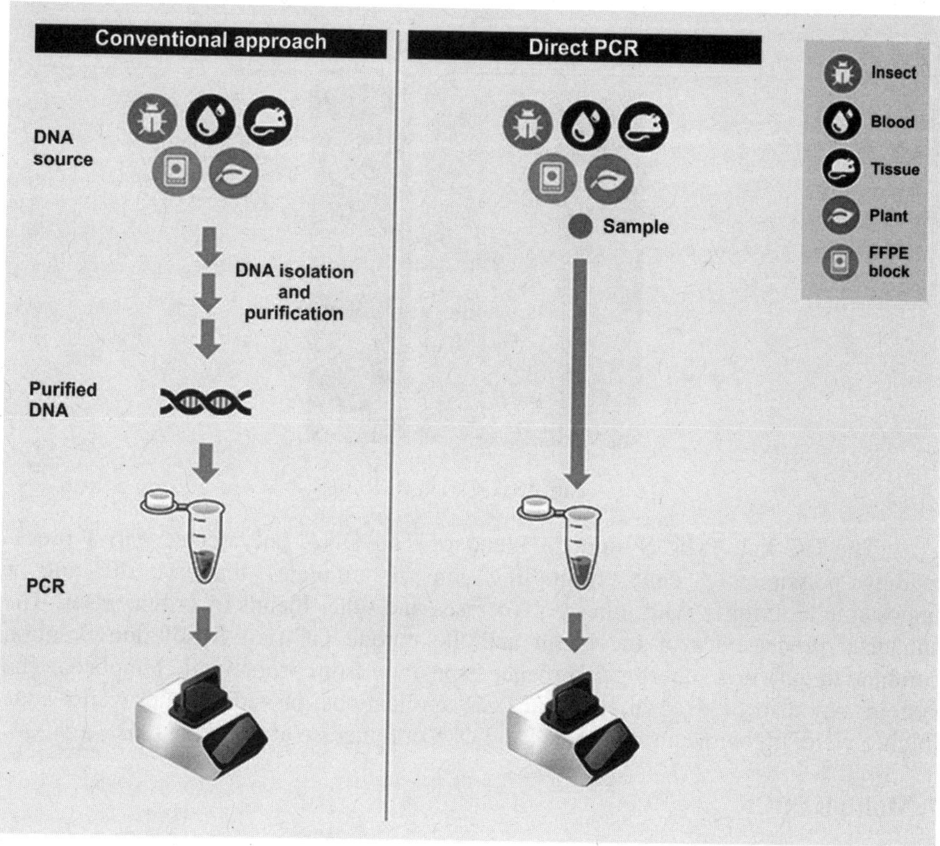

Fig. 12: Differences between Conventional PCR and Direct PCR

6. GC-rich PCR

The templates of DNA that contain high GC content (>65%) can be difficult to amplify because of the stronger hydrogen bonds between G and C bases. Such sequences can also be involved in secondary structures. Hence, these can cause DNA polymerases to stutter along templates and interrupt synthesis of DNA. To amplify G-C rich targets, the double-stranded template must be separated for the primers to bind and DNA polymerase to read through the sequence. To overcome the strong G-C interactions, the most common approach relies on PCR additives or co-solvents like DMSO to help DNA denature. These reagents often lower the primer T_m and so the annealing temperature should be adjusted accordingly. The highly processive DNA polymerases are beneficial for G-C rich PCR because of their strong binding to the templates during the extension of the primer. Hyper-thermo-stable DNA polymerases are also advantageous for this method since a higher denaturing temperature (for example, 98^0C instead of 95°C) may facilitate the strand separation and PCR amplification.

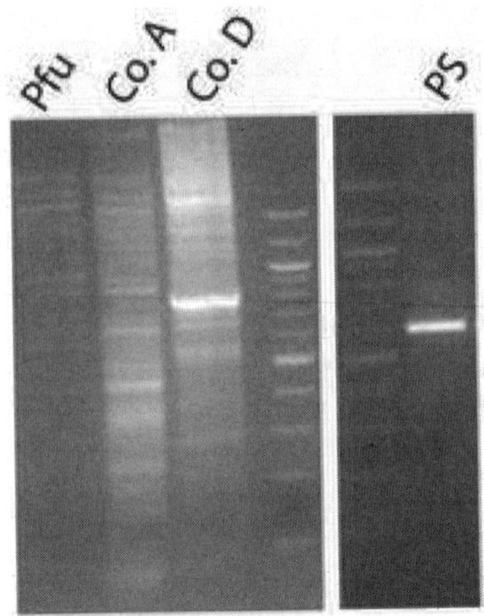

Fig. 13: GC-rich PCR

The GC-rich PCR System (a blend of Taq DNA polymerase and a proof-reading polymerase), enables amplification of templates that are difficult or impossible to amplify with other polymerases and other blends of polymerases. The enhanced processivity of the blend and the unique GC-rich Resolution Solution combine to deliver a superb performance especially from problematic templates. The system may also be used in standard PCR applications, providing improved results (higher yield, higher accuracy) over Taq DNA polymerase alone (Fig. 13).

7. Multiplex PCR

This method allows concurrent amplification of different targets in a single PCR tube. This method not only saves time, reagents, and samples but also makes simultaneous comparison of multiple amplicons possible.

When there are multiple primer pairs in a single tube, non-specific amplification and reduced efficiency are the concerns because the reaction cannot be optimized for a single primer pair and target but rather for all of the primers and targets (Fig. 14). Hence, the design of the primer is critical to minimize mis-priming that would lead to non-specific amplification. The primer sequences should be as unique to their target as possible, and the T_ms of the primers should be within 5°C of each other. Each primer set should be validated in a singleplex reaction for specificity and efficiency prior to multiplexing. Furthermore, the amplicons should be of distinct sizes that can be resolved by gel electrophoresis for identification. In addition to primer design, and amplicon size, a hot-start DDNA polymerase and a PCR buffer specially formulated for multiplexing can help PCR success and specificity. Though this method is performed routinely as endpoint reactions, the approach is more popular with real-time PCR because of capabilities in multi-labelling and detection of target amplification. Multiplex real-time PCR is also frequently employed in the detection of genetic markers for human identification.

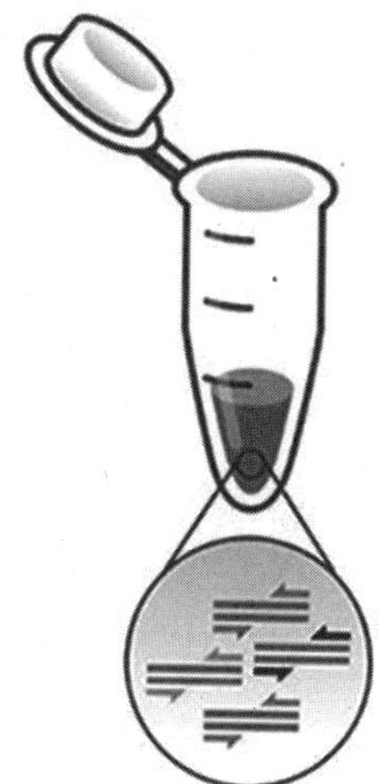

Fig. 14: Multiplex PCR where multiple primer pairs are used to amplify Multiple targets in one reaction.

8. Long PCR or Long-range PCR

This generally refers to the amplification of DNA targets longer than 5kb. This method has traditionally been performed with a blend of Taq DNA polymerase (for fast elongation) and a high-fidelity enzyme (for accuracy) (Fig. 15). Long PCR can now be performed with greatly improved accuracy in a shorter time with the invention of highly processive, high-fidelity DNA polymerases. High processivity is achieved by designing a strong DNA-binding domain, which enables the amplification of long fragments (for example, >20 kb from gDNA) in a few hours. In addition, extremely high-fidelity (for example >100x fidelity relative to Taq polymerase) helps to ensure a low error rate in replication of long fragments. High processivity of a DNA polymerase shortens the reaction times of long PCR significantly (by half) while high-fidelity reduces the effort required to screen for colonies with error-free inserts.

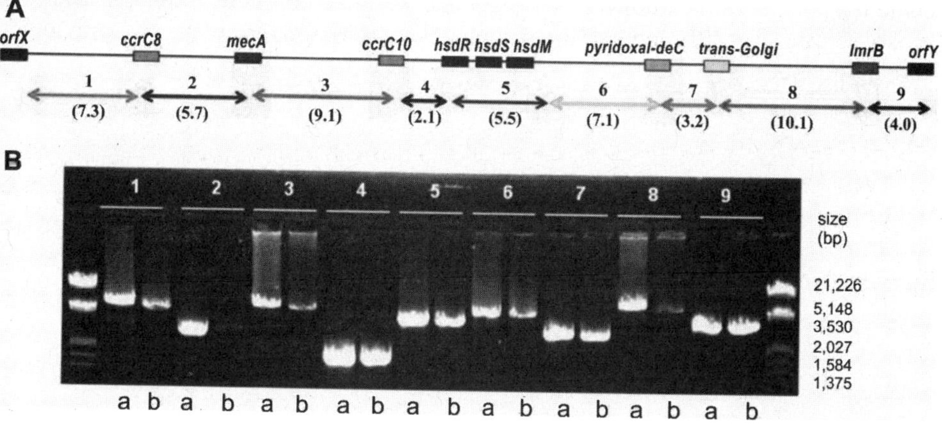

Fig. 15: Long-range PCR

When amplifying the targets of >10kb, the PCR protocol may be optimized as follows:

- to ensure that the DNA samples are of good quality and purity,
- to use higher amounts of DNA polymerase if it is of low thermo-stability, to compensate for loss of activity over extended cycling times,
- to decrease the temperatures of the annealing and extension steps to help with the primer binding,
- to increase the durations of the PCR steps to promote complete separation of the template DNA and binding the primers, and
- to extend PCR elongation times accordingly to ensure full length replication of the target regions.

9. Inverse PCR

This method was originally designed to determine the sequences of adjacent unknown regions. This is helpful for investigating the promoter sequence of a gene; oncogenic chromosomal re-arrangements like gene fusion, translocation, and transposition, and viral gene integration. This is known as inverse PCR because the primers are designed to extend away from each other rather than toward each other as in regular PCR. As on today, this is routinely employed in site-directed mutagenesis to replicate a target plasmid, while introducing the desired mutations.

Conventionally, for studying the unknown sequences of genomic DNA, restriction digestion and ligation precede inverse PCR, which is then followed by sequencing of the PCR amplicon. A restriction enzyme is chosen to generate the fragments of suitable lengths that can self-ligate for the digestion of gDNA. The selected restriction enzyme should not cleave the known sequence, so ligation occurs between the flanking unknown sequences. Ligation is optimized by using low concentrations of digested DNA fragments to favour self-ligation over multi-fragment ligation (that is, concatemer formation). After self-ligation, inverse PCR is performed by priming from the known region of DNA. The resulting amplicons contain a portion of the known DNA sequence at each end. These amplicons can then be sequenced from the end(s) to examine the regions adjacent to the previously known sequence (Fig. 16).

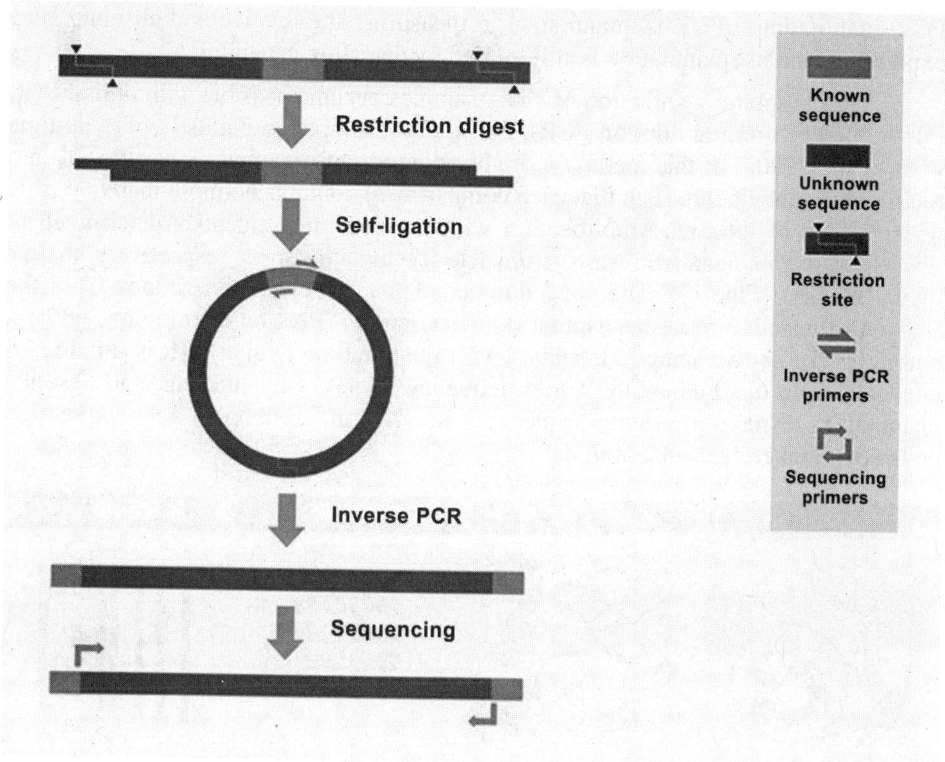

Fig. 16: Inverse PCR for adjacent unknown sequences

10. Quantitative PCR or Quantitative Real-time PCR (Q-RT PCR)

This is also called as Kinetic Polymerase Chain Reaction. As the yield (extent) of amplification of a sequence depends on the amount of the input of the template, PCR is commonly employed to quantify the amount of DNA present in a sample, of which the most common application is the quantitation of gene expression. Endpoint PCR is one possible method, but it has major disadvantages since the yield is determined by gel electrophoresis, where the detection sensitivity is limited. More important is that quantitation would be at the end of PCR when amplification has reached the plateau phase where intensities of DNA gel staining do not correlate linearly with amount of DNA input. Nevertheless, for semi-quantitative analysis of gene expression by endpoint PCR before reaching the plateau phase, serially diluted DNA samples can be used as inputs, or amplicons at specific PCR cycles can be collected for the estimation of expressed genes via the intensity of the gel.

The limitations of endpoint PCR-based quantitation were overcome by the real-time monitoring of PCR amplification using fluorescent signals (Higuchi *et al.*, 1993). This technique became the basis of quantitative PCR (qPCR). The first qPCR machines were introduced to the market in 1997 making accessible the power of PCR to accurately quantify the gene expression and copy numbers. This method relies upon real-time monitoring of fluorescence intensity associated with the target amplification during the exponential phase, thus avoiding the pitfalls of endpoint

PCR quantitation. qPCR is quantitative in measuring the relative and absolute gene expression, and its quantitation is still limited by detection capabilities.

True absolute quantitation of DNA samples became possible with digital PCR (also called as limiting dilution PCR), a method developed in parallel with real-time PCR in the 1990s. In this method, a highly diluted DNA sample is partitioned in a multi-compartment chip such that each compartment contains no more than one copy of the target of interest. Amplification within each compartment is determined to obtain positive or negative results (from 1 to 0 template copies, respectively, that is, digital results) (Fig. 17). The copy number of the sample is determined from the fraction of negative reactions using a statistical model (Poisson distribution) without a need for the known samples (standards) for quantitation. Digital PCR is suitable for applications in discrimination of low-frequency alleles, viral tittering, and absolute quantitation of next-generation sequencing libraries, in addition to gene expression and copy number determination.

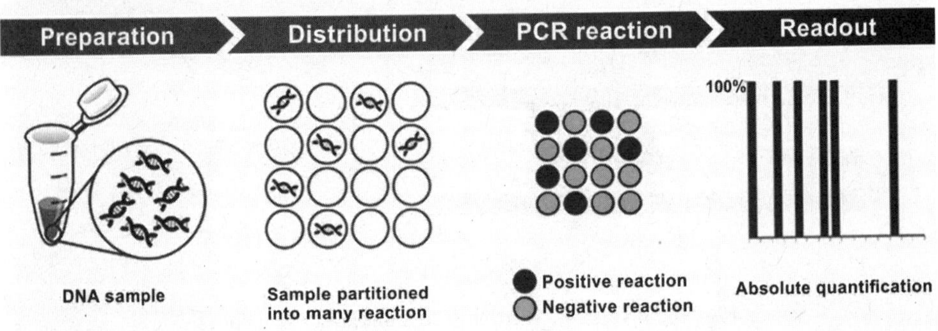

Fig. 17: Work flow of digital PCR

11. Anchored PCR

This is yet another variation wherein at times, when one has the knowledge about sequence at only one of the two ends of the DNA sequence to be amplified, anchored PCR can be used and this will utilize only one primer instead of two. In this method, due to the use of only one primer, only one strand of the template will be copied first, after which a poly-G tail will be attached at the end of the newly synthesized strand. This new strand with poly-G tail at its 3' end will then become template for the synthesis of daughter strand utilizing an anchor primer with which a poly-C sequence is linked to complement with poly-G of the template. In the next cycle, both the original primer and the anchored primer will be used for gene amplification (Fig. 18).

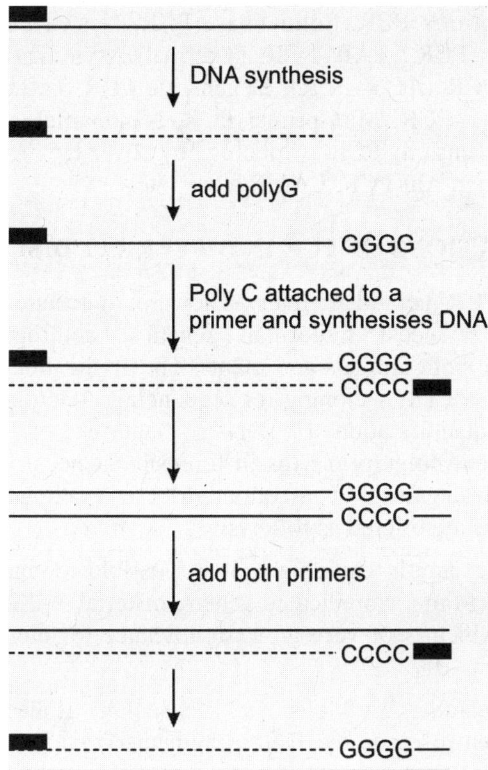

Fig. 18.: Anchored PCR – Single primer in the first cycle

12. Single Sided PCR

This is a slight variant of conventional PCR, where only one type of oligonucleotide primer is included in the reaction mixture. This primer anneals with the 3`end of the DNA strand which is complementary to the DNA strand to be amplified. The other strand does not involve in the polymerization. Taq polymerase adds complementary nucleotides to the primer and synthesizes a full DNA strand, and thus, a duplex DNA is formed. The new strand never acts as a template and it gets accumulated in the reaction mixture. The parent strand is used as a template for several times to amplify its complementary strand. This method is used to propagate the single stranded-DNA to make hybridization probes, which can be used in the clinical diagnosis of the genetic diseases and pathogenic diseases and also in detecting the genes (DNAs) in a sample.

In addition to the amplification of a target DNA sequence by the typical PCR procedures described above, several specialized types of PCR have been developed for specific applications. They are: Real-time PCR, Reverse Transcriptase PCR (RT-PCR), Single-cell PCR, Methylation-specific PCR, High-fidelity PCR, *In situ* or In colony PCR, Variable Number of Tandem Repeats (VNTR) PCR, Asymmetric PCR, Repetitive sequence-based PCR, Overlap extension PCR, Assembly PCR (Polymerase Cycling Assembly - PCA), Allele - Specific PCR, Inter-sequence-specific PCR (ISSR), Convective PCR, Ligation-mediated PCR, Methylation-specific PCR, Miniprimer PCR, Solid Phase PCR, Nested PCR, Digital PCR (In Suicide PCR, Cold PCR, TAIL-PCR (Thermal Asymmetric Interlaced PCR), Arbitrarily Primed PCR (AP – PCR), Degenerate PCR, Differential Display PCR (DD – PCR), Dial-out PCR, Mini-primer PCR, Nanoparticle Assisted PCR (Nano PCR), Over-lap Extension PCR (Splicing by over-lap extension), PAN-AC, Universal Fast Walking, Alu-PCR, LAMP-assay, etc.

APPLICATIONS OF PCR IN DIFFERENT DISCIPLINES

The applications of PCR method are many. They are: in genetic engineering, medical sciences, forensic sciences, molecular genetics, anthropology, archaeology, palaeontology, etc, in both animals and plants. The application of PCR include: the selective DNA isolation, DNA cloning for sequencing, functional analysis of genes, amplification and quantification of DNA, diagnosis of hereditary diseases, identification of genetic fingerprints (as in forensic sciences) and paternity testing, and the detection and diagnosis of infectious diseases (pathogens) etc. Some of the important applications are briefed as follows:

- By using simple equipment, it is possible to multiply a given DNA segment from a complicated genetic material, millions of times in a few hours, which is of very great significance for biochemical and genetic research.
- The researchers in Molecular Archaeology (Palaeontology), PCR has been used to clone the DNA fragments from the mummified (buried) remains of the humans and extinct animals (such as Woolly Mammoths and the Dinosaurs) can also easily produce DNA from the animals that became extinct millions of years ago by using this method on fossil material. The movie – Jurassic Park – by Michael Crighton, epitomised dinosaurs by DNA amplification and their life in ancient times, Jurassic period.
- It is also possible to re-programme the genetic code by replacing the specific amino acids in the proteins.
- This is useful for the propagation of DNA for manipulation of the genes and also for the construction of the c-DNA libraries.
- This method is used to amplify the fragments of DNA isolated from the organisms, which can be used for the chemical diagnosis of the genetic and pathogenic diseases.
- The method is also used to propagate DNAs in *in vitro* fertilized eggs, to detect baby's sex (whether male or female in humans) even at 8–60 celled stage of development.
- Single-sided PCR is useful to make the gene probes to detect the genetic diseases, pathogenic diseases and also to detect the presence of a particular gene in the sample.
- Inverse PCR is used in chromosome walking to detect the correct order of gene arrangement in a chromosome.

- Random PCR is useful to amplify a particular DNA present in a pool of DNAs.
- RT-PCR is useful to make copies of c-DNA from a single species m-RNA from a pool of m-RNAs.
- RACE amplifies a specific portion of c-DNA to reduce the sequence of amino acids in a protein.
- The technique provides DNA that is useful for cloning in PCR-based expression vectors.
- By using the PCR, DNA polymorphism can be studied at loci with known DNA sequence (using the primers on the known sequence) or at random sites (using the random primers). This involves the use of RAPD or Rapid.
- Using PCR, the genetic and physical chromosome maps can also be prepared, both in plants and animals. After the detection of RAPD`s, the recombinant inbreds (RIs) or F_2/backcross segregating populations or doubled haploids (DHs) derived from haploids in plants can be used to detect the linkage and recombination frequencies.
- Through PCR, the molecular markers that have been identified can also be assigned to specific chromosomes using aneuploids or alien addition lines (such as: wheat–barley additions = 21 pairs of wheat chromosomes+1 pair of a specific barley chromosome). A number of molecular markers have been mapped in maize, soybean, barley, mouse, humans etc using PCR. In addition, STMS or SSR loci have also been examined and mapped in mouse (and as many as 200 microsatellites have been detected in genome database) using this technique. PCR can be profitably utilized for genetic mapping for screening RIs in mouse, somatic hybrid cell lines in humans and F_2 individuals in plants. This allows the detection of linkage and calculation of recombination frequencies that lead to the preparation of genetic maps.
- PCR has also been utilized for the development of molecular markers closely linked to specific genes of economic importance in plants. For example, in tomato, 144 random primers were used to produce 625 primer products from a set of near isogenic lines (NILs which differ only for the presence and/or absence of a specific gene) for *Pseudomonas* resistance. Three of these products were present in one and not in the other of a pair NIRs showing linkage with *Pseudomonas* resistance gene (*Pto*). These PCR products could be mapped close to *Pto*, and such tagging of the genes through PCR will be used for plant breeding in future (the selection of desirable plants in segregating populations) and also for the isolation of the genes.
- In the gene therapy experiments, the transfer and the presence of a marker gene for neomycin resistance (*NeoR*) could be detected in the patient's blood even after 60 days, although the frequency of marked lymphocytes gradually decreased necessitating a second transfusion of such marked cells. Similarly, ADA gene (adenosine deaminase

deficiency) was transferred for therapeutic purposes, and PCR was used to detect the presence of this gene up to 6.5 months, so that the 4 year old girl patient was then placed on a programme of maintenance infusions at 3–5 months interval. As such PCR proves to be of immense help in gene monitoring (in genetic engineering or gene therapy experiments).

- The technique can be used for DNA amplification, utilizing individual sperms (the products of meiosis and recombination). Hence, by studying the PCR products from a large number of individual sperms, one can calculate the proportion of sperms that are recombinant for linked markers. This allows the construction of the genetic maps at a resolution which is not possible with pedigree analysis. The micro-dissected chromosome bands can also be used for PCR to allow the physical location of the genetic lesions.
- The sex of the human or the livestock embryos (sexing), fertilized *in vitro*, can be determined before implantation as PCR can be used to study the DNA sequences in single cells. For this purpose, PCR primers and probes that are specific for the sex chromosomes (like ZFY or ZFX sequences) can be used. The Y-specific primers and probes in humans are available for this use. The method can also be used to detect the sex-linked disorders in the fertilized embryos.
- Some specific characterized sequences of microsatellites are used for primer designing so that can be easily achieved through this technique, which is employed in forensic medicine for the easy identification of the criminals, rapists, uniting the lost children to their parents and/or relatives after confirming their identity, and also disputed parentage. This method allows the amplification of DNA from the little amounts of the tissues, such as individual hairs, stains of blood or the seminal fluid containing partially degraded DNA at the site of crime.
- There are many pathogens that normally grow very slowly as such they are found in lesser number in the infected cells and/or the tissues. Due to this, it is difficult to culture them on the artificial medium. PCR-base assays can detect the presence of certain specific sequences of the pathogens in the infected cells/tissues. This method is also useful in the detection of the viral infection even before they produce the symptoms (or serious diseases).
- Genetically speaking, in human beings, there are 1000s of genetic diseases and the mutations are also related to these diseases. By using PCR, the presence of a faulty DNA sequence can be easily detected (even before the establishment of the diseases). The very first application is the prenatal diagnosis of sickle cell anaemia, and was also useful for the diagnosis of PKU (phenylketonuria), β-thalassemia, haemophilia and muscular dystrophy. The test is carried out in less than one day in contrast to several weeks when Southern blots were used for hybridization with a probe, and in all these cases, the PCR product is examined using a probe (labelled) to suggest whether or not the mutant

sequence that causes the disease is present or not; and in all these cases, the PCR product is examined using a probe (labelled) to suggest whether or not the mutant sequence that causes the disease is present or not. If the genetic diseases are not curable, it is generally recommended to go for abortion. In the present time, PCR-based diagnostic tests (rapid, sensitive and specific) for AIDS (Acquired Immune Deficiency Syndrome), Chlamydia, TB (Tuberculosis), Hepatitis, HPV (Human Papilloma Virus) and many other infectious agents and diseases are being developed.

- In research, the DNA fingerprinting of new microbes isolated from different extreme environments (such as soil, water, sediments, air, extreme habitats) is carried out to confirm their identity by comparing with the DNA of the known microbes. The DNA and RNA of the new microbes can be amplified, and it is also useful to determine the orientation and location of the restriction fragments relative to one another.
- PCR technique is also applied in the diagnosis of plant diseases; and a large number of plant pathogens in different hosts and environmental samples are detected by this technique. The examples are: Viroids (such as: hops, apple, pears, grapes, citrus species etc.); Viruses (such as: TMV [Tobacco Mosaic Virus], CMV [Cauliflower Mosaic Virus], Bean Yellow Mosaic Virus, Plum Pox Virus and Poty Viruses etc.); Mycoplasmas bacteria (such as: *Agrobacterium tumefaciens, Pseudomonas solanacearum, Rhizobium leguminosarum, Xanthomonas compestris* etc.); fungi (such as: *Colletotrichum gloeosporioides, Glomus* species, *Laccaria* species, *Phytophthora* species, *Verticillium* species etc.); and nematodes (such as: *Meloidogyne incoginta, M. javanica* etc.).
- In research, PCR is used to create copies of DNA for introduction into host organisms (like *E. coli*) in genetic engineering and also to amplify the stretches of genetic material for Sanger sequencing.
- PCR-based technologies are used every day, in addition to basic research, in clinical diagnostics, forensic investigations, and agricultural biotechnology. These require reliable performance, superb sensitivity, and stringent specifications. Hence, thermal cyclers and the reagents must be compliant to and specially designed for such purposes. The examples of molecular diagnostics include genetic testing, detection of oncogenetic mutations, and testing for infectious diseases. In forensics, human identification by PCR relies on the amplification of unique short tandem repeats (STRs) on gDNA to differentiate the individuals. In agriculture, PCR plays an integral role in food-pathogen detection, plant genotyping for breeding and GMO testing. Since its introduction (1980s) this method continues to prove to be a useful tool with many broader applications in discovery biology, medical diagnostics, forensics, and agriculture.

- This method has been used to identify and to explore the relationships among the species in evolutionary biology. For example, in anthropology, it is used to understand the ancient human migration patterns, and in archaeology, it has been used spot the ancient human races. PCR is used in this field to identify human or animal remains (including the insects trapped in amber), and also to track the human migration patterns, degraded DNA samples may be able to reconstructed during the early cycles of PCR. This is used to differentiate between similar organisms (ticks) or work out the relationships between different animal and plant species. This is used by palaeontologists to amplify DNA from the species which are extinct or fossils of millions of years that are cryo-preserved for further study to elucidate on.
- Real-time PCR has been shown to enable the quantitative detection of *Ralstonia solanacearum* in a rapid and reliable manner, thus providing an improved alternative assay of potato brown rot agent, *Ralstonia solanacearum.* The food contamination by mycotoxins is of great concern; Real-time PCR detection assays targeted at genes involved in toxinogenesis have been developed for trichothecene-producing *Fusarium* and aflatoxin-producing *Aspergillus* species. A refinement of this technique recently implemented is based on the quantification of the *nor1* mRNA, which directly addresses aflatoxin biosynthesis in infected wheat. The implementation of real-time PCR tests to quantify a number of pathogens on *Arabidopsis* demonstrated that they are a very interesting alternative to classical tests.
- The absence of gluten in baby food can be controlled by an amplification test of cereal genes by real-time PCR. This technique has been shown to be an ideal tool for assessing common wheat (*Triticum aestivum*) adulteration in durum wheat pasta.
- Real-time PCR is able to meet this technical requirement, as the expression of numerous genes may be tested on the same RNA preparation. This method has been successfully used to investigate the expression of 8 genes involved in brassinosteroid (BR) biosynthesis, degradation or signal transduction (Shimada *et al*, 2003). They were able to correlate the expression of the BR synthesizing enzymes with the presence of BR *in planta*, and provided a more complete view of BR metabolism at the whole plant level.
- The appearance of genetically modified organisms on the food market a few years ago, and the demand for more precise and reliable techniques to detect foreign (transgenic or pathogenic) DNA in edible plants, have been the driving force for the introduction of real-time PCR techniques in plant research. This was followed by numerous fundamental research applications aiming to study the expression profiles of endogenous genes and multigene families. Since then, the interest in this technique in the plant scientist community has increased exponentially.

PCR –Benefits, Drawbacks and Limitations

PCR is at present, a common and often an indispensable technique used in medical and biological research laboratories for a variety of applications. This technique can provide many advantages over the traditional techniques. Many of the tests can be rapidly performed and interpreted the same day as sample submission. A large number of samples can be simultaneously completed. Differential PCR assays can determine if an isolate is non-pathogenic or contains toxigenic properties (genes) necessary to induce a disease. It can amplify very small amounts of genetic material and thus, it can detect very low numbers of organisms in a sample. This testing is valuable for detecting specific pathogens that are difficult to culture *in vitro* or require a long cultivation period. The test is significantly more rapid in providing the results compared to culturing – as it enables earlier informed decision making, and rapid diagnosis of *bacteremia*, in particular the lower levels of bacteria in the specimens. This technique also becomes useful in detecting the cases in extra pulmonary specimens which may be missed by smear and/or culture. This is a valuable tool and is still considered as an adjunct test for certain diagnostic tests that still rely on the smears and cultures (like tuberculosis).

The major drawbacks are: lack of antimicrobial sensitivity data, complexity of the assay, and the price of equipment and the kits.

The limitations include: this testing alone may be limited as a diagnostic tool and still it needs culture for testing for drug/antibiotic susceptibility and gene typing. The non-availability of the information related to the knowledge of the nucleotide sequence of the DNA segment or the gene to be amplified, becomes a limitation in the applications of PCR, though it has been overcome through the use of random DNA primers. The post-treatment diagnosis may become challenging because this test detects dead organisms that may be shed for weeks after the patient stops showing the symptoms. This is also unclear regarding the persistence of infection, as detection of dead organisms at this stage may have no clinical relevance. The results of PCR should not be used as the sole basis of a patient treatment management decision. All the results should be interpreted by a trained professional in conjunction with the review of the patient's history and clinical signs and symptoms. The false negative results can arise from: improper sample collection/transport; insufficient amount of the specimen; degradation of nucleic acids (particularly RNA) during shipping or storage; specimen collected prior to the onset of symptoms or late in illness; quantity of the organisms is below the detection limits; non-homogeneous distribution of the organism of interest; the presence of amplification inhibitors in the specimen; and laboratory processing/testing errors. Similarly, the false negative results can arise from: detecting the contaminants introduced during the collection of the specimen, its transport or processing; detection of the organism representative of normal flora near the specimen collection site, acid-fast bacilli in water, and the contaminants in the laboratory; mislabelling; and specimen mix-up.

Conclusion

The PCR revolutionized molecular biology to a similar extent as the discovery of plasmids and restriction endonucleases. This is a simple and widely used process in which the minute amounts of DNA can be amplified into multiple copies. In addition to the rapidity with which this assay will work, it is able to quantitatively demonstrate how much of a particular sequence is present. Similar to all other methods, the validity of the results should be compared with the specificity associated with this method. The future of this method is most promising, combining various assays and approaches to produce a greater insight into many gene combinations. For example, in a study to link distinct taxa within the microbial community to specific metabolic processes, stable isotope probing can also be combined with qPCR, and similarly, microarray experiments can be validated by qPCR approaches due to its rapidity.

■■■

CHAPTER - 19

PHOTON COUNTING AND PHOTO-MULTIPLIER TUBES (PMTs)

Introduction

Photon counting is a collective name for all the techniques that use single photon detection for the total measurements of light flux, occurring under extremely low-light conditions. At present, this approach can be applied to a variety of radiation wavelengths, from X-ray and deep ultraviolet ending with far-infrared of the spectrum. The astronomers were the first to use PMTs in the photon counting mode for the measurements on the stars of a very low magnitude in the early 1950s. The PMT is an extremely sensitive light detector providing a current output proportional to light intensity. These are used to measure any process which directly or indirectly emits light.

The operating principle of this technology is that – the photons that strike a photocathode at the entrance window of a PMT produce the electrons, which are accelerated by a high voltage field and get multiplied in their number with a chain of dynodes by secondary emission.

PMTs are vacuum phototubes which are extremely sensitive detectors of light in the UV, visible and near IR ranges of the electromagnetic spectrum. PMT ia also a vacuum tube that detects the light from dim sources through the use of photoemission and successive instances of secondary emission to produce enough number of electrons to generate a useful current. The first known use of PMT was in 1941.

PMT also called as a photomultiplier, is – a photo emissive device in which the absorption of a photon results in the emission of an electron. This is useful for the detection of light of very weak signals. The photomultiplier is a very sensitive light detector which provides a current output proportional to light intensity. As such, these are used to measure any process which directly or indirectly emits light. These can multiply the current that is produced by the incident light by as much as 100 million times.

PMTs are of a very essential use in low light level spectroscopy, confocal microscopy, Raman spectroscopy, fluorescence spectroscopy, nuclear physics, particle physics, astronomy, in addition to medical diagnostics (including blood tests, medical imaging), telecine (motion picture film scanning), radar jamming, and drum scanners etc. This technology is also the basis of night vision devices. In research, this is also useful in the study of polymers in solution.

The different applications of PMTs are as follows: The very first electric eye devices were the PMTs used to measure the interruptions in the light beams. These are also used in conjunction with scintillators (to detect ion ionizing radiation) as well as particle radiation (in experiments).These are also used in research labs as compound semiconductors and quantum dots to measure the intensity and the spectrum of light-emitting materials; and are also used as the detector in spectrophotometers. In medical field, these are used to design flow cytometers (for blood analysis) to determine the relative concentration of different components in blood samples; and in incandescent lamps (in combination with optical fibres); in a gamma camera (as an array) and as detectors in flying-spot scanners.

The following details explain how to use PMTs in photon counting (low-light-level measurement with highest sensitivity and higher accuracy), in addition to the principle of photon counting, operation of PMTs, characteristics of PMTs and their selection for experimentation.

Photon Counting

This is a technique in which the individual photons are counted by using a SPD (Single Photon Detector), and the efficiency is determined by the quantum efficiency and any electronic losses that are present in the system.

The photodetectors are configured to detect the individual photons, with their own relative advantages and disadvantages. The best known examples are: Photomultiplier (PM), Geiger counter (GC), Single-photon Avalanche Diode, Super-conducting Nanowire Single-photon Detector, Transition Edge Sensor, or Scintillation counter. Sometimes, charge-coupled devices (CCDs) can also be used.

PHOTOMULTIPLIER TUBE (PMT)

Photomultipliers (German), sometimes also called photon multipliers, are a type of photo-emissive detectors which have a very high sensitivity due to an avalanche multiplication process, and also exhibit a high detection bandwidth. These are defined as: photo-detection devices based on the photo-electric effect and charge multiplication by secondary emission of electrons. PMT is a light detector that is useful in low intensity applications like fluorescence spectroscopy. PMTs are very sensitive detectors because of high internal gain. The invention of the PMT is predicated upon two prior achievements, the separate discoveries of the *photoelectric effect* (associated with Albert Einstein in 1905) and of the *secondary emission* (first reported by Villard in 1899).

Design – Construction and Operation Principle

These are similar to phototubes, and they consist of a photocathode and a series of dynodes in an evacuated glass enclosure. The light is absorbed on a photo-emissive cathode and generates free photo-electrons due to external photoelectric effect. Instead of collecting these few electrons at an anode as in the phototubes, the electrons are subsequently accelerated with a high voltage (100s of volts) to a first dynode (an electrode) where they generate several secondary electrons. These are accelerated towards other dynodes (a series of additional electrodes) maintained each at a more positive potential (Fig. 2). The voltage between two subsequent dynodes is (for example, 100 V) and also between the last dynode and the anode. The dynodes have a surface which is optimized for the emission of secondary electrons (usually a semi-conductor like AgMgO, $SbCs_3$, GaP, GaAsP, BeO or oxidized copper with some beryllium). The additional electrons are generated at each dynode where the number of electrons is getting several times larger each time. Such a cascading effect creates 10^5 to 10^7 electrons for each photon hitting the first cathode depending on the number of dynodes and the accelerating voltage. Finally, a strong amplified photocurrent is collected with an anode near the last dynode (Fig. 1). This amplified signal is finally collected at the anode where it can be measured. Typical specifications are: the range of the wavelength – 110 to 1100 nm – the quantum efficiency (Q. E) – 1 to 10% - and the response time – 1 to 15 ns.

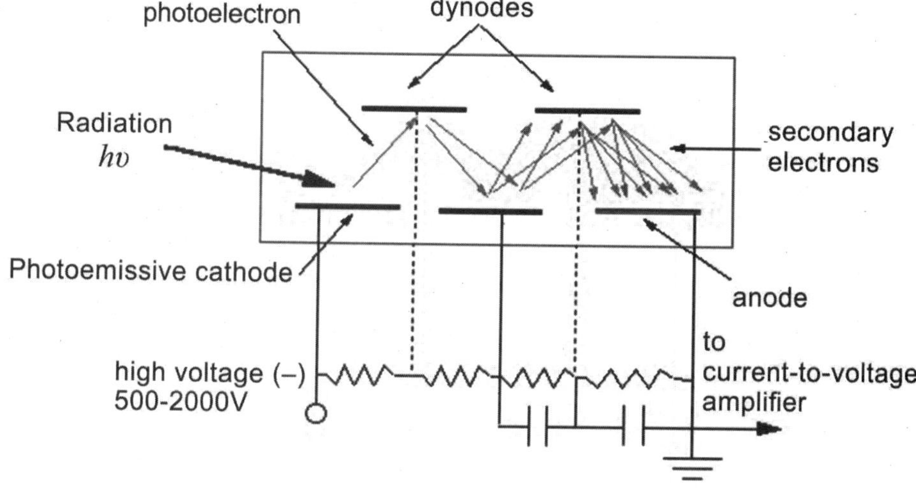

Fig. 1: PMT Design – Schematic

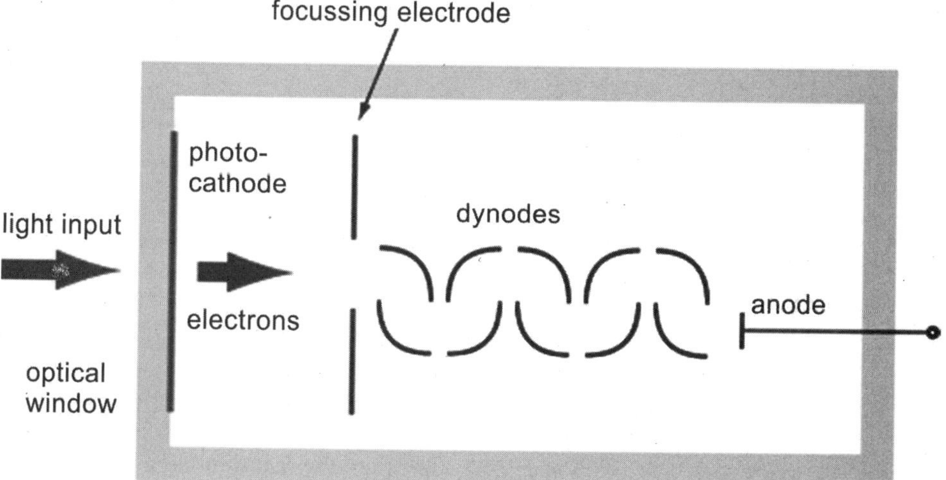

Fig 2: Setup of a head on PMT

PMTs are used generally as photodetectors because of their fast response, high gain, and high signal-to-noise ratio. They are sensitive to light detection in the ultra-violet, visible and near-infrared region of ~200 to 900 nm. The selection of a specific PMT used is based on the spectral region of interest. For the detection of longer wavelengths in the range 800–1800 nm, other detectors such as germanium or InGaAs photodiodes etc, are required.

The electrode designs can be optimized concerning different aspects like: the electron collection efficiency, maximum output current, uniformity of response over a large active area, detection band width and sensitivity to magnetic fields etc. For example, a researcher can use specially curved focusing electrodes, which reduce the spread of electron transit time for a minimum rise time and maximum bandwidth. Therefore, one can use different designs like box-in-grid, circular, linear-focussed, venetian blind and mesh etc, depending on the requirements for a specific application.

Some multipliers are *head-on photomultipliers* where the light enters the roughly cylindrical tube on one of the flat faces, for which transmissive photo-cathodes are needed. A reflective photo-cathode is to be used in *side-on photomultipliers*. The glass tube of a photomultiplier has a substantial number of contact pins for all the electrodes on one side, with which it can be inserted into a suitable socket.

Micro-channel photomultipliers (MCPs or micro-channel plates) are a very compact form of photomultiplier tubes which contain a large number of very narrow channels (for example, 1 mm long), in which the amplification occurs, and these exhibit a particularly high detection bandwidth. In all cases, one needs to maintain a high vacuum (typically between 10^{-6} Pa and 10^{-4} Pa) in a PMT (Fig. 3).

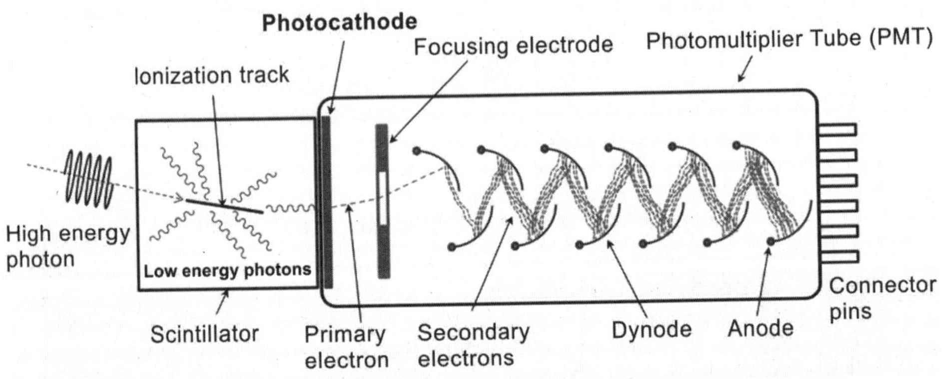

Fig. 3: PMT – Schematic Diagram

The Supply of Voltage

A PMT requires a whole sequence of different electrical potentials in order to have an appropriate voltage of the order of 100 V between the photocathode and the first dynode, between each pair of dynodes, and between the last dynode and the anode. Normally, a resistive voltage divider is used for generating the required voltages, and it is sufficient to supply a single high voltage of the order of 1 kV to the voltage divider chain. This can be generated from a low supply voltage with a DC-to-DC converter, or alternatively with a moderately complex power supply at the grid. The operation voltage should be very stable, with little temperature dependence since it sensitively affects the gain. It should not exhibit any significant ripple or other noise. The device draws some current due to the resistors even without any light input, where the output current is close to zero. The current through the voltage divider train should be at least 10 to 20 times larger than the maximum possible output current for good signal linearity. Modified voltage dividers can also contain Zener diodes which are most frequently applied for the last few dynodes which can be used to stabilize a voltage level. They also produce some noise in case of low currents which can be mitigated with a parallel capacitor. The capacitors can also be used in parallel to resistors in order to stabilize the voltages in case of pulse detection. The voltage divider may be place conveniently into the socket into which the multiplier tube is inserted, which is detrimental as it may heat the tube. A more power-efficient solution is a Cockcroft-Walton voltage multiplier which consists of two sequences of capacitors connected with a ladder of silicon diodes, which is driven with an alternating voltage and naturally provides the required set of stable dynode voltages. It can also be electronically regulated to produce a precisely constant voltage, and the only disadvantage is that it cannot be as compact is a simple voltage divider chain with resistors (Fig. 4).

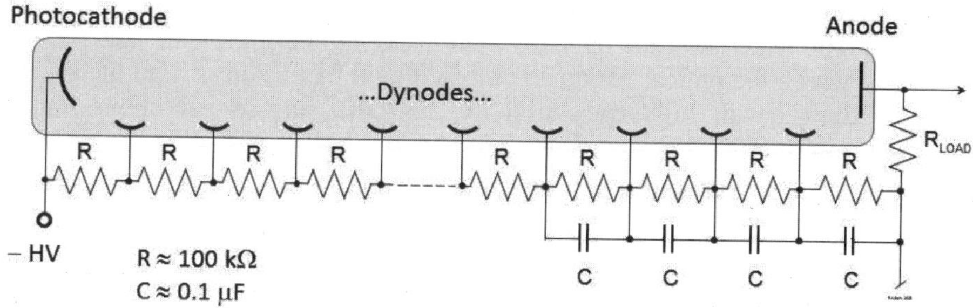

Fig. 4: PMT Voltage Divider Circuit - Diagrammatic

Output Signal – Cathode and Anode Potentials

The anode current is sent into a load resistor in order to convert it to an output voltage, and this should be a voltage against the ground potential. Therefore, the electrical circuit is often made as such the anode stays close to the ground potential. The photocathode needs to be at a strongly negative potential, which becomes problematic (when grounded parts need to be placed close to the cathode). Hence, a photomultiplier with grounded cathode is sometimes used, and the anode is operated at a high positive potential. In such a case, a capacitor is inserted into the output signal line, which allows only for AC detection. A current amplifier instead of a load resistor may be used on the basis of an operational amplifier. The details of the signal processing are less critical (than for photodiodes) and the output signal is strong and less sensitive to noise influences.

Gated Operation

In some cases, a gated operation is required, where the detector is made sensitive only at certain times where a signal is expected. Then, it should be completely insensitive to much stronger signals coming at all times. This can be accomplished by switching the cathode potential, and when it is made more positive than that of the first dynode, the electron flow to that dynode is blocked.

Photomultipliers – Properties

1. Gain and Responsivity

The photocurrent can be orders of magnitude higher than that of a photodiode. The responsivity can also be orders of magnitude larger than for an ordinary phototube without electron multiplication. Hence, photomultipliers can be used for photon counting, because the current pulse created by a single photoelectron is large enough to be easily tested.

2. Quantum Efficiency and Noise

The quantum efficiency of a photomultiplier depends on 3 important factors, namely:

a. The quantum efficiency of the photocathode (key factor) – it may be several 10s of percent or much lower, sometimes below 1% depending on the material and the optical wavelength of the detected light,

b. The material of the optical window and its thickness – the material with which the optical window is made of and its thickness limit the spectral range with high transmissivity, and the surface reflections can be minimized with an anti-reflection coating,

c. The electron collection efficiency – some electrons may be lost on their way to the first dynode (more than for later stages), and the efficiency depends upon the electrode geometry and operation voltage, which can be close to 100% for optimized designs, but can be well below 80% for others.

3. Dark Current

This refers to the output current at the anode without any light input. The dark current is usually very small, but limits the ability to detect very weak light. This is achieved when the device has been kept in complete dark condition for a while (for an hour or so), and it may be substantially higher for sometime after exposure of the photocathode to brighter light. The different effects contribute to the dark current. Some of them are:

a. The dominant (very weak) contribution is the leakage current due to non-perfect electrical isolation at very low operation voltages.

b. The dark current from thermionic emission, mainly from the photocathode and the first dynode, is dominant with normal voltages. The low temperature operation is a very effective measure for reducing the dark current as it is strongly temperature dependant.

c. There can be a steeper rise of the dark current at high operation voltages due to field emission at different locations in the bulb. As it leads to accelerated ageing, the regime should be avoided.

d. Some dark current is also contributed by the ionization of the residual gas.

e. When the electrons hit the glass tube, some minimum amount of light can be produced by scintillation.

f. The weak flashes of light are caused by cosmic rays and radio active substances at low level in the glass tube or its surroundings.

4. Dynamic Range and Linearity

Generally PMTs have a very large dynamic range (of 80 dB) and it is limited at the lower end by the dark current or by electronic noise and at the higher end by a loss of linearity. Though the linearity of response can be very good over several orders of magnitude, it can be limited by the following factors:

- The voltages supplied by the voltage divider network to the single electrodes will not stay stable when the output current gets large. The linearity starts to suffer when the output current exceeds 5 to 10%of the current through the voltage divider network (as it consists of resistors). A possible counter measure is to use smaller resistors (and increasing the current which increases the power consumption to lead to heating of the tube). Otherwise, one may use Zener diodes (with parallel capacitors for reducing the noise) for the last few dynodes.
- One can have space charge effects for the pulse detection: for example, the accumulated electrons on the last few dynodes repel further electrons and reduce the output current; while in some other cases, the finite conductivity of the photocathode is the limiting factor – for the operation at very low temperatures, where the conductivity of a semi-conductor gets very low.

5. Active Area and Uniformity of Response

The active area of a photo multiplier can be fairly large, and a special property is the combination of a large active area with a high detection bandwidth and low noise, compared to other photodetectors like photodiodes. The uniformity of response differs very much between different designs. Even if the photocathode has a very uniform response, the collection efficiency of the electrons can depend on the position. At present, there are optimized electrode designs like venetian blind type for high and uniform collection efficiency. But the head-on devices tentatively offer better uniformity of response.

6. Detection bandwidth

The transit time of electrons is fairly small between a few nanoseconds and some 10s of nanoseconds, and it exhibits a relatively small spread depending on the electrode design. With a standard design, the spread of transit time and the rise time of the output signal may be larger than 10 ns, but with optimized electrode designs, it can be far below 1 ns. Hence, the detection bandwidth can be several gigahertz. The transit time scales inversely with the square root of the applied voltage, and high operation voltages also reduce the transit time spread.

7. Signal Drifts

The signal drifts over shorter time scales include: if the device has seen some high light levels for a while, and if the level is suddenly reduced, the output current may also change with minutes. The dark current may also be significantly higher than after a while of keeping the device in the dark.

8. Secondary Pulses

Slightly after the actual signal pulses, one may get some additional weak after-pulses, which can result from the electrons which bounce off the first dynode or from positive ions, which are generated from the residual background gas by the intense output current, and these are accelerated to open electrode and generate secondary electrons there.

9. Operation Voltage

The required operation voltage as an input to the voltage divider chain is usually of the order of 1 kV. This influences many performance parameters, in particular, the gain, but also the bandwidth and the noise level, and also the lifetime.

10. Sensitivity to the Magnetic Fields

The photomultipliers are relatively sensitive to the magnetic fields because in some environments, they influence the electron trajectories. The sensitivity depends on the design of the device. A relatively low sensitivity is achieved with micro-channel plates which may be used even in strong fields of several teslas, while the more conventional designs are disturbed by the fields well below 1 mT.

11. Lifetime

The lifetime of a PMT in principle can be very long, 10s of 1000s of hours. But, premature death may be caused by some circumstances, such as:

- PMTs can be easily damaged by mechanical shock (for example, at the high voltage feed-through through the glass).
- A PMT should not be exposed to too much intense input light (for example, daylight) even when it is switched off.
- The degradation of the dynode system can occur for operation with high anode currents, which essentially means a loss of gain over time.
- Due to the bombardment with the positive ions which it attracts, the photocathode may exhibit some ageing, which means a drop of cathode sensitivity.

- Depending on the used glass material, for example silicate glass, helium can leak into the tube. The building up of helium pressure inside the tube leads to a performance degradation (formation of after pulses to complete damage).
- The operation at a reduced air pressure (between 100 Pa and a few 100 Pa) may lead to electric discharges at the socket, which can damage the tube.
- There can also be some problems with moisture (cooled operation) causing the electric discharge, or with damage related to thermal expansion, and
- There are various additional degradation mechanisms, apart from the direct performance degradation (concerning the dark current) for operation in high radiation environment (for example in nuclear reactors or in spacecrafts).

12. Housing

Placing of a PMT in some housing can have several important functions, namely,

- Mechanical protection
- Shielding against unwanted light
- Shielding against electric and magnetic fields
- Cooling of the tube for reduced dark current and keeping the moisture away from it to avoid condensation
- Providing space for additional electronics (for example, a high voltage power supply and a signal amplifier).

The photomultiplier modules come with a housing which also contain a power supply, additional amplifier electronics etc.

Photomultipliers – Modification

The photomultipliers are being increasingly replaced with the cheaper and more robust and more compact type of semiconductor detectors in different applications (such as spectro-photometers). These are sometimes even more superior in terms of quantum efficiency. These still occupy the application areas where the combination of performance parameters are critical (for example, high responsivity, low noise, large active area and higher bandwidth). The best examples are as follows: There are neutrino detectors, where a large volume of liquid is surrounded by many sensitive photodetectors. A photomultiplier can be used in conjunction with a scintillator crystal for gamma or neutron radiation (ads in scintillation counters). The incoming gamma quanta or the neutrons can be easily detected within the tube. The modified devices without a conversion dynode can be used for the detection of the electrons, ions, vacuum ultra-violet or X-ray radiation.

Photon Counting: SPCMs, CPMs and CCDs: Which Technology Works Best?

About 20 or more years ago, who would have thought that photon counting would play such an important and a critical role in many of the cutting-edge industrial and scientific endeavours of today? Photon counting seems to be a very limited strategy needed and/or used by the researchers (molecular biologists, biochemists and biophysicists) to use when the emissions are so weak (e.g., 10–14 watts approximately) that the ordinary detectors operating in linear mode cannot distinguish the light above the noise.

The 3 - Technologies

The researchers/scientists can choose one of the 3 different devices for photon counting, namely SPCMs (Single Photon Counting Modules), CPMs (Channel Photo Multipliers), and CCDs (Charged Coupled Devices) for getting the best results during their experimentation.

1. SPCMs (Single Photon Counting Modules)

SPCMs count the photons with an APD (Avalanche Photo Diode), which is a solid-state, silicon-based device that can convert the photons into the electrons than an ordinary photodiode in a different way. In the ordinary photodiodes, the photon/electron ratio is 1:1. However, in APDs, the photon-generated electrons can

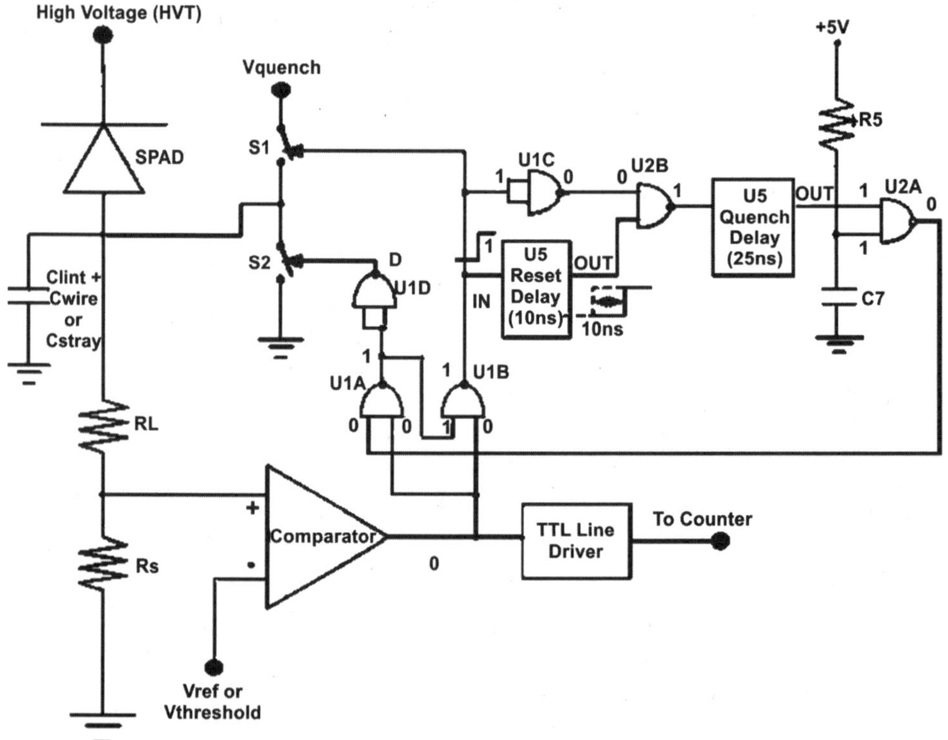

Fig. 5: SPCM – Schematic diagram

excite more number of electrons (creating an avalanche effect) resulting in an internal gain that can be adjusted with the bias (for example, 1 to 1000) that results in a signal which is proportional to the intensity of the light (Fig. 5, 6).

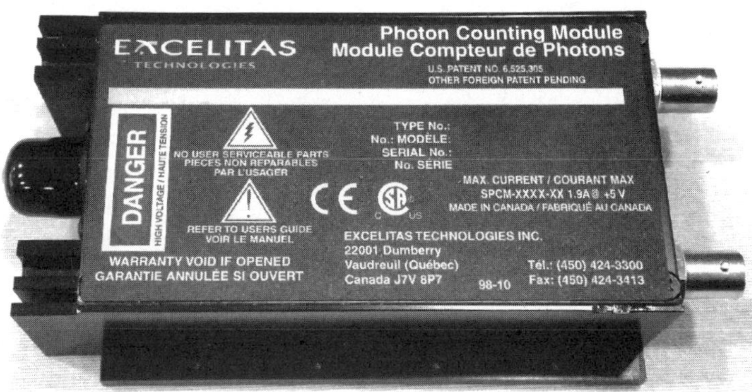

Fig. 6: SPCM (Single Photon Counting Module)

2. CPMs (Channel Photo Multipliers)

CPM is an improvement on the classical PMT that conserves the advantages of the PMT and avoids the disadvantages.

In PMT, the photons are allowed to enter via a transparent face- plate, where some are absorbed by the photocathode through which the photoelectrons are emitted into the vacuum. Several secondary electrons are created at each collision of an electron with the dynode coating. These secondary electrons are accelerated on to the next dynode and more number of secondary electrons gets generated. Such multiplication is repeated up to 10 times, to produce the secondary electrons on the order of one million. Such a gain is possible when each dynode is electrically isolated from others and has been supplied with its own voltage.

The photons that enter the CPM also produce photoelectrons that are accelerated into the vacuum towards the anode by an electrical field. But, a single hollow glass tube that is coated internally with a semi-conductor (which replaces the complicated multi-element dynode that is needed in a PMT). The secondary electrons that are generated continue to travel down the channel. By subsequent collisions, finally an electron multiplication of 100 million or more is obtained. The resultant current can be read out at the anode. The CPM channel requires only one applied voltage (unlike PMT dynode); and the CPM is more robust mechanically than the PMT, and is also much more resistant to the influence of the external magnetic fields compared to PMT. The CPM is manufactured under ultra-high vacuum and presents a smooth glass surface for out gassing during the process of manufacture. Both the ultra-high vacuum and the bent-tube configuration successfully prevent the generation of ion currents (responsible for the after-pulsing in PMTs). This technology also produces a very low background counts because is the channel is virtually silent (unlike the dynode assembly in PMT). There is no external glass bulb to charge and discharge in CPM as it is manufactured as a monolithic semiconducting channel structure; as such the background is extremely stable and free of noise bursts that are common in PMT (Fig. 7, 8).

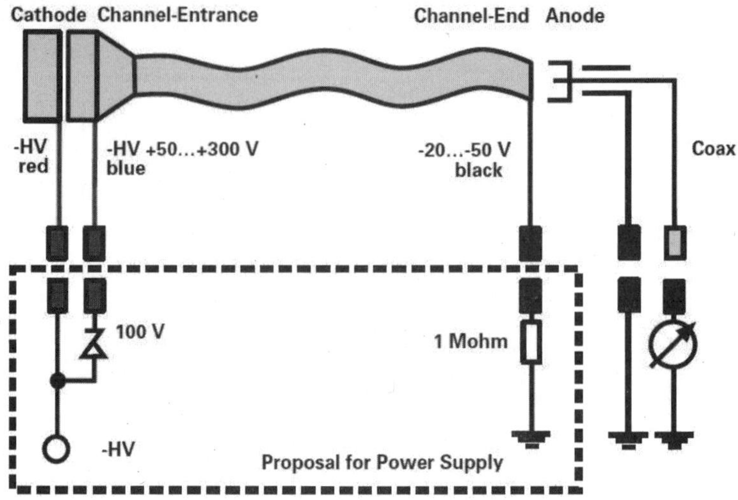

Fig. 7: CPM – Schematic

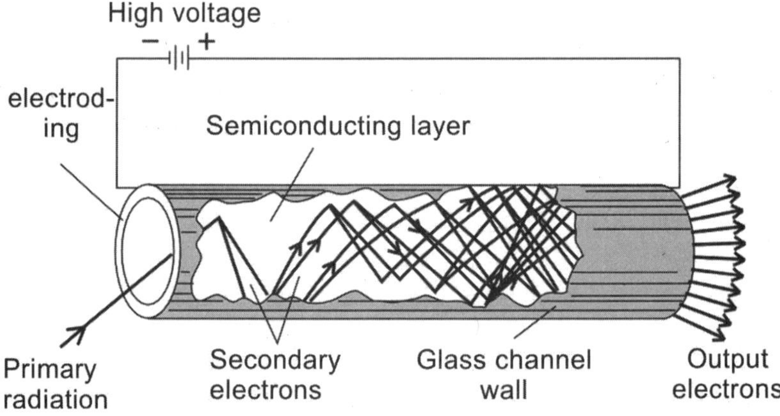

Fig. 8: Single-Channel Electron Multiplier - Diagrammatic

3. CCDs (Charged Coupled Devices)

The CCD was invented by George Smith and Willard Boyle (1969) at Bell Labs, at present a part of Lucent Technologies. A CCD is a highly light-sensitive photon detector or an IC (integrated circuit) that stores and displays the data for an image, in such a way that each pixel (picture element – potential wells) in the image is converted into an electrical charge the intensity of which is related to a colour in the spectrum. The CCD is divided up into a large number of light-sensitive small areas (called the pixels) which can be used to build up an image of the scene of interest. A photon which falls within an area defined by one of the pixels will be converted into one (or more) electrons, and the number of electrons collected will be directly proportional to the intensity of the scene at each pixel. The number of electrons in each pixel is measured and the scene can be reconstructed when the CCD is clocked out (Fig. 9, 10).

The digital cameras replace the sensitized film with a CCD photon detector, a thin silicon wafer that is divided into a geometrically regular array of 1000s (or millions) of light-sensitive regions that capture and store image information as localized electrical charge which varies with incident light intensity. The generation of the image with a CCD camera can be divided into four primary stages (or functions), namely, charge generation through photon interaction with the photosensitive region of the device, the collection and storage of the liberated charge, transfer of the charge and the measurement of the charge.

At present CCDs are commonly included in digital still cameras and video cameras, and are also used in astronomical telescopes, scanners, and bar code readers. These also find use in machine vision for robots, in OCR (optical character recognition), in satellite photograph processing and in the enhancement of radar image (in meteorology).

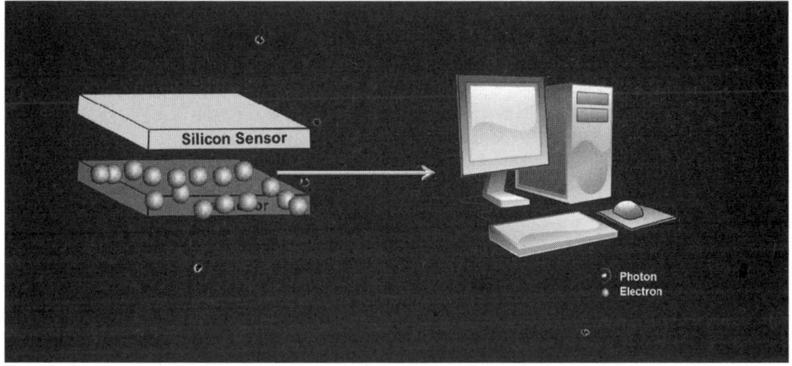

Fig. 9: CCD – Diagrammatic

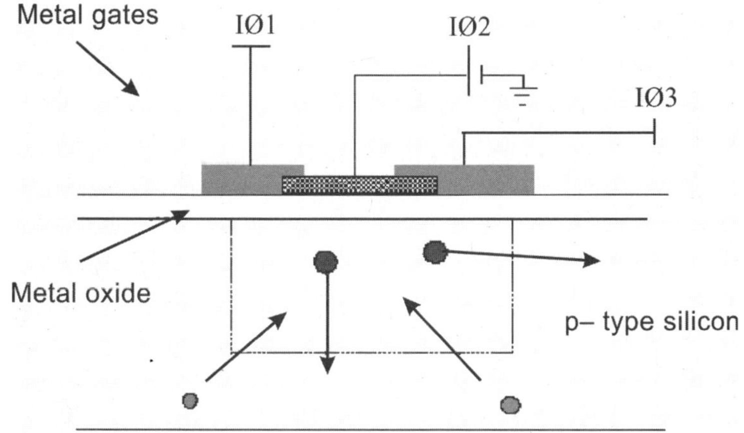

Fig. 10: CCD _ Operation - Diagrammatic

SPCM (Single Photon Counting Modules)

The SPCM uses an APD (Avalanche Photodiode) that is 200 microns in its diameter, and operates at 20 volts above the breakdown voltage to generate a gain of 108. The SPCM comes with an electronic circuitry that monitors and repeatedly quenches the current with a TTL pulse, since a gain at this level cannot be sustained without burning the device. In this photon-counting mode, the operation is very fast and this allows ther user to count a photon every 50 nanoseconds. At present, SPCMs are inte4grated into OEM devices for their application, and they are being used to detect and measure environmental toxins, proteins, particles in emulsions and solutions, for drug development, cancer, genetic research and clinical diagnosis. SPCMs are ultra-sensitive and can identify and count single photons.

Considerations

The design engineers will need to base their selections on what precisely they want to achieve, and on how best the detector complements other instrumentation as the devices vary in their capability and limitation. The most important considerations are as follows:

1. Size of the Sample

The SPCMs can easily detect the light emitted from extremely small volumes (centi-litres in volume) and count the photons at the molecular level. The SPCMs have very small detectors (0.2 to 0.3 mm) too small for the sizes of the sample that are much larger than a millimeter. The SPCM will not work with large volume as it is difficult to focus the sample's light on to such a small detector. The CPM is the detector of choice, if the size of the sample is bigger than several tenths of a milli-meter. CPM's detector size ranges from a few millimeters to centimeters, and enables the researchers to examine the samples which extend form several tenths of a millimeter to several centimeters. The CCDs have much smaller collection areas (in general, as small as 7μm), even then they can still look at extended sources, as long as these can be imaged on the CCD.

2. Speed

When the speed becomes an issue, SPCMs and CPMs are the detectors of choice. CCDs are far slower because of the time they take to charge the capacitor, which becomes important when the researchers need to make some time-resolved measurements. For example, let us suppose that the researchers choose between a SPCM, CPM and a CCD to examine the fluorescence in a microarray; all these devices can count the photons to measure the intensity, which, in case of CCD, is proportional to the charge of the capacitor. Even then, there are some events that occur within the time frame required to charge the capacitor that a CCD fails to detect (for example, the time it takes for fluorescence to decay or for the molecules to diffuse through a volume). Both the SPCM and a CPM can measure the decay and diffusion, as they can count photons every 50 nanoseconds or 30 nanoseconds, respectively.

3. Electronic Shutter

Very often, the use of a high intensity flash lamp or laser pulse to excite the sample is common in the applications in fluorescence and luminescence. In such cases, in general, the light level of the excitation light is far beyond the range of the detector. The filters can be used if there is a significant wavelength shift between the excitation and the emission light. If both ranges of the wavelength (excitation and emission) are close or even overlap, filter techniques are just not sufficient. In such a case, as an alternative, the effective method that can be used to suppress the excitation light is to employ the electronic shutter function of the CPM, which is a unique feature of the CPM that is not available either on a PMT or SPCM. The MP series CPM modules have an incorporated electronic gate function with rise times in the nanosecond region.

The different types of instruments for different tasks are built by the engineers, and the design parameters will invariably determine which photon counting technology works at its best. A CCD is the best and more appropriate, if the engineers want to perform spatial-resolved analyses on the basis of fluorescence alone. A CPM is well suited, if a company wants to analyze the sample size between less than a millimeter and up to a few centimetres in ultra-violet or the visible regions of the spectrum. In a CPM, the fast internal electronic shutter will enable the possibility to suppress effectively short-time highlight events such as the excitation of a flash lamp. Such a fast-time response enables the time-resolved fluorescence and luminescence applications. One should choose a SPCM, of the researchers want to use a laser, in the visible light range, to examine the intensity of single molecules from appoint-like source, and also to measure the decay and diffusion. Every technology has its own niche, and the design engineers will need to select a device on the basis of their requirements.

4. Range of the Wavelength

The different tasks require different wavelength sensitivities, for example, the researchers would most likely choose a SPCM if they want to excite fluorescence on a microarray with a laser. This is because these devices are the most sensitive from the visible spectrum, the red, green and yellow regions, which are the wavelength range of the lasers, up to the infrared. Similarly, the sensitivity of SPCM extends from 500 nm to 1100 nm (approximately), with a peak of 70% quantum efficiency at 650 nm. But, SPCMs are virtually blind below 400nm. The best detector for the researchers to use would be a CPM if they want to use a flash lamp or any another different light source to excite a sample in the ultra-violet or visible wavelength area. Though a CPM is most sensitive in the UV and blue region of the spectrum, the spectral sensitivity covers the entire range from the vacuum UV at 115nm up to the near infra-red at 900 nm. The CCD is a special case, as it is an array, and the sensitivity depends on the specific photodiodes in the device; and the sensitivity corresponds to the capabilities of silicon, which allows for a range of 300 nm to 1000 nm.

PHOTON COUNTING MODE (ANALOG AND DIGITAL)

A PMT (photomultiplier tube) consists of a *photocathode*, an *electron multiplier* (composed of several dynodes) and an *anode*. When light enters the photocathode, the photoelectrons are emitted from it. These photoelectrons get multiplied by secondary electron emission through the dynodes and are collected by the anode as output pulses. Such output pulses are not handled as individual pulses, but dealt with an analog current created by a multitude of pulses (called *analog mode*).

In this case, a number of photons incident on the PMT per unit time. The resulting photoelectrons are emitted from the photocathode. The photoelectrons multiplied by the dynodes are then derived from the anode as output pulses. When the pulse-to-pulse interval is narrower than each pulse width, or the signal processing circuit is not fast enough, the actual output pulses overlap each other, and finally can be considered as electric current with shot noise fluctuations. When the light intensity becomes so low, the incident photons are separated. The output pulses obtained from the anode are discrete. This condition is called a single *photoelectron state*. The number of output pulses is in direct proportion to the amount of the incident light and this pulse counting method has advantages in signal-to-noise ratio and stability over the analog mode in which an average of all the pulses is made. This pulse counting technique is known as the *photon counting method*. This is also referred to as the *digital mode*, since the detected pulses undergo binary processing for digital counting.

PHOTON COUNTING METHODS

This is an effective technique which is used to detect very-low-level-light (like Raman spectroscopy, fluorescence analysis, and chemical or biological luminescence analysis) where the absolute magnitude of the light is extremely low.

DETECTORS FOR HIGH SPEED PHOTON COUNTING

The detectors of photon counting must have sufficient gain to deliver a useful output pulse for a single detected photon. This output pulse must be short enough to resolve the individual photons at high count rate, and the transit time jitter in the detector should be small to achieve a good time resolution. A wide variety of commercially available photomultipliers and a few avalanche photodiode detectors meet all these general requirements. These different detectors are applicable to TCSPC (Time-correlated Single Photon Counting), SPC (Steady-state Photon Counting), multichannel-scaling, and FCS (Fluorescence Correlation Measurements).

PHOTON COUNTING TECHNIQUES

In a detector with a gain of the order of 10^6 to 10^8, and a pulse response width of the order of 1 ns, each detected photon yields an output current pulse of some mA peak amplitude. The output signal for a low level signal is then a train of random pulses, the density of which represents the light intensity. Hence, counting the detector pulses within defined time intervals – that is, photon counting is the most efficient way to record the light intensity with a high gain detector.

1. Steady State photon Counting

The simple intensity measurement of slow signals can be easily accomplished by a high-gain detector; a discriminator, and a counter that is read in equidistant time intervals. Simple photon counting heads that are connected to a PC via an RS232 interface can be used to collect the light signals with photon rates up to a few 10^6/s within time intervals from a few ms to minutes or hours.

2. Gated Photon Counting

These counters (GP Counters) use a fast gate in front of the counter, and the gate is used to count the photons only within defined, usually short time intervals. Gating in conjunction with pulsed light sources can be used to reduce the effective background count rate or to distinguish between different signals components. Several parallel counters with different gates can be used to obtain the information about the fluorescence life time. This method is used for life time imaging in conjunction with laser scanning microscopes. The count rate within the short gating interval can be very high, and therefore, the gated photon counters can have maximum count rates of 800MHz.

3. Multichannel Scalers or Multiscalers

These help to count the photons within subsequent time intervals and store the results in subsequent memory locations of a fast data memory. Each sequence, also called '*sweep*' is started by a trigger pulse, and therefore, the waveform of the repetitive signals can be accumulated over many signal periods. Two different versions of multiscalers with different accumulation technique exist. The photons can either be directly counted and accumulated in a large and fast data memory, or the detection times of the individual photons are stored in a FIFO memory, and the waveform is reconstructed when the measurement is finished. The direct accumulation achieves higher continuous count rates and higher sweep rates, with the FIFO principle; it is easier to obtain a short time channel width. Multiscalers for direct accumulation are available with 1 ns channel width and 1 GHz continuous count rates. Multiscalers with FIFO principle are available for 500 ps channel width. But, this is not fast enough for the measurements of fluorescence lifetimes or most of the organic dyes. But, these can be an excellent solution for phosphorescence, delayed fluorescence, and luminescence lifetime measurements of the inorganic samples. In addition, multiscalers are also used for LIDAR applications.

Benefits of Multiscaler Techniques

The important benefits of multiscaler techniques are:

- These have a near-perfect counting efficiency and as such one can achieve optimum signal-to-noise ratio for a given number of detected photons.
- These are able to record several photons per signal period.

- These can exploit extremely high detector count rates, and
- These cover extremely long time intervals with high resolution in one sweep.

4. TCSPC (Time-Correlated Single Photon Counting)

This method is based on the detection of single photons of a periodical light signal, the measurement of the detection times of the individual photons and the reconstruction of the wave from the individual time measurements.

This method makes use of the fact that for low level, high repetition rate signals the light intensity is usually so low that the probability to detect one photon in one signal period is much less than one. Hence, the detection of several photons can be neglected. The detector signal consists of a train of randomly distributed pulses due to the detection of the individual photons. There are many signal periods without photons, other signal periods contain one photon pulse (Fig. 11). The periods with more than one photon is very unlikely.

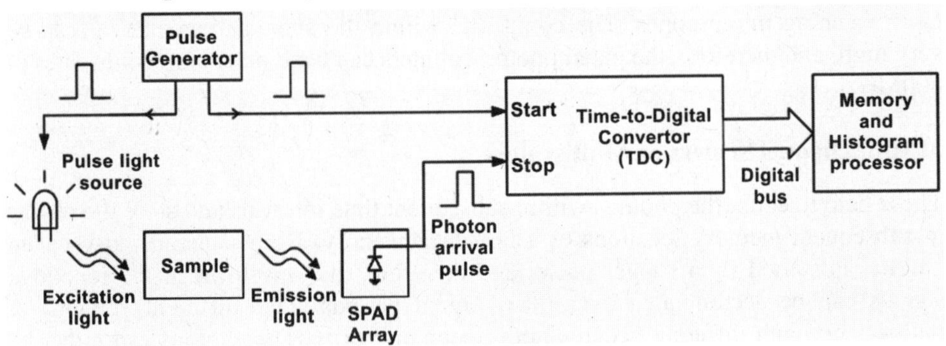

Fig. 11: TCSPC – Block Diagram

When a photon is detected, the time of the corresponding detector pulse is measured. The events are collected in a memory by adding a '1' in a memory location with an address proportional to the detection time. After many photons, in the memory, the histogram of the detection times, that is, the waveform of the optical pulse builds up.

Benefits of TCSPC

This method has a number of benefits though this looks complicated at first glance. They are:

- The time resolution is limited by the transit time spread, not by the width of the output pulse of the detector. With fast MCP PMTs, an instrument response width of less than 30 ps is achieved.
- This method has a near-perfect counting efficiency and achieves optimum signal-to-noise ratio for a given number of detected photons.
- This method is able to record the signals from several detectors simultaneously.

- This method can be combined with a fast scanning technique and can be used as a high resolution, high efficiency lifetime imaging (FLIM) technique in confocal and two- photon laser scanning microscopes.
- This method is able to acquire fluorescence lifetime and fluorescence correlation data simultaneously, and
- The state-of the-art TCSPC devices achieve the count rates in MHz range and acquisition times down to a few milliseconds.

5. Multi-Detector TCSPC

Multi-detector TCSPC operation makes use of the fact that the simultaneous detection of the photons in several detector channels is unlikely, and therefore, single photon pulses from several detector channels, either individual detectors or the anodes of a multi-anode PMT, can be combined in a common timing pulse line. If a photon is detected in one of the channels, the pulse is sent through a normal time-measurement circuitry of a single TCSPC channel. In the mean time, an array of discriminators connected to the detector outputs generates a data word that indicates in which of the channels the photon was detected. This information is used to store the photons of the individual detector channels in separate blocks of the data memory.

The multi-detector TCSPS can be used to simultaneously obtain time and wavelength resolution or to record photons from different locations of the sample. This does not involve any multiplexing or scanning process. Hence, the counting efficiency for each detector channel; is still close to one, meaning that the efficiency of a multi-detector TCSPC system can be considerably higher of a single channel TCSPC device.

6. Fluorescence Correlation Spectroscopy – Photon Counting

FCS (Fluorescence Correlation Spectroscopy) exploits the intensity fluctuations in the emission of a small number of chromophore molecules in a femtoliter sample volume. The FCS is the auto-correlation function of the intensity fluctuation and this yields the information about the diffusion processes, conformational changers of chromophore-protein complexes, an intra-molecular dynamics. FC spectra can be obtained directly by hardware correlators or by recording the detection times of the individual photons and by calculating the FCS curves by software. The second technique can be combined with TCSPC to obtain FCS and lifetime data simultaneously. The multi-detector capability of TCSPC can also be used to detect the photons in different wavelength intervals or of different polarisation simultaneously. For each detector, an individual correlation spectrum and a fluorescence decay curve can be calculated; and if several detectors are used to record the photons from different chromophores, the signals of these can be cross-correlated. The FCS shows whether the molecules of both chromophores and the associated protein structures are linked or diffuse independently.

Detector Principles

Photomultiplier tubes (PMTs) are the most common detectors for low level detection of light. A conventional PMT is a vacuum device which contains a photocathode, a number of dynodes (amplifying stages) and an anode which delivers the output signal. The electrons are accelerated from the cathode to the first dynode D1, then to the next dynodes, and from D8 to the anode by the operating voltage that builds up an electrical field. When a photoelectron emitted by the photocathode hits D1, it releases several secondary electrons; and the same happens for the electrons emitted by D1 when they hit D2. The overall gain reaches the values of 10^6 to 10^8. The secondary emission at the dynodes is very fast, and the secondary electrons resulting from one photoelectron arrive at the anode within a few nanoseconds or less. Such a high gain and the short response, a single photoelectron yields easily detectable current pulse at the anode.

A wide variety of dynode geometries have been developed in recent time, and of special interest for photon counting are: the '*linear focused*' type dynodes which can yield a fast single electron response, and the '*fine mesh*' and '*metal channel*' types that offer position-sensitivity when used with an array of anodes.

A similar gain effect can be achieved in the channel PMT and MCP (Microchannel PMT) as in the conventional PMTs. These use the channels with a conductive coating, the walls of which work as secondary emission targets. MCPs are the currently available fastest photon counting detectors. Further, the microchannel plate technique allows building position-sensitive detectors and image intensifiers. The position sensitivity, the single anode can be replaced with an array of individual anode elements. By detecting the pulses individually from the anode elements, the position of the corresponding photon on the photocathode can be easily determined. Multi-anode PMTs in conjunction with TCSPC can process the photon pulses from several detector channels in only one time-measurement channel.

The gain systems in PMTs can also be used to detect the electrons or ions, and such '*electron multipliers*' are to be operated in the vacuum, and the particles are fed directly into the dynode system, the multiplier channel or on to the multichannel plate.

The cooled avalanche photodiodes are used to detect single optical photons if operated close to or slightly above the breakdown voltage as the generated electron-hole pairs initiate an avalanche breakdown in the diode. To restore normal operation after each photon an active or passive quenching circuits must be used. SPADs (Single-Photon Avalanche Diodes) have high quantum efficiency in the visible and near infra-red range of the spectrum.

PIN diodes can detect X-ray photons. A single high energy X-ray photon generates many electron-hole pairs in the diode so that the resulting charge pulse can be detected by an ultra-sensitive charge amplifier. These detectors have a time resolution in the ultra-sensitive range and do not reach high count rates due to the limited speed of the amplifier. But, they can distinguish photons of different energy by the amount of the charge generated.

The detector parameters include: single electron response, transit time spread and timing jitter, cathode efficiency, pulse height distribution, dark count rate, and after pulses.

The detectors to be selected for efficient performance of photon counting include: R3809U MCP-PMT, H7422, H7421, H5783 and H5773 Photo-sensor Modules, PMH-100, R7400 and R5600 TO-8 PMTs, R5900-L16 Multi-channel PMT and PML-16 Multi-channel Detector Head, Side-Window PMTs, CP 944 Channel Photomultiplier, and SPCM-AQR Single-Photon Avalanche Photodiode Module.

Photon Detection - Uses

Single photon detection is useful in different fields like: Fibre-optic Communication, Quantum Information Science, Quantum Encryption, Medical Imaging, Light Detection and Ranging, DNA Sequencing, Astrophysics and Materials Science. In radiology (X-ray imaging in medicine), photon counting detectors could help (due to their ability to reject the noise more easily) in such a way that the lower doses of radiation can be used for the same diagnostic image quality, and other advantages compared to conventional integrating detectors. In addition, commercial photon counting mammography machines have been produced because of their ability to produce comparable images at lower doses than other digital mammography systems (with flat panel detectors). Another key area of interest is Photon-counting Computed Tomography where its ability to discriminate between the photon energies could improve their ability to distinguish the tissue types when an image is being reconstructed.

Photon Counting – Applications

The applications are so many and the number is growing day by day, because these applications have a decisive impact on the advances in different fields of science, from astronomy and metallurgy to the environmental protection, health diagnostics, medical analysis and medical imaging, and to be more specific in biomedical research. Much of the dramatic progress in recent times in genomics has been made possible by microarray technology that depends on the ability to detect the 1000s of reactions of DNA strands that are distributed across a micro-array slide. The researchers/scientists employ fluorescence detection techniques to achieve this and often detect fluorescence by counting the photons. As such, this has played an important and an essential role in mapping the human genome.

Photon Counting - Benefits

This is a digital method of measuring the light intensity with time. Unless the bandwidth of the change of intensity is very low, and the high intensity, this is the best method of quantifying changing light intensity. Until recently, this method has dominated light measurement for low intensity, but with the increase of the bandwidth in new devices, the counting of much higher light intensities, is possible.

Conclusion

There is no detector that meets all requirements of photon counting: high quantum efficiency, low dark count rate, short transit time spread, narrow pulse height distribution, high peak count rate, high continuous count rate, and low after-pulse probability. Perkin Elmer SPCM-AQR is the detector with highest efficiency as it has a broad and count-rate dependent transit timer spread. The R7400 miniature PMTs and the H5783 and H5773 photo-sensor modules (Hamamatsu) have a short transit-time spread and work well for TCSPC, steady state photon counting and multi-scaler applications, but cannot be used for correlation experiments below 1.5 ultra-sensitivity because of their high after-pulse probability. The H7422 modules offer a high efficiency combined with an acceptable transit time spread, and if they are operated at reduced gain, the after-pulse probability can be kept low. The two really remarkable detectors are: R3809U MCP (Hamamatsu) and the R5900 multi-anode PMT, and both these tubes are free of after-pulses. These two detectors appear to be fully applicable for simultaneous fluorescence correlation and lifetime experiments.

■■■

References

Andrews AT 1986 *Electrophoresis Theory, Techniques and Biochemical and Clinical Applications.* Oxford: Oxford University Press.

Attwood TK, Gisel A, Eriksson NE and Bongcam-Rudloff E 2011 *Concepts, Historical Milestones and the Central Place of Bioinformatics in Modern Biology: A European perspective. Bioinformatics – Trends and Methodologies.* InTech.

Attwood TK and Parry-Smith DJ 2002 *Cell and Molecular Biology In Action Series: Introduction to Bioinformatics,* Delhi: Pearson Education.

Baker HG 1957 Origin of the Vasculum. *Bulletin Botanical Survey of India* **2** 324, 41-43.

Baker JR 1958 *Principles of Biological Microtechnique; A Study of Fixation and Dyeing.* New York: Wiley.

Barz W, Reinhard E and Zenk MH 1976 Plant Tissue Culture and its Bio-technological application: *Proceedings of the First International Congress on Medicinal Plant Research, Section B* - University of Munich, Germany. Springer Science & Business Media.

Bhadur PIN and Majumdar B R 1955 *An improved Pollen Grain Smear Method, BIO-331 Cell biology – Lab 6.2 Plant histology & microtechnique.* pdf

Bhojwani SS and Razdan MK. 1996 *Plant Tissue Culture: Theory and Practice.* Elsevier.

Bhojwani SS and Dantu PK 2013 *Plant Tissue Culture: An Introductory Text.* Springer Science & Business Media.

Biotechnology – Myths and Facts 2002 *Produced by the Global Knowledge Center on Crop Technology,* ISAAA, SE Asia Centre, IRRI, Metro Manila, Philippines.

Birnie GD and Rickwood D (eds.) 1978 *Centrifugal Separations in Molecular and Cell Biology.* London: Butterworths.

Bradbury S 1967 *The Evolution of the Microscope.* New York: Pergamon Press.

Bradbury S and Bracegirdle B 1998 *Introduction to Light Microscopy.* Oxford UK: BIOS Scientific Publishers.

Bridson D and Forman L (eds) 1998 *The Herbarium Handbook.* Kew: Royal Botanic Gardens.

British Columbia Ministry of Forests 1996 *Techniques and Procedures for Collecting, Preserving, Processing, and Storing Botanical Specimens.* Res. Br., C. Min. For., Victoria, B.C. Work. Pap. 18/1996

Brown RW 2009 *Histologic Preparations: Common Problems and Their Solutions.* Northfield: College of American Pathologists

Carl Zeiss: *Microscopy from the Very Beginning 1997 – Brochure,* 44 pages. Jena, Germany: Carl Zeiss Mikroskopie.

Carson FL and Hladik C 2009 *Histotechnology: A Self-instructional Text.* Chicago: ASCP Press.

Chandra V and Ravichandran K 2011 Herbaria In: *A Handbook on Plant Taxonomy:* FRI, Dehradun 321-358 pages

Chanut F 2012 *Writing Research Articles – How to Write-in Introduction section.* Office of Career & Professional Development, University of San Francisco.

Clark G 1973 *Staining Procedures.* Baltimore: Williams and Wilkins Company.

Clydesdale FM and Ahmed EM 2009 *Colorimetry - Methodology and Applications* - Paper No. 2197: From the Massachusetts Agricultural Experiment Station, University of Massachusetts, Amherst. 243-301 pages

Coleman C 1938 *Preparation of Leuco-basic Fuchsin for use in the Feulgen Reaction.* College Press.

Conn EE and Stumpf PK 1978 *Outlines of Biochemistry.* New Delhi: Wiley Eastern Limited, New Delhi.

Connor O and Woodford FP 1977 *Writing Scientific papers in English.* Amsterdam: Elsivier.

Craig R and Wilson C 1935 A Microtome Knife Holder for Safety Razor Blades. *Science* **81** 404-405

Cramer CS 1999 Laboratory Techniques for determining ploidy in plants. *Hortechnology* **9(4)** 594-596

Creager ANH 2004 The Industrialization of Radioisotopes by the U.S. Atomic Energy Commission. In: K. Grandin, N. Wormbs, & S. Widmalm (Eds.), The Science – Industry Nexus: History, Policy, Implications. *Proceedings of Nobel Symposium 123* (pp. 141–167). Sagamore Beach, MA: Science History Publications.

Crisp MD & Cummings DJ 1977 *How to Use Field Note Books?* Australian National Herbarium, Centre for Australian National Biodiversity Research, Canberra.

Crow JF and Dove WF 1995 (ed). In: *Perspectives - Anecdotal, Historical and Critical Commentaries on Genetics. Early Days of Gel Electrophoresis.* O. Smithies (1995), Genetics Society of America, Genetics **139**: 1-4

Day RA 1979 *How to Write and Publish Scientific Paper*? Philadelphia, ISI Press, 96-100 pages

De Duve C 1971 Tissue Fractionation: Past and Present. *Journal of Cell Biology* **50** 20D –55D

De Duve C 1975 Exploring Cells with a Centrifuge. *Science,* **189** 186–194

De Tomasij A 1936 Improving the Technique of the Feulgen Stain. *Stain Technololgy* **11** 137-144

DeWolf GP 1968 "Notes on Making a Herbarium". *Arnoldia.* **28** 69-111

Dwivedi JN and Singh RB 1990 *Essentials of Plant Techniques.* Jodhpur, India: Scientific Publishers.

Edwards M and Morse DR 1995 The Potential for Computer-aided Identification in Biodiversity Research. *Trends in Ecology and Evolution* **10** 153-158

Evans DG and Gruba P 2002 *How to write a better thesis?* Carlton: Melbourne University Press.

Fisher RA 1958 *Statistical Methods for Research Workers.* New York: Hafner Publishing Co.

Fox JH 1958 Criteria of Good Research, *Phi Delta Kappa* **39** 284-286

Freedman P 1960 *The Principles of Scientific Research.* New York: Pergamon Press.

Fristensky B and Wang Y 2008 *Lab Manual for Introductory Cytogenetics.*

Funk VA 2002 The Importance of Herbaria. *Plant Science Bulletin* **49** 94-95

Gaddagimath RB 1988 Technical Report Writing. Foundation Course Manual, NAARM, Rajendranagar, Hyderabad.

Gilbert D 2004 Bioinformatics Software Resources. *Briefings in Bioinformatics* **5 (3)** 300–304

Gray P 1952 *Handbook of Basic Microtechnique,* New York: McGraw-Hill.

Gray P 1954 *The Microtomist's Formulary and Guide,* New York.

Gray P 1964 *Handbook of Basic Microtechnique*, New York.

Gupta PK 1998 *Elements of Biotechnology*, Meerut: Rastogi Publications.

Hernan A, Retamales and Tanya Scharaschkin 2014 A Staining Protocol for Identifying Secondary Compounds in Myrtaceae: *Appl. Plant Sci.* Oct, 2 (10)

Howat WJ and Wilson BA 2014 Tools and Techniques in Biology 33 Some Common Preservatives. Transactions of American Micro. Society 66 379-382

Hildebrand EM 1938 Technique for the Isolation of Single Micro-organisms. *Bot. Rev.* **4** 628

Holiday R 2006 Physics and the origins of Molecular Biology. *J. Genetics* **85(2)** 93-97

Johansenn DA 1940 *Plant Microtechnique*, New York: McGraw-Hill Book Company, Inc.

Johnston I.M 1939 *The Preparation of Botanical Specimens for the Herbarium.* Jamaica Plain, MA: Arnold Arboretum of Harvard University.

Kapitza HG 1994 *Microscopy from the Very Beginning.* Carl Zeiss, Oberkochen, Germany.

Keener K, Hoban T and Balasubramanian R 2014 *Biotechnology and its Applications*, The North Carolina Cooperative Extension Service, North Carolina State University, NC, USA.

Kiernan JA 2009 *Histological and Histochemical Methods: Theory and Practice.* Scion, Bloxham.

Kothari CR 1985 *Research Methodology: Methods and Techniques*, New Delhi: Wiley Eastern.

Kothari CR 1984 *Quantitative Techniques,* New Delhi: Vikas Publishing House.

Kraft A 2006 Between Medicine and Industry: Medical Physics and the rise of the Radio-isotope, 1945–65 *Contemporary British History,* **20** 1–35

Kuhr WG 1990 Capillary Electrophoresis. *Anal. Chem.* **62** 403R–414R

Lavoie C 2013 Biological collections in an ever changing world: Herbaria as tools for biogeographical and environmental studies. *Perspectives in Plant Ecology, Evolution and Systematics* **15** 68–76

Lee AB 1913 *A Handbook of the Methods of Microscopic Anatomy.* Philadelphia: P. Blakiston's Son & Co.

Li SFY 1992 Capillary Electrophoresis: Principles. Practice and Applications, Amsterdam: Elsevier.

Luna L (ed) 1968 *AFIP Manual of Histological Staining Methods*, 3rd ed, New York: McGraw Hill Publications

Macfarlane RB 1985 *Collecting and Preserving Plants*, New York: Dover Publications

Maden, K. 2004 Plant Collection and Herbarium Techniques, *Our Nature,* **2**, 53-57

Mani A, Narayanan LM, Dusly Fatima, Selvaraj AM and Arumugam N 1999 *Immunology & Microbiology*, Nagarcoil: Saras Publication

Martinsson A 1983 *Guide for the Preparation of Scientific Papers for Publication.* Paris: General Information Programme and UNISIST, UNESCO.

Matthews JR, Bowen JM and Matthews RW 2000 *Successful Scientific Writing.* Cambridge: Cambridge University Press.

McMullan D 1953 An improved Scanning Electron Microscope for opaque specimens; *Proc. Inst. Electr. Engrs* 100, Part II, 245-249

Minot CS 1903 The History of the Microtome, *Journal of Applied. Micro. and Lab. Meth*ods **6** 2157-2160

Mowshowitz D 2012 Writing a Laboratory Report: Notes to Student Experimenters. Adapted from an article in: Biochemical. ed. by Blackman, D. 7: 82 (1979)

Murthy BR 2011 A simple and illustrative method of Ion-exchange chromatography, *Resonance,* **8(12)** 77–82

Nair AJ 2008 *Introduction to Biotechnology and Genetic Engineering*, New Delhi: Laxmi Publications.

Nair AS 2007 *Computational Biology & Bioinformatics – A gentle overview,* Communications of Computer Society of India.

Narayanan LM, Mani A, Selvaraj AM, Arumugam N and Padmalatha Singh 1996 *Molecular Biology & Genetic Engineering,* Nagercoil: Saras publication.

Narayanaswamy S 1994 *Plant Cell and Tissue Culture,* New Delhi: Tata McGraw-Hill.

Needham GH 1958 *The Practical Use of the Microscope*, Springfield, IL: Charles C. Thomas.

Oregon State University 2004 *Using Paper Chromatography.* Hydroville Curriculum Project, http://www.ehsc.orst.edu/outreach.htm

Östergren G and Heneen WBK 1962 A Squash Technique for Chromosome Morphological Studies, *Hereditos* **18**, 332-341

Ouzounis CA and Valencia A 2003 Early Bioinformatics: The Birth of a Discipline - A Personal View. *Bioinformatics* **19(17)** 2176–219

Perrett D 2010 *200 Years of Electrophoresis Chromatography Today.*

Primrose SB 1991 Part V: Plant Biotechnology: Plant Cell, Tissue and Organ Culture, In: *Biotechnology*, New Delhi: Panima Publishing Corporation, 153-164 pages

Pullaiah T 2007 *Taxonomy of Angiosperms*. New Delhi: Regency Publications.

Pullaiah T and Subba Rao MV 2009. *Plant Tissue Culture*, Jodhpur: Scientific Publishers.

Ramakrishnana P 1999 *Biostatistics*, Nagercoil: Saras Publications.

Sabesh M 2007 Computer Applications in Agricultural Research Model Training Course on "*Cultivation of Long Staple Cotton (ELS)*", December 15-22, 2007. Coimbatore: Central Institute for Cotton Research, Regional Station.

Saravanavel P 1987 *Research Methodology,* Allahabad : Kitab Mahal.

Satyanarayana U 2007 *Biotechnology.* Kolkata.: Books and Allied (P) Ltd.

Seboyar GE 1929 *Manual for Report and Thesis Writing.* New York: F.S. Crofts & Co.

Sharma BK 2000 *Chromatography*, 3rd edition. (ed) Manjula Sharma and Piyush Sharma, Meerut: Goel Publishing House.

Sheehan DC and Hrapchak BB 1980 *Theory and Practice of Histotechnology.* St.Louis, MO: C.V.Mosby Co.

Singh, B. D. 2000 *Biotechnology.* Ludhiana: Kalyani Publishers.

Smith EE Jr 1971 *Preparing Herbarium Specimens of Vascular Plants.* St Louis, MO: U.S.D.A. Agriculture Information Bulletin No. 348. Mosby.

Smith KCA. and Oatley CW 1955 The Scanning Electron Microscope and its fields of application. *British Journal of Applied Physics* **6** 391-399

Smith Jr CE 1971 *Preparing Herbarium Specimens of Vascular Plants,* Washington D.C.: USDA Agriculture Information Bulletin No. 348

Spectrophotometry (Handbook), GE Healthcare UK Limited Amersham Place Little Chalfont, Buckinghamshire HP7 9NA, UK.

Srinivas A (ed.) 2006 *A Handbook of Computational Molecular Biology.* Chapman & Hall/CRC Computer and Information Science Series.

Stuessy TF and Sohmer SH (eds.) 1996 *Sampling the Green World: Innovative Concepts of Collection, Preservation, and Storage of Plant Diversity.* New York: Columbia University Press.

Subramanian N 1986 *Introduction to Computers.* New Delhi: Tata McGraw-Hill, New Delhi.

Sundberg MD 1981 Making the Most of Onion Root tip Mitosis. *Amer. Biol. Teacher* **43** 386–388

Svenson HK 1935 The Preparation of Herbarium Specimens. Brooklyn Botanic Garden Leaflets, series XXIII, no. 4. New York.

Taiz L and Zeiger E 2012 *A Companion to Plant Physiology,* Fifth Edition Topic 7.1 Principles of Spectrophotometry.

Thiers BM 2001 *Herbaria, Plant Sciences.* www.encyclpedia.com

Tucker AO and Calabrese L 2005 *The Use and Methods of Making a Herbarium/plant specimens.* An Herb Society of America Guide.

University of Auckland, School of Graduate Studies. *Guide to Theses and Dissertations,* Last updated: September 2014, 34 pages. www.postgrad.auckland.ac.nz

Utter B and Huber W 2011 *Introduction into Colorimetry and Spectrophotometry* Heidelberger Druckmaschinen AG Germany.

Venkatasubramanian V 1999. *Introduction to Research Methodology in Agricultural and Biological Sciences.* Chennai: New Century Book House.

Vesterberg O 1989 History of Electrophoretic Methods. *J. Chromatography.* **480**, 3-19

Victor JE, Koekemoer M, Fish L, Smithies SJ and Mössmer M 2004 *Herbarium Essentials: The Southern African Herbarium User Manual.* Pretoria: Southern African Botanical Diversity Network Report No. 25. SABONET.

Weissberg R and Buker S 1990 *Writing Up Research: Experimental Research Report. Writing for Students of English.* NJ, USA : Prentice Hall Regents, Englewood Cliffs.

Westermeier, R. 1997 *Electrophoresis in Practice*. Weinheim: VCH.

Whiley E 1938 Aids to section cutting and mounting. *The Microscope*, **2** 138-140

White PR 1942 Plant Tissue Cultures. *Annu Rev Biochem* **11** 615–628

Wilkins AJ 1995 *Visual Stress.* Oxford University Press.

Whelan WJ 1995 The Advent of Chromatography. *The FASEB Journal.* **9**: 287-288

Wilson K and Walker JM 1994 *Principles and Techniques of Practical Biochemistry.* Cambridge University Press.

Wixom, R.L. and Gehreke, C.W. (ed). 2010 Chromatography - A Science of Discovery. John Wiley & Sons, New Jersey.

www.encyclopaediabritannica.com

www.encyclopedia.com

www.webopedia.com

www.wikipedia.com

Zworykin VA, Hillier J and Snyder RL 1942 A Scanning Electron Microscope. *ASTM Bull.* **117** 15-23

■■■

Related Book

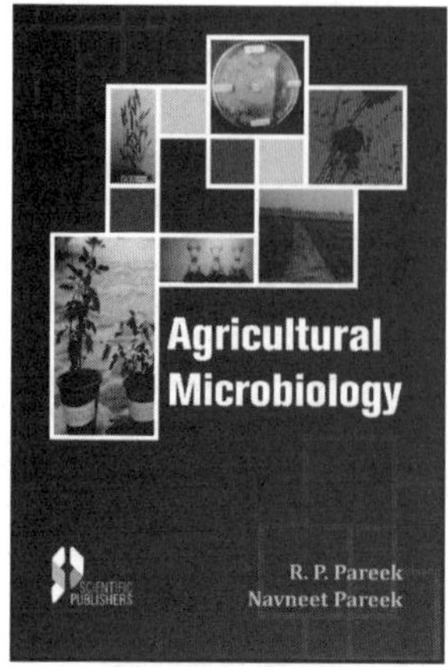

ISBN: 9789386652287

ISBN: 9789388172004

ISBN: 9788172339340

ISBN: 9789387991446

ISBN: 9788172337513

ISBN: 9789386652089

ISBN: 9788172333836

ISBN: 9789386102638

ISBN: 9789386652362

For Full Contents & Sample Pages Visit our Website